Y VII 75

THE
INTERNATIONAL SERIES
OF
MONOGRAPHS ON PHYSICS

GENERAL EDITORS
W. MARSHALL   D. H. WILKINSON

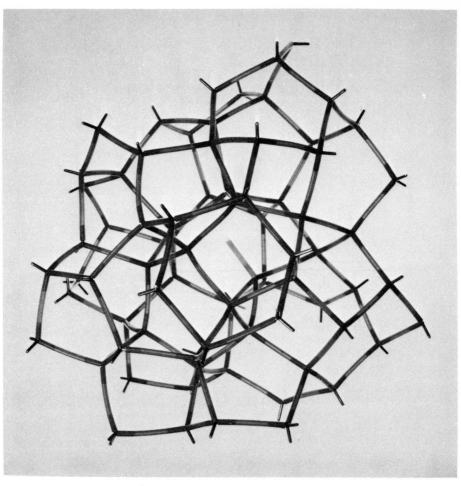

A dangling bond in a random network simulating the structure of amorphous germanium or silicon.

ELECTRONIC PROCESSES IN NON-CRYSTALLINE MATERIALS

BY

N. F. MOTT

Emeritus Cavendish Professor of Physics in the University of Cambridge

AND

E. A. DAVIS

Lecturer in Physics in the University of Cambridge

SECOND EDITION

CLARENDON PRESS · OXFORD

1979

Oxford University Press, Walton Street, Oxford OX2 6DP

OXFORD LONDON GLASGOW
NEW YORK TORONTO MELBOURNE WELLINGTON
IBADAN NAIROBI DAR ES SALAAM LUSAKA CAPE TOWN
KUALA LUMPUR SINGAPORE JAKARTA HONG KONG TOKYO
DELHI BOMBAY CALCUTTA MADRAS KARACHI

© Oxford University Press 1971, 1979

First Edition 1971
Second Edition 1979

British Library Cataloguing in Publication Data

Mott, *Sir* Nevill Francis
   Electronic processes in non crystalline materials.—
   2nd ed.—(The international series of monographs on physics).
   1. Amorphous substances—Electric properties
   I. Title   II. Series   III. Davis Edward Arthur
   530.4'1   QC176.8.E4   78-40236

ISBN 0-19-851288-0

Printed in Great Britain
by J. W. Arrowsmith Ltd.
Bristol

TO
RUTH
AND
CHRISTINE

PREFACE TO SECOND EDITION

IN THE seven years since we completed our first edition, there have been very substantial advances in our knowledge of non-crystalline materials based on experiment and in their theoretical interpretation, and we have therefore found it necessary to rewrite the greater part of this book. Many review articles now exist, and because of these and so as to keep the book within a reasonable size we have not attempted as full a review of the literature as we did in the first edition. Our purpose, however, remains the same: to present a theoretical framework and to relate it to the experimental material.

Some of the outstanding advances since our first edition seem to us to be the following: the very detailed understanding of the electrical properties of glow-discharge-deposited silicon obtained notably at Dundee and Marburg, and in particular the realization that this material can be doped; experiments on conduction in an inversion layer at the interface between silicon and silicon oxide, which combined with numerical calculations of the minimum metallic conductivity in two dimensions provides strong evidence for the existence of this quantity; real-space calculations of the energy spectra of both electronic and vibrational states from model structures, and increasing realization for chalcogenide and silicate glasses of the importance of the polaron concept and of the distortion produced by a trapped carrier, leading to Anderson's concept of a 'negative Hubbard U' and to detailed models of photoluminescence, states in the gap, and other properties of these materials. These are some highlights; many other advances may be of comparable importance.

Finally, it is a pleasure to thank many colleagues who have helped us in the writing of this book, Dr S. R. Elliott for reading the galley proofs, and Miss Shirley Fieldhouse for her invaluable work in the preparation of the bibliography.

Cambridge
August 1977

N.F.M.
E.A.D.

PREFACE TO FIRST EDITION

TEN YEARS ago our theoretical understanding of electrons in non-crystalline materials was rudimentary. The classification of materials into metals, semiconductors, and insulators was based on band theory, and band theory starts from the assumption that the material is crystalline. According to band theory, an insulator is a material with an energy gap between the conduction and valence bands, and a transparent insulator is one in which the gap is greater than the quantum energy of visible light. Ordinary soda glass is an insulator and transparent; a gap seemed to exist but we did not know how to describe the gap. Even now we do not know how to calculate it, but the concepts that we have to use are fairly clear.

A milestone in the development of the subject was Ziman's quantitative explanation of the electrical properties of liquid metals, put forward in 1960. This was a weak-interaction theory, the effect of each atom being treated as small. The success of this theory prompted investigations of what happens when the interaction is large, as it must be when an energy gap exists. The keys to our present understanding have been the principle of Ioffe and Regel (1960) that the mean free path cannot be less than the distance between atoms, and the concept of localization introduced by Anderson in his paper 'Absence of diffusion in certain random lattices', published in 1958. In a sense, our book is written around these two themes. We have built a theoretical edifice on them, and since mathematical rigour is anything but easy in this subject we have not hesitated to guess at the approximate solutions of problems that at present are unsolved. Our aim is to suggest models that can be compared with experiment. We have chosen the experimental material, too, with a view to comparing it with our theory and our conjectures. Thus we have given a rather full account of what is known in October 1970 about the electrical and optical properties of certain amorphous semiconductors, in particular silicon, germanium, chalcogenide glasses, and selenium, which we think relevant. We have said much less about conduction in glasses containing transition-metal ions as they would in our view fit better into a book about polarons. Our chapter on impurity conduction is not meant to be exhaustive; we include it because impurity conduction is the most fully understood process of conduction in a random field. We have said rather little about the phenomenon of switching, fearing that anything we could write would be out of date too quickly.

Finally it is a pleasure to thank our many colleagues who are interested in non-crystalline materials and who have helped us to write this book. We are particularly indebted to Dr. T. E. Faber for making available to us some tables from his forthcoming book, to Dr. R. S. Allgaier for the table in the Appendix, to Dr. L Friedman, Mr. C. H. Hurst, and Dr. F. Stern for help in correcting the proofs, and to Miss Shirley Fieldhouse for her help in preparing the bibliography.

Cambridge   N.F.M.
October 1970   E.A.D.

ERRATUM

Note that on pp. 35 and 135 in the discussions of two-dimensional problems $N(E)$ refers to the number of states per unit area, while in §§7.4 and 7.5 for thin films it is always the bulk density, the two-dimensional density being $N(E)d$, where d is the film thickness.

CONTENTS

1. INTRODUCTION   1

2. THEORY OF ELECTRONS IN A NON-CRYSTALLINE
   MEDIUM   7
   2.1. Introduction   7
   2.2. The Kubo–Greenwood formula   11
   2.3. Anderson localization   15
   2.4. Situation in which states are localized in one range of energies and
         not localized in another   22
   2.5. Photon-activated hopping; the ω^2 law   27
   2.6. The minimum metallic conductivity   28
   2.7. Hopping and variable-range hopping   32
   2.8. The Anderson transition   37
   2.9. Mobility and percolation edges   39
   2.10. Conductors, insulators, and semiconductors   42
   2.11. Semimetals and pseudogaps   50
   2.12. Some calculations of the density of states   51
   2.13. Thermopower   52
   2.14. Hall effect   56
   2.15. Hopping conduction for alternating currents   59
   2.16. One-dimensional problems   62

3. PHONONS AND POLARONS   65
   3.1. Introduction   65
   3.2. Distortion round a trapped electron   66
   3.3. Dielectric and acoustic polarons   69
   3.4. Rate of loss of energy by free carriers   73
   3.5. Transitions between localized states   75
         3.5.1. Introduction   75
         3.5.2. Single-phonon hopping processes   76
         3.5.3. Multiphonon processes   78
   3.6. Motion of a polaron in a crystal   89
   3.7. Trapped and localized polarons   90
   3.8. Thermopower due to polarons   91
   3.9. Hall effect due to polarons and other forms of hopping   92
   3.10. Examples of hopping polarons   93
   3.11. Degenerate gas of polarons   94
   3.12. Charge transport in strong fields   95

4. THE FERMI GLASS AND THE ANDERSON
   TRANSITION   98
   4.1. Introduction   98

  4.2. Metal-insulator transitions in crystals   101
        4.2.1. Band-crossing transitions   101
        4.2.2. Hubbard bands and the Mott transition   104
        4.2.3. Wigner crystallization   109
        4.2.4. Effect of disorder on Mott transitions   109
        4.2.5. Effect of correlation and distortion on Anderson tran-
                sitions   110
  4.3. Doped semiconductors   111
        4.3.1. Impurity conduction; direct current   111
        4.3.2. Impurity conduction; alternating currents   117
        4.3.3. Metal–insulator transitions in doped semiconductors   119
        4.3.4. Metal–insulator transitions in metal–rare-gas systems   126
        4.3.5. Metal–insulator transitions in compensated semiconduc-
                tors   127
  4.4. Anderson transition in a pseudogap; magnesium–bismuth alloys   130
  4.5. Anderson transition of type II; Amorphous films of Fe–Ge   132
  4.6. Two-dimensional conduction in an inversion layer   135
  4.7. Fermi glasses where lattice distortion is important   139
        4.7.1. Impurity conduction in nickel oxide   140
        4.7.2. Conduction in glasses containing transition-metal ions   142
        4.7.3. Lanthanium strontium vanadate ($La_{1-x}Sr_xVO_3$)   144
        4.7.4. Pyrolytic carbons   145
        4.7.5. Cerium sulphide   146
        4.7.6. Metal–insulator transitions in tungsten bronzes   148
        4.7.7. Vanadium monoxide (VO_x)   150
  4.8. Impurity conduction in magnetic semiconductors   152
  4.9. Granular metal films   157
  4.10. Polycrystalline aggregates   160

5. LIQUID METALS AND SEMIMETALS   161
  5.1. Introduction   161
  5.2. Scattering of electrons by a random distribution of centres;
        degenerate semiconductors   163
  5.3. Resistivity of liquid metals; Ziman's theory   165
  5.4. Resistivity of liquid alloys   170
  5.5. Thermoelectric power of liquid metals   172
  5.6. Hall effect in liquid metals   172
  5.7. Density of states   173
  5.8. Change of Knight shift on melting   174
  5.9. X-ray emission spectra and photoemission   175
  5.10. Liquid transition and rare-earth metals   176
  5.11. Amorphous metals and grain boundaries   177
  5.12. Injected electrons in liquid rare gases   179
  5.13. Liquid semimetals and semiconductors   181
  5.14. Liquid systems in which the depth of the pseudogap changes with
        volume   185
        5.14.1. Mercury   185
        5.14.2. Caesium   189
  5.15. Liquid alloys with pseudogaps in which the composition is varied   190

5.16. Liquid systems in which the depth of the pseudogap changes with
      temperature   194
      5.16.1. Tellurium   194
      5.16.2. Some liquid alloys of tellurium   197

6. NON-CRYSTALLINE SEMICONDUCTORS   199
   6.1. Introduction   199
   6.2. Preparation and classification of materials   200
   6.3. Methods of determining structure   204
   6.4. Electrical properties of non-crystalline semiconductors   209
         6.4.1. Density of states   210
         6.4.2. The mobility edge   215
         6.4.3. Temperature dependence of the d.c. conductivity   219
         6.4.4. Behaviour in the liquid state   222
         6.4.5. a.c. conductivity   223
         6.4.6. Thermopower   235
         6.4.7. Hall effect   240
         6.4.8. Magnetoresistance   242
         6.4.9. Field effect   243
   6.5. Drift mobility and photoconduction   247
         6.5.1. Drift mobility   247
         6.5.2. Photoconductivity   254
         6.5.3. Quantum efficiency   265
   6.6. Conduction at a mobility edge versus hopping   270
   6.7. Optical absorption   272
         6.7.1. Absorption edges and Urbach's rule   273
         6.7.2. Effect of externally applied fields   285
         6.7.3. Interband absorption   287
         6.7.4. Absorption at high energies   293
         6.7.5. Intraband absorption   297
         6.7.6. Photoluminescence   300
         6.7.7. Vibrational spectra. Density of phonon modes   301
   6.8. Other measurements   305
         6.8.1. Specific heat and thermal conductivity   305
         6.8.2. Photoemission and density of states   310
         6.8.3. Electron spin resonance   314

7. TETRAHEDRALLY-BONDED   SEMICONDUCTORS—
   AMORPHOUS GERMANIUM AND SILICON   320
   7.1. Methods of preparation   320
   7.2. Structure of amorphous Ge and Si   322
   7.3. Voids, impurities and other defects in amorphous Ge and Si   333
   7.4. Electrical properties of amorphous germanium   345
   7.5. Electrical properties of amorphous silicon   362
   7.6. Optical properties of amorphous germanium   375
   7.7. Optical properties of amorphous silicon   384
   7.8. Density of states in the valence and conduction bands of amor-
         phous germanium and silicon   396

8. ARSENIC AND OTHER THREE-FOLD
   CO-ORDINATED MATERIALS   408
   8.1. Introduction   408
   8.2. Forms and preparation of arsenic   408
   8.3. Structure of amorphous arsenic   410
   8.4. Electrical properties of amorphous arsenic   419
   8.5. Optical properties of amorphous arsenic and the density of states
         in the bands.   426
   8.6. States in the gap of amorphous arsenic   434
   8.7. Amorphous antimony, phosphorus, and related materials   439

9. CHALCOGENIDE AND OTHER GLASSES   442
   9.1. Introduction   442
   9.2. Structure   445
   9.3. Electrical properties of chalcogenide glasses   452
         9.3.1. Introduction   452
         9.3.2. d.c. conductivity   457
         9.3.3. Thermopower   458
         9.3.4. Hall effect   459
   9.4. States in the gap   460
         9.4.1. Introduction   460
         9.4.2. Screening length   468
         9.4.3. Field effect   469
   9.5. Drift mobility   470
   9.6. Luminescence   474
   9.7. Photoconductivity   483
   9.8. Effect of alloying on the dark current   484
   9.9. Numerical values in the model of charged dangling bonds   488
   9.10. Charge transport in strong fields   490
   9.11. Density of electron states in conduction and valence bands   491
   9.12. Optical properties   497
   9.13. Switching in alloy glasses   507
   9.14. Oxide glasses   512

10. SELENIUM, TELLURIUM, AND THEIR ALLOYS   517
    10.1. Structure of amorphous selenium and tellurium   517
    10.2. Optical properties of amorphous selenium and tellurium   521
    10.3. Electrical properties of amorphous selenium   530
           10.3.1. Electrical conductivity   530
           10.3.2. Drift mobilities   534
           10.3.3. Carrier lifetimes and ranges   539
           10.3.4. Photogeneration in amorphous Se; xerography   540
    10.4. Some properties of liquid selenium and Se–Te alloys   543

REFERENCES   548

INDEX   583

1

INTRODUCTION

THE principal subject matter of this book is those properties of non-crystalline materials that are due to the movement of electrons, particularly electrical conduction and optical absorption. Among non-crystalline materials are liquid metals and semiconductors, glasses, and amorphous films evaporated or deposited in other ways. A closely related subject also described in this book is the phenomenon of impurity conduction in semi-conductors, in which an electron moves directly (by tunnelling) from one impurity atom or point defect to another. Whether the material surrounding the impurities is crystalline or not, the impurity atoms are distributed at random, so that impurity conduction provides a particularly simple example of the movement of an electron in a non-periodic field of force.

The book starts with a description of the theoretical concepts necessary to describe these phenomena. Chapter 2 sets out a theory of non-interacting electrons in a rigid non-crystalline array of atoms. By a rigid array, we mean a model that neglects the effect of phonons and of distortions of the lattice, such as polarons, produced by an electron. For many of the phenomena described in this book, this is legitimate; for instance, the resistivity of a liquid metal is determined mainly by the scattering of electrons that results from the disordered arrangement of the atoms, and the resistance of a disordered alloy is normally calculated without considering the energy that may be transferred to an atom when an electron is scattered.†

Confining ourselves then to a rigid array of atoms, we have to ask first which of the concepts appropriate to crystalline solids can be used in non-crystalline materials. The first concept, equally valid for crystalline and for non-crystalline materials, is the density of states, which we denote by $N(E)$. The quantity $N(E)\,dE$ denotes the number of states in unit volume available for an electron with given spin direction with energies between E and $E+dE$. As in crystalline solids, the states can be occupied or empty, and $N(E)f(E)\,dE$ is the number of occupied states per unit volume, where f is the Fermi distribution function. The density of states can in principle be determined experimentally, for instance by photoemission. In general, the available evidence suggests that the form of the density of states in a liquid or non-crystalline material does not differ

† See Chapter 5.

greatly from the corresponding form in the crystal, except that the finer features may be smeared out, and some localized states may appear in the forbidden energy range in semiconductors. In contrast, the description of individual electron states used for electrons in crystalline materials is not always appropriate for the non-crystalline case. In crystalline materials, assuming a perfect crystal and neglecting the effect of phonons, we describe each electron by a Bloch wavefunction

$$\psi = u(x, y, z) \exp(i\mathbf{k} \cdot \mathbf{r}) \qquad (1.1)$$

where $u(x, y, z)$ has the periodicity of the lattice. The wavevector $\mathbf{k}$ is a quantum number for the electron. Because of phonons or impurities, scattering takes place, and a mean free path L is introduced; for instance, if there are N impurities per unit volume each with a differential scattering cross-section $I(\theta)$, the mean free path is given by

$$1/L = N \int_0^\pi I(\theta)(1 - \cos \theta) 2\pi \sin \theta \, d\theta. \qquad (1.2)$$

This formula assumes that the Fermi surface is spherical, so that $I(\theta)$ is independent of the initial direction of motion of the electrons. But it is, of course, characteristic of the conduction and valence bands of many crystalline solids that the energy $E(\mathbf{k})$ corresponding to the wavefunction (1.1) does depend on the direction of $\mathbf{k}$.

In non-crystalline materials there are two possibilities. One is that the mean free path is large, so that $kL \gg 1$. This is the case in most liquid metals, and in the conduction band of liquid rare gases and of some glasses such as SiO_2. The wavevector $\mathbf{k}$ is then still a good quantum number, and for metals a Fermi surface can still be defined; but, since the liquid or amorphous solid has no axis of symmetry, the Fermi surface must be spherical. In fact, since the mean free path is large, the deviation of the density of states from the free-electron form must be small. This will be shown in Chapter 5, which deals with liquid metals and other related problems. But if in a liquid or amorphous material the atomic potential (or pseudopotential) is strong enough to produce a band gap, or any large deviation from the free-electron form, then it must give strong scattering and a short mean free path ($kL \sim 1$). This in our view is the most important difference between the theories of crystalline and non-crystalline materials. In the latter case, phenomena frequently occur in which electrons have energies for which $kL \sim 1$. This is so, as we shall see in subsequent chapters, for the carriers in most amorphous and liquid semiconductors. Under such conditions the $\mathbf{k}$-selection rule[†] breaks down in optical transitions (Chapter 6). For conduction processes when this is the case, there is much

---

[†] This says that $\mathbf{k} - \mathbf{k}' \pm \mathbf{q} = 0$ when $\mathbf{k}, \mathbf{k}'$ are the wavenumbers before and after the transition and $\mathbf{q}$ is the wave vector of the light.

evidence that the Hall coefficient R_H is less than that predicted by the usual formula ($R_H = 1/nec$), and may even have the wrong sign.

It was first emphasized by Ioffe and Regel (1960) that values of L such that $kL < 1$ are impossible; this leads us to expect that, when the interaction of the carrier with atoms is sufficiently strong, something new ought to happen. It was first conjectured by Gubanov (1963) and by Banyai (1964) that near the edges of conduction or valence bands in most noncrystalline materials the states are *localized*, and the concept of localization will play a large part in this book. There is nothing unfamiliar about the concept of localized states; they are simply 'traps', and the most direct evidence for their existence in amorphous materials is provided by measurements of the transit time for injected carriers (Chapter 6); if this shows an activation energy, a trap-limited mobility can be inferred. The new concept for amorphous materials is that a continuous density of states, $N(E)$, can exist in which for a range of energies the states are all traps, or in other words localized, and for which the mobility at the zero of temperature vanishes, *even though the wavefunctions of neighbouring states overlap*. Moreover, at the bottom of a conduction band or top of a valence band, such localized states must necessarily occur in a disordered material.

The first phenomenon for which this was generally recognized was impurity conduction in doped and compensated semiconductors, which was first fully understood in the early 1960s. The centres in these materials are located at random positions, and in addition there is a random potential at each centre. This is discussed in Chapter 4. Our understanding of localization in this case derives from Anderson's paper (1958) on the absence of diffusion in certain random lattices which is central to our theme and is discussed in Chapter 2. In impurity conduction, each time an electron moves from one centre to another, it emits or absorbs a phonon; processes in which it absorbs a phonon are rate determining, and in consequence the conductivity contains an activation energy, so that it takes the form (cf. §4.3.1)

$$\sigma = \sigma_3 \exp(-\varepsilon_3/kT) \tag{1.3}$$

and tends to zero at low temperatures where, as we shall see, it behaves like $A \exp(-B/T^{1/4})$ instead of like (1.3). We call this form of charge transport *thermally activated hopping*, or just hopping. Hopping can also be responsible for an a.c. conductivity $\sigma(\omega)$ at frequency ω proportional to ω^s, where $s \simeq 0.8$. In this process an electron hops between pairs of localized states, absorbing or emitting a phonon each time.

In solid-state theory of crystalline materials a distinction is always made between those in which the density of states $N(E)$ is finite at the low-temperature Fermi energy E_F, and those in which E_F lies in a band gap so that $N(E_F)$ vanishes. The former are metals or metallic compounds, the conductivity σ tending to a finite value at low temperatures; the latter are semiconductors or insulators, the conductivity behaving at low temperatures like $\exp(-w/kT)$. In metals the conductivity depends only on those electrons that have energies near E_F and on their interaction with phonons and impurities. The same distinction can be made in non-crystalline systems. If $N(E_F)$ vanishes, the material is a semiconductor, but if $N(E_F) \neq 0$, it does not *necessarily* follow that σ tends to a finite value as T tends to zero. If the states with energy E_F are localized, conduction will be by hopping and σ will tend to zero with T, normally as $\exp(-\mathrm{const}/T^{1/4})$. Materials in which this is so have been called by Anderson (1970) 'Fermi glasses' and a discussion of them occupies a large part of this book. A doped and compensated semiconductor behaves like a Fermi glass at low temperatures. These, in common with many other materials, show a conductivity of Fermi-glass type at low temperatures, while at higher temperatures conduction is due to thermally excited electrons in non-localized (extended) states.

If in any non-crystalline material the states are localized at a band edge, then unless they are localized throughout the band there must exist an energy E_C separating localized and non-localized states, as first pointed out by Mott (1967). As we shall argue in the next chapter, they cannot coexist at the same energy. In several non-crystalline materials for which $N(E_F) \neq 0$, it is possible by altering the composition or some other parameter such as stress, to make E_F pass through E_C. In this case a transition occurs from metallic to hopping conduction, with a discontinuous change in the conductivity at zero temperature, which jumps from a finite value, which we write σ_{min}, to zero. We call this an Anderson transition, depending as it does on Anderson's (1958) localization theorem, and many examples will be described in this book (§ 2.8 and Chapter 4). The electrical behaviour of Fermi glasses at the Anderson transition gives the clearest evidence for the existence of the energy E_C and the minimum metallic conductivity σ_{min}.

In the conduction band of an amorphous semiconductor, an energy E_C will also separate the localized states at the band edge from the non-localized ones. This was first realized by Cohen, Fritzsche, and Ovshinsky (1969), who called this energy a 'mobility edge'. For energies below the mobility edge, an electron moves by hopping; the diffusion coefficient is of the form

$$D = \tfrac{1}{6}\nu_{\mathrm{ph}} a^2 \exp(-w/kT) \qquad (1.4)$$

where a is the distance between localized states, ν_{ph} depends on the phonon frequencies and is in some cases of order $10^{12}\,\mathrm{s}^{-1}$ and w will probably tend to zero with decreasing T. The mobility μ is given by the Einstein relation

$$\mu = eD/kT.$$

As the energy E of the electron approaches E_C, w will tend to zero (§ 2.7); but above E_C, as we shall see, the diffusion coefficient is of the form

$$\tfrac{1}{6}\nu_{el}a^2$$

where ν_{el} is an electronic frequency of order $\hbar/ma^2$. Once again μ is given by the Einstein relationship, so that a discontinuity in μ as a function of energy of order 10^2–10^3 is expected at the mobility edge. In semiconductors we shall find that the main part of the current is sometimes carried by electrons above, and sometimes below, the mobility edge.

In Chapter 2, where the general theory of the mobility edge is developed, we start from the simplest model of a disordered material, that of a crystalline array of potential wells with random depths, introduced by Anderson in (1958). This is the only case that has proved at all tractable mathematically, and we use it for discussions of localization, the minimum metallic conductivity σ_{min}, Hall coefficient, and so on. The model is fairly directly applicable to impurity conduction, but the use of results obtained from it for non-crystalline materials in general involves a certain amount of guesswork, which we shall try to justify by appeal to experiment.

Chapter 3 describes the effects of phonons. These are of three kinds.

(a) Phonons can scatter an electron with a non-localized wavefunction, making a contribution to the resistance just as in a crystalline metal or semiconductor.

(b) As we have seen, they can, by exchanging energy with an electron, enable it to hop from one localized state to another, as for instance in impurity conduction. They are also responsible for multiphonon transitions occurring when an electron and hole recombine without emission of radiation.

(c) They can be trapped by the electrons to form a *small polaron*. Effects such as polaron formation in which several phonons are trapped, so that the interaction between electron and phonon is not to be treated as a small perturbation, play a part in impurity conduction in all polar semiconductors and in other phenomena. In Chapter 3, therefore, a description is given of polaron behaviour in crystalline narrow-band semiconductors as an introduction to the related problems in non-crystalline materials.

In Chapter 4 we describe many phenomena involving a degenerate electron gas in a non-crystalline solid medium, particularly those in which states at the Fermi energy are localized and those in which a metal–insulator transition (the Anderson transition) occurs if the Fermi energy can be made to cross the mobility edge. This chapter also describes, as far as possible, the effect of electron–electron interaction on the behaviour of degenerate electron gases in a non-crystalline medium. Chapter 5 extends the discussion to liquids, starting with a brief account of Ziman's theory of liquid metals, which has been described in detail in other books (e.g. Faber 1972).

Chapter 6 and subsequent chapters describe the electric and optical properties of amorphous semiconductors. Here we face the problem of the nature of an amorphous film or a glass. We shall suppose, though this cannot be taken as certain, that a fully co-ordinated non-crystalline structure is possible, but that real materials, just like crystals, contain defects such as vacancies and dangling bonds. These in many materials are thought to produce deep donor and acceptor states which determine the position of the Fermi energy. Moreover, evaporated films frequently contain voids of diameter greater than an atomic distance; surface states on these voids will also affect the position of the Fermi energy.

The possibility of long-range fluctuations of potential, either due to charges on these voids or to other reasons, must always be kept in mind. If present, they may necessitate a treatment more akin to classical percolation theory than to the concepts developed in Chapter 2 (see § 2.9). In Chapter 6 and subsequent chapters we shall examine both possibilities.

2

THEORY OF ELECTRONS IN A
NON-CRYSTALLINE MEDIUM

2.1. Introduction
2.2. The Kubo–Greenwood formula
2.3. Anderson localization
2.4. Situation in which states are localized in one range of energies and not localized in another
2.5. Photon-activated hopping; the ω^2 law
2.6. The minimum metallic conductivity
2.7. Hopping and variable-range hopping
2.8. The Anderson transition
2.9. Mobility and percolation edges
2.10. Conductors, insulators, and semiconductors
2.11. Semimetals and pseudogaps
2.12. Some calculations of the density of states
2.13. Thermopower
2.14. Hall effect
2.15. Hopping conduction for alternating currents
2.16. One-dimensional problems

2.1. Introduction

This chapter introduces some of the theoretical concepts appropriate to the discussion of electronic processes in non-crystalline materials, particularly electrical conduction and optical absorption. Except where otherwise stated, the discussion will be in terms of the same approximation as that normally used in the elementary band theory of crystalline materials, the interaction energy e^2/r_{12} between electrons being neglected except in so far as it can be included in the averaged Hartree–Fock field. The effect of this interaction, which can assume considerable importance, is discussed in Chapter 4.

In crystalline materials the wavefunction of each electron is of the Bloch form (1.1). In non-crystalline materials the wavefunctions $\psi_E(x, y, z)$ do not necessarily have this form. Nevertheless, solutions of the Schrödinger equation must exist, and therefore the first concept that can be carried over from the theory of crystals to the theory of non-crystalline materials is the density of states $N(E)$, defined so that $N(E)\,dE$ is the number of eigen-states in unit volume for an electron in the system with given spin direction and with energy between E and $E+dE$. Then at a temperature T the number of electrons in the energy range dE is, for each spin direction,

$$N(E)f(E)\,dE$$

where $f(E)$ is the Fermi distribution function

$$f(E) = \frac{1}{\exp\{(E - E_\mathrm{F})/kT\} + 1}. \tag{2.1}$$

The Fermi energy E_F is a function of T and tends to a limiting value as $T \to 0$, E_F then separating occupied from non-occupied states.

For the calculation of $N(E)$ and the corresponding wavefunctions, there are two possibilities. The first is that the free-electron approximation is a good one and that the electrons are not strongly scattered. We may then write

$$E = \hbar^2 k^2/2m \tag{2.2}$$

where k is the wavevector and m the effective mass. In this case the Fermi surface is spherical, and the density of states for the electrons is given for each spin direction by the free-electron formula

$$N(E) = \frac{4\pi k^2}{8\pi^3} \bigg/ \frac{dE}{dk} = \frac{km}{2\pi^2\hbar^2}$$

$$= \frac{(Em^3/2)^{1/2}}{\pi^2\hbar^3}. \tag{2.3}$$

In crystalline materials the interaction with the field of the lattice can lead to large deviations from eqn (2.2), because the energy depends on the direction of k and also because of the formation of band gaps. Small deviations from a perfect lattice, such as those due to phonons, impurities, or defects, lead to a finite mean free path L (cf. eqn (1.2)), but unless L is small $(kL \sim 1)$ the changes in the density of states are not large. In non-crystalline materials, however, the disorder is responsible both for the finite mean free path and for deviations from eqn (2.3) for the density of states, and large deviations will occur if the scattering is strong. The following situations may therefore arise.

(i) The scattering by each atom is weak. The wavevector k is then a good quantum number, the uncertainty Δk in k is such that $\Delta k/k \ll 1$, the surfaces of constant energy are spherical and eqns (2.2) and (2.3) are valid. This is the situation in most liquid metals for values of E near the Fermi energy (Chapter 5).

(ii) The scattering by each atom is strong, so that $\Delta k/k \sim 1$. In this case k is not a good quantum number for describing the eigenstates, and the concept of a Fermi surface (for metals) is no longer valid. Ioffe and Regel

(1960) pointed out that under these conditions the mean free path is of the order $\sim 1/k$, and that it cannot be shorter than this. When $\Delta k/k \sim 1$, considerable deviation from (2.3) can occur for the density of states. (iii) If the interaction becomes yet stronger, a new phenomenon occurs which is absent in crystalline materials, namely that for a given energy E all† the wavefunctions ψ_E are localized. This means that each wavefunction ψ_E is confined to a small region of space, falling off exponentially with distance as $\exp(-\alpha R)$ and with a quantized energy value. This was first recognized by Anderson (1958) in his paper on 'The absence of diffusion in certain random lattices' and this form of localization is known as Anderson localization. There is of course nothing new in the concept of a localized wavefunction for an electron in a trap below the conduction band. What is new is the concept that one can have a finite and continuous density of states $N(E)$ in which all states are localized, although there can be strong overlap between the wavefunctions of neighbouring states. Moreover, as we shall show, if states are filled up to a limiting Fermi energy E_F in the range where states are localized, the conductivity, which we denote by σ_E, vanishes as the temperature tends to zero.‡ In crystals, the condition for insulating behaviour is the vanishing at E_F of $N(E)$; in non-crystalline materials insulating behaviour is compatible with a finite value of $N(E_F)$. Materials where this is so, states at E_F being localized, have been called 'Fermi glasses' (Anderson 1970).

We argue that, if for a given energy some states are localized, then for that energy they must *all* be localized.§ If a non-localized state exists, it will have the result that any state with the same energy which might otherwise be localized will become a 'virtual bound state' in the sense described by Friedel (1954). Localized and non-localized states cannot coexist at the same energy for a given configuration (Cohen 1977). In contrast, in any non-crystalline system, for instance a liquid, a great many configurations of the atoms are possible; we call the totality of such configurations an *ensemble*. Any calculated quantity that is to be compared with experiment must be averaged over all configurations of the ensemble. These are bound to include some for which states are non-localized, for instance the crystalline configuration. Our statement, then, that σ_E can

---

† In terms of an ensemble average, very nearly all; see below.

‡ It was many years after Anderson's original work before this was generally accepted. Intuitively one might think that an electron could find a state with its own energy if it tunnelled far enough. This turns out to be wrong, as we shall see. Lloyd (1969), Brouers (1970), and Ziman (1969) assumed, we believe incorrectly, that, if G is the Green's function for the system, a finite value of $\langle G \rangle$ was sufficient to yield non-localized states. Anderson (1970), Thouless (1970, 1974), and Mott (1974b) gave contrary arguments.

§ An experimental proof that localized and extended states do not coexist at a given energy in one typical system, the impurity band of crystalline silicon doped with phosphorus, is due to Geschwind, Romestain, and Devlin (1976) and is described in Chapter 4.

vanish must mean that $\langle \sigma_E \rangle = 0$, where the angular brackets denote a configurational average. The contribution to $\langle \sigma_E \rangle$ for non-localizing configurations must tend to zero as the volume of the specimen tends to infinity.

For this reason we think that the most satisfactory definition of localization for wavefunctions of energy E is that

$$\langle \sigma_E \rangle = 0. \tag{2.4}$$

We need therefore a method for calculating this quantity applicable to the case of localization, and also for short mean free paths $\Delta k/k \sim 1$, as well as to the normal case when $\Delta k/k \ll 1$ and the Boltzmann formulation is applicable. We do this by considering an electromagnetic wave of small frequency ω acting on a medium in which the states are full up to a limiting energy E, and deducing the conductivity $\sigma_E(\omega)$ at zero temperature. The d.c. conductivity for such a system is given by

$$\sigma_E = \sigma_E(0) = \lim_{\omega \to 0} \sigma_E(\omega). \tag{2.5}$$

If this vanishes even though $N(E)$ is finite, the system is a Fermi glass.

For a semiconductor, that is to say a system in which the current is carried by excited electrons rather than being determined by those at E_F, the quantity σ_E can also be used to describe the conductivity at a finite temperature, if interaction with phonons is neglected. This conductivity is

$$\sigma = -\int \sigma_E \frac{\partial f}{\partial E}\, dE$$

where f is the Fermi distribution function (2.1). We shall see later that an energy E_C always exists which separates energies where states are localized and non-localized in a conduction or valence band. The contribution to the conductivity of states in a conduction band above E_C (extended states) is thus

$$\sigma_{\min} \exp\left\{ -\frac{(E_C - E_F)}{kT} \right\} \tag{2.6}$$

where $\sigma_{\min}$ is the value† of σ_E at $E = E_C$. If this is written

$$N(E_C)kTe\mu \, \exp\left\{ -\frac{(E_C - E_F)}{kT} \right\}$$

where μ is the mobility for electrons with energy E_C, then

$$\mu = \frac{\sigma_{\min}}{N(E_C)ekT}. \tag{2.7}$$

† The reason for this notation is explained in § 2.5.

The mobility for electrons with energies below E_C, which must hop from one state to another, involves interaction with phonons and cannot be determined from σ_E. This will be the subject of later sections.

An expression for σ_E obtained from (2.5), known as the Kubo–Greenwood formula,† will be the basis of many of the considerations of this book, and will be derived in the next section.

2.2. The Kubo–Greenwood formula

We shall now deduce formulae for the quantities $\sigma_E(\omega)$ and $\sigma_E(0)$ introduced in the last section. The calculation will be carried out for a degenerate electron gas at zero temperature, states being occupied up to an energy E_F. Suppose that the eigenfunctions for an electron with energy E in the non-periodic field, with any appropriate boundary conditions, are $\psi_E(x, y, z)$, and that these are normalized to give one electron in a volume Ω. Suppose that an alternating field $F \cos \omega t$ acts on an electron so that the potential energy is $exF \cos \omega t$. Then the chance per unit time that an electron makes a transition from a state with energy E to any of the states with energy $E + \hbar\omega$ is

$$\frac{1}{4}e^2F^2\frac{2\pi}{\hbar}|x_{E+\hbar,E}|^2_{\text{av}}\Omega N(E+\hbar\omega). \tag{2.8}$$

The matrix element $x_{E',E}$ is defined by

$$x_{E',E} = \int \psi_{E'}^* x \psi_E \, d^3x$$

and the suffix av represents an average over all states having energy near $E' = E + \hbar\omega$. It is convenient to write

$$x_{E+\hbar\omega,E} = \frac{\hbar}{m\omega} D_{E+\hbar\omega,E} \tag{2.9}$$

where

$$D_{E',E} = \int \psi_{E'}^* \frac{\partial}{\partial x}(\psi_E) \, d^3x.$$

Thus (2.8) becomes

$$\frac{\pi e^2 \hbar \Omega}{2m^2\omega^2}F^2|D|^2_{\text{av}}N(E+\hbar\omega). \tag{2.10}$$

We now introduce as in the last section the conductivity for frequency ω, written $\sigma_E(\omega)$ and defined so that $\sigma_E(\omega)\frac{1}{2}F^2$ is the mean rate of loss of energy per unit volume. To obtain this, we must multiply (2.10) by

† Kubo (1956), Greenwood (1958).

$N(E)f(E)\,dE$, the number of occupied states per unit volume in the energy range dE, by $\{1-f(E+\hbar\omega)\}$, the chance that a state with energy $E+\hbar\omega$ is unoccupied, by $\hbar\omega$, the energy absorbed in each quantum jump, and finally by 2 for the two spin directions. We find, integrating over all energies,

$$\sigma_E(\omega)=\frac{2\pi e^2\hbar^2\Omega}{m^2\omega}\int [f(E)\{1-f(E+\hbar\omega)\}$$

$$-f(E+\hbar\omega)\{1-f(E)\}]|D|^2_{av}N(E)N(E+\hbar\omega)\,dE. \qquad (2.11)$$

The second term in the square brackets gives the energy absorbed in stimulating downward jumps. $|D|^2$ is now averaged over all initial and final states. The quantity in the square brackets simplifies to

$$f(E)-f(E+\hbar\omega)$$

so that (2.11) reduces to

$$\sigma_E(\omega)=\frac{2\pi e^2\hbar^3\Omega}{m^2}\int \frac{\{f(E)-f(E+\hbar\omega)\}|D|^2_{av}N(E)N(E+\hbar\omega)}{\hbar\omega}\,dE. \qquad (2.12)$$

When $T=0$, (2.12) becomes

$$\sigma_E(\omega)=\frac{2\pi e^2\hbar^3\Omega}{m^2}\int \frac{|D|^2_{av}N(E)N(E+\hbar\omega)}{\hbar\omega}\,dE. \qquad (2.13)$$

The lower limit of integration is $E_F-\hbar\omega$, this being the lowest energy of an electron that can absorb a quantum; the upper limit is E_F; E_F is now the Fermi energy at zero temperature.

To obtain the d.c. conductivity we take the limit of $\sigma(\omega)$ when $\omega\to 0$. At $T=0$ this depends only on the values of the quantities in the integral when $E=E_F$. We define $\sigma_E(0)$ by

$$\sigma_E(0)=\frac{2\pi e^2\hbar^3\Omega}{m^2}|D_E|^2_{av}\{N(E)\}^2 \qquad (2.14)$$

where

$$D_E=\int \psi^*_{E'}\frac{\partial}{\partial x}\psi_E\,d^3x \qquad (E=E').$$

The av represents an average over all states E and all states E' such that $E=E'$, so that at $T=0$ the conductivity $\sigma(0)$ is given by

$$\sigma(0)=\{\sigma_E(0)\}_{E=E_F}$$

where E_F is the Fermi energy. Eqn (2.14) is the Kubo–Greenwood formula.

If states with energy E are localized, all the functions D_E vanish, because $\int \psi_E (\partial/\partial x) \psi_E \, d^3x$ is zero, and overlap between two localized functions ψ_1, ψ_2 with the same energy is impossible. This is because, if the overlap is finite, the degeneracy will be removed by forming two functions of the form $A_1\psi_1 + A_2\psi_2$ and $B_1\psi_1 + B_2\psi_2$ with an energy separation depending on the overlap.

If scattering is weak and the mean free path L is long ($kL \gg 1$), the Kubo–Greenwood treatment should give the same result as the Boltzmann formulation. This is, for a spherical Fermi surface,

$$\sigma_E = \frac{S_E e^2 L}{12\pi^3 \hbar} \tag{2.15}$$

where S_E is the area $4\pi k^2$ enclosed by the surface in k-space of energy E. If $E = E_F$, S_E is the area of the Fermi surface, which we denote by S_F. Eqn (2.15) is easily shown to be equivalent to

$$ne^2\tau/m \qquad (\tau = L/u) \tag{2.16}$$

where n is the number of electrons per unit volume and $u(=\hbar k/m)$ is the velocity at the Fermi surface.

A rough demonstration that the Kubo–Greenwood formula yields (2.15), using a method that we shall employ when kL is not large, is as follows (Mott 1970). We define the mean free path as the distance L in which the phase of ψ loses all memory of its value. If we introduce a volume v equal to that of a sphere with radius L, so that

$$v = 4\pi L^3/3$$

the phases of the wavefunctions in any two of these volumes will be uncorrelated. Thus, if δ is defined by

$$\delta = \int^v \psi_{k'}^* \frac{\partial}{\partial x} \psi_k \, d^3x$$

then D is equal to the sum of Ω/v contributions, each equal to δ but with random signs. We may thus write

$$D = (\Omega/v)^{1/2} \delta.$$

To evaluate δ we write

$$\delta = \int^v \exp\{i(\mathbf{k} - \mathbf{k}') \cdot \mathbf{r}\} \frac{d^3x}{\Omega}$$

and, setting

$$|\mathbf{k} - \mathbf{k}'| = 2k \sin \tfrac{1}{2}\theta \simeq k\theta$$

where θ is the scattering angle, we approximate by writing

$$\delta = kv/\Omega \quad \text{if } kL\theta < 1,$$

$$= 0 \quad \text{otherwise.}$$

Thus, averaging over all angles θ between 0 and $1/kL$

$$|D|^2_{\text{av}} = \frac{\Omega}{v}\frac{k^2v^2}{\Omega^2}\int^{1/kL}\frac{2\pi \sin \theta}{4\pi}\,\mathrm{d}\theta = \pi L/3\Omega. \tag{2.17}$$

Substituting from (2.3) for $N(E)$ we see from (2.14) that $\sigma = e^2k^2L/6\pi^2\hbar$, which is the same as (2.16) apart from a factor of 2.

This approximate method does not give the correct numerical factor; a method due to Thouless (1975) is exact. We assume that the eigenfunctions of energy $k_i^2/2m$ are superpositions of plane waves with random phases and that the mean square amplitude of a plane-wave component is the same as one would obtain if one made a state with complex wavenumber $k_i - i/2L$, where L is the mean free path. That is, we take the state of energy $k_i^2/2m$ to have the form

$$|i\rangle = \sum_k a_k^i |k\rangle \tag{2.18}$$

where the amplitudes a_k^i are independent Gaussian random variables whose variance is given by

$$\langle a_k^{i*} a_{k'}^i\rangle = \delta_{ij}\,\delta_{kk'}\frac{\pi}{L\Omega kk_i}\left\{\frac{1}{(k-k_i)^2+\tfrac14 L^2}-\frac{1}{(k+k_i)^2+\tfrac14 L^2}\right\}$$

$$\simeq \frac{\delta_{ij}\,\delta_{kk'}\pi}{Lk_F^2\{(k-k_i)^2+\tfrac14 L^2\}} \tag{2.19}$$

where the second form is a suitable approximation for a degenerate Fermi gas with $k_F L \gg 1$.

Using (2.18) and (2.19) we can work out the quantities $|D|^2$ as

$$\hbar^2 \sum\sum a_k^{i*} a_k^j a_{k'}^{j*} a_{k'}^i k_x k_x'$$

and, since $i \neq j$, only the terms with $k = k'$ contribute to the average, and we obtain, writing $p = (\hbar/i)D$

$$\langle|p_{ij}|^2\rangle = \frac{1}{3}\frac{\hbar^2 k_F^2\pi^2}{L^2\Omega^2 k_F^4}\sum\frac{1}{\{(k-k_i)^2+\tfrac14 L^2\}\{(k+k_i)^2+\tfrac14 L^2\}}$$

$$= \frac{2\hbar^2\pi^3 L}{3\Omega\{1-(k_i-k_j)^2 L^2\}}. \tag{2.20}$$

Substitution of (2.20) in eqn (2.13) gives

$$\sigma(\omega) = \frac{e^2 k_F^2 L}{3\pi^2 \hbar} \frac{1}{\{1 + m^2 \omega^2 L^2 / \hbar^2 k_F^2\}}. \tag{2.21}$$

For $\omega = 0$ this gives (2.16), and in addition gives the Drude term $1/(1 + \omega^2 \tau^2)$ for a.c. conduction, since $\tau = L/u = L/(\hbar k/m)$.

The value of the mean free path L depends on the scattering mechanism. Edwards (1958), taking weak-scattering potentials, also derived the Boltzmann formula starting from the Kubo–Greenwood formalism. He proved directly that, with $I(\theta)$ given by the Born approximation, eqn (2.15) follows from (2.14).

If the density of states differs from the free-electron value, we may introduce the factor g, defined (Mott 1967) by

$$g = \frac{N(E_F)}{N(E_F)_{\text{free}}}. \tag{2.22}$$

(2.14) then becomes

$$\sigma = \frac{S_F e^2 L g^2}{12\pi^3 \hbar} \tag{2.23}$$

L being defined as the distance in which electrons lose phase memory. If, however, we write L_0 for the value of L calculated by first-order perturbation theory, then as shown by Edwards (1961)

$$L = L_0/g^2 \tag{2.24}$$

so (2.15) remains true if L_0 is written for L. The relation (2.24) is valid because, if τ is the time of relaxation, $1/\tau$ must be proportional to the density of the states into which the electron is scattered; also

$$L = u\tau$$

and a high density of states implies a low value of dE/dk (eqn (2.3)) and hence of the velocity u. As we shall see later, however, the factor g^2 in (2.23) becomes important if L has its minimum value a, which in the model to be introduced in the next section is the distance between atoms.

2.3. Anderson localization

We stated in the last section that sufficient disorder can produce characteristic solutions of the Schrödinger equation which are localized in space. The first paper which proved this, and gave a quantitative criterion for localization, was that by Anderson (1958). His result will now be described.

Starting from the Schrödinger equation

$$\nabla^2 \psi + \frac{2m}{\hbar^2}(E - V)\psi = 0 \qquad (2.25)$$

he uses the *tight-binding approximation*, in which a crystalline array of potential wells produces a narrow band of levels, as in Fig. 2.1(a). Applications could be to the d band of a transition metal or to donors producing

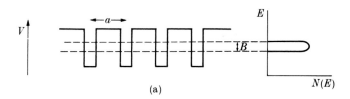

(a)

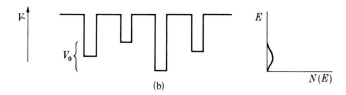

(b)

Fig. 2.1. (a) Potential wells for a crystalline lattice. (b) Potential wells for the Anderson lattice. The density of states $N(E)$ is also shown.

a metallic impurity band in a semiconductor (Chapter 4). We suppose that the wells are so far apart that the overlap between the atomic wavefunctions $\phi(\mathbf{r})$ on adjacent wells is small. If the suffix n describes the nth well and $\mathbf{R}_n$ its lattice site, the Bloch wavefunction for an electron in the crystal is

$$\psi_k(x, y, z) = \sum_n \exp(i\mathbf{k} \cdot \mathbf{R}_n)\phi(\mathbf{r} - \mathbf{R}_n). \qquad (2.26)$$

We take the functions ϕ to be spherically symmetrical (s functions). If then W_0 is the energy level for an electron in a single well, the energies for an electron in a simple cubic lattice corresponding to the wavefunctions (2.26) are

$$E = W_0 + W_k$$

where

$$W_k = -2I(\cos k_x a + \cos k_y a + \cos k_z a).$$

Here, I is the transfer integral given by

$$I = \int \phi^*(\mathbf{r} - \mathbf{R}_n) H \phi(\mathbf{r} - \mathbf{R}_{n+1}) \, d^3x \qquad (2.27)$$

where H is the Hamiltonian. The transfer integral occurs many times in this book. It depends on the shape of the wells, but for our purpose it will be sufficient to write it as

$$I = I_0 \exp(-\alpha R) \qquad (2.28)$$

Here α is defined so that $\exp(-\alpha r)$ is the rate at which the wavefunction on a single well falls off with distance $(\alpha = (2mW_0)^{1/2}/\hbar)$. For hydrogen-like wavefunctions I_0 can be evaluated and is (Slater 1963)

$$I_0 = \{\tfrac{3}{2}(1 + \alpha R) + \tfrac{1}{6}(\alpha R)^2\} e^2 \alpha. \qquad (2.29)$$

The effective mass m^* at the bottom of a band is

$$m^* = \hbar^2/2Ia^2 \qquad (2.30)$$

and the bandwidth B is

$$B = 2zI \qquad (2.31)$$

where z is the co-ordination number.

Our problem is to consider what happens to this band of energies when the potential-energy function V is non-periodic. A non-periodic potential can be formed in two ways:

(a) By the displacement of each centre by a random amount, as for instance by lattice vibrations or by the destroying of the long-range order (lateral disorder) as in a liquid.

(b) By the addition of a random potential $\tfrac{1}{2}V$ to each well (vertical disorder); Anderson supposed that V took all values at random between $\pm V_0$, so that V_0 is the spread of energies. Other distribution functions, such as the Gaussian, are of course admissible.

We consider case (b) first. If V_0 is small, a large mean free path is introduced. An application of the Born approximation (Mott and Massey 1965, p. 86) gives

$$\frac{1}{L} = \frac{2\pi}{\hbar} \tfrac{1}{2}(\tfrac{1}{2}V_0)^2 a^3 \frac{N(E)}{u} \qquad (2.32)$$

E and the velocity u being taken at the Fermi energy. Using (2.3) for $N(E)$, (2.32) reduces to

$$a/L = \frac{(V_0/I)^2}{32\pi}. \qquad (2.33)$$

In Chapter 1 the rule of Ioffe and Regel was introduced, according to which a mean free path such that $kL < 1$ is impossible. With the potential energy illustrated in Fig. 2.1 and no disorder, $ka = \pi$ in the middle of the

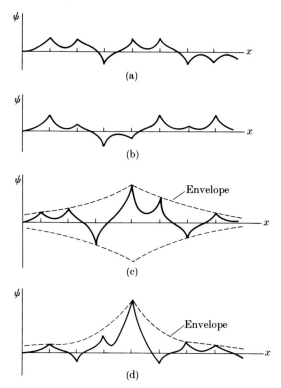

Fig. 2.2. Form of the wavefunction in the Anderson model: (a) when $L \sim a$; (b) when states are just non-localized $(E \gtrsim E_C)$; (c) when states are just localized $(E \lesssim E_C)$; (d) strong localization.

band. The Ioffe–Regel rule then means that the shortest possible mean free path arises when the wavefunction loses phase memory in going from atom to atom, so that instead of (2.26) it has the form

$$\sum_n A_n \phi(\mathbf{r} - \mathbf{R}_n) \qquad (2.34)$$

where the A_n have random phases and amplitudes, as in Fig. 2.2(a). In such a case we write for the magnitude of the mean free path

$$L \sim a.$$

Eqn (2.33) leads us to suppose that this occurs when

$$V_0/I \simeq \sqrt{(32\pi)} \sim 10$$

or if the co-ordination number z is 6,

$$V_0/B \simeq 0.83. \qquad (2.35)$$

To see what happens when V_0 exceeds this value, we consider a pair of wells at a distance R from each other with energies shifted from the mean by amounts V_a, V_b. The two wavefunctions for a pair of electrons in these states are

$$\psi_1 = A\phi_a + B\phi_b$$
$$\psi_2 = B\phi_a - A\phi_b \qquad (2.36)$$

ϕ_a, ϕ_b being the atomic wavefunctions. The values of A, B and the energies E_1, E_2 can be found by minimizing the energy integral; the results are rather complicated (see, for instance, Miller and Abrahams 1960), and we need quote only the following limiting cases:

(a) If $|V_a - V_b| \ll I$, then $A \sim B$ and $E_1 - E_2 \approx 2I$. It cannot be less than $2I$.

(b) If $|V_a - V_b| \gg I$, then

$$\frac{A}{B} \simeq \frac{|V_a - V_b|}{2I}.$$

Wavefunctions for the two cases, and also a plot of $E_1 - E_2$ as a function of $|V_a - V_b|$, are shown in Fig. 2.3.

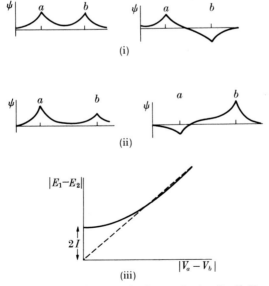

Fig. 2.3. Wavefunctions of odd and even parity for a pair of wells: (i) $V_a = V_b$; (ii) $V_a < V_b$ and $V_b - V_a \gg 2I$; (iii) plot of the difference in the energies of the two states.

If we turn again to the infinite array of wells, the form of ψ for a pair suggests that there should be random fluctuations of the amplitude (as well as of the phase) of ψ in going from well to well, and that as V_0/B increases, these fluctuations become larger (Fig. 2.2(b)). This undoubtedly occurs. However, if V_0/B is very large, one would expect intuitively that the wavefunctions for each isolated well would be little perturbed by all the other wells, and so would fall off exponentially with distance as in Fig. 2.2(c) and 2.2(d). The important questions are: Does this in fact occur, and if so at what value of V_0/B? To answer this question, Anderson took the potential of Fig. 2.1(b) and asked the following question. If at time $t = 0$ an electron is placed on one of the wells, what happens at $t \to \infty$? Is there a finite probability at the absolute zero of temperature that the electron will have diffused to large distances or does the chance that an electron will be found at a large distance r vary as $\exp(-2\alpha r)$, in which case there is no diffusion? Anderson found that there is no diffusion if V_0/B is greater than a constant that depends on the co-ordination number z. This means that, if V_0/B is greater than this constant, all the wavefunctions for an electron in the system are of the type shown in Fig. 2.2(c), decaying with distance r from the neighbourhood of some well n. The wavefunctions of the localized states are of the form (instead of (2.34))

$$\sum A_n \phi(\mathbf{r} - \mathbf{R}_n) \exp(-\alpha r) \qquad (2.37)$$

the coefficients A_n having random phases as before.

Another and perhaps preferable way of expressing the Anderson condition is to take for the initial state an electron localized in a large volume of diameter Δx, the energy being defined as closely as allowed by the uncertainty principle. Then, if the energy E is taken at the centre of the band, there will be no diffusion if the condition is satisfied; if not the amplitude within Δx will tend to zero as the time t increases. This description allows us to ask whether states are localized at a given energy E.

There is now an extensive literature on the critical value of V_0/B for Anderson localization. Anderson (1958) found for co-ordination number 6, a value of 5·5; Edwards and Thouless (1972), as a result of numerical calculations, found a much smaller value, about $\frac{4}{3}$ in two dimensions and about 2 in three. Herbert and Jones (1971) also find that a smaller value is likely. Schönhammer (1971), using a probability density for the random energies V_i of the form

$$P(V_i) = \frac{2}{\pi B^2}(B^2 - V_i^2)^{1/2} \qquad (2.38)$$

obtained the value 2·2. Numerical calculations by Schönhammer and Brenig (1973) confirm this. The value is roughly proportional to the co-ordination number z (Economou and Cohen 1970a, b, c, 1972), though for

$z = 2$ it becomes zero, *all* states in a one-dimensional random chain being localized (§ 2.16). Economou and Cohen (1972), using a Lorentzian distribution const./$(V_i^2 + \Gamma^2)$, found $\Gamma/B = \frac{1}{2}$, a much smaller value; Kikuchi (1970) found $V_0/B \approx 4$. Abou-Chacra, Anderson, and Thouless (1973) found by analytical methods a value closer to that of Anderson (1958), but conclude that the numerical results are more reliable. A review of all this work has been given by Thouless (1974).

Many applications, for instance to impurity conduction, are for a random array of impurities, and for these we do not know the appropriate value of z. Experimental estimates of the value come from measurements of the minimum metallic conductivity (§ 2.6 and Chapter 4) and suggest $V_0/B \sim 2$ for s functions and about 1 for d functions for which the effective coordination number is smaller.

For lateral (non-diagonal) disorder, that is for a random distribution of centres in space, the condition for Anderson localization has been estimated roughly by Mott (1973a, 1977a) as follows. Following Lifshitz (1964), we pair each atom with its nearest neighbour, which we suppose to be at a distance r_1 from it, given by

$$\frac{4\pi}{3} r_1^3 = \frac{1}{N}$$

where N is the number of centres per unit volume. The energy of an electron located on any two centres in a pair is $I_0 \exp(-\alpha r_1)$ and, since r_1 varies from pair to pair, we take this quantity as V_0 in the Anderson formula (a *very* rough approximation). The mean distance between pairs is

$$r_2 = 1/(\tfrac{1}{2}N)^{1/3}$$

so we suppose the band width B to be

$$2zI_0 \exp(-\alpha r_2)$$

and, putting in the Anderson criterion in the form $V_0 = 2B$, we have

$$\exp(-\alpha r_1) = 4z \exp(-\alpha r_2).$$

Thus

$$\alpha(r_2 - r_1) = \ln(4z) \approx 3\cdot 2 \qquad \text{if } z = 6$$

whence

$$\alpha N^{-1/3}\{2^{1/3} - (\tfrac{3}{4}\pi)\}^{1/3} \approx 3\cdot 2.$$

The term in braces is equal to $0\cdot 63$, so if we write $\alpha^{-1} = a_H$, we find

$$N^{1/3}a_H = 0\cdot 2. \qquad (2.39)$$

Mott (1977f) suggests that, because of the directional properties of the wavefunction for a pair, z should be nearer 2 and $V_0 = B$ would be more appropriate, which leads to 0·4 instead of 0·2 in (2.39). Debney (1977), in a computer-based study, finds $1/(N^{1/3} a_H)$ to be 4 in three dimensions and 3 in two, but concludes that these are upper bounds, the true value being up to 25 per cent lower. If so, the right-hand side of (2.39) should be in the range 0·3–0·35. Economou and Antoniou (1977), however, find that pure Anderson localization is not possible in the centre of a band with only off-diagonal disorder. If this is correct, for a half-full impurity band as discussed in Chapter 4 the additional effect of electron–electron interaction must allow it. Probably for the same reason, Hoshino and Watabe (1977) find that the relationship $\sigma \propto \{N(E)\}^2$ does not hold for off-diagonal disorder; it would not be expected unless $L \sim a$ (see also p. 124).

Turning now to diagonal disorder, as the value of V_0/B approaches the critical value, the quantity α in the middle of the band in eqn (2.37) will tend to zero. According to Abram and Edwards (1972), Anderson (1972a), and Lukes (1972) it behaves like

$$\text{const}\{(V_0/B) - (V_0/B)_{\text{crit}}\}^{0\cdot6}. \tag{2.40}$$

Freed (1972) obtains a slightly different value of $\frac{2}{3}$ for the exponent. This behaviour is discussed in more detail in connection with the behaviour of the wavefunctions at the mobility edge, and these indices are perhaps open to doubt.

The Anderson criterion in two dimensions is discussed by Licciardello and Thouless (1975) and by Yoshino and Okazaki (1977), who obtain similar results, $V_0/B \sim 1$. In one-dimension, all states are localized (§ 2.16); in two and three, as we have seen, a criterion exists for localization; in four according to Toulouse (1975) and Toulouse and Pfeuty (1975) localization cannot occur.

2.4. Situations in which states are localized in one range of energies and not localized in another

This can occur for an electron with the potential energy illustrated in Fig. 2.1(b), if the Anderson criterion is not satisfied. σ_E will then be finite in the middle of the band, but zero for energies near its extremities. In fact, any form of random potential, however small, will introduce a range of localized states at the band tail. This being the case, a critical energy E_C must separate localized from non-localized states, defined so that

$$\langle \sigma_E \rangle = 0 \qquad E < E_C$$
$$\neq 0 \qquad E > E_C. \tag{2.41}$$

This is illustrated for a density of states resulting from the Anderson potential in Fig. 2.4; E'_C separates localized and non-localized states at the

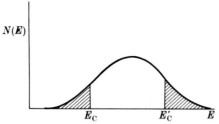

$N(E)$

E_C E'_C E

Fig. 2.4. Density of states in the Anderson model when states are non-localized in the centre of the band. Localized states are shown shaded. E_C, E'_C separate the ranges of energy where states are localized and non-localized.

top edge of the band. There is no discontinuity in $N(E)$ nor in any of its derivatives at E_C, as shown by Thouless (1970).

The existence of an energy E_C was first pointed out by Mott (1966, 1967) and follows rigorously if we define localization at a given energy E by the equation $\sigma_E = 0$, because σ_E cannot be both zero and non-zero. The concept was introduced by Cohen *et al.* (1969) for the conduction band of an amorphous semiconductor; E_C is then referred to as the 'mobility edge'. Determination of the position of a mobility edge on the Anderson model has received considerable attention, calculations being made by Economou and Cohen (1970a,b, 1972), Schönhammer (1971), and Abou-Chacra and Thouless (1974). The latter authors calculate the position of the mobility edge with the same square distribution function for V_0 as used by Anderson (1958)

$$P(V) = 1/V_0 \qquad V < \tfrac{1}{2}V_0$$
$$= 0 \qquad \text{otherwise.}$$

Fig. 2.5 is reproduced from their paper, in which E_C is plotted against V_0/B for $z \sim 4$. For weak disorder they find that the distance of E_C from

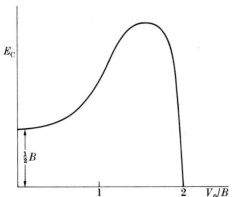

E_C

$\tfrac{1}{2}B$

1   2 V_0/B

Fig. 2.5. Plot of E_C against V_0/B measured from mid-band. (From Abou-Chacra and Thouless (1974).)

the bottom of the band is given by

$$E_C - E_A \simeq \tfrac{1}{4} V_0^2 / KI \tag{2.42}$$

where $K = z - 1$, so that B (the bandwidth without disorder) is $2zI$. The term $\tfrac{1}{2} V_0^2$ could be replaced by the mean square of the deviation of the site energy whatever its form. As V_0 increases, the band broadens and E_C moves into its tail, so (E_C being measured from mid-gap) the numerical value increases. Eventually, when $V_0/B \sim 2$, E_C moves to mid-gap. Similar behaviour is obtained by Schönhammer (1971).

A calculation of the band form with this model is given by Mookerjee (1973), and discussed further in § 2.12. The bottom of the band is at $E_A - \tfrac{1}{2} V_0$; this occurs when statistical fluctuations give a large number of wells with maximum depth close together. This will of course be very improbable, so an exponential tail occurs. E_C lies in the tail unless V_0/B is quite near to, say, half the Anderson value.

For lateral disorder the problem has been attacked by Kikuchi (1974) using the Lifshitz method of pairs as in the last section and the formula of Abou-Chacra and Thouless (eqn (2.42)). An interesting result is that localization is stronger at the top of the band than in the tail, because of the weak overlap integral I between pairs in antibonding states.

The behaviour of the wavefunctions for energies near E_C is of great importance, and has been the subject of considerable controversy. The form of the wavefunction for E just on the localized side of E_C is shown in Fig. 2.2(c). It can be written for large r in the form (2.37), where $\exp(-\alpha r)$ represents the envelope shown by the dotted line. α must tend to zero as $E \to E_C$, just as it must in the middle of the band when $V_0 \to (V_0)_{crit}$. Therefore the spatial extent of the wavefunctions becomes large and they overlap strongly. An early conjecture on dimensional grounds (Mott 1969a) was that

$$\alpha \propto \frac{\{2m(E_C - E)\}^{1/2}}{\hbar}$$

where m is the effective mass. Later investigations by Lukes (1972), Abram and Edwards (1972), Anderson (1972a), and Freed (1972) predict that, near the mobility edge, α behaves like $(E_C - E)^s$, where $s = 0.6$ (or $\tfrac{2}{3}$ according to Freed), so a dimensionally correct formula would be

$$\alpha \propto \alpha_0 \left(\frac{E_C - E}{E_C} \right)^s \tag{2.43}$$

where α_0 is the value of α far from the mobility edge. It follows that near the mobility edge each wavefunction strongly overlaps a number of other wavefunctions, the number being of order $(4\pi/3)(\alpha a)^3$. Hopping, therefore, is possible without tunnelling through any region in which $\exp(-\alpha r)$

decays appreciably. In two dimensions, according to Abram (1973), the index in (2.43) should be 0·75. The values of these indices are, however, open to doubt,† and in § 2.9, following considerations due to Mott (1976c), we shall propose $s = \frac{2}{3}$ (Freed's value) in three dimensions and $s = 1$ in two. There is some experimental indication that the latter value is correct in two dimensions (§ 4.6), while in three dimensions the experimental work of Sayer *et al.* (1975) discussed in § 4.7.3 indicates a value near 0·6; one cannot, however, distinguish between this and $\frac{2}{3}$.

When $E = E_C$, the quantity α becomes zero and the wavefunction has the form (2.34) illustrated in Fig. 2.2(b). These wavefunctions spread through all space and are called 'extended'. The conductivity σ_E can be calculated using the Kubo–Greenwood formula (2.14) if it is assumed that the phases of A_n in (2.34) vary in a random way from atom to atom, but that fluctuations in the magnitude of A_n do not have a major effect. This calculation is carried out in the next section; we call the result the 'minimum metallic conductivity' and denote it by $\sigma_{\min}$. If, as we believe, such a quantity exists, the conductivity σ_E must show a discontinuity‡ at E_C, as illustrated in Fig. 2.6(a). In systems of the Fermi glass type, in which E_F (the Fermi energy

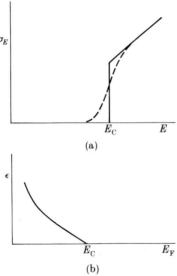

(a)

(b)

Fig. 2.6. (a) Conductivities σ_E at zero temperature as a function of E; the full line is d.c. and the broken line is $\sigma_E(\omega)$. (b) Activation energy ε for excitation to a mobility edge as a function of E_F.

† Last and Thouless (1971; see also Thouless 1974, p. 133) suggest that for energies near the mobility edge α may fall off as a power law. Another important point is that, even when $\alpha \to 0$, a finite amount of $\int \psi^2 \, d^3x$ may lie near the centre of the localized wavefunction; this is so in one dimension (see § 2.16).

‡ The criticisms of this concept due to Cohen and others are discussed in § 2.9.

at zero temperature) can move from below to above E_C, consequent on change of composition or some other parameter, there will thus be a sharp change in the d.c. conductivity at $T = 0$ from zero to a finite value. Such a change has been called an 'Anderson transition'.

If the Fermi energy E_F lies on the localized side of E_C, then two forms of d.c. conduction are possible: thermally activated hopping from one localized state to another, which is described further in § 2.7, or excitation to the mobility edge E_C, which will have an activation energy ε varying as shown in Fig. 2.6(b).

Fig. 2.7(a) shows the expected behaviour of the resistivity as a function of T if E_F and E_C are varied in such as way that $|E_F - E_C|$ changes sign. Fig. 2.7(b) shows as a function of $1/T$ the behaviour of the resistivity for charge

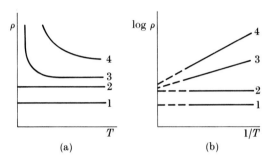

Fig. 2.7. Plots of resistivity ρ against T and $\log \rho$ against $1/T$ for values of V_0/B increasing from curves 1 to 4, conduction for the Fermi glass being by excitation to E_C. Curve 2 shows the value of ρ for E_F at E_C, so that $1/\rho$ for this curve is the minimum metallic conductivity.

transport due to excitation to the mobility edge, the conductivity thus being of the form

$$\sigma = \sigma_{\min} \exp\{-(E_C - E_F)/kT\}. \tag{2.44}$$

$\sigma_{\min}$ is here the minimum metallic conductivity, i.e. σ_E when $E = E_C$, as may be seen from eqn (2.6). The behaviour in the hopping region is described in § 2.7.

As in Fig. 2.6, we do not expect a discontinuity in $\sigma(\omega)$ when $E = E_C$ unless $\omega = 0$. The behaviour of this quantity is described in the next section.

The description given here of Anderson localization and of a mobility edge is for a particle. It is applicable also for any excitation, for example an exciton. Koo, Walker, and Geschwind (1975) have explored the form of the exciton band in ruby, which is a line due to absorption in one of the Cr^{3+} ions located at random sites on the Al_2O_3 lattice. Using a ruby laser beam with width much smaller than the exciton band, they found that only for a range of energies in the middle of the band could the exciton diffuse away, allowing it to reach traps from which radiative emission occurred at small frequency. For a low concentration of Cr, too, there was no diffusion for any frequency of the laser beam. A somewhat similar phenomenon seems to occur in $C_{10}H_8$–$C_{10}D_8$ mixtures, the exciton-transfer rate changing discontinuously as the concentration is varied (Kopelman *et al.* 1975). According to Klafter and Jortner (1977), the change in the zero-point energy gives sufficient fluctuations to cause an Anderson transition.

2.5. Photon-activated hopping; the ω^2 law

In this section we calculate $\sigma(\omega)$, the conductivity for frequency ω, for a Fermi glass, that is to say a degenerate electron gas in which the states at E_F are Anderson localized. We start with eqn (2.12) and suppose that $\hbar\omega$ is small so that $N(E)$ can be taken as constant. For D we have to sum over pairs of occupied and empty states differing in energy by $\hbar\omega$, and from (2.9) write

$$D = \frac{m\omega}{\hbar} \int \psi_{E+\hbar\omega} x \psi_E \, d^3x. \tag{2.45}$$

The functions ψ fall off with distance as $\exp(-\alpha R)$ and within the radius $1/\alpha$ change their sign in a random way from atom to atom. If the two states are at a large distance R from each other, we may approximate D by writing

$$D = \frac{m\omega R}{\hbar} \exp(-\alpha R)(\alpha a)^{3/2}. \tag{2.46}$$

The last term arises because of the random change of sign from atom to atom.

Now if the two centres are close together, the wavefunctions for an electron resonating between them will have the forms (2.36) and the states will be separated in energy by

$$2I_0 \exp(-\alpha R) \quad \text{or} \quad W_D \tag{2.47}$$

whichever is the larger. Here W_D is the difference in the energies of the zero-order functions ϕ. I_0 will be given by (2.29), but should be multiplied by $(\alpha a)^{3/2}$ if this quantity is less than unity. For large values of R and under these conditions

$$I_0 \approx \tfrac{1}{6}(\alpha R)^2 (\alpha a)^{3/2} e^2 \alpha.$$

If R is large, transitions will take place between states for which $W_D = \hbar\omega$, but if R is not large, so that

$$2I_0 \exp(-\alpha R) > \hbar\omega$$

the separation between the states is larger than $\hbar\omega$ and no transition can occur. Thus significant contributions due to a given localized state come from other states at a distance from it between R and $R + \alpha^{-1}$, where

$$R_\omega = \ln(2I_0/\hbar\omega).$$

Moreover, at this distance, since ψ resonates between the two wells as in Fig. 2.3(i), we may write

$$D = \tfrac{1}{2}(m\omega R/\hbar)(\alpha a)^{3/2}.$$

It follows then from (2.12) that

$$\sigma(\omega) = \frac{\pi e^2}{2\hbar}\{N(E_F)\}^2 (\hbar\omega)^2 a R_\omega^4. \tag{2.48}$$

This formula was first given by Mott (1970). It shows that, as $\omega \to 0$, $\sigma(\omega)$ tends to zero as ω^2, apart from the logarithmic term.

Tanaka and Fan (1963) were the first to obtain an ω^2 law with a model of this kind; they considered the case when $kT > V_0$. With this condition our formulae should be multiplied by V_0/kT.

2.6. The minimum metallic conductivity

The purpose of this section is to calculate the value of σ_E when E has the value E_C at which localization occurs. A similar problem is to obtain a value for σ_E in the centre of the band when V_0 takes the critical value for localization. We call either the 'minimum metallic conductivity', denoted by $\sigma_{\min}$, because, for a system at zero temperature in which electron states are occupied up to the Fermi energy E, it is the smallest non-zero value that the conductivity at $T = 0$ can have. Our calculation (Mott 1972c), which is for the Anderson potential of Fig. 2.1, starts from the Kubo–Greenwood formula (2.14) and makes use of a method due to Hindley (1970) and to Friedman (1971). Each pair of atoms is treated as a bond, and the matrix element D_E is the sum of the matrix elements δ_E for each pair of atoms, summed with the assumption that they have random signs. Thus

$$|D_E|^2_{\text{av}} = \tfrac{1}{2}Nz|\delta_E|^2$$

where N is the number of atoms and z the co-ordination number, so that $\tfrac{1}{2}Nz$ is the number of pairs (bonds). We can transform δ_E into the overlap integral I (eqn (2.27)) using a formula due to Holstein and Friedman (1968), namely

$$\delta = maI/\hbar^2$$

where a is the interatomic distance; then from eqn (2.14) we find

$$\sigma = \frac{\pi e^2}{\hbar a} z a^6 I^2 \{N(E_F)\}^2. \tag{2.49}$$

This formula is valid in the range of energies for which $L \sim a$. If the density of states were unaltered from the form for a crystal, we could set in the middle of the band for a simple cubic lattice (Mott and Jones 1936, p. 85)

$$N(E_F) \simeq \frac{1 \cdot 75}{2 z I a^3} \qquad (z = 6)$$

so that

$$\sigma = \text{const. } e^2 / \hbar a \tag{2.50}$$

with the constant equal to $3\pi/4z$ or, with $z = 6$, equal to $\pi/8$; we may note that starting with eqn (2.15) and putting $L = a$, we obtain the same formula with the constant equal to $\frac{1}{3}$, which is perhaps fortuitously good agreement.

If the random potential V_0 is much larger than the crystalline bandwidth B, an approximation to the density of states in the middle of the band will be

$$N(E) \simeq 1/a^3 V_0 \tag{2.51}$$

substituting into (2.49) and writing $B = 2zI$, we find

$$\sigma = \frac{\pi}{4z} \frac{e^2}{\hbar a} \left(\frac{B}{V_0}\right)^2. \tag{2.52}$$

For the conductivity, therefore, we obtain the following regimes for increasing values of the disorder parameter V_0.

(i) The regime where $L > a$ and L is given according to the Born approximation by eqn (2.33), which may be written

$$\frac{a}{L} = \frac{z^2}{8\pi} \left(\frac{V_0}{B}\right)^2.$$

from (2.51)

$$\sigma = \frac{1}{6} \frac{e^2}{\hbar a} \left(\frac{B}{V_0}\right)^2. \tag{2.53}$$

L and a become comparable when $V_0 \simeq 0 \cdot 6B$. A value of the conductivity of this magnitude, $0 \cdot 5 \, e^2/\hbar a$ (or $e^2/3\hbar a$ if we use eqn (2.15) with $L = a$), which occurs when $L \sim a$ but when the disorder has little effect on $N(E)$, plays a role in the theory. For instance, we think it is appropriate for the resistance of liquid transition metals (§ 5.11). It is of order[†] $3000\ \Omega^{-1}\,\mathrm{cm}^{-1}$.

(ii) The regime where $L \sim a$ and $\sigma \propto \{N(E_\mathrm{F})\}^2$. At first as V_0 increases we may write $N(E) \sim 1/a^3(B^2 + V_0^2)^{1/2}$, so that σ does not decrease rapidly as V_0/B increases; for large V_0, (2.51) becomes a good approximation. The conductivity thus behaves as shown in Fig. 2.8.

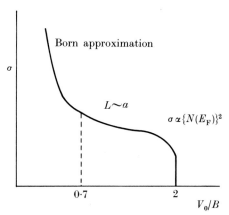

Fig. 2.8. Plot of conductivity for half-filled Anderson band as a function of V_0/B at zero temperatures. (From Mott 1973a.)

(iii) When V_0/B reaches the Anderson criterion for localization, the conductivity has fallen to the value (2.52) with the critical value of B/V_0,

$$\sigma_{\min} = \frac{\pi e^2}{4z\hbar a}\left(\frac{B}{V_0}\right)^2_{\mathrm{crit}}. \tag{2.54}$$

$(B/V_0)_{\mathrm{crit}}$, as we have seen, may depend on the co-ordination number.[‡] If we take

$$(V_0/B)_{\mathrm{crit}} = 2$$

then the minimum conductivity is

$$\sigma_{\min} = 0 \cdot 026\ e^2/\hbar a = 610/a\quad \Omega^{-1}\,\mathrm{cm}^{-1} \tag{2.55}$$

when a is in ångströms. When V_0/B exceeds this value, σ_E drops discontinuously to zero, as shown in the figure.

[†] In § 5.16.2 we comment on the conclusion of Andreev (1976) that for liquid Se and Te alloys the value 1000–$3000\ \Omega^{-1}\,\mathrm{cm}^{-1}$ is appropriate in this regime, and that, for these materials with a considerable number of electrons outside a closed shell, $L \sim 0 \cdot 4a$ fits the observed results. We note that $L \sim a$ is correct only for the tight-binding model of Fig. 2.1, and that a critical value of $k_\mathrm{F}L$ of order unity might be a better approximation in these liquids.

[‡] Thouless (1974) finds that the dependence is weak.

If a is 3 Å, σ_{min} given by (2.55) has the value $\sim$200 Ω^{-1} cm^{-1}. For an impurity band with a, say 50 Å, σ_{min} is $\sim$12 Ω^{-1} cm^{-1} (see Chapter 4).

If E_C lies far away from the centre of the band, a in eqn (2.55) should be the distance a_E in which the wavefunction loses phase memory; we conjecture that this is of order

$$a_E^3 \simeq 1/BN(E_C). \tag{2.56}$$

An investigation by Lukes (1975) confirms this, giving $N(E)$ of the form

$$N(E) = \frac{B}{\pi a^3 (B^2 + E^2)}$$

and

$$a_E/a_0 = \left\{\frac{(B^2 + E^2)}{B^2}\right\}^{1/3}$$

a_0 being the hopping distance in the middle of the band and E being measured from the middle of the band.

In this book, particularly in Chapters 4 and 5, we shall describe the experimental work which determines the magnitude of σ_{min}. We summarize here some of our findings.

(1) The constant 0·026 in (2.55) is not far wrong for impurity bands, though some empirical evidence suggests that 0·05 would be a better value. In this connection we note that Thouless (1974) maintains that the average hopping distance should be rather greater than to nearest neighbours, which may account for the discrepancy as may also the many approximations in the calculation.

(2) If the atomic orbitals are d states, values of σ_{min} of about 10^3 Ω^{-1} cm^{-1} are usual, suggesting that the constant is $\sim$0·1. This may be due to the smaller effective co-ordination number for wavefunctions that are large only in certain directions.

(3) For random positions of the centres in space (lateral disorder), σ_{min} is about three times larger than (2.55), possibly because of the smaller co-ordination number between pairs.

(4) If long-range fluctuations in potential exist, the concept of a minimum metallic conductivity is still valid as long as tunnelling through the potential maxima is possible; σ_{min} in this case may, however, be much smaller (§ 2.8).

(5) In two dimensions, the minimum metallic conductivity is of the form const. $e^2/\hbar$; the distance a between wells or a_E does not occur. According to Licciardello and Thouless (1975) the constant has a universal value $\sim$0·1 independent of the kind of disorder; the conductance is thus

$$0·1 \, e^2/\hbar = 3 \times 10^{-5} \, \Omega^{-1}$$

though Pepper (1977a) gives experimental evidence to suggest that long-range fluctuations can lead to a higher value; also lower values for Landau orbitals induced by a magnetic field are predicted by Aoki and Kamimura (1977); these are discussed in § 4.6.

(6) In one dimension all states are localized and there is no 'metallic' conduction (§ 2.16).

The concept of a minimum metallic conductivity, either in two or three dimensions, has proved controversial, as will be shown in § 2.9. At this point, however, we need only say that in non-crystalline materials values of the conductivity tending to a finite value as $T \to 0$ seem always to lie above $\sim 300 \ \Omega^{-1} \ \text{cm}^{-1}$ or the corresponding value $0 \cdot 05 \ e^2/\hbar a$ for impurity bands. The evidence for its existence is therefore strong. Our discussion of the 'mobility edge' in amorphous semiconductors (§ 6.4.2) also depends on this concept.

2.7. Hopping and variable-range hopping

In this section we continue our discussions of conduction in a 'Fermi glass', that is a degenerate electron gas in a highly disordered medium, and consider what happens to the d.c. conductivity when the Fermi energy lies in the range of energies where states are localized. Then two mechanisms for conduction are possible.

(i) Excitation of electrons to E_C; the contribution to the conductivity is

$$\sigma = \sigma_{\text{min}} \exp\left(-\frac{\varepsilon}{kT}\right) \tag{2.57}$$

where σ_{min} is as before the value of σ_E at $E = E_C$, and

$$\varepsilon = E_C - E_F.$$

This form of conduction is normally predominant at high temperatures or when ε is small. In an alloy or other system in which the composition x changes in such a way that E_F passes through E_C, ε should tend linearly to zero, as already illustrated in Fig. 2.6(b).

(ii) Thermally activated hopping conduction by electrons in states near the Fermi energy. This is illustrated in Fig. 2.9; the rate-determining

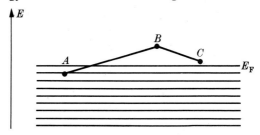

Fig. 2.9. The mechanism of hopping conduction. Two hops are shown, from A (an occupied state) to B and from B to C.

process is the hopping of an electron from a state (A) below the Fermi energy to one above (B). The probability p per unit time that this occurs is the product of the following factors.

(a) The Boltzmann factor $\exp(-W/kT)$, where W is the difference between the energies of the two states.

(b) A factor ν_{ph} depending on the phonon spectrum, discussed in Chapter 3.

(c) A factor depending on the overlap of the wavefunctions.

If localization is very strong (Fig. 2.2(d)), an electron will normally jump to the state nearest in space because the term $\exp(-2\alpha R)$ falls off rapidly with distance. This we call 'nearest-neighbour' or 'Miller–Abrahams' hopping, after the work of these authors (Miller and Abrahams 1960; cf. § 4.3) on impurity conduction. The conductivity is obtained as follows. The number of electrons jumping a distance R in the direction of the field will be made up of the following two factors.

(i) The number of electrons per unit volume within a range kT of the Fermi energy, namely $2N(E_F)kT$.

(ii) The difference of the hopping probabilities in the two directions, which are

$$\nu_{ph} \exp\left\{-2\alpha R - \frac{W \pm eRF}{kT}\right\}$$

where F is the field. The current j is obtained by multiplying by e and R and is thus

$$j = 2eRkTN(E_F)\nu_{ph} \exp(-2\alpha R - W/kT)\sinh(eRF/kT). \qquad (2.58)$$

For weak fields, $eRF \ll kT$, the conductivity is

$$\sigma = j/F = 2e^2 R^2 \nu_{ph} N(E_F) \exp(-2\alpha R - W/kT). \qquad (2.59)$$

The sinh in (2.58) for strong fields will be noted and is discussed further in § 3.12.

Nearest-neighbour hopping is only expected if $\alpha R_0 \gg 1$, where R_0 is the average distance to a nearest neighbour. The hopping energy W is of the order of the bandwidth; we write

$$W \sim 1/R_0^3 N(E_F).$$

Thus nearest-neighbour hopping with an exponential factor $\exp(-W/kT)$ can be observed only if states are Anderson localized throughout the whole band, so that any mobility edge is in a higher band.

If αR_0 is comparable with or less than unity, or in all cases at sufficiently low temperatures, the phenomenon of variable-range hopping is *always* to be expected, the hopping distance R increasing with decreasing temperature. This was first pointed out by Mott (1968, 1969a) and gives a conductivity that depends in the limit of low T on T as $\exp(-B/T^{1/4})$, or in two dimensions as $\exp(-B/T^{1/3})$. The proof given was as follows. We consider that at a temperature T the electron will normally hop to a site at a distance smaller than a value R which depends on the temperature. This implies that it will have available $4\pi(R/a)^3/3$ sites. It will normally jump to a site for which the activation energy W is as low as possible, and for this site

$$W = \frac{3}{4\pi R^3 N(E_F)}. \tag{2.60}$$

The average hopping distance is[†]

$$\bar{R} = \frac{\int^R r^3 \, dr}{\int r^2 \, dr} = \frac{3R}{4}$$

so the probability of a hop is, per unit time,

$$\nu_{ph} \exp\left(-2\alpha\bar{R} - \frac{W}{kT}\right).$$

Assuming (cf. Chapter 3) that ν_{ph} varies little with R or T, we have to take the maximum value of this quantity, which occurs when

$$\frac{3}{2}\alpha = \frac{9}{4\pi R^4 N(E_F)kT} \tag{2.61}$$

giving for the optimum value of R

$$R = \frac{3^{1/4}}{\{2\pi\alpha N(E_F)kT\}^{1/4}} \tag{2.62}$$

with the hopping distance $\frac{3}{4}$ of this. The hopping probability therefore becomes

$$\nu_{ph} = \exp(-B/T^{1/4}) \tag{2.63}$$

with

$$B = B_0\left\{\frac{\alpha^3}{kN(E_F)}\right\}^{1/4} \qquad B_0 = 2\left(\frac{3}{2\pi}\right)^{1/4} = 1.66.$$

The conductivity will be obtained by multiplying (2.63) by $e^2 N(E_F)\bar{R}^2$.

The calculation of ν_{ph}, depending on assumptions made about the electron–phonon interaction, is reviewed in Chapter 3. The value of the

---

† In the papers quoted this was taken to be R.

constant B_0 varies considerably as between various treatments, as we shall see below. It should lie in the range 2·5–1·7.

A treatment by Apsley and Hughes (1974, 1975) carries out the averaging in a different way, again obtaining the form (2.63). Hamilton (1972), Pollak (1972), Apsley and Hughes (1974, 1975, 1976), and Overhof and Thomas (1976) have discussed the effect on variable-range hopping of a non-constant form for the density of states. In the limit when $T \to 0$, (2.63) is always expected if $N(E_F)$ is finite, but deviations will occur at higher temperatures. Overhof and Thomas have applied their analysis to the case where E_F lies in a sharp minimum of $N(E)$, as may be the case in amorphous silicon, and find $1/T^{1/2}$ behaviour rather than $1/T^{1/4}$ over a significant temperature range.

Percolation theory applied to these problems gives a more rigorous treatment. The essence of such an approach is that the spheres of radius R on sites along a most favoured path must join up to form a percolation channel through the material. R must be chosen so that it does, and a value of R comes out from the analysis varying as $T^{-1/4}$. Shklovskii and Efros (1971) and Ambegaokar, Halperin, and Langer (1971) were the first to apply percolation theory to the problem. These authors find a value of B_0 differing little from that in (2.63). Ambegaokar, Cochran, and Kurkijärvi (1973), Brenig, Döhler, and Wölfe (1971, 1973), Jones and Schaich (1972), Pollak (1972, 1974), Kurkijärvi (1974), Maschke, Overhof, and Thomas (1974), Pike and Seager (1974), Seager and Pike (1974), and Butcher (1974a,b, 1976a,b) have developed the theory. Seager and Pike give a table of values of B_0 (eqn 2.63) obtained by different methods; they range from 1·78 to 2·48. The pre-exponential factor has also been studied by these and other authors.† We quote the result of Kirkpatrick (1974); if the hopping probability between a pair of sites at distance R is

$$C(\alpha R)^2 \frac{\Delta E}{kT} \exp\left\{-2\alpha R - \frac{\Delta E}{kT}\right\}$$

then

$$\sigma = 0 \cdot 022 (C\alpha) \left(\frac{T_0}{T}\right)^{0 \cdot 35} \exp\left\{-\left(\frac{T_0}{T}\right)^{1/4}\right\}.$$

In two dimensions, in the equation $\sigma = A \exp(-B/T^{1/3})$, the expected value of B is (Mott, Pepper, Pollitt, Wallis, and Adkins 1975, Arnold 1974)

$$B = \frac{3\alpha^2}{N(E_F)k}. \tag{2.64}$$

The effect of high fields is considered by Shklovskii (1973a,b), Apsley and Hughes (1975), and Pollak and Riess (1976). For moderate fields F,

† See also p. 353.

according to Pollak and Riess, the conductance must be multiplied by $\exp(eF\gamma R/kT)$, where R is the low-field hopping distance and $\gamma = 0\cdot17$ in three dimensions and $0\cdot18$ in two dimensions. Pollak and Riess compare their results with those of Elliott, Yoffe, and Davis (1974) in amorphous germanium (cf. Chapter 7). At strong fields a form of conduction in which the electron hops downwards without thermal activation gives, according to all three authors,

$$\sigma \propto A \exp(-B/F^{1/4}) \qquad (2.65)$$

though they give different values for the constant. Some of their results are presented in greater detail in § 7.4.

Comparison with experiment is given in various chapters of this book. $1/T^{1/4}$ behaviour was first observed by Walley (1968a,b) in amorphous germanium. Many experiments on this material and on silicon are reviewed in Chapter 7, including the transition to $T^{1/3}$ behaviour observed by Pollak *et al.* (1973) and Knotek (1975a,b) for films of amorphous silicon and germanium of thickness less than the hopping distance R. This form of conductivity is via 'states in the gap' which are probably due to divacancies, dangling bonds, or other defects of the nature of which there is some uncertainty; the creation of defects by bombardment with heavy ions increases this form of conduction (Olley 1973, Stuke 1976). Another uncertainty in the interpretation is the possible temperature dependence of ν_{ph} (Chapter 3). Also $1/T^{1/4}$ behaviour should not be thought *necessarily* to imply variable-range hopping; in fact other explanations have been given (cf. Alder, Flora, and Sentura 1973) based on percolation theory.

We think that the clearest evidence for $1/T^{1/4}$ behaviour comes from low-temperature measurements in impurity bands, where the nature of the disorder and of the electron–phonon interaction is well understood, and in two dimensions ($1/T^{1/3}$ behaviour) from conduction at a Si/SiO$_2$ interface. Other examples are amorphous pyrocarbons (Bücker 1973), VO$_x$ where the disorder is due to random vacancies, and La$_{1-x}$Sr$_x$VO$_3$. All these are reviewed in Chapter 4.

Klinger (1976) has discussed variable-range hopping for small polarons. These are described in Chapter 3.

Voget-Grote, Stuke, and Wagner (1976) have shown that the temperature-dependent part of the linewidth of the e.s.r. signal of glow-discharge-deposited silicon after bombardment by helium ions varies as $\exp(-B/T^{1/4})$; the interpretation is that the electron in its excited state at a point r_1 can interact with another at a point r_2 through the term

$$\frac{(\sigma_1 r_1)(\sigma_2 r_2)}{|r_1 - r_2|^3}$$

leading to a spin flip. σ_1, σ_2 are here the spin vectors.

The effect of correlation on hopping has been considered by various authors. Pollak (1971b) and Srinivasan (1971) first pointed out that Coulomb repulsion between electrons in occupied centres could lead to ordering (similar to Wigner crystallization) with a reduction in the density of states at E_F or perhaps a gap. In the same spirit Knotek and Pollak (1974) show that the field in which the localized states exist is itself partly determined by the electrons, and a given electron can raise the energies of nearby states above the Fermi level. It thereby creates for itself a sort of electronic polaron, and the electron can only jump to a distant state of very nearly the same energy by taking the 'polaron' with it. This means that simultaneously with its jump, other electrons with energies near the final state must jump into new positions. This reduces the hopping probability by a factor

$$\prod_i \int \psi_1(x_i)\psi_2(x_i)\, d^3x_i \tag{2.66}$$

where for each electron ψ_1, ψ_2 are the two relevant states in the fields before and after hopping. The effect on the $1/T^{1/4}$ law at low T was not discussed. According to Mott (1976b), however, this will affect the pre-exponential factor and the coefficient B in (2.63), but not the limiting $T^{1/4}$ behaviour at low T. We consider, following Knotek and Pollak (1976), that a range of temperature may exist in certain cases in which a transition from single-particle to multiparticle hopping gives σ proportional to $\exp(-B/T^s)$, with $s > \frac{1}{4}$.

It is not in our view true, however, as claimed by some authors (e.g. Efros and Shklovskii 1975, Efros 1976) that correlation introduces an energy gap or $1/T^{1/2}$ behaviour (see Mott 1975d), or that deviations at low T are expected as proposed by Kurosawa and Sugimoto (1975).

2.8. The Anderson transition

A Fermi glass has been defined as a material in which there is a degenerate electron gas, with a finite density of states at the Fermi energy E_F, but in which there is enough disorder for states there to be localized. An Anderson transition occurs if the position of the Fermi energy or of the mobility edge E_C (or both) is varied in such a way that $E_C - E_F$ changes sign; a transition will then occur from semiconducting behaviour, the resistivity ρ tending to infinity as $T \to 0$, to metallic behaviour with ρ tending to a finite value. The transition can occur because of a change of composition, or in some cases change of stress, magnetic field, or for two-dimensional conduction at the Si/SiO$_2$ interface by varying the gate voltage. Many cases are reviewed in Chapter 4. The considerations of the last section show that the resistivities of three-dimensional systems behave as in Fig. 2.10(a); for activated conduction the relationship $\sigma = \sigma_{\min} \exp\{-(E_C - E_F)/kT\}$ is

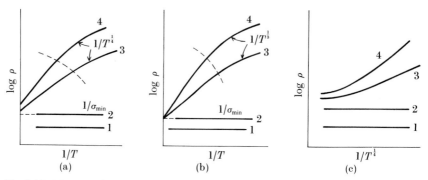

Fig. 2.10. Plot of log (resistivity), with V_0/B increasing from curve 1 through to curve 4 (a) as a function of $1/T$ (three dimensions), (b) as a function of $1/T$ (two dimensions), and (c) as a function of $1/T^{1/4}$ (three dimensions).

expected at high temperatures, variable-range hopping at low. The behaviour of a two-dimensional system is shown in Fig. 2.10(b). The difference between two and three dimensions, namely that in the former all activated curves extrapolate to the same point as $1/T$ tends to zero, is because in the former $\sigma_{min} \simeq 0.1\, e^2/\hbar$ and does not contain a, so the pre-exponential factor is the same for all curves. At sufficiently low temperatures a plot against $1/T^{1/4}$ (or $1/T^{1/3}$) should appear as in Fig. 2.10(c).

In some cases the range in which there is excitation to the mobility edge seems to be absent. This form of Anderson transition may occur for a *half*-filled band in which, as disorder is increased, localization occurs everywhere else before E_F. In this case only $T^{1/4}$ behaviour may be observed near the transition, since when states are localized at E_F there is no mobility edge in the band (as in the Ge–Fe alloys discussed in § 4.5).

Measurements of thermopower and of Hall coefficient show a striking difference between conditions when charge transport is due to electrons excited to a mobility edge and those when it is due to electrons with energies at E_F; these differences are set out in subsequent sections and in Chapter 4, where the experimental evidence is reviewed. This gives clear evidence for the existence of a mobility edge and a minimum metallic conductivity in Fermi glasses, and it is on the basis of this evidence that we can confidently apply the concept to the conduction and valence bands in amorphous semiconductors (Chapter 6).

A detailed description of these phenomena is given by Mott, Pepper, Pollitt, Wallis, and Adkins (1975). Since the conductivity of a degenerate electron gas depends in the low-temperature limit only on the behaviour of the wavefunctions at E_F, the Anderson transition gives a method of locating the mobility edge with more precision than is possible for materials and ranges of temperature where a non-degenerate gas is involved.

No discontinuity in the electronic specific heat is expected at an Anderson transition, because $N(E)$ shows no discontinuity (cf. § 2.4). Experimental evidence that this is so is cited in § 4.3.3.

2.9. Mobility and percolation edges

Classical percolation theory has been applied to electrons in non-crystalline materials by many authors. A wavy potential energy function such as that illustrated in Fig. 2.11 is envisaged, and it is supposed that the

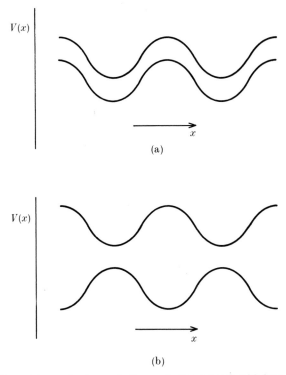

Fig. 2.11. Potential energies in conduction and valence bands with long-range fluctuations due to (a) electrostatic charges and (b) fluctuations in density.

conductivity can be calculated by considering the fraction P_1 of the volume available to an electron of energy E and the fraction P_2 in connected channels. Schematically the two volumes will appear as functions of E as in Fig. 2.12; $P_1(E)$ represents the total volume available for the electron of energy E. Curve P_2, representing the volume available for conduction, has been shown† to behave like $(E - E_P)^{1.6}$ near the percolation edge. If the

† Last and Thouless (1971), Stinchcombe (1973, 1974), Kirkpatrick (1971, 1973a); for a review see Esser (1972) and Kirkpatrick (1973b).

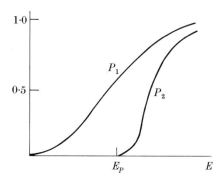

Fig. 2.12. P_1 is the volume available to an electron of energy E in the potential of Fig. 2.11; P_2 is the volume in connected channels.

conductivity in the allowed regions is constant, then we expect (at $T = 0$) that the d.c. conductivity σ_E would behave like

$$\sigma_E \propto (E - E_P)^{1 \cdot 6}. \qquad (2.67)$$

It need hardly be stated that this classical theory is only valid if the 'mountains' of Fig. 2.11 are both high and wide enough effectively to prohibit tunnelling and, if the valleys are wide enough, to include many electron wavelengths. Even if they are high enough, as in the problem of the mobility of electrons in dense helium vapour, the probability of the electron being able to get through one of the hills of Fig. 2.11 cannot be calculated classically.

Cohen[†] and co-workers in several papers have maintained that near a mobility edge large long-range fluctuations of potential will always introduce a range of energies such that classical percolation theory can be applied, so that $\langle \sigma_E(0) \rangle$, the conductivity when $T = 0$, goes continuously to zero as E approaches E_C. Arguments to show that this is not so have been advanced (Mott 1972b, 1974a,b, 1976c), by Thouless (1974), and by Licciardello and Thouless (1975). Since, however, the existence of a discontinuity in σ_E is basic to a great many of the arguments of this book, we point out here the strength of the experimental evidence that the former conjecture is wrong. If it were right, the resistivity–temperature curve for an 'Anderson transition' would appear as in Fig. 2.13, in contrast to the observed behaviour shown in Figs. 2.10 and 2.11. One of the few cases known to us where the behaviour of Fig. 2.13 is observed is that of $NiS_{1-x}Se_x$ investigated by Jarrett and co-workers[‡] and discussed in detail by Mott (1974a,c). In this case, where a metal–insulator transition occurs, a phase

[†] Cohen (1970a,b, 1973), Cohen and Jortner (1973, 1974a,b), Cohen and Sak (1972), Economou *et al.* (1974), Webman, Jortner, and Cohen (1975).
[‡] Jarrett *et al.* 1973, Bouchard, Gillson, and Jarrett 1973.

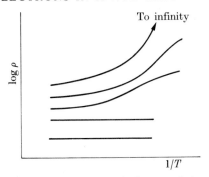

Fig. 2.13. Resistivity–temperature curves expected when percolation theory is applied to the potentials of Fig. 2.11.

separation is certainly to be expected (§ 4.2), so we think macroscopic regions of metallic and non-metallic type may have formed.

Conduction by granular metals embedded in an insulator are described in § 4.9. For these, metallic behaviour starts at the percolation limit and the conductivity varies as $(x - x_0)^{1.9}$, but on the non-metallic side tunnelling between particles can occur, and the concept of minimum metallic conductivity may be relevant. As we shall see in § 4.9, when the temperature coefficient of resistance changes sign, $\sigma \sim 50 \ \Omega^{-1} \, \text{cm}^{-1}$, which could correspond to $\sigma_{\min}$ if a is the diameter of the grains.

Materials may exist in which fluctuations of potential are large and long-range, but not quite large enough to prevent tunnelling altogether. In this case we argue that a sharp mobility edge E_C must exist, at which σ_E is discontinuous as in Fig. 2.14. The argument is that localized and extended

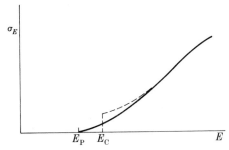

Fig. 2.14. σ_E (conductivity at zero temperatures) for different energies E in the presence of long-range fluctuations: full curve, classical percolation theory; broken curve, when tunnelling is allowed.

states cannot coexist at the same energy, and that, if extended states exist, σ_E must be finite. However, in such a case $\sigma_{\min}$ is likely to be much smaller than the values previously estimated and E_C should lie near the percolation edge E_P. Physical situations which may be described in this way are discussed in Chapters 4 and 5.

A percolation treatment has been applied to many fluids, fluctuations in density or composition being assumed great enough to prevent tunnelling. It is uncertain under what conditions this is so. The papers by Cohen and Jortner on expanded fluid mercury, liquid Te, and metal ammonia solutions are examples, as is the work of Hodgkinson (1974) on some liquid semiconductors. The effective medium theory of Landauer (1952) is used to calculate the conductivity, which should give the same result as percolation theory except near the percolation threshold (Kirkpatrick 1971).

The important point in our view is whether *statistical* fluctuations will lead to a value of σ which tends to zero at the mobility edge. This problem has been considered by Mott (1976*d*), who supposes that, if a localized wavefunction obtained from the Anderson potential of Fig. 2.1 behaves like

$$\psi = \exp(-\alpha r) \sum A_n \phi_n$$

then (cf. eqn 2.43)

$$\alpha a = \text{const.} \left(\frac{E_C - E}{B} \right)^s. \tag{2.68}$$

He finds that values of s greater or equal to $\frac{2}{3}$ in three dimensions and 1 in two dimensions are necessary if large-scale fluctuations in the amplitude of *extended* wavefunctions are not to determine σ as $E - E_C \to 0$. Such fluctuations would lead to a continuous change in σ and no minimum metallic conductivity. Since as shown in Chapter 4 there is evidence for a minimum metallic conductivity in a wide variety of materials in both three and two dimensions, it seems likely that $s = \frac{2}{3}$ and $s = 1$ are the correct values.

It has been suggested by some authors (cf. Wegner 1976) that, whereas a value $\sigma_{\min} \sim 0 \cdot 1 \, e^2/\hbar$ may exist in two dimensions so that a discontinuous transition will occur, in three dimensions the quantity a in const. $e^2/\hbar a$ may tend to infinity as $E - E_C \to 0$; if this is so, no $\sigma_{\min}$ should be observable. The absence of metallic conductivities below $\sim 200 \, \Omega^{-1} \, \text{cm}^{-1}$ is, however, strong evidence against this possibility.

2.10. Conductors, insulators, and semiconductors

In crystalline materials the distinction between metallic conductors and non-metals has been understood since the work of Wilson in 1930. In his model each electron is described by a Bloch wavefunction as in eqn (1.1) associated with a wavenumber k. The allowed energy states E_k fall into bands, which may be separated by energy gaps. If at zero temperature all bands are either full or empty, the material is a non-metal. If an energy gap ΔE separates occupied from empty bands, the material is an intrinsic semiconductor with the number of current carriers proportional to

$\exp(-\Delta E/2kT)$ and is transparent for frequencies below $(\Delta E - E_{ex})/h$, E_{ex} being the binding energy of an electron and hole in an exciton. Extrinsic semiconductors contain impurities which provide shallow donors or acceptors. If at zero temperatures one or more bands are partly occupied, the material is metallic, the conductivity tending to a finite value as $T \to 0$.

In non-crystalline materials we have already seen that the wavefunctions corresponding to eigenstates are not Bloch functions, k not being a good quantum number. None the less, the first and most striking fact about many non-crystalline materials is that they are transparent, either in the infrared or in the visible; SiO_2, borosilicate glasses, and chalcogenide glasses are examples. Therefore an energy gap must certainly exist. The band theory of solids had depended so firmly since about 1930 on the concept of Bloch functions and a periodic lattice that in the early days of the theory of non-crystalline materials there was some surprise that energy gaps could exist in them too (Ziman 1970). However, the Anderson tight-binding model (Fig. 2.1) makes it clear that they can, if the variation of the random potential V is limited, because the Anderson band cannot have a width greater than $V_0 + B$, and if there are two atomic levels in each well separated by an energy greater than this, there will be a gap between them. If, however, the random potential has for instance a Gaussian distribution with no upper limit, there must always be *some* states at all energies in the gap, though often a very small number. Thus in all liquids this must be so, since thermal fluctuations in density are not limited in magnitude, though in transparent liquids such as water the number must be negligibly small.

The glasses mentioned are either semiconductors or insulators, in the sense that the conductivity tends to zero with T. This does not *necessarily* mean that there is a gap, with E_F in mid-gap so that $N(E_F) = 0$. If $N(E_F)$ is finite, but states at E_F are localized, we expect a value of the conductivity behaving as $\exp(-B/T^{1/4})$, which tends to zero with T and may in certain cases be quite small even if $N(E_F)$ is appreciable. In fact most of the non-crystalline semiconductors to be described in this book do have a finite value of $N(E_F)$, but this appears to be due to point defects or voids of one kind or another. We must, then, in discussing glasses and amorphous silicon and germanium, start with the concept of a continuous random network† and, as in a crystal, admit the presence of certain point defects. In silicon and germanium these may be vacancies, divacancies, or more complicated clusters. In SiO_2 non-bridging oxygens, bonded to a single Si, are well known. In chalcogenides in the same way a Se or Te atom can be bonded to a single atom. In glasses which can be quenched from the melt, an equilibrium concentration of such defects should be present above the

---

† The idea goes back to Zachariesen (1932), Polk (1971), and Duffy, Boudreaux, and Polk (1974), who showed that such a structure could be constructed without adding excessive distortion energy as the number of atoms increased (cf. Chapter 6).

glass transition temperature. In sputtered or evaporated films, their concentration may depend on the rate of deposition. They have many roles, which are described in Chapters 6–10; they can pin the Fermi energy, give an e.s.r. signal, give variable-range hopping as the dominant conductivity mechanism, act as recombination centres, and so on.

The conductivity of crystalline non-metals is very sensitive to impurities which can form shallow donors or acceptors. This is not so for the semi-conducting glasses, which are insensitive to composition and have the Fermi energy pinned near mid-gap by the defects referred to above. The insensitivity of the conductivity to composition for these materials was first established by the Leningrad school under Kolomiets (1964) and is described in Chapters 6 and 9. Moreover, evaporated Si and Ge also have conductivities much less sensitive to purity than the crystals. The explanation given by various authors (Mott 1967, Haisty and Krebs 1969a,b) is that the positions of atoms in a glass or amorphous material will normally be such that all available electrons are taken up in bonds. Thus, while phosphorus in crystalline germanium is placed as in Fig. 2.15(b), the fifth

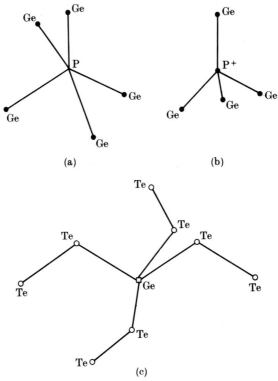

(a)   (b)

(c)

Fig. 2.15. Suggested position of the phosphorus atom in (a) amorphous and (b) crystalline germanium; (c) the configuration in Ge–Te according to Adler *et al.* (1970).

electron being loosely bound, in amorphous germanium it is suggested that a phosphorus atom would be surrounded by *five* or *three* germanium atoms (Fig. 2.15(a)). Pentavalent as well as trivalent phosphorus compounds exist (Pauling 1960, p. 62). The same result would be obtained and is expected for arsenic in amorphous Ge using the three p electrons, the two s electrons being pushed down into the valence band. As we shall see in Chapters 5 and 6, structures of this kind may break down at high temperatures in the liquid state; the co-ordination number then increases and the material becomes metallic.

An alternative explanation, which in principle goes back to Gubanov (1963), is that in the amorphous material there are always deep localized states in the gap which act as acceptors for the electrons from any shallow donors or donors for shallow acceptors.

For chalcogenide glasses, and indeed probably for all glasses, there is experimental evidence that the former is the correct model, though it is less certain for deposited layers of silicon (see below). For the former Betts, Bienenstock, and Ovshinsky (1970) for example have determined the radial distribution function of several Ge–Te alloys and have concluded that Te–Te chains are cross-linked by a Ge atom which thus has four Te neighbours. Their model is shown in Fig. 2.15(c). Adler *et al.* (1970) have interpreted n.m.r results of Senturia, Hewes, and Adler (1970) and come to a similar conclusion. Evidence from EXAFS (extended X-ray absorption fine structure) is discussed in § 6.3. More recent work by Bienenstock and co-workers includes $Ge_x S_{1-x}$ (Rowland, Narasimhan, and Bienenstock 1972), GaS, GeSe, and GeTe (Bienenstock 1973), and is reviewed by Bienenstock (1974). Of interest is the possibility of threefold co-ordination in GeTe, each atom behaving as if it has five electrons of which three p electrons form bonds. A quite different system showing the same effect is amorphous boron. Carbon increases the conductivity of the crystals, but decreases it in the amorphous state (Moorjani and Feldman 1975). Thus in glassy substances we do not in general expect shallow donors or acceptors due to impurities or lack of stoichiometry.

This is true only for materials where covalent bonds are formed; amorphous films of for instance $Mg_{3-x}Bi_{2+x}$ behave quite differently (§ 4.4), the Fermi level being shifted out of a deep gap into the conduction or valence bands as x, and with it the electron concentration, is varied. Moreover, Le Comber and Spear (1976) first showed that by the glow-discharge decomposition of PH_3 together with SiH_4 it is possible to form shallow donors (fourfold co-ordinated phosphorus) in amorphous Si. These donors lose their electrons to gap states at defects; it has not yet proved possible in amorphous Si to produce unionized donors. Thus for these materials Gubanov's hypothesis is correct. The matter is discussed further in Chapters 6 and 7.

Non-crystalline materials, then, have conduction and valence bands and also gap states which pin the Fermi energy. A form of density of states appropriate to some forms of amorphous silicon and germanium is shown in Fig. 2.16(b). The gap states are due to deep donors with energies lying below those of acceptors; if the two ranges of energy due to these overlap, then $N(E_F)$ is finite, giving free spins and an e.s.r. signal.

In chalcogenide glasses unpaired electrons (giving e.s.r.) are rarely found in the annealed material (except those due to magnetic impurities such as iron), though there is evidence to be reviewed in Chapter 9 that the Fermi energy is pinned.[†] We there present a model of 'charged dangling bonds', and pinning of the Fermi energy by electron pairs.

We turn now to the nature of the conduction or valence band. It is generally considered that any form of disorder will lead to a range of 'intrinsic' localized states at the band edges, also shown in Fig. 2.16 beyond E_C and E_V. The conjecture that they exist goes back a long way (Fröhlich 1947, Banyai 1964), and they are indeed little different from the electron or hole traps found in many crystalline semiconductors. What is new for amorphous semiconductors are the concepts of a continuous range or band of *intrinsic* localized states and of charge transport by activated hopping between them.

At the time of writing little progress has been made in calculating the range of these localized states for covalent semiconductors; localization is thought to be due either to variation of bond angle or to fluctuations in density $\Delta\Omega/\Omega$ which, in conjunction with a deformation potential E_0, would lead to a random potential. In such a case the random potential would be $E_0\Delta\Omega/\Omega$. Stern (1971) has given a treatment in which random fluctuations and their radial extensions are arbitrary parameters; for Si he obtains the position of the mobility edge, the mobility there (6 cm^2 V^{-1} s^{-1}) and the density of states $N(E_C)$.

The absence of long-range order is not likely to give an appreciable range of localization for s states, as the considerations of § 2.3 show; an example is the conduction band of liquid argon (§ 5.13). However, for glassy materials with a high proportion of p states, overlap integrals depend on the orientation of the two orbitals concerned. Although short-range order is preserved, the strongly directional overlap integral for pairs which are next-nearest neighbours should vary strongly. It is known that second-nearest neighbours contribute to the band energy as much as first-nearest neighbours, so perhaps we may use eqn (2.42) with $V_0 \simeq \frac{1}{2}z^{1/2}I$. If so

$$E_C - E_A \sim \frac{1}{16}I(z-1) \sim \frac{B}{32(z-1)}. \qquad (2.69)$$

[†] By 'pinned' we mean that if a small number δn of shallow donors is introduced, the zero-temperature Fermi energy is not shifted as $\delta n \to 0$ as it would be in an intrinsic semiconductor.

With $z = 4$ we see that ~ 1 per cent of a band may be localized. A range of localized states of $0 \cdot 1$ eV would thus correspond to a bandwidth of 10 eV. Naturally no weight can be given to such an estimate except to indicate that a localized range of a few per cent of the bandwidth is reasonable.

In a semiconductor in which there is a range of localized states at the band edge, two terms in the conductivity are expected. These are as follows,

(a) A term due to electrons at E_A. This gives a term of the form

$$(eN_C\mu) \exp\left\{ -\frac{(E_A - E_F)}{kT} \right\} \qquad (2.70)$$

where N_C is the number of states available in the conduction band, given by

$$N_C \simeq \int_0^{kT} N(E)\,dE \qquad (2.71)$$

and μ is the hopping mobility

$$\mu = \left(\nu_{ph} \frac{ea^2}{kT} \right) \exp\left(-\frac{w}{kT} \right). \qquad (2.72)$$

w is a hopping energy which may decrease at low temperatures because variable-range hopping is possible. A discussion of this due to Grant and Davis (1974) is described in Chapter 6. Also some distortion of the surrounding material will occur near a trapped electron, which may mean that part of w is of polaron type.

(b) A term due to electrons at the mobility edge, and thus equivalent to that given by eqn (2.57), namely

$$\sigma_{\min} \exp\left\{ -\frac{(E_C - E_F)}{kT} \right\}. \qquad (2.73)$$

It is convenient to write (2.73), for comparison with (2.70), in the form

$$e^2 N(E_C)\nu_{el}a^2$$

where

$$\nu_{el} = \frac{\sigma_{\min}}{e^2 N(E_C)a^2}$$

$$= \frac{1}{\hbar a^3 N(E_C)} \sim \frac{\Delta E}{\hbar} \qquad (2.74)$$

where $\Delta E = E_C - E_A$. The quantity ν_{el}, an electronic frequency, is expected to be considerably larger than ν_{ph}, so a 'kink' in the conductivity temperature curve is expected. The kink observed by Le Comber and Spear (1970) for high-resistivity amorphous silicon is described in § 6.4.2. The kink is also observed in the measurements of the drift mobility and provides perhaps the strongest evidence for a sharp change in μ at a mobility edge in an amorphous semiconductor.

In a given range of temperature the question of whether the main part of the current is carried by electrons at E_A or E_C (or the equivalent energies for holes) is often of interest. The following methods are available to distinguish them,

(i) If conduction is by hopping, the activation energies which determine the conductivity (eqn (2.72), see also § 2.13) and the thermopower

$$S = \frac{k}{e}\left(\frac{E_C - E_F}{kT} + \text{const.}\right)$$

should differ by w. If conduction is at E_C, they should have the same activation energies. This method has been used in crystals to establish polaron hopping (Chapter 3). Its use in amorphous semiconductors is discussed in Chapter 6.

(ii) If conduction is at E_C, the Hall mobility (see § 2.14) μ_H should be independent of T and of order $0 \cdot 1 \, \text{cm}^2 \, \text{V}^{-1} \, \text{s}^{-1}$. If it is at E_A, it is probably small.

(iii) The quantity σ_0 in the conductivity ($\sigma = \sigma_0 \exp(-E/kT)$) should be smaller in the hopping case; this is discussed in Chapters 3 and 6 but has not proved a very reliable criterion.

For amorphous semiconductors the 'CFO' model, put forward by Cohen, Fritzsche, and Ovshinsky (1969), has played an important part in the development of the subject and is illustrated in Fig. 2.16(a). Apart from the concepts of mobility edges, identical with the 'E_C' first proposed by Mott (1967) but applied by them for the first time to semiconductors, they envisaged tails of localized states pulled out of the conduction and valence bands by the disorder and some overlap between these tails. Where they overlap equal numbers of *charged* states of either sign are formed. Present thinking is that such overlapping tails do not exist in most amorphous semiconductors, though doubtless they do in other cases, for instance expanded fluid mercury (see the next section and § 5.14.1). The overlapping bands that give rise to a finite value of $N(E_F)$ are due to defects of acceptor and donor type (Fig. 2.16(b)).

(c) When $N(E_F)$ is finite, we expect a third form of conduction, namely variable-range hopping by carriers with energies near the Fermi level, leading to a term in the conductivity proportional to $\exp(-B/T^{1/4})$, which gives the major contribution to the current at sufficiently low temperatures. These three terms are also discussed in Chapter 6.

States in the gap may be of various kinds. Vacancies and divacancies have been postulated, divacancies particularly for amorphous silicon

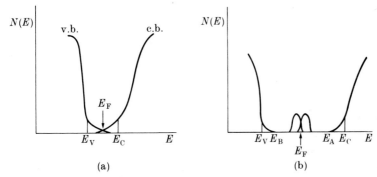

Fig. 2.16. Density of states in amorphous non-crystalline semiconductor: (a) model of Cohen *et al.* (1969) (CFO model); (b) with states in the gap due to dangling bonds acting as deep donors below acceptors.

(Chapter 7). These will be spinless deep donors when neutral, and they can also act as acceptors. When charged they will carry a spin. If the bands of energy levels in their two capacities overlap as in Fig. 2.16(b), *charged* states with spin are expected. Another form of defect, not expected in crystals except at dislocations and surfaces, is the 'dangling bond' (frontispiece). A chain end in amorphous selenium is the clearest example, but they probably exist in all amorphous materials, but in crystals only when an atom (as As in As_2Se_3) has odd co-ordination. When neutral a dangling bond should carry a free spin; when charged it will not. A third form of defect is the void. In most crystals the number of dangling bonds on a void surface is even, and the spins are believed to form pairs by a rearrangement of the atomic positions (Chapter 7) so that there is no resultant spin. This is not so in glasses or crystals with odd co-ordination, however, and an odd number is possible.

For point defects, in order to give a finite value of $N(E_F)$, a mixture of charged states is essential. In such a case, optical excitation from E_F to the conduction band is possible, as well as variable-range hopping and a.c. conductivity varying as $\omega^{0.8}$ (§ 2.15).

For states deep in the gap the positions of atoms round the defect will depend on whether the state is occupied or not. For shallow states the effect is small, but not necessarily so for deep states. In fact for dangling bonds in chalcogenides the effect is very large, particularly for positively charged dangling bonds which can form a strong chemical bond with a neighbour. This can lead to the consequence that all such bonds are positively or negatively charged and that the Fermi energy is 'pinned' with a finite density of states, even though $N(E_F)$, the *one-electron* density of states, vanishes. This is discussed further in Chapters 3, 6, 8, and 9.

There remains the question of whether significant long-range fluctuations in potential exist in amorphous semiconductors sufficient to confine the carriers to preferred percolation channels as in Fig. 2.11. There is some evidence† that they may in evaporated or sputtered films containing voids which may be charged. The evidence is reviewed in Chapter 6. For materials without voids, however, such as glow-discharge-deposited silicon and chalcogenide glasses, there is as far as we know no evidence for their presence.

2.11. Semimetals and pseudogaps

In some amorphous and liquid materials the tails of a conduction and valence band may overlap, much as in the CFO model and as illustrated in Fig. 2.16(a), and the degree of overlap and sometimes the Fermi energy can be varied by changing the composition or temperature. The density of states then shows a minimum, often called a 'pseudogap'. Examples are Mg–Bi alloys (§ 4.4), expanded fluid mercury, and some liquid tellurium alloys (§ 5.17). Overlapping Hubbard bands are also considered in § 4.2.4. It is a natural extension of the idea of localized states in band tails to introduce localized states in the pseudogap. An Anderson transition can occur when, owing to decreasing overlap, the states at E_F become localized.‡ An estimate of the density of states at which this occurs can be given as follows. If we write, as in eqn (2.23), $g = N(E_F)/N(E_F)_{\text{free}}$, then the conductivity is given by

$$\sigma = \frac{S_F e^2 a g^2}{12\pi^3 \hbar}. \tag{2.75}$$

Writing $S_F = 4\pi k^2$ and (for a divalent metal) $k \simeq \pi/a$, (2.75) gives

$$\sigma = \frac{g^2 e^2}{3\hbar a}.$$

Localization should occur when this is equal to the minimum metallic conductivity, which we write from (2.54)

$$\sigma_{\min} = \frac{\pi}{4z}\frac{e^2}{\hbar a}\left(\frac{B}{V_0}\right)^2_{\text{crit}}.$$

Thus for localization

$$g = \left(\frac{3\pi}{4z}\right)^{1/2}\left(\frac{B}{V_0}\right)_{\text{crit}}$$

$$= 0{\cdot}31 \tag{2.76}$$

† Barna and co-workers (see for instance Barna *et al.* 1976) maintain on the basis of results with the transmission electron microscope that evaporated amorphous Si and Ge show fluctuations in density with a scale of ~100Å.

‡ The considerations of the CFO model for semiconductors (§ 2.11) suggest that states should be negatively and positively charged.

if $(B/V_0)_{crit} = \frac{1}{2}$. This estimate, empirically rather successful, is subject to the assumption that the expression for minimum metallic conductivity can be used for diverse kinds of disorder.

Thus if g is equal to or greater than $\sim\frac{1}{3}$, the conductivity is metallic and equal to or greater than σ_{min}. If g is less than $\sim\frac{1}{3}$, the current at high temperatures will be carried by electrons excited to the mobility edge E_C, or holes at E_V, and at lower temperatures by hopping by electrons at E_F. Much of the evidence for this behaviour comes from work on liquids, to be discussed in Chapter 5; for these there is some doubt as to whether hopping can occur, rather than a drift of the localized carriers through the liquid, though excitation to E_C is certainly observed. In solids the case of amorphous Mg–Bi alloys is discussed in Chapter 4.

2.12. Some calculations of the density of states

For non-crystalline systems few exact calculations of the density of states exist. For covalent semiconductors much work has been done, reviewed in later chapters, giving forms of the density of states which agree with optical and photoemission studies. These at present have not successfully treated the behaviour at the band edge (E_A or E_B in Fig. 2.16) which is all important for the electrical properties.

Among studies which treat band tails the following are available. For one-dimensional disordered chains of hydrogen-like wells, Frisch and Lloyd (1960) gave a treatment showing a tail with a density of states

$$N(E) = \text{const. } |E| \exp(-\tfrac{4}{3}|E/E_0|^{3/2});\qquad (2.77)$$

here E_0 is the ionization energy of one of the wells. The three-dimensional aspects of the problem are relevant to the tails to the bands in heavily doped semiconductors which were first discussed by Kane (1963); Lifshitz (1964), Halperin and Lax (1966), and Zittartz and Langer (1966) extended the one-dimensional calculations to three.† An application to the luminescence properties of GaAs:Si is described by Redfield, Wittke, and Pankove (1970).

For the Anderson model, Lloyd (1969) gave an exact expression for $N(E)$ using a Lorentzian form for the distribution function $P(V_i)$, defined in §2.3. Mookerjee (1973) carried out a calculation for a square form of $P(V_i)$ and a simple cubic lattice. This shows a marked tail. The width of the band is $B + V_0$, but such states occur where a statistical fluctuation gives a large volume in which all wells have the lowest (or highest) value of V_i. In this model, therefore, the radius of the lowest localized state is infinite. This would of course not be so for a Gaussian distribution.

---

† The latter authors gave a band form varying in d dimensions as $\exp(-\text{const. }|E|^{2-1/2d})$. Evidence for the behaviour as $\exp(-\text{const. }|E|)$ in two-dimensional systems is given by Pollitt (1976).

For a random distribution of hydrogen-like centres, the problem appropriate to an uncompensated impurity band with neglect of correlation, calculations are due to Matsubara and Toyozawa (1961), Lifshitz (1964), and Cyrot-Lackmann and Gaspard (1974; references to several earlier papers by these workers are given in this paper). The form of $N(E)$ is as in Fig. 2.17. The low-energy tail, similar to that of Halperin and Lax (1966),

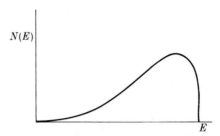

Fig. 2.17. Density of states for random array of centres.

arises because pairs of centres close together give low states, with energy $-4e^4 m/2\kappa^2 \hbar^2$ if they coincide, while its absence at the high-energy limit is because the corresponding state of highest energy turns into a p state of energy $-e^4 m/2\kappa^2 \hbar^2$. Lifshitz found that for low densities $N(E)$ should show a minimum; this was obtained by dividing the centres into nearest-neighbour pairs, with molecular wavefunctions of odd and even parity and separation $2I$, so that two bands occur which probably overlap. This result was confirmed by Lukes, Nix, and Suprapto (1972). The minimum occurs, however, only at densities at which a half-full band would not be metallic when correlation is included, and the minimum in impurity bands, for which there is some evidence (Chapter 4), is probably due to correlation.

2.13. Thermopower

In this section we summarize the formulae needed for the interpretation of the thermoelectric power. This can be expressed in terms of σ_E; in § 2.2 we have defined this quantity for a disordered lattice at zero temperature and have shown that at a finite temperature where f is the Fermi function (2.1)

$$\sigma = -\int \sigma_E \frac{\partial f}{\partial E} \, dE.$$

The thermoelectric power S is then given by (Cutler and Mott 1969)

$$S\sigma = \frac{k}{e} \int \sigma_E \frac{E - E_F}{kT} \frac{\partial f}{\partial E} \, dE. \tag{2.78}$$

The proof is as follows. If F is the field, then the current dj due to electrons with energies between E and $E + dE$ is given by

$$dj = -\sigma_E \frac{\partial f}{\partial E} F \, dE.$$

The free energy carried by this current is $-(E - E_F) \, dj/e$, which becomes

$$\frac{1}{e} \frac{\partial f}{\partial E} \sigma_E (E - E_F) F \, dE.$$

Integrating this equation we obtain the total electronic heat transport, which is equal to $j\Pi$, where Π is the Peltier coefficient, so that

$$\Pi j = \frac{F}{e} \int \sigma_E \frac{\partial f}{\partial E} (E - E_F) \, dE.$$

Since $S = \Pi/T$, eqn (2.78) follows.

We may deduce the following expressions. For metals the current is determined by electrons with energies in the neighbourhood of E_F, so we obtain the familiar equation

$$S = \frac{\pi^2}{3} \frac{k^2 T}{3} \left\{ \frac{d(\ln \sigma)}{dE} \right\}_{E = E_F}. \tag{2.79}$$

If the gas is non-degenerate in a parabolic band,

$$S = -\frac{k}{e} \frac{3}{2} \ln T + \text{const.} \tag{2.80}$$

If kT is greater than the bandwidth,

$$S = \frac{k}{e} \ln\left(\frac{c}{1-c}\right) \tag{2.81}$$

where c is the ratio of the number of electrons to the number of sites. Eqn (2.81) is due to Heikes and Ure (1961). It is in agreement with experiment for glasses containing the vanadium ions V^{4+} and V^{5+} (Chapter 3) if the temperature is not too low.

For semiconductors in which a mean free path L can be defined, we obtain the usual formula

$$S = \frac{k}{e} \left(\frac{E_C - E_F}{kT} + \frac{5}{2} + r \right)$$

where $r = d(\ln \tau)/d(\ln E)$ and τ is the relaxation time L/v. For amorphous semiconductors, in which a discontinuity in $\sigma(E)$ is assumed at E_C, a short calculation gives (Cutler and Mott 1969, Fritzsche 1971)

$$\frac{k}{e}\left(\frac{E_C - E_F}{kT} + 1\right). \tag{2.82}$$

Additional terms are of order T and depend on $d \ln \sigma/dE$. It should be noticed that, if $E_C - E_F$ varies with T and is of the form

$$E_C - E_F = \varepsilon_0 - \gamma T$$

then†

$$S = \frac{k}{e}\left(\frac{\varepsilon_0}{kT} - \gamma + 1\right). \tag{2.83}$$

A relationship between σ and S is

$$\ln\left(\frac{\sigma_{min}}{\sigma}\right) = \frac{kS}{e} - 1 \tag{2.84}$$

which is used further in Chapter 5.

In the neighbourhood of an Anderson transition (where $\varepsilon = E_C - E_F$ is near zero), we have (Mott 1975c)

$$\sigma = \sigma_{min}f(\varepsilon)$$

tending to $\frac{1}{2}\sigma_{min}$ when $\varepsilon = 0$, and

$$S = \frac{k}{e}\left\{\frac{\varepsilon}{kT} + \left(1 + \exp\frac{\varepsilon}{kT}\right)\ln\left(1 + \exp-\frac{\varepsilon}{kT}\right)\right\} \tag{2.85}$$

tending to $(k/e)2 \ln 2$ when $\varepsilon = 0$.

For conduction at E_A (or E_B) in amorphous semiconductors, it is generally supposed that (2.82) is valid with E_A substituted for E_C, but with $n + 1$ substituted for 1 within the parentheses if $N(E) \propto E^n$ at the bottom of the valence band. In the hopping regime, therefore, the activation energies for σ and S differ by the hopping energy w. This is amply confirmed for hopping of polaron type (Chapter 3), and for hopping between Anderson localized states (Chapter 6). An important point to be brought out in § 6.4.8 is that in a semiconductor, as the temperature is raised and charge transport goes over from hopping at E_A to extended-state conduction at E_C, the thermopower may *rise* for a short range of T, or if the transition is not sharp there may be a region in which the *apparent* activation energy ε_S in (2.83) is less than the conduction energy ε_σ.

---

† It has been queried in the literature (Emin 1977b) whether γ should appear in this equation. We believe it to be correct, though γ disappears from the Thomson coefficient (Butcher and Friedman 1977).

Another possible reason for a difference between ε_S and ε_σ often observed in amorphous semiconductors is inhomogeneity, if random fields lead to a change in the Fermi energy over regions too wide for tunnelling. According to Overhof (1975) one then expects

$$\varepsilon_\sigma - \varepsilon_S \simeq \tfrac{1}{6}\Delta E.$$

where ΔE is the energy range between the top and bottom of any potential fluctuations as in Fig. 2.11.

When the conductivity is determined by the motion of electrons at the Fermi level, if states are non-localized we can use eqn (2.79), and if $L \sim a$ so that $\sigma \propto \{N(E_F)\}^2$ we may write

$$S = \frac{2\pi^2}{3}\frac{k^2 T}{e}\frac{\mathrm{d}\ln N(E_F)}{\mathrm{d}E}. \tag{2.86}$$

Where states are localized and conduction is by hopping it was assumed by Mott (1967) and in the first edition of this book that (2.79) can still be used, but this is not correct. Treatments of the thermopower for this case have been given by Zvyagin (1973), Kosarev (1975), Overhof (1975), and Butcher (1976b). At low temperatures below the amorphous Néel temperature the material will have no free moments and negligible magnetic entropy. In this case, starting from eqn (2.78), we integrate over the range of energies W which contribute to the thermopower and suppose this to be the hopping energy, which in the case of variable-range hopping is

$$W \sim k(T_0 T^3)^{1/4}$$

where T_0 is defined by the relation $\sigma \propto \exp\{(-T_0/T)^{1/4}\}$. We set for the conductivity

$$\sigma = e^2 N(E) D$$

where D is the diffusion coefficient, whatever form of hopping obtains. Then (2.75) gives

$$S = \frac{1}{2}\frac{k}{e}\frac{W^2}{kT}\left(\frac{\mathrm{d}\ln N}{\mathrm{d}E}\right)_{E=E_F} \tag{2.87}$$

For variable-range hopping

$$W^2/kT = k(T_0 T)^{1/2}$$

so S varies as $T^{1/2}$. For hopping to nearest neighbours, $S \to \infty$ as $1/T \to 0$.

Overhof (1975) and Zvyagin (1973) express the constant in S as $\tfrac{1}{3}\xi^2$ and give different values of ξ, namely $0 \cdot 453$ and $0 \cdot 35$.

These equations are essentially for spinless particles, and are valid below the Néel temperature† of the amorphous antiferromagnet which any 'Fermi glass' must form. Above the Néel temperature there should be an additional term $(k/e)\ln 2$ in the thermopower, frequently observed in amorphous semiconductors in the hopping regime (Butcher 1976b). The thermopower plotted against T should thus appear as in Fig. 2.18.

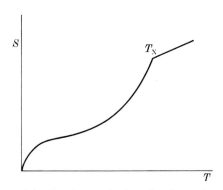

Fig. 2.18. Thermopower S for hopping conduction, showing conjectured behaviour near a Néel temperature (T_N).

2.14. Hall effect

The classical formula for the Hall coefficient

$$R_H = 1/nec$$

where n is the number of free electrons per unit volume, appears to be valid for a degenerate gas even in the regime $L \sim a$, as for instance in liquid metals of high resistivity, as long as g (cf. eqn (2.22)) does not drop below unity. In all other cases to be discussed in this book the equation is *not* valid. These are as follows.

(a) Conduction by electrons excited to a mobility edge.

(b) Conduction in a degenerate electron gas when $g < 1$ but states are not localized.

(c) Conduction by hopping, either by non-degenerate electrons or by electrons with energies near E_F. This depends *essentially* on the mechanism of interaction with phonons, and the main part of the discussion of hopping is therefore in the next chapter (§ 3.9).

The treatment of cases (a) and (b) is due to Friedman (1971, 1973); with an Anderson model, using the assumption that phases of the wavefunctions

---

† The existence of a Néel temperature for an amorphous antiferromagnetic, for any kind of interaction, is not yet proved (see § 4.2.5), but will be assumed here.

on the atoms are random and that the three-site model (§ 3.9) is appropriate, he finds for the Hall mobility[†]

$$\mu_H = \frac{2\pi\eta\bar{z}}{z^2} \frac{ea^2}{\hbar} a^3 BN(E_C).$$ (2.88)

Here as before B is the bandwidth without disorder, z the co-ordination number, and $\bar{z}$ the average number of closed three-site paths about an arbitrarily chosen site. For η, a projection of the area of a three-site path in the direction perpendicular to the field, Friedman takes $\frac{1}{3}$ in three dimensions and $\frac{3}{4}$ in two dimensions. For $a^3 BN(E_C)$ a rough approximation will be

$$\{1+(V_0/B)^2\}^{-1/2}$$

and thus, with V_0/B having a value of about 2 in three dimensions and about 1 in two dimensions, it is $1/\sqrt{5}$ and $1\sqrt{2}$ respectively. Friedman assumes $\bar{z} = z$, and we suppose $z = 6$ in three dimensions and $z = 4$ in two. This gives in three dimensions

$$\mu_H = ea^2/7\hbar = 0\cdot23 \text{ cm}^2 \text{ V}^{-1} \text{ s}^{-1}$$ (2.89)

if $a = 3$ Å. The corresponding value for two dimensions is

$$\mu_H = 2ea^2/3\hbar = 1 \text{ cm}^2 \text{ V}^{-1} \text{ s}^{-1}.$$ (2.90)

A further result due to Friedman is that the Hall coefficient should always be negative, whether the carriers are electrons or holes. This result is dependent on the three-site assumption and would be appropriate in the Anderson model where a random potential is superimposed, for instance, on a close-packed structure. For a simple cubic one would have to use four sites (ABCD) at the corners of a square, and in this case one expects a *positive* effect for holes. For the realistic situation, namely sites at random positions in space, it is likely that a three-site model is appropriate. In amorphous semiconductors for which the sign of the thermopower establishes p-type behaviour, the Hall coefficient is normally negative. However, cases exist of a positive value of R_H with n-type thermopower, a sign reversal not predicted by Friedman. These are described in Chapters 6, 7, 8, and 9. Emin (1977a,b) has proposed that hopping of an excess electron between *antibonding* orbitals on a three-site ring could give p-type behaviour. If this is the correct explanation, it must mean that the wavefunction loses phase memory in going from one pair orbital to another; the quantity a_E of § 2.6 must be equated to a.

[†] A similar result is found by Brinkman and Rice (1971) for an electron at the bottom of the conduction band in a magnetic semiconductor, where it is strongly scattered by random moments (see § 4.8 and a discussion by Friedman 1973). Also Kaneyoshi (1972, 1974), using a treatment due to Matsubara and Kaneyoshi (1968), obtain a similar result. Kaneyoshi (1976) gives a critical review of the theory.

The behaviour of R_H as the mean free path increases, when its sign (for holes) should change to a positive value (p-type behaviour), has not been investigated theoretically or observed experimentally, except perhaps for the change of sign of R_H (positive to negative) in Li-doped (p-type) NiO near the Néel temperature where the strong spin scattering may produce the situation for which $L \sim a$ (Austin and Mott 1969, p. 60).

It will also be noticed that, if we write for the conductivity mobility $\mu_C = (0.05\ e/\hbar a)/N(E_C)kT$, then

$$\mu_H/\mu_C \simeq 3a^3 N(E_C)kT$$

and unless the range of energies up to the mobility edge is small, this is expected to be considerably less than unity (say $\sim \frac{1}{10}$).

Friedman's formula (2.88) is applied in this book to the following.

(a) The current due to electrons or holes at mobility edges in amorphous semiconductors (§ 6.4.7).

(b) Electrons at a mobility edge in expanded fluid mercury (§ 5.14.1).

(c) Electrons at a mobility edge in Si:P (§ 4.3.5).

(d) Electrons at a mobility edge in two-dimensional conduction at the Si/SiO$_2$ interface (§ 4.6).

It is fair to say that the prediction of a temperature-independent Hall mobility of the order given by eqn (2.89) is in agreement with experiment, though the numerical factor as well as the sign is open to doubt.

Turning now to the case of a degenerate gas, we consider the Hall coefficient rather than the mobility. For a Fermi glass, where current is carried by electrons at the mobility edge, we expect

$$R_H = \frac{1}{c} \frac{\mu_H}{\sigma_{\min}} \exp\left(\frac{\varepsilon}{kT}\right)$$

where $\varepsilon = E_C - E_F$. Convenient formulae to evaluate $\mu_H/\sigma_{\min}$ are (2.49) for $\sigma_{\min}$ and (2.88) for μ_H; we find

$$R_H = \tfrac{4}{3} zcIN(E) \qquad (E = E_C). \qquad (2.91)$$

When $\varepsilon \to 0$, that is at the Anderson transition, there is no discontinuity in R_H and (2.91) remains valid with $E = E_F$. If we express I in terms of the free-electron density of states using eqns (2.3) and (2.28), then (2.91) reduces for a divalent metal to

$$R_H = C/necg \qquad (2.92)$$

with $C \simeq 0.7$. Here g is the ratio $N(E_F)/N(E_F)_{\text{free}}$, where $N(E)_{\text{free}}$ means the value that $N(E_F)$ would have without disorder. $N(E_F)$ will normally be reduced by disorder for a single band as in Fig. 2.1(b), or g may refer to the drop in $N(E)$ below the free-electron value in a pseudogap (Fig. 2.16(a).

Eqn (2.92) has been applied with reasonable success to the following cases.

(1) Two overlapping bands in expanded fluid mercury.

(2) Two overlapping bands in metal–ammonia solutions, particularly with caesium and lithium as solutes where the transition is far from a critical point (Thompson 1976).

(3) Tungsten bronzes (§ 4.7.6).

(4) Si:P, on the assumption that the electron gas is not highly correlated (§ 4.2.4).

It does not correspond, however, to what is observed for liquid Te and Te–Se alloys (§ 5.16), where $\mu_H \simeq$ const. and $R_H \propto 1/g^2$, or to conduction in an impurity band in Si:P, where $R_H = 1/nec$ and seems independent of g (§ 4.33). We believe these to be unsolved problems.

For hopping, whether by electrons at E_F or at a band edge, the behaviour depends on the nature of the electron–phonon interaction. For polarons in a crystalline lattice $\mu_H \propto \exp(-\frac{1}{3}W_H/kT)$, where W_H is the polaron hopping energy (Chapter 3). For hopping where the activation energy is primarily due to disorder, we shall give reasons in § 3.9 for supposing that R_H is small.

2.15. Hopping conduction for alternating currents

In a very wide variety of materials, crystalline and non-crystalline, a value of the a.c. conductivity $\sigma(\omega)$ varying as $C\omega^s$ with s of the order 0·8 is observed; a survey is given by Jonscher (1975). C is in general only weakly dependent on temperature. This behaviour is predicted if the material contains dipoles which can point in two or more different directions, with energies W_1 and W_2 ($\Delta W = W_1 - W_2$) and with a jump time τ from the lower to the upper state, if both ΔW and τ vary over a wide range including zero. The analysis is due to Debye (see Fröhlich 1958). A review is given of various mechanisms applicable to amorphous semiconductors in Chapter 6. Here we shall survey the behaviour for a 'Fermi glass'. In § 2.5 we have already evaluated a term in $\sigma(\omega)$ due to direct interaction with photons of energy $\hbar\omega$ and have shown that for this $\sigma(\omega) \propto \omega^2$. At low frequencies a term involving interaction with phonons giving $\sigma \propto \omega^s$ was first obtained for a 'Fermi glass' by Pollak and Geballe (1961) in their treatment of impurity conduction (§ 4.3.2). The development given here follows that of Austin and Mott (1969).

In Debye's analysis, we may suppose that a material contains n sites per unit volume at each of which a dipole D can point in two opposite directions with different energies, so that $\Delta W = W_1 - W_2$. If the dipole is inclined at an angle θ to a field F, the polarization if the field has frequency ω is

$$\frac{(nDF\cos^2\theta/kT)\{1+\exp(\Delta W/kT)\}^{-1}}{1+\omega^2\tau^2}$$

where τ is the mean jump time from the lower to the upper state. If we average $\cos^2\theta$ over all directions, we obtain $\frac{1}{3}$. For the conductivity at frequency ω, this yields

$$\sigma(\omega)=\frac{1}{3}\frac{(nD^2/kT)\{1+\exp(\Delta W/kT)\}^{-1}\omega^2\tau}{1+\omega^2\tau^2}. \tag{2.93}$$

In amorphous materials, whatever the nature of the dipole, we may expect to average over ΔW and τ. If, near $\Delta W=0$, there are $N\,\mathrm{d}W$ dipoles with ΔW in the range $\mathrm{d}W$, then the integration over ΔW gives

$$\int N\{1+\exp(\Delta W/kT)\}^{-1}\,\mathrm{d}(\Delta W).$$

Supposing N to be fairly constant in the range near $\Delta W=0$, this integral becomes $nkT\ln 2$, so

$$\sigma(\omega)=\frac{1}{3}\frac{(\ln 2)ND^2\omega^2\tau}{1+\omega^2\tau^2}.$$

If the relaxation process involves an electron surmounting a barrier of height U, so that

$$\frac{1}{\tau}=\nu_{\mathrm{ph}}\exp\left(-\frac{U}{kT}\right)$$

and $B(U)\,\mathrm{d}U$ is the probability that a barrier has height between U and $\mathrm{d}U$, then writing $\mathrm{d}\tau/\tau=\mathrm{d}U/kT$, the average of $\omega^2\tau/(1+\omega^2\tau^2)$ is

$$kT\omega^2\int_0^{\infty}B(U)(1+\omega^2\tau^2)^{-1}\,\mathrm{d}\tau=\tfrac{1}{2}\pi kTB\omega$$

if B is constant. It follows that

$$\sigma(\omega)=(\pi/6)(\ln 2)NBkT\omega. \tag{2.94}$$

Thus a value of σ proportional to ω and to kT is expected for mechanisms involving thermally activated rotation of a dipole.

Applying these concepts to an electron hopping between two localized states at a distance R from each other, we write

$$\frac{1}{\tau}=\nu_{\mathrm{ph}}\exp(-2\alpha R)\exp\left(-\frac{\Delta W}{kT}\right). \tag{2.95}$$

The number of electrons taking part in the hopping is $N(E_F)kT$ per unit volume. Supposing as before that only hops of energy $\sim kT$ make an important contribution, then the last factor in (2.95) is of order unity. The number of vacant states into which an electron can jump is then $N(E_F)kT$. The important hops are those for which $\omega\tau \sim 1$, that is those for which the hopping distance is R, where

$$2\alpha R = \ln(\nu_{ph}/\omega) \tag{2.96}$$

and we may take a range of ΔR as significant where

$$2\alpha\,\Delta R = 1.$$

Thus, multiplying by 2 to take account of both spin directions, we find

$$\sigma(\omega) = \frac{\pi}{3}\ln 2\; e^2\omega\{N(E_F)\}^2(kT)^2\frac{R_\omega^2}{kT}4\pi R_\omega^2\,\Delta R.$$

This gives

$$\sigma(\omega) = A\left(\frac{e^2}{\alpha^5}\right)\{N(E_F)\}^2 kT\omega\left\{\ln\left(\frac{\nu_{ph}}{\omega}\right)\right\}^4 \tag{2.97}$$

where $A = (\pi^2/24)\ln 2 \approx 0.3$. A more accurate way of averaging over the electrons gives $\pi^4/96$ (Pollak 1977).

The analysis here supposes that the electron goes through the barrier between two sites, not over it. Pike (1972) and Elliott (1977) have considered the opposite case, which also gives $\sigma(\omega)\propto \omega^s(s\leqslant 1)$. In our view this can only occur when there is considerable distortion round each occupied site, as shown in Chapters 3 and 6.

It should be noted that (2.97) is deduced subject to two assumptions.

(a) $kT\ll E_F$. Pollak and Geballe (1961) treated impurity conduction with very low compensation, so in this case eqn (2.97) is not valid. We must multiply it by E_F/kT, and $\sigma(\omega)$ is then independent of T, apart from any dependence of ν_{ph} on T (Chapter 3).

(b) The resonance energy I of centres of distance R_ω apart is less than kT. Pollak (1964) has given a discussion of the very-low-temperature case when this is not so; $\sigma(\omega)$ will always tend to zero with T.

In impurity conduction in germanium ν_{ph} is of the order $10^{12}\,\text{s}^{-1}$, and then the factor $\{\ln(\nu_{ph}/\omega)\}^4$ varies approximately as $\omega^{-0.2}$ for frequencies in the neighbourhood of 10^4 Hz, so $\sigma\propto\omega^{0.8}$, a form of behaviour which is often observed. However, much smaller or larger values of ν_{ph} are possible, so smaller or larger powers of ω can occur.

It is interesting to compare eqns (2.97) and (2.48), the conductivity (proportional to ω^2) due to optical transitions; one would expect the second formula ($\sigma \propto \omega^2$) to be predominant at high frequencies. The two contributions are equal when

$$\frac{kT}{\hbar\omega} = \left\{\frac{\ln(I_0/\hbar\omega)}{\ln(\nu_{ph}/\omega)}\right\}^4.$$

The transition temperature is very sensitive to ν_{ph}.

Eqn (2.97) is valid whether R_ω is greater or less than the mean distance between centres. In the latter case we may expect a.c. conduction but no d.c. variable-range hopping.

2.16. One-dimensional problems

The problem of the motion of electrons in one dimension in the kind of random field discussed in this chapter is of considerable theoretical interest. It is perhaps not yet certain whether it is applicable to any real case, such as polymers. Only a brief account will be given here.

Early papers (James and Ginzbarg 1953, Dyson 1953, Landauer and Helland 1954, Lax and Phillips 1958, Frisch and Lloyd 1960) deal with the density of states, a quantity accessible to machine calculation. If ψ_μ is any solution of the Schrödinger equation defined in the range $0 < x < l$ with cyclic boundary conditions, the quantum number μ may be taken to denote the number of zeros. If we write $k = 2\pi\mu/l$, then the density of states is given by

$$N(E) = \frac{l}{2\pi(dE/dk)}.$$

These authors calculate $N(E)$. More recent reviews are due to Lieb and Mattis (1966) and Hori (1968).

A minimum (pseudogap) will replace an energy gap for an array of scatterers separated by a distance $a \pm \delta$ if the quantities have (for instance) a Gaussian distribution. The question of whether a gap can ever exist in a random lattice was discussed by Landauer and Helland (1954), by Makinson and Roberts (1962), and by Halperin (1967). The result appears to be that a gap can exist if limits are set on the magnitude of δ, but that if no limits are set, as for instance with a Gaussian distribution, $N(E)/l$ will not become zero for infinite l anywhere. A similar theorem may well be true in three dimensions.

Finally we come to the question of localization. Mott and Twose (1961), using a random Kronig–Penney model, first gave arguments to suggest that *all* states in the one-dimensional lattice were localized. Borland (1963) considered a random array of scatterers, and proved for this model that, considering all configurations of the ensemble, the expectation fraction of the number of states that are not localized tends to zero as $l \to \infty$. Halperin

(1967) and Matsuda and Ishii (1970) have examined in detail the mathematical rigour of the argument. By a localized wavefunction is meant one that decays exponentially in space. This surprising result is valid for all energies of the electron and for all strengths of the scattering potential, however weak.

It must follow from the same argument as for the three-dimensional case that $\langle \sigma_E(\omega) \rangle$ vanishes for all values of E in the limit when $\omega \to 0$. This is considered by Halperin (1967, p. 173) to be not rigorously proved, because Borland's argument fails to establish that the rate of decay of the exponential wavefunction is independent of l and does not tend to zero as $l \to \infty$. It seems intuitively likely that localized wavefunctions do not depend on what happens a long way away, so it is very probable that $\langle \sigma_E(0) \rangle$ always vanishes.

Landauer (1970) has given a treatment of the zero-temperature conductivity for a finite one-dimensional array, and finds that it is finite but tends to zero exponentially with l. Cohen (1970b) and Economou and Cohen (1970c) come to the same conclusion. This behaviour, we believe, should be shown also in three dimensions when states are localized (§ 2.4).

The remainder of this section will show in an elementary and nonrigorous way how this localization arises. Consider a wave e^{ikx} falling on the random array of scatterers illustrated in Fig. 2.19; by familiar methods

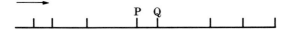

Fig. 2.19. A random array of scatterers in one dimension.

a mean free path L can be calculated, and at a distance of some multiples of L the wavefunction should be

$$A\, e^{ikx} + B\, e^{-ikx}$$

with $|A| \sim |B|$. But conservation of current gives

$$|A|^2 - |B|^2 = 1.$$

If $|A|$ and $|B|$ are nearly equal, this is only possible if $|A|$ and $|B|$ are large. This suggests that the solution of the equation that varies as e^{ikx} to the left of the array *increases* exponentially with x as $\exp(x/L)$.

We now look at real solutions of the form $\sin(kx + \eta)$ to the left of the array. In general, such a solution will increase with x, but Borland's analysis shows that there is one value of η for which ψ decreases exponentially. The solution then corresponds to a beam of electrons incident on the array and totally reflected, just as they would be at a potential step.

Localized states are obtained as follows. In a given interval PQ of an infinite one-dimensional lattice, any solution will have the form $\sin(kx + \eta)$. We can choose η uniquely so that ψ will decay exponentially for values of x to the right of PQ; let the value be η_1. We can also choose η so that ψ decreases exponentially for decreasing values of x to the left of PQ; let the value be η_2. In general $\eta_1 \neq \eta_2$. But quantized values of the energy E exist such that $\eta_1 = \eta_2$. The two solutions fit together, and we have a bounded solution with ψ continuous everywhere. This is the localized eigenstate that we require, the localization being in the neighbourhood of PQ.

Papatriantafillou and Economou (1976), Papatriantafillou, Economou, and Eggarter (1976), and Economou and Papatriantafillou (1972, 1974) have examined the localization properties in some detail. At large distances the wavefunction of a localized state falls off $\exp(-x/L)$ and L is equivalent to the 'mean free path' if calculated from a formula such as eqn (1.2). Their most interesting result is that most of $|\psi|^2$ is located in a much smaller length than L.

Bloch, Weisman, and Varma (1972) proposed that in a degenerate gas in one dimension the conductivity will behave like $A \exp(-B/T^{1/2})$. Kurki-järvi (1973), however, pointed out that in one dimension the resistance is determined by the most opaque obstacles; he finds a variation as $A \exp(-B/T)$, but for an infinite chain B should be infinite. A detailed discussion of the interaction of electrons with phonons in a one-dimensional disordered chain is given by Gogolin, Mel'nikov, and Rashba (1975).

3

PHONONS AND POLARONS

3.1. Introduction
3.2. Distortion round a trapped electron
3.3. Dielectric and acoustic polarons
3.4. Rate of loss of energy by free carriers
3.5. Transitions between localized states
       3.5.1. Introduction
       3.5.2. Single-phonon hopping processes
       3.5.3. Multiphonon processes
3.6. Motion of a polaron in a crystal
3.7. Trapped and localized polarons
3.8. Thermopower due to polarons
3.9. Hall effect due to polaron and other forms of hopping
3.10. Examples of hopping polarons
3.11. Degenerate gas of polarons
3.12. Charge transport in strong fields

3.1. Introduction

This chapter treats the interaction of current carriers (electrons or holes) with phonons and the distortion of the material round a trapped or free carrier. For the subject matter of this book the effects to be considered are the following.

(a) A carrier trapped by a defect in a crystal or in an Anderson localized state in a non-crystalline material will always distort its surroundings to some extent. This distortion will lower its energy, and can lead to a Stokes shift in the frequency of photoluminescence if this occurs.

(b) For a carrier in the conduction band of a crystal, if the interaction energy between the carrier and the phonons is strong in comparison with the bandwidth, a 'small polaron' may form. A small polaron behaves, as we shall see, like a free particle with enhanced mass (m_p) at low temperatures, but moves by thermally activated hopping at high temperatures $(kT > \frac{1}{2}\hbar\omega$, where ω is a phonon frequency). In such cases, when there is disorder, it may be appropriate to apply the Anderson localization criterion (§ 2.3) using for B the polaron bandwidth $(2z\hbar^2/m_p a^2)$.

(c) In non-crystalline materials, as in crystals, scattering of electrons by phonons will increase the resistivity. Also the rate of loss of energy to

phonons by injected electrons is of importance, for instance in considerations of quantum efficiency (§ 6.5.3) and other problems.

(d) Interaction with phonons will normally determine the rate of hopping between one localized state and another (§ 2.7).

(e) The radiationless (multiphonon) recombination between electrons
and holes produced by the absorption of radiation or in other ways in
amorphous or crystalline non-metals is due to interaction with phonons.
These matters will be considered in turn in this chapter.

3.2. Distortion round a trapped electron

The wavefunction of an electron in a donor centre in crystalline silicon or
germanium has a radius large compared with the lattice parameter, and the
distortion of the lattice which it produces is therefore small and is usually
neglected. For electrons in deep traps this neglect may not be justified, and
this could be true also for localized states at the band edges in amorphous
semiconductors, though we know of no phenomenon where this is
important. In order to see what happens, within the limitations of the
Born–Oppenheimer approximation, we consider first a system in which a
bound electron has an energy which depends on some configurational
co-ordinate q. Holstein (1959), in a classic paper on small polarons, considered a one-dimensional molecular lattice, q being the change in the
distance between nuclei in each molecule (see Fig. 3.4).

The configurational co-ordinates that we shall use in this chapter are as
follows.

(a) The displacement of an atom or anion containing a hole, when it
forms a bond with a neighboring atom as in solid rare gases or for the V_K
centres in alkali halides or dangling bonds in chalcogenides (§ 3.3).

(b) For band-edge states in amorphous silicon and similar materials q
may represent the dilatation, driven by the deformation potential, within
the volume a^3 of the trap.

(c) In polar materials q will be a measure of the polarization of the
medium outside the trap.

In all these cases the lattice energy is of the form Aq^2, and the energy of
the electron varies linearly† with q for small q, and we write it $-Bq$. The
total energy is then

$$Aq^2 - Bq$$

which has a minimum when $q = q_0$, where

$$q_0 = B/2A.$$

† We argue in § 3.3 that for *weakly* bound states this may be modified.

The energy of the electron $(-Bq)$ is lowered by Bq_0; the distortion of the system requires energy equal to Aq_0^2, and

$$Aq_0^2 = \tfrac{1}{2}Bq_0; \tag{3.1}$$

the energy of the whole system is thus lowered by W_p, where

$$W_p = \tfrac{1}{2}Bq_0 = B^2/4A. \tag{3.2}$$

These energies are illustrated in Fig. 3.1.

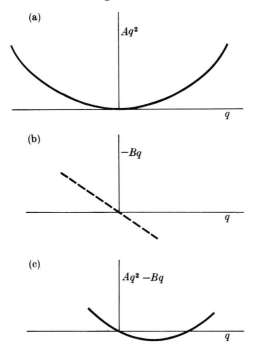

Fig. 3.1. Terms in the energy of an electron as a function of a configurational parameter q.

For case (a), an example is a chain end in amorphous selenium; here the centre, if neutral, contains a single electron in a lone-pair orbital which can form a bond with a neighbouring selenium, but if it is negatively charged, it cannot. This type of defect is discussed in Chapters 6 and 9. For case (b), we consider band-edge localized states in amorphous silicon or germanium. If q is the dilatation of the volume a^3 occupied by the electron, B is the deformation potential E_1 (Bardeen and Shockley 1950) and $A = \tfrac{1}{2}Ka^3$, where K is the bulk modulus. The polarization energy is then†

$$W_p = \tfrac{1}{2}E_1^2/Ka^3. \tag{3.3}$$

† For large values of a, the kinetic energy of an electron in the defect is $\sim \hbar^2/ma^2$, so W_p must be small in comparison in the limit as $a \to \infty$.

As regards magnitudes, if $E_1 = 1$ eV, $K = 10^{11}$ c.g.s., and $a = 10^{-7}$ cm, this gives $W_p = 0.01$ eV. Smaller values of a or larger values of E_1 will give larger values.

For a polar lattice, we suppose an electron to be trapped in a centre of radius r_0, which is neutral when empty. When the electron comes into the trap, then, before the surrounding ions are displaced, the potential energy of (another) electron at a distance r is

$$e^2/\kappa_\infty r \qquad (r > r_0)$$

where κ_∞ is the high-frequency dielectric constant. After the ions are displaced, the potential energy becomes $e^2/\kappa r$. Therefore an electron forms a potential well for itself given by

$$V_p(r) = -e^2/\kappa_p r \qquad (r > r_0)$$
$$= -e^2/\kappa_p r_0 \qquad (r < r_0) \tag{3.4}$$

where

$$1/\kappa_p = 1/\kappa_\infty - 1/\kappa. \tag{3.5}$$

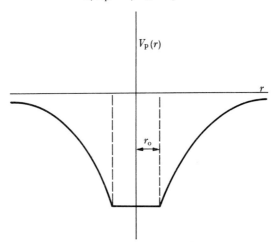

Fig. 3.2. Potential well due to the polarization of an ionic lattice round a trapped electron.

The well is illustrated in Fig. 3.2. The energy of the electron† is lowered by $e^2/\kappa_p r_0$. At the same time the polarization energy of the surrounding medium, in terms of electric field E and polarization P, is

$$\tfrac{1}{2} \int_{r_0}^{\infty} EP4\pi r^2 \, dr = (8\pi)^{-1} \int (e^2/\kappa_p r^4) 4\pi r^2 \, dr$$
$$= \tfrac{1}{2} e^2/\kappa_p r_0.$$

† The energy $\sim h^2/mr_0^2$ of the electron above the bottom of the well is here neglected.

Therefore the energy of the system is lowered by W_p, where

$$W_p = \tfrac{1}{2}e^2/\kappa_p r_0. \tag{3.6}$$

In practical cases W_p may be of order up to one-half of an electronvolt if r_0 is of atomic dimensions.

It is convenient to introduce a co-ordinate q by denoting the polarization potential energy near the centre by $qe^2/\kappa_p r$ when $r > r_0$; the considerations illustrated in Fig. 3.1 can then be applied.

3.3. Dielectric and acoustic polarons

All the forms of distortion discussed in the last section can, under certain conditions, occur for free carriers, electrons or holes, though only types (a) and (c) are known to be of practical importance. We start with (c), partly for historical reasons. The idea that a free electron in a polar lattice can be trapped by 'digging its own potential hole' goes back to Landau (1933, see Mott and Gurney 1940, p. 116). The further development of the concept is due to Fröhlich (1954), Allcock (1956) Holstein (1959), and many subsequent workers. An elementary exposition within the limitations of the Born–Oppenheimer approximation is as follows. As for an electron in a localized state, we introduce a distance r_p from the electron beyond which the medium is fully polarized. Before the ions are displaced, the potential energy of another electron in the field of the electron considered would be

$$V(r) = e^2/\kappa_\infty r$$

and after displacement of the ions

$$V(r) = e^2/\kappa r,$$

so that as in eqn (3.4) the electron 'digs a potential well' for itself, for which

$$V_p(r) = -e^2/\kappa_p r \qquad (r > r_p)$$
$$= -e^2/\kappa_p r_p \qquad (r < r_p). \tag{3.7}$$

However, unlike the case of a state localized because of a defect or disorder, instead of putting r_p equal to the radius of the defect we must here determine r_p by minimizing the total energy including the kinetic energy of the electron, which it has by virtue of its localization within a sphere of radius r_p. This to a first approximation is

$$\pi^2\hbar^2/2m^* r_p^2$$

where m^* is the effective mass in the undisturbed lattice. The energy of the electron is $-e^2/\kappa_p r_p$, the polarization energy is $\tfrac{1}{2}e^2/\kappa_p r_p$, and the total energy is therefore

$$\frac{\pi^2\hbar^2}{2m^* r_p^2} - \frac{e^2}{2\kappa_p r_p}. \tag{3.8}$$

Minimizing the quantity (3.8), we find

$$r_p = 2\pi^2\hbar^2\kappa_p/m^*e^2 \tag{3.9}$$

and

$$W_p = e^2/4\kappa_p r_p. \tag{3.10}$$

Fröhlich (1954) and Allcock (1956) gave a self-consistent calculation leading to a modified value of r_p, namely

$$r_p = 5\hbar^2\kappa_p/m^*e^2 \tag{3.11}$$

which is smaller than the value given above, and of course to be preferred to it. If $m^* = m$ and $\kappa_p = 10$, r_p is 25 Å.

This calculation is not correct if the value of r_p given by eqns (3.9) or (3.11) is comparable with or smaller than the distance between ions in the solid. This can happen only if the effective mass m_{eff} in the undistorted lattice is somewhat larger than m. We may therefore think in terms of the tight-binding approximation and a narrow band. The situation is illustrated in Fig. 3.3. We cannot consider the polarization well as extending nearer to

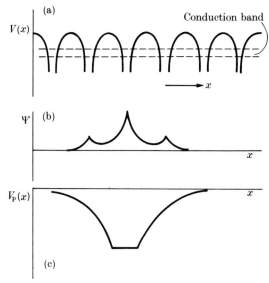

Fig. 3.3. (a) The potential energy of the electron in the undistorted lattice; (b) the wavefunction of a large or intermediate polaron; (c) the polarization well.

the electron than the radius of the ion or atom on which it is placed. The radius of this ion is then a rough approximation to r_p; a treatment by Bogomolov, Kudinov, and Firsov (1968), in which the polarization well is analysed into the normal modes of the lattice vibrations, gives for r_p

$$r_p = \tfrac{1}{2}(\pi/6N)^{1/3} \tag{3.12}$$

where N is the number of wells per unit volume. The formulae that we have used for localized states are now applicable; the energy of the polaron is $-W_p$ where

$$W_p = \tfrac{1}{2}e^2/\kappa_p r_p. \tag{3.13}$$

It will be seen that, as the coupling becomes weaker and the radius of the polaron increases, there is no sharp limit at which the potential well disappears. However, the use of the Born–Oppenheimer approximation breaks down when the radius is large and a dynamic treatment is necessary leading to the concept of the 'large' or 'Fröhlich' polaron. On this there is a large literature (for references see Devreese 1972). In this book we need only consider situations in which the polaron radius is at any rate near to the minimum radius (3.12) and the Born–Oppenheimer approximation can be used.

Self-trapping due to polarization of an ionic lattice is not the only kind that can occur, and we now consider the analogy of case (a) of the last section. Holstein (1959) in his pioneering work on small-polaron movement considered a linear array of diatomic molecules as in Fig. 3.4; the

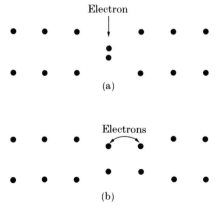

Electron

(a)

Electrons

(b)

Fig. 3.4. (a) A molecular polaron and (b) the excited state which must be formed before an electron can hop.

electron was supposed to distort one of the molecules, and to move carrying along its distortion in the way that will be explained in § 3.6. In three dimensions there are many examples of this kind of motion. Thus holes in alkali halides are self-trapped, the Cl^- ion and Cl atom attracting each other and forming a 'molecular ion' Cl_2^- known as a V_K centre (Känzig 1955, Castner and Känzig 1957, Stoneham 1975). The same is true of solid and liquid argon (Howe, Le Comber, and Spear 1971) (for other examples, see § 3.6 and Spear (1974a)). Problems in three dimensions are different in an important way from those in one, as shown by

Toyozawa (1961), Emin (1973a), Anderson (1972b), Sumi and Toyozawa (1973), and Mott and Stoneham (1977). These authors show that, just as a potential well of depth H and radius a falling off with distance faster than $1/r$ is only capable of trapping an electron if $2mHa^2/\hbar^2 > \frac{1}{4}\pi^2$, so a molecular distortion such as that in a V_K centre can only trap an electron if q is greater than some critical radius q_C. Thereafter the energy of the electron will increase first quadratically and then linearly and we may write the energy of the system, measured from the bottom of the conduction band, as

$$\tfrac{1}{2}Aq^2 - E_1(q - q_C) \qquad q > q_C. \tag{3.14}$$

The minimum value occurs when $q = E_1/a$, giving

$$E_1 q_C - \tfrac{1}{2}E_1^2/A.$$

Self-trapping will only happen if this is negative. The energy of the system is shown in Fig. 3.5. For a strongly localized defect (Fig. 3.5(c)) q_C is zero and energy will always be gained by deformation. Mott and Stoneham point out, however, that for weak localization ($\alpha a_0 \ll 1$) this is not so and both of the situations illustrated in Figs. 3.5(a) and (b) can occur. This concept is used in § 9.14 in our discussion of the mobility of holes in SiO_2.

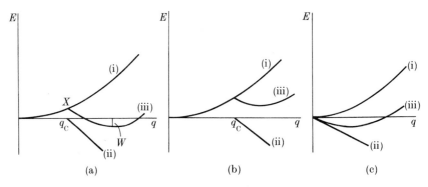

Fig. 3.5. (a, b) The energy of a carrier in a three-dimensional lattice, where trapping of V_K type is possible: (i) the elastic energy; (ii) the carrier's electronic energy; (iii) the total energy. W is the polaron energy and a stable polaron is only formed if this is positive (case (a)). (c) The same quantities for a carrier trapped at a defect. (From Mott and Stoneham 1977.) For a short-range potential, curve (ii) will vary as $(q - q_C)^2$ for small values of this quantity.

Mott and Stoneham show that, in the situation of Fig. 3.5(a), the lifetime τ of a carrier before it forms a polaron should be such that $1/\tau \sim \omega \exp(-w/\frac{1}{4}\hbar\omega)$ at low T, or $\sim \omega \exp(-w/kT)$ at higher temperatures (cf. eqn (3.35)), where w is the height of the potential barrier at X. There is thus a delay in the self-trapping process for a free or weakly bound carrier. This is not so for a dielectric polaron.

Emin (1973*a*) points out that in a system of alloys in which these parameters vary with composition a 'metal–insulator transition' will occur at the value where trapping sets in. The polaron energy and hopping energy will then increase continuously from zero, but there is a discontinuity in the electronic energy and thus in the Stokes shift.† Evidence for this kind of transition is presented by Sumi and Toyozawa (1973) for the silver halides. For these an electron is not self-trapped, while a hole is free in AgBr and trapped in AgCl (Höhne and Stasiw 1969, Känzaki and Saguraki 1973). Sumi and Toyozawa quote this as an example of the critical condition for self-trapping; also the more dramatic observation by Känzaki, Sakuragi, and Sakamoto (1971) of a sudden appearance in $AgBr_{1-x}Cl_x$ of a Stokes shift in the emission spectrum due to exciton annihilation, which occurs at $x \sim 0.43$, is in accordance with the predicted discontinuity in the Stokes shift. This applies rather to the self-trapping of an exciton, to which similar considerations apply. Another example is presented by the findings of Weinberg and Pollak (1975) that holes have a large drift mobility in glassy Si_3N_4, while in SiO_2 the mobility is $\sim 10^{-6} \, cm^2 \, V^{-1} \, s^{-1}$ at room temperature, indicating self-trapping (§ 9.14).

3.4. Rate of loss of energy by free carriers

In crystals the mean free path of a carrier is due to scattering by phonons and by impurities. In metals and semiconductors each time a carrier is scattered by a phonon of frequency ω, it exchanges energy $\hbar\omega$ with it. Collisions with impurities are usually taken as elastic, though an electron can also suffer an inelastic collision. In metals the effect of such collisions has been observed by Panova, Zhernov, and Kutaïtsev (1968) and Kagan and Zhernov (1966) as a pronounced maximum at 55 K in the resistivity of Mg–Pb alloys; below that temperature there are too few electrons with energy sufficiently above the Fermi level for inelastic collisions to make a significant contribution.

In pure crystalline non-polar semiconductors the effect of scattering by acoustic phonons gives the well-known variation of the carrier mobility with $T^{-3/2}$. This behaviour can be derived as follows. Each electron has mean energy $\frac{3}{2}kT$ and thus a wavenumber $(3mkT)^{1/2}/\hbar$. In the absence of a many-valley structure, phonons of this wavenumber are those involved in the scattering, and these have energy small compared with $\hbar\omega_0$. The probability per unit time that an electron is scattered will be of the form

$$(2\pi/\hbar)\langle H\rangle^2 N(E) \tag{3.15}$$

where $\langle H\rangle$ is the matrix element of the electron–phonon interaction and $N(E)$ is the density of electron states. To obtain H we consider the

† For definition of the Stokes shift see § 3.5.3.

phonons in a block of the material of volume Ω and density ρ_0. The Schrödinger equation for a normal mode of wavenumber q is

$$\frac{d^2\phi}{dX^2} + \frac{2\rho_0\Omega}{\hbar^2 q^2}(E - \tfrac{1}{2}\Omega\rho_0 s^2 X^2)\phi = 0 \qquad (3.16)$$

where X is the dimensionless dilatation, $\phi(X)$ is the vibrational wavefunction, and s is the velocity of sound. The interaction between an electron and a phonon is

$$XE_1 \exp(i\mathbf{q} . \mathbf{r}) \qquad (3.17)$$

where E_1 as before is the deformation potential. The matrix elements of X for transitions upwards or downwards are respectively

$$|X_{n,n+1}|^2 = (\hbar q/2\rho_0\Omega s)\begin{cases} n_q \\ n_q + 1. \end{cases} \qquad (3.18)$$

where n_q is the number of phonons in the initial state, given by

$$n_q = \{\exp(\hbar\omega/kT) - 1\}^{-1}. \qquad (3.19)$$

The term $\exp(i\mathbf{q} . \mathbf{r})$ ensures that q is equal to the change in wavenumber of the electrons, which is small. For $\hbar\omega$ we set $\omega = qs$, and (3.19) gives

$$n_q = kT/\hbar\omega = kT/qs\hbar.$$

We see therefore that q cancels from (3.18), and $|X|^2$ is proportional to T. The electronic density of states is proportional to $\sqrt{E}$ and thus for occupied states to $\sqrt{T}$; therefore if τ is the time between collisions, $1/\tau$ is proportional to $T^{3/2}$.

Of more importance for the considerations of this book is the rate of loss of energy to phonons. In crystals this is $\hbar\omega/\tau$, where $1/\tau$ is the difference in the probabilities per unit time for scattering downwards and upwards. This is, by (3.18),

$$\frac{2\pi}{\hbar}\hbar\omega E_1^2 \frac{\hbar q}{\rho_0 s}N(E)$$

where we have taken Ω to be unity. For q we may write as before ω/s, so this reduces to

$$E_1^2 N(E)\frac{2\pi\hbar\omega^2}{\rho_0 s^2}. \qquad (3.20)$$

Here ω will be the maximum frequency that can interact with an electron of energy E, so that $\omega/s = (2mE)^{1/2}/\hbar$; (3.20) thus becomes

$$4\pi E_1^2 mEN(E)/\hbar\rho_0. \qquad (3.21)$$

This does not depend on the temperature of the material and varies as $E^{3/2}$, that is as $T^{3/2}$ where T is the temperature of the electrons.

For amorphous semiconductors and disordered systems generally a treatment is given by Hindley (1970). For this problem it is not correct to consider separately each scattering process due to the disorder and ask whether it is elastic or inelastic, as for impurities in metals; this is because the scattering processes are too frequent to be considered independently. Instead Hindley uses the random-phase model of §§ 2.6 and 2.16. The consequence of this is that there is no selection rule; the electron can lose energy to any phonon and (3.21) is valid with ω equal to a mean phonon frequency (ω_0). Thus if M is the atomic mass the rate of loss of energy is†

$$\hbar\omega_0^2\left\{\frac{\pi E_1^2 N_0(E)}{\frac{1}{2}Ms^2}\right\} \qquad (3.22)$$

where N_0 is the density of electronic states per atom. The term in braces contains only electronic energies and is likely to be of order unity, except near the bottom of a band where $N_0(E)$ is small. If it is much greater than unity, perturbation theory will not be applicable. A treatment of this strong-coupling case is given by Thornber and Feynman (1970) and applied to the rate of loss of energy in Al_2O_3, which is much greater than $\hbar\omega_0^2$.

Eqn (3.22) for the rate of loss of energy should be valid when the energy crosses E_C, the mobility edge, electrons falling from extended to localized states. Here again the treatment is somewhat different from that applicable to capture by individual shallow traps in crystals (Gummel and Lax 1957, Ascarelli and Rodriguez 1961; for a review see Milnes 1973, Stoneham 1975).

3.5. Transitions between localized states

3.5.1. Introduction

In this section we discuss transitions of an electron or hole between one localized state and another, the difference W_D between their energies being obtained from or given up to phonons. There are two main applications of this analysis.

(a) The thermally activated hopping process discussed in § 2.7, where the rate-determining step in charge transport is the jump from one localized state to one of higher energy.

(b) The converse process in which an electron jumps down from a localized state at the conduction-band edge, or a deep state, to one at the valence-band edge. In § 6.5.2 we show that such transitions can be responsible for recombination in amorphous photoconductors.

† Hindley finds for the numerical factor $\pi^3/4$ instead of our π.

If the transition rate upwards is $p_0 \exp(-W_D/kT)$, the rate downwards is p_0 by the principle of detailed balancing.

We can distinguish three cases.

(i) W_D is less than the maximum phonon energy $\hbar\omega_0$, and the polarization energy (3.3) is small compared with $\hbar\omega_0$. The latter condition implies, for a deformation potential E_1 and localized state diameter a,

$$\frac{E_1^2}{\frac{1}{2}Ka^3\hbar\omega_0} < 1 \tag{3.23a}$$

or for a polar lattice

$$\frac{(e^2/r_0)(1/\kappa_\infty - 1/\kappa)}{\hbar\omega_0} < 1. \tag{3.23b}$$

In either case this means in practice that $a \sim 10^{-7}$ cm or more. We can then treat the process as a single-phonon transition, as in § 3.5.2.

(ii) W_D is still smaller than $\hbar\omega_0$, but the inequality (3.23a) or (3.23b) is not valid. The hopping is then essentially of polaron type, the main part of the activation energy being derived from the term (3.6) when $T > \frac{1}{2}\Theta_D$.

(iii) W_D is greater than $\hbar\omega_0$, so that the transition involves the simultaneous emission or absorption of many phonons. The main application here is to recombination.

3.5.2. Single-phonon hopping processes

This process (number (i) of those enumerated above) was first treated by Miller and Abrahams (1960) in their discussion of impurity conduction (§ 4.4.2). We suppose, following the analysis of § 2.5, that the two localized states a, b are at a distance R from each other, that their wavefunctions ψ_a, ψ_b fall off as $\exp(-\alpha r)$, and that the difference between their energies is W_D. Distortion of the lattice is neglected. Then the eigenfunctions for an electron resonating between the two states are

$$\Psi_i = \psi_a + (I/W_D)\psi_b$$
$$\Psi_j = \psi_b - (I/W_D)\psi_a$$

where

$$I = I_0 \exp(-\alpha R), \qquad I_0 = \frac{e^2\alpha}{\kappa}\left\{\frac{3}{2}(1+\alpha R) + \frac{1}{6}(\alpha R)^2\right\}.$$

It is supposed that $I/W_D \ll 1$, so that the energy difference is still, to a first approximation, equal to W_D. We have to calculate the transition probability p between these states. For transitions upwards, p is of the form

$$p_0 \exp(-W_D/kT)$$

and (by the principle of detailed balancing), p is equal to p_0 for transitions downwards. If first-order perturbation theory is valid, the transition probability will be given by the usual formula

$$\frac{2\pi}{\hbar}\langle H_{ij}\rangle^2 N(E) \tag{3.24}$$

where, in contrast to eqn (3.15), $N(E)$ is the density of states for the *phonons* in the final state, so that

$$N(E)=\frac{1}{2\pi^2}\frac{\Omega q^2}{\hbar s} \tag{3.25}$$

where s as before is the velocity of sound. The transition probability, using eqn (3.18), is

$$\frac{q^3 E_1^2 |\mathcal{M}|^2 n_q}{2\pi\hbar p_0 s^2} \tag{3.26}$$

where

$$\mathcal{M} = \int \Psi_i^* \exp(i\mathbf{q}.\mathbf{r})\Psi_j \, d^3x.$$

We have next to evaluate $\mathcal{M}$. We consider four cases.

(i) The limit of small values of W_D, so that $qR \ll 1$. Then

$$\mathcal{M} = 2qR(I_0/W_D)\exp(-\alpha R).$$

(ii) Also for small W_D, the limit for large R, so that $qr \gg 1$. Then

$$\mathcal{M} = (I_0/W_D)\exp(-\alpha R).$$

These are the two cases considered by Miller and Abrahams (1960) and relevant to impurity conduction.

(iii) The case when W_D is comparable with $\hbar\omega_0$. This is applicable to hopping between band-edge localized states and variable-range hopping between states at the Fermi level in semiconductors. In either case

$$\mathcal{M} = (I_0/W_D)\exp(-\alpha R)\int |\Psi|^2 \exp(i\mathbf{q}.\mathbf{r}) \, d^3x \tag{3.27}$$

the integral being over a single site. For hopping at a band edge, localized states will have radius a larger than the lattice parameter, and the integral is small. The simplest way to evaluate it is to take for ψ

$$\psi = \text{const. } \exp\{-(r/a)^2\}$$

so that the integral contains the exponential factor $f = \exp(-\frac{1}{4}q^2 a^2)$. Therefore, if we set $q = \pi/a_0$, (3.27) contains the factor

$$f = \exp(-\pi^2 a^2/4a_0^2) \tag{3.28}$$

which, if $a \sim 3a_0$, is 10^{-7}. Owing to the small values of f, single-phonon hopping by electrons in states of large radius, for values of W_D comparable with $\hbar\omega$, have very low probability. Emin (1974) shows that the multiphonon process considered in the next section may make the major contribution. Since the multiphonon process gives a pre-exponential factor dependent on T, it has to be explained why variable-range hopping with $\sigma \propto A \exp(-B/T^{1/4})$ is observed over such a large temperature range in amorphous silicon and germanium (Chapter 7), where optical modes (for which the factor f is absent) are weak. A suggestion due to Overhof (private communication) is that, for localized states centred on a defect, a localized phonon mode coupled with the acoustic spectrum is expected; in this case, too, the integral in (3.27) can remain large even if α is small.

(iv) For weakly localized states at the Fermi energy ($\alpha a_0 \ll 1$), the random-phase behaviour can occur within the radius occupied by a wavefunction and f is then $(a_0/a)^{3/2}$.

The approximations of this section are only valid if (3.23a) is satisfied; the strongest interaction allowed, without going over to the approximations of the next section, is when $E_1^2 \sim Ka^3\hbar\omega_0$. Assuming also that $Ka_0^3 \simeq Ms^2$ (both are a few electronvolts), the hopping probability is

$$\sim\omega\left[\left(\frac{I_0}{W_D}\right)^2\left(\frac{a}{a_0}\right)^3\right]f^2\exp\left(-2\alpha R-\frac{W_D}{kT}\right). \qquad (3.29)$$

The term in the square brackets is less than or comparable with unity, I_0 and W_D being in many cases of the same order. A pre-exponential factor of order ω is often assumed in discussions of hopping, and we think that such a value is probably near the upper limit for the situation when single-phonon hopping can occur. It seems likely, however, that, because of the factor f, the pre-exponential factor can be much smaller.

Single-phonon hopping appears not to give rise to any observable Hall coefficient (cf. § 6.4.7). Holstein (1961) has considered a two-phonon process, of a higher order in the coupling than the single-phonon process, in which an electron jumping from a site A to a site B can interfere with the path ACB via a third site C. He finds, for the coupling constant used, a significant value of R_H with $\mu_H > \mu_c$. A Hall constant of this kind has not been observed; the reason is discussed in § 3.9.

3.5.3. Multiphonon processes

A transition in which several phonons are simultaneously emitted or absorbed may give a larger transition probability than a single phonon hop; for acoustic phonons, as emphasized by Emin (1974), f in eqn (3.9) is small for high-energy phonons if the radius of the localized state is large, and then multiphonon processes involving many low-energy phonons may be rate determining. In addition multiphonon transitions must be considered in the following two situations.

(a) When for a polar lattice the inequality (3.23) is not satisfied, or for other kinds of distortion an equivalent condition holds. This means that the energy of distortion round each centre is greater than $\hbar\omega$. We must then use a multiphonon treatment even when $W_D < \hbar\omega$.

(b) When $W_D \gg \hbar\omega$, for instance in the problem of recombination between electrons and holes in two band-edge states.

In either case we then use the notation of § 3.2 and denote by q_1, q_2 configurational co-ordinates for the ions or atoms surrounding each centre, typically local dilatations, changes in a bond distance, or polarization round each site. Then we argue that if an electron is to jump from one centre to another, one mechanism which will allow this is for q_1, q_2 to attain values such that the two electronic energies are identical; thus

$$B(q_1 - q_2) = W_D.$$

If the electron is initially on site 2, the energy required to produce such a state is

$$A(q_2 + W_D/B)^2 + A(q_0 - q_2)^2 \tag{3.30}$$

which is a minimum when

$$q_2 = \tfrac{1}{2}q_0 - W_D/2B \qquad q_1 = \tfrac{1}{2}q_0 + W_D/2B.$$

Substituting in eqn (3.30) we find that the minimum energy required to produce a configuration of the required kind is

$$W = W_H + \tfrac{1}{2}W_D + W_D^2/16W_H \tag{3.31}$$

where

$$W_H = \tfrac{1}{2}W_p.$$

As we shall see, it is only at temperatures above $\sim\tfrac{1}{2}\hbar\omega/k$ that W appears as an activation energy; if it does, the transition probability per unit time upwards is of the form $P_0 \exp(-W/kT)$, and that downwards is $P_0 \exp\{-(W - W_D)/kT\}$. In the case when $W_H \gg W_D$, appropriate to impurity conduction in polar materials, the last term in (3.31) can be neglected.

The analysis leads to the important result that the term W_H in the hopping activation energy is approximately half the energy of polarization W_p. This is true only in a model in which the electron on one molecule does not affect the value of q for the other. It is not true, for instance, in polar materials, in which two polarization wells overlap and can affect each other. For these we picture the process as follows. Initially the electron is

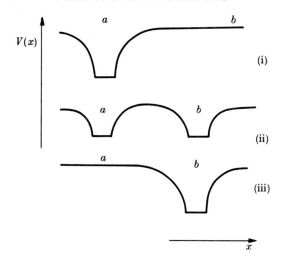

Fig. 3.6. Potential wells on a pair of ions a and b during the hopping process: (i) before hopping; (ii) thermally activated state when electrons can move; (iii) after hopping.

trapped in a potential well as in Fig. 3.6. If the electron is to be transferred, thermal fluctuations must ensure that the wells have the same depths. If W_D is zero or negligible, it is obvious that the smallest activation energy that can produce such a configuration is that when both wells have half the original depth. The energy required to produce this configuration consists of the following terms.

Energy to raise the electron in well a; W_p.

Polarization energy released in well a; $W_p - \frac{1}{4}W_p = \frac{3}{4}W_p$.

Energy to form well b; $\frac{1}{4}W_p$.

These give a total activation energy $\frac{1}{2}W_p$. If $W_D \neq 0$, formula (3.31) can be used.

For polar lattices, if the distance R through which the electron must be transferred is not large compared with r_0, the formula

$$W_H = \tfrac{1}{2}W_p = e^2/4\kappa_p r_0$$

is no longer valid and must be replaced by (Killias 1966, Austin and Mott 1969)

$$W_H = \frac{e^2}{4\kappa_p}\left(\frac{1}{r_0} - \frac{1}{R}\right). \tag{3.32}$$

This occurs for the reason already mentioned: the wells overlap and the energy required to produce the intermediate configuration of Fig. 3.6(ii) is diminished. For the adiabatic case the exponential term $\exp(-2\alpha R)$ does *not* occur.

We turn now to the term outside the exponential in the expression for the probability per unit time for a jump from site a to site b. For the case $W_D = 0$ this has been investigated in detail by Holstein (1959) and Emin and Holstein (1969) in their work on the behaviour of polarons; the extension to the case when $W_D \neq 0$ is due to Schnakenberg (1968). We distinguish two cases.

(a) The 'adiabatic' case in which, during the time of the order of 10^{-12} s in which the activated state of Fig. 3.6 persists, the electron makes several transitions backwards and forwards between the two wells. In this case the analysis shows that the jump rate is of the form

$$p \exp(-W_H/kT) \tag{3.33}$$

where p is equal to the frequency ω_0 of an optical phonon (there is no dispersion in the simple model used).

(b) The non-adiabatic case, where in each fluctuation the chance of an electron jumping is small. Then

$$p = \frac{\pi^{1/2} I^2}{\hbar (W_H kT)^{1/2}}. \tag{3.34}$$

Since $I = I_0 \exp(-\alpha R)$, this exponential factor is now included. In this case the factor $I_0^2 / \hbar (W_H kT)^{1/2}$ outside the exponential $\exp(-2\alpha R - W/kT)$ may be greater than ω_0. Perhaps the simplest way of considering this is to say that the frequency with which the system reaches the situation in which tunnelling can occur is $\omega \exp(-W_H/kT)$, the time it stays in the energy range where tunnelling is possible is $\omega^{-1}\{I/(W_H kT)^{1/2}\}^{-1}$ and the chance per unit time of tunnelling is $I/\hbar$.

To consider what happens at low temperatures, we draw the configurational co-ordinate diagram for the two states of the electrons as functions of the co-ordinate q already defined (Fig. 3.7(a)). The transition probability upwards is

$$C \exp\left(-\frac{W_H}{\frac{1}{4}\hbar\omega}\right) \exp\left(-\frac{W_D}{kT}\right). \tag{3.35}$$

For discussions of the constant C, of order ω in the adiabatic case, see Schnakenberg (1968) and Emin (1975). The factor $\exp(-w/\frac{1}{4}\hbar\omega)$ for the transition probability through a potential barrier of height w is of widespread applicability and has already been used in the last section. It is simply the value of $|\chi|^2$ at X in Fig. 3.7(a), χ being the vibronic wavefunction of the system. This shows that the same term (with a numerical factor slightly different from $\frac{1}{4}$ depending on the shape of the barrier) applies for the tunnelling of a heavy particle from one potential well to another, ω being its vibrational frequency in the well.

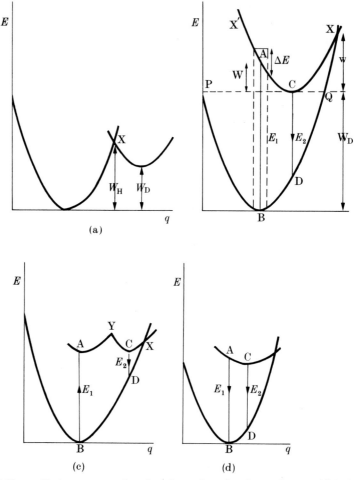

Fig. 3.7. Energy E of a system as a function of a configurational co-ordinate q: (a) for hopping between two states separated by energy W_D, when polarons are formed; (b) for recombination of excited defect or for two electrons in localized states (note that W_H is now written w); (c) for free exciton; (d) when ω is different for excited and normal states. The kink at Y may not be sharp.

At intermediate temperatures, W_H in (3.33) should be replaced by

$$\frac{W_H \tanh(\tfrac{1}{4}\hbar\omega/kT)}{(\tfrac{1}{4}\hbar\omega/kT)} \sim W_H\left(1 - \frac{1}{3}\frac{\hbar\omega}{kT}\right)^2.$$

At $T \sim \tfrac{1}{2}\Theta$, the hopping energy has dropped by 8 per cent, and at $\tfrac{1}{4}\Theta$ by 30 per cent.

An important application is made in § 6.4.5 to the passage of a trapped electron (or electron pair) between two positively charged sites fairly close to each other (Pike 1972, Elliott 1977). The barrier is then lowered by $4e^2/\kappa R$ (or $8e^2/\kappa R$ for an electron pair), where R is the distance between the sites. However, if the trapping is of the V_K (molecular) type, the trapping energy is not affected by proximity. Thus, if $T > \sim\frac{1}{2}\Theta_D$, excitation over the barrier should be more probable than tunnelling through it. This has important consequences for a.c. conduction in chalcogenides.

We now consider the case when $W_D \gg \hbar\omega$. Here the main application is to the radiationless recombination of an electron and a hole. They may be in band-edge localized states, such that their wavefunctions overlap; in this case q will be defined as above. The analysis will, however, also apply to trapped excitons, that is to an excited defect, where both states are functions of some co-ordinate q. We shall discuss also free excitons.

For a trapped exciton, that is an excited defect, we draw the energies of the two states as in Fig. 3.7(b). In this case radiation is possible from the upper to the lower state; by the Franck–Condon principle the frequency of the radiation is given by $E_2/\hbar$, while $E_1/\hbar$ is the corresponding absorption frequency. The quantity

$$(E_1 - E_2)/\hbar$$

is called the 'Stokes shift'. At low temperatures the width, broadened by zero-point motion, is

$$\Delta E = 2(W\hbar\omega)^{1/2} \qquad (3.36)$$

where W is the interval A to C. At temperatures above the Debye temperature $\hbar\omega$ should be replaced† by kT. ΔE for absorption is shown in the figure, the two vertical broken lines enclosing the width of the vibration.

The quantity W_D is the zero-phonon absorption frequency. If the Stokes shift is small, absorption lines due to excitation of one, two, or more phonons are observed in crystals. In amorphous materials it is likely that phonon lines are sufficiently broadened by interaction with other phonons for the effect of quantization of phonon energies to be negligible (see Robertson and Friedman 1977).

The multiphonon transition probability downwards when the system is in equilibrium at C at zero temperature can be calculated as follows. We make use of the Born–Oppenheimer approximation, according to which any state of the system can be represented by a product

$$\psi(x; q)\chi(q)$$

where $\psi(x; q)$ is the wavefunction of the electrons (with co-ordinates x) when the configurational co-ordinate has the value q and $\chi(q)$ is the vibrational wavefunction. We can calculate the transition probability

† See Street 1976 (eqn 12) for intermediate temperatures.

downwards between the state when $\chi(q)$ is the ground state of the upper curve and $\chi_2(q)$ that of the lower curve vibrating between the points PQ. This is a transition with no change of energy; it is envisaged that the large vibrational energy after the transition is dissipated by coupling with other phonons.† The probability per unit time will be given by

$$(2\pi/\hbar)|M|^2 N(E) \qquad (3.37)$$

where $N(E)$ is the phonon density of states ($\sim 1/\hbar\omega$) and M is the matrix element of the terms omitted from the Schrödinger equation in the Born–Oppenheimer approximation. In general, if $\frac{1}{2}Aq^2$ is the potential energy, this is $\frac{1}{2}A(\partial^2/\partial q^2)$, not acting on χ. So the matrix element contains terms involving the electronic functions, which are difficult to estimate, and the terms

$$\int \chi_1 \chi_2 \, dq \quad \text{and} \quad \int \chi_1 \left(\frac{\partial}{\partial q}\right) \chi_2 \, dq. \qquad (3.38)$$

Provided both upper and lower states have the same frequency ω, these can be calculated exactly (see Englman and Jortner 1970, Robertson and Friedman 1976). Mott, Davis, and Street (1975) give a calculation due to Rees. For instance, if the ground-state wavefunction is $\exp(-\beta^2 q^2)$, we find

$$\left| \int \chi_1 \left(\frac{\partial}{\partial q}\right) \chi_2 \, dq \right|^2$$

$$= \frac{\beta^2}{2n!} \exp\left(-\frac{W_D - E_2}{\hbar\omega}\right)\left(\frac{W_D - E_2}{\hbar\omega}\right)^{n-1}\left(\frac{E_2}{\hbar\omega}\right)^2$$

where $n = W_D/\hbar\omega$ and is the number of phonons emitted. Using Stirling's formula for $n!$, this is of the form const. $\exp(-\gamma W_D/\hbar\omega)$ with

$$\gamma = \ln\left(\frac{W_D}{W_D - E_2}\right) - \frac{E_2}{W_D}. \qquad (3.39)$$

The quantity γ is often as great as 2 for materials with small Stokes shift or about 1 for materials with large Stokes shift. It is here assumed that ω is the same in both states. The matrix element without $\partial/\partial q$ has the same exponential term, so at zero temperature P_0, the transition probability downwards, is given by‡

$$C \exp\left(-\frac{\gamma W_D}{\hbar\omega}\right) \qquad C \sim \omega. \qquad (3.40)$$

---

† Or possibly by ejecting an anion (Pooley 1966).
‡ A short calculation shows that, if the barrier height w is small, (3.40) becomes

$$P_0 = C \exp(-w/\tfrac{1}{4}\hbar\omega)$$

indicating that in this limit we may consider the transition as tunnelling through the barrier.

At finite temperatures integrals of the type (3.37) can also be evaluated, χ_1 referring to an excited state. Robertson and Friedman find† that (3.37) should be multiplied by

$$\{1-\exp(-\hbar\omega/kT)\}^{-n} \qquad n = W_D/\hbar\omega.$$

This is nearly constant until kT approaches $\hbar\omega$ and then increases as $(kT/\hbar\omega)^n$. The approximation is valid only in the weak coupling limit (small Stokes shift).

At higher temperatures the transition rate downwards will be of the form $\omega \exp(-w/kT)$, as first proposed by Mott (1938). The transition probability downwards is thus independent of T only below $\hbar\omega/nk$, and then increases as T^n and eventually as $\exp(-w/kT)$.

There is a large literature on multiphonon transitions of this kind, of which early papers are those of Huang and Rhys (1951), Kubo (1952), and Kubo and Toyozawa (1955). The treatment given above derives from Englman and Jortner (1970) and Robertson and Friedman (1976, 1977). A review is given by Struck and Fonger (1975). Other workers who derive equations of the type (3.39) are Hagston and Lowther (1973). The exponential form (3.40) is found to be in very good agreement with experiment for rare-earth ions in various matrices, where the luminescence and non-radiative transition probabilities can be compared. Figs. 3.8 and 3.9 show some results on P_0 and its temperature variation obtained by

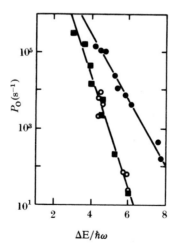

Fig. 3.8. Spontaneous multiphonon transition rate P_0 of trivalent rare earth ions in LaCl$_3$, LaBr$_3$ (lower curve) and LaF$_3$ (upper curve) as a function of $\Delta E/\omega_{\mathrm{max}}$. (From Moos 1970.)

† This factor was first given by Kiel (1964).

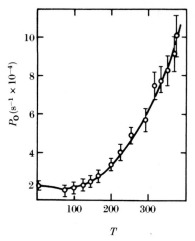

Fig. 3.9. Temperature dependence of the multiphonon transition rate P_0 in LaF_3HO^{3+}. (From Moos 1970.)

Moos (1970). More recent work by Layne *et al.* (1977) for rare earths in glasses confirms (3.40) and its temperature dependence in detail. Fig. 3.10 shows some non-radiative lifetimes for the polymers C_xH_y and C_xD_y. Since deuterium (D) gives a smaller phonon energy $\hbar\omega$ than hydrogen, one expects a longer lifetime; this is well shown.

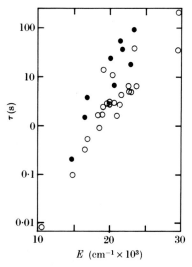

Fig. 3.10. The dependence of the non-radiative lifetime τ on the radiative energy gap E for the hydrocarbon C_xH_y (○) and C_xD_y (●). The exponential dependence on E and particularly the effect of the smaller value of $\hbar\omega$ for deuterium is well shown. (From Moos 1970.)

If Fig. 3.7(b) refers to an optically allowed transition, in which BA is the absorption energy (E_1) and CD (E_2) that for the emission frequency, there is another process by which the system can return to the ground state without the emission of radiation. It was first suggested by Dexter, Klick, and Russell (1955) that if A lies above X, the system will not come to a temporary equilibrium at C through interaction with other phonons but will normally cross over at the point X and then oscillate on the curve PBQX until its energy is dissipated. It is envisaged that the excited system loses energy through anharmonic coupling with other phonons (heat conduction) until the vibration is from X to X′ in Fig. 3.7(b); then the interaction between the two electron wavefunctions, which will cause the splitting between the two curves, will lead to a large transition probability onto the lower curve. If for symmetry reasons this term vanishes, a small excitation of another configurational co-ordinate (called in the literature a 'promoting mode') will cause one.

Bartram and Stoneham (1975) have made detailed calculations for F centres which show that radiationless recombination occurs with high probability when the condition of Dexter *et al.* is valid. Parke and Webb (1973) apply the idea to the optical properties of thallium, lead, and bismuth in oxide glasses. If after irradiation in the band due to a defect there is no photoluminescence and if the energies (as in SiO_2) are several electronvolts, so that (3.40) is small compared with the probability of radiation, this is the most likely mechanism.

For a free exciton formed from a hole subject to trapping by the V_K mechanism and an electron, a configurational diagram as in Fig. 3.7(c) has been proposed by Mott and Stoneham (1977). There is a barrier which must be surmounted if the exciton is to be self-trapped, for a value q_C which depends on the exciton's effective mass. Here again self-trapping is only possible if the minimum C lies below A and occurs after a delay τ such that

$$1/\tau \sim \omega \exp(-w/\tfrac{1}{4}\hbar\omega) \qquad kT < \tfrac{1}{2}\hbar\omega$$

$$\sim \omega \exp(-w/kT) \qquad kT > \tfrac{1}{2}\hbar\omega.$$

A consequence is that free excitons should give narrow Lorentzian absorption lines, and bound excitons should give a broad Gaussian form.[†] In solid and liquid rare gases there is much evidence that this is the case (cf. Jortner 1974).

An important conclusion from (3.40) for variable-range hopping or a.c. conduction by electrons with energies at the Fermi level is that α, the tunnelling term in $e^{-2\alpha R}$, will normally be only slightly increased because the electron has to tunnel between the two states of Fig. 3.6(ii), so that

$$\alpha^2 = \alpha_0^2 + mW_p/\hbar^2$$

† See for instance Street 1976, p. 402.

where α_0 is the value without polaron formation. However, ν_{ph} is reduced by $\exp(-4W_H/\hbar\omega)$. This can greatly decrease the magnitude of the pre-exponential term in variable-range hopping, but since ν_{ph} occurs for a.c. conduction (cf. eqn 2.96) in the term $\{\ln(\nu_{ph}/\omega)\}^4$ the effect here will be much less. This is possibly one reason why a.c. conductivity according to eqn (2.97) is observed for chalcogenides, but d.c. conduction by variable-range hopping is very weak if present at all. More recent work (Elliott 1977) suggests, however, that a.c. conduction is due to motion of electron pairs between very close centres (§ 6.4.5).

These concepts are applied to defects in non-crystalline materials in Chapters 6–10. We can summarize their behaviour as follows.

(a) For materials with fourfold co-ordination (silicon, germanium) dangling bonds (not possible in crystals except at dislocations and grain boundaries), divacancies, voids, and impurities such as Si—H bonds have been proposed as producing states in the gap. It is known (Chapter 7) that at crystalline silicon surfaces there is a reconstitution of the atomic arrangement, the atoms forming pairs. However, observations of photoluminescence (§ 6.7.6) due to transitions between gap states show that the Stokes shift, in certain cases at any rate, is not large, perhaps because of the rigidity of the lattice. Also deep states due to impurities in crystalline silicon normally show rather small Stokes shift, as has been shown in a number of papers by Grimmeiss and co-workers (Braun and Grimmeiss 1974, Grimmeiss et al. 1974, Engstrom and Grimmeiss 1975, 1976, and for a review Grimmeiss 1977). Energies of optical edges from the valence band to such levels and from the level to the conduction band add up to the band gap, showing that a zero-phonon line is involved. For such levels (e.g. gold or zinc in Si, and also for deep levels in GaP) the capture cross-section is temperature dependent, tending to a constant value at low temperature (Henry and Lang 1976). The configuration diagram may, then, appear as in Fig. 3.7(d), with a frequency ω_1 for the excited state smaller than the ground-state frequency ω_0. For this case the transfer integrals were calculated by Hutchisson (1930) and Siebrand (1967). In eqn (3.40) ω is now ω_0 and $\gamma = \log\{(\omega_0 + \omega_1)/(\omega_0 - \omega_1)\}$, in the limit of zero Stokes shift.

(b) At the other extreme are SiO_2 and the chalcogenide glasses, in which the chalcogen is bonded to only two metal atoms. These glasses show a term of the form γT in the specific heat, thought to be due to vibrations of this chalcogen when the bond angle approaches 180° (§ 6.8.1). This is absent in amorphous arsenic with threefold co-ordination (Phillips and Thomas 1977). Photoluminescence in chalcogenides shows a very large Stokes shift (§ 9.6), as it does in silicate glasses containing thallium (Parke and Webb 1973). For radiationless recombination, eqns (3.36, 3.37) are then likely to be applicable with $\gamma \sim 1$. A plausible model for the defect is then analogous to the 'non-bridging oxygen' in SiO_2, namely a chalcogen bonded to a single metal. A feature of these materials is that the upper valence band is formed from lone-pair orbitals, so that the dangling bond on the non-bridging oxygen can form a strong bond with a lone-pair orbital on another chalcogen. This will be strongest when the centre is positively charged since there are then two bonding electrons only and no antibonding one. The hypothesis is made in Chapters 6 and 9 that the bond is then so strong that all centres are either positively or negatively charged (D^+ and D^-). The co-ordinate q in this case is the distance from the singly co-ordinated chalcogen to the other chalcogen atom with which it is bonded, and a very large Stokes shift is expected (Street and Mott 1975).

3.6. Motion of a polaron in a crystal

In a crystal the motion of a polaron from site to site in the adiabatic approximation will be thermally activated and occur with frequency

$$\omega \exp(-W_H/kT).$$

Here, for dielectric polarons, W_H is given by (3.32); the only difference from the treatment of § 3.5.2 is that W_D now vanishes. For polarons of any kind, dielectric or otherwise, W_H is the energy of the intermediate stage illustrated in Figs. 3.4 and 3.6.

However, at low temperatures ($T \ll \Theta_D$) a polaron in a crystal moves with a well-defined wavevector k and effective mass m_p given by

$$\frac{m_p}{m} = \frac{\hbar}{2m\omega R^2} \exp\frac{W_H}{\frac{1}{2}\hbar\omega}. \tag{3.41}$$

R is here the hopping distance. The occurrence of this exponential factor comes because the wavefunction of the moving polaron must be of the form

$$\Sigma \exp(ika_n)\psi(x; q_n)\chi_n(q_n)$$

where $\chi_n(q)$ is the vibrational function in the ground state at atom n. The energy will then be of the form $\hbar^2 k^2/2m_p$, and will contain the integral

$$\int \chi_n(\partial/\partial q)\chi_{n+1}\, dq$$

which as we have seen is determined by the term $\exp(-W_H/\frac{1}{2}\hbar\omega)$. For transition metal oxides, for which $R \sim 4$ Å, the pre-exponential term in (3.41) will be of order 3–5. It is important to notice that for quite small values of W_H of order $\hbar\omega$, which would not give an observable temperature activation, mass enhancement of order 10 or more is possible. We refer to such pseudoparticles as 'heavy polarons'—they are identical with the 'nearly small polaron' treated by Eagles (1966, 1969a, b).

The optical properties of polarons have been considered in many papers. For molecular polarons, in which the distorted molecules are reasonably far apart, the transition of lowest frequency is normally one in which the electron jumps to a neighbouring site (photon-assisted hopping); for this a treatment of the temperature dependence is due to Reik (1970). For polarons in a polar medium, however, the potential well formed on the site where the carrier is, by virtue of its Coulomb form, will always have excited states; absorption processes similar to those in an F centre are expected (Austin 1972).

A mathematical treatment of bound piezoelectric polarons is given by Hattori (1975).

3.7. Trapped and localized polarons

We have seen that a charge carrier (electron or hole) trapped in a localized state will always distort its surroundings to some extent, and that under certain conditions this distortion may determine the activation energy for hopping. Particularly in materials of high dielectric constant, it may be a useful approximation to treat the polaron in a crystal as moving in the field of a trapping centre or a random field, so that the polaron bandwidth $B_p = 2z\hbar^2/m_p a^2$ can be used in the Anderson localization criterion (§ 2.3) when the material is amorphous.

If a polaron is bound to a charged donor or acceptor, its potential energy at a distance R from the donor or acceptor is $-e^2/\kappa R$, where κ is the *static* dielectric constant, so long as R is greater than the polaron radius r_p. We can therefore distinguish two cases for small polarons.

(a) If the radius $\hbar^2\kappa/m_p e^2$ is large compared with R (the distance of the nearest metal ion from the centre), the polaron is described by a hydrogen-like function and the energy required to remove it from the centre is $e^4 m_p/2\hbar^2\kappa_p^2$. This probably occurs only for $\kappa \gg 100$, as for example in $SrTiO_3$ (see § 3.11).

(b) If this is not the case, then the carrier must be thought of as located on the metal lattice site next to the donor, so that the energy required to remove it is $e^2/\kappa R$, where R is the distance between the two sites. This is probably a good approximation† for NiO and TiO_2.

The resistivity of a semiconductor in which the carriers are small polarons with W_H great enough to show hopping conduction will therefore appear as in Fig. 3.11. At low temperatures the electrons are bound to the donors; when they are all freed, if that occurs at a temperature below $\frac{1}{2}\Theta_D$, they are capable of transport as heavy particles scattered by phonons, so

---

† The direct measurement of drift mobility in undoped NiO by Spear and Tannhauser (1973) suggests that the carriers are *large* polarons, with effective mass enhanced by only 1·5 and $m_{eff} \sim 1·5\, m_e$. This surprisingly low value led these authors to conjecture that they may have been measuring the mobility of holes in the 2p band.

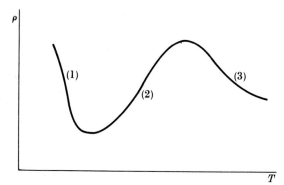

Fig. 3.11. Typical plot of resistivity against temperature for a crystalline semiconductor where the carriers are polarons. In regime 1 electrons are bound to donors; in regime (2) they are free but $kT \ll \frac{1}{2}\hbar\omega$ so that they behave as heavy particles increasingly scattered by phonons as the temperature is raised; in regime (3) they move by hopping.

the resistivity rises with increasing temperature. At the temperature $\frac{1}{2}\Theta_D$ the mean free path satisfies the relation $L \sim a$ and σ is of order

$$ne^2 a/mv.$$

Here $mv^2 \sim kT$, so the denominator is $m^{1/2}(kT)^{1/2}$. At temperatures above $\frac{1}{2}\Theta_D$ conduction is by hopping, and the conductivity decreases.

3.8. Thermopower due to polarons

Determination of the thermopower of a semiconductor is the most direct way of finding out whether the mobility contains an activation energy, either due to polarons or to disorder. In extrinsic compensated semiconductors the conductivity varies as $\exp(-E/kT)$, and the thermopower S as $(k/e)(E/kT + \text{const.})$; the two values of E, the donor ionization energy, are identical. If, however, conduction is by small polarons and the temperature range is such that thermally activated hopping occurs, the conductivity varies as

$$\exp\left\{-\frac{E + W_H}{kT}\right\}.$$

W_H does not occur in the expression for the thermopower. Various authors have used this criterion to determine whether or not thermally activated hopping is present in a given material. For instance, in polycrystalline lithium-doped nickel oxide Bosman and Crevecoeur (1966) found that the activation energies deduced from conductivity and thermopower were identical, as shown in Fig. 4.37 at temperatures above that for which impurity conduction sets in. Keem, Honig and van Zandt (1978), however, found that in single crystals of NiO of high purity they differ by $\sim 0 \cdot 1 \, \text{eV}$, indicating a polaron hopping energy. In this case the acceptors are Ni vacancies. These authors believe that in polycrystalline Li-doped specimens a lithium-rich region exists near grain boundaries, that this carries the current, and that the contraction of the lattice parameter in this region so broadens the d-band that no polarons are formed. Crevecoeur and de Wit (1968) have made similar measurements for MnO; for this material the activation energy deduced from the conductivity is greater than that deduced from the thermopower. Polaron hopping seems to be present, indeed down to temperatures below $\frac{1}{2}\Theta_D$. Other examples are given in § 3.10.

3.9. Hall effect due to polarons and other forms of hopping

For polarons moving by band motion and when kT is small compared with the polaron bandwidth, the Hall coefficient is given by the usual formula ($R_H = 1/nec$) and its sign depends on the sign of the carrier. This is not so in the hopping regime, and the calculation of the effect follows quite different principles. It depends on the interference between the wavefunction due to a direct jump and an indirect jump via another site, on more than one site (Friedman and Holstein 1963, Friedman 1963, Holstein and Friedman 1968, Holstein 1973). For three-site configurations R_H is always negative, whether the moving particles are electrons or holes. The interference can only take place when the electrons in the three sites have the same energy, and an elementary calculation (Austin and Mott 1969) shows that this requires an activation energy $\frac{4}{3}W_H$. Therefore the Hall mobility has the form

$$\mu_H \sim T^s \exp(-\tfrac{1}{3}W_H/kT); \tag{3.42}$$

s is believed to be $-\frac{3}{2}$.

The occurrence of the factor $\frac{1}{3}$ in the activation energy for the Hall mobility in crystalline materials has at the time of writing been verified in only one material ($LiNbO_3$); this is described in the next section. One expects also in disordered materials, when $W_D > kT$ and W_H is the largest term in the activation energy, that (3.42) should still be valid; an example may possibly be cerium sulphide (§ 4.7.5). However, much evidence suggests that, in disordered materials when $W_D \gtrsim kT$, the Hall effect is small, whether W_H is large (multiphonon processes) or small (single-phonon Miller–Abrahams hopping). The evidence from impurity conduction and from conduction at band edges is reviewed in Chapters 4 and 6. The reason for this behaviour is probably that preferred percolation paths only give a Hall voltage at the relatively few points where they come close together (see the discussion given by Mott, Davis, and Street 1975). No fully quantitative theory exists at the time of writing for the Hall effect under these conditions and our belief in its absence must be based mainly on experiment, through the analysis of Böttker and Bryksin (1977) gives it some support.

3.10. Examples of hopping polarons

As we have stated, in all narrow-band ionic materials and many that are not ionic there must be some enhancement of the mass of the carriers due to polaron formation. Evidence for thermally activated hopping in crystalline materials is, however, limited. There are three ways in which it could be obtained.

(a) It can be obtained from a discrepancy between the activation energies in the conductivity and thermopower, as in MnO (§ 3.8). This method has been extensively used with the aim of establishing the existence of hopping (not of polaron type) in amorphous semiconductors (Chapter 6) and in vanadium glasses.

(b) Direct measurements of the drift mobility for injected carriers can be used. Ghosh and Spear (1968) established a hopping mobility with activation energy $0 \cdot 24$ eV for electrons in crystalline sulphur, which they interpreted as due to hops from one S_8 ring to another. The mobility increases with pressure (Dolezalek and Spear 1970). Also Le Comber et al. (1974) have measured an activated mobility for holes in solid rare gases; the polaron is of molecular type, an argon atom and ion forming an Ar_2^+ molecule. This behaviour was predicted by Druger and Knox (1969) and Song (1971). The self-trapped hole, known as a V_K centre, found in alkali halides is similar (Castner and Känzig 1957, Stoneham 1975). Polaron transport (of the molecular or acoustic type) is frequently observed in molecular liquids, as in the work of Ghosh and Spear (1968) (see also Spear 1974a) on liquid sulphur. The activated mobility of electrons in liquid hydrocarbons observed by many authors (Robinson and Freeman 1974, Dodelet, Shinsaka, and Freeman 1973) can probably be intepreted in this way.

Schein et al. (1978) measure the electron drift mobility in naphthalene. Above 100 K this is almost temperature-independent and of order $0 \cdot 5$ cm^2 V^{-1} s^{-1}; this in our view indicates the hopping motion of a 'nearly small' polaron, of the form $T^{-1} \exp(-W_H/kT)$ with small W_H. Below 100 K there is a sharp rise in μ_H, indicating, these authors claim, the transition to polaron-band motion.

A solvated electron in ammonia or water forms for itself a sort of polaron (Jortner 1959, Catterall and Mott 1969, Thompson 1976), giving the well-known absorption spectrum in the red. It is also likely that in liquids such as NaCl plus a few per cent of Na (Nachtrieb 1975), or CsAu with Cs (Freyland and Steinleiter 1976), the electron is self-trapped, forming the liquid analogy of an F centre.

(c) Comparison between the activation energies in the Hall and conductivity mobilities can be used. For crystals detailed evidence is provided by the work of Nagels, Callaerts, and Denayer (1975) on conduction in (slightly reduced) single crystals of $LiNbO_3$. Here, between 325 K and 750 K, the activation energy for conduction is 0·65 eV, and for the (n-type) thermopower 0·25 eV. The latter is thought to represent the binding energy of an electron to an oxygen vacancy and the difference (0·40 eV) a polaron hopping energy. The n-type Hall effect shows a small Hall mobility with activation energy 0·13 eV, which is $\frac{1}{3}$ of 0·4 eV as predicted. The results of these authors for conductivity and thermopower are shown in Fig. 3.12.

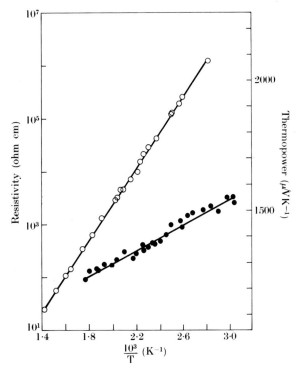

Fig. 3.12. Temperature dependence of the electrical resistivity (circles) and thermopower (dots) of a single crystal of slightly reduced $LiNbO_3$. (Nagels *et al.* 1975.)

3.11. Degenerate gas of polarons

In this book we give many examples of a 'Fermi glass' (Chapters 2 and 4), that is to say a degenerate gas of electrons in which states at the Fermi energy are Anderson localized. Earlier sections of this chapter have shown that if the localization radius $1/\alpha$ of a state is such that $\alpha a_0 \sim 1$, a_0 being the lattice parameter, strong deformation of the lattice is likely, and motion of the carriers involves dragging a polarization cloud along with the carrier. This term is not important for problems of impurity conduction in silicon and germanium because $\alpha a_0 \ll 1$; in deep states in these materials, whether amorphous or otherwise, the effect may or may not be large (§ 3.5), but it is certainly important in chalcogenides as explained in Chapter 9.

There may be cases in which a useful approximation is to consider the carriers as a degenerate gas of polarons *before* Anderson localization, so that the bandwidth is contracted and localization occurs more easily. In the next chapter it is suggested that crystalline $La_{1-x}Sr_xVO_3$ and VO_x may be treated in this way, the disorder being due to the random positions of La and Sr in the lattice or to vacant sites in VO_x.

It is of interest to ask whether a degenerate gas of polarons is a suitable treatment for any substance that shows metallic behaviour. It is well known that in metals interaction with phonons can increase the effective mass of electrons at the Fermi surface by a considerable factor. In calculations of this effect the interaction between the electron gas and the phonons is treated as a perturbation. If the small polaron energy is greater than the Fermi energy, a better approximation may be to consider the metal as a degenerate gas of small polarons; this will of course be possible only if the number of carriers is small compared with the number of atoms at any rate for dielectric polarons. The condition that a gas of polarons will be metallic will be that the hydrogen radius a_H ($=\hbar^2\kappa/m_p e^2$) should be greater than the interatomic distance, so that the binding energy is given by condition (a) of § 3.7; this will be so only for large values of κ. We can then apply eqn (4.1) of Chapter 4, namely $n^{1/3}a_H > 0\cdot2$, as the condition for metallic behaviour.

A degenerate gas of small polarons is therefore only likely to occur in materials of high static dielectric constant κ, for values of κ of 100 or more. The slightly reduced crystalline materials $SrTiO_3$ and $KTaO_3$ have been described in this way by Mott (1967, 1974c). The former provides a degenerate electron gas with conductivity tending to a finite value as $T \to 0$, if the concentration of carriers is greater than 3×10^{18} cm^{-3}. This would be quite impossible if the large *static* dielectric constant were not the quantity determining the Bohr radius. Eagles (1969a), in a detailed discussion of the properties of $SrTiO_3$, describes the carriers as 'nearly small' with a mass in the range $7-10m_e$; for our description what is necessary is that they are small enough for κ_∞ not to play any role in the binding energy, which means that the polaron radius is smaller than a.

3.12. Charge transport in strong fields

The discussion can be divided into (a) the mobility of a single carrier forming part of a non-degenerate electron gas, and (b) the conductivity of a degenerate gas when states at the Fermi energy are localized, and variable-range or nearest-neighbour hopping occurs. The latter is treated in § 2.7.

For a single carrier, we may envisage the following possibilities.

(i) The carrier is hopping between localized states at a band edge. If R is the hopping distance and F the field, we should expect the activation energy to be reduced by eRF. If eRF becomes comparable with the range of localized states, it seems likely that the activation energy will disappear; this case has not been treated quantitatively.

(ii) The carrier forms a small polaron. Its drift velocity in a field F is then expected to be

$$\frac{\omega eRF}{kT} \exp(-2\alpha R) \sinh \frac{eRF}{2kT} \tag{3.43}$$

as has been shown by many authors (Reik 1970, Austin and Mott 1969, Austin and Sayer 1974, Austin 1976). R is here the hopping distance. Here it is assumed that the particle hops to the nearest site. If $2\alpha < eF/2kT$, then this will not be so, and the electron will jump to a site where the activation energy disappears. This case has not been treated in the literature.

Austin and Sayer (1974) and Austin (1976) have discussed charge transport in vanadium phosphate glasses, and found that the conductivity varies with field according to (3.43) but that R is considerably greater than the hopping distance. They consider that this is probably due to enhancement of the field by structural irregularities at the rate-determining steps.

(iii) Trap-limited mobility. A treatment was first given by Bagley (1970). If the traps are neutral when empty, the escape time will be increased by $\exp(\pm eaF/kT)$ for jumps along and against the field. A one-dimensional model would suggest a release time proportional to $\cosh(eaF/kT)$, but Ieda, Sawa, and Kato (1971) find that in three dimensions it is proportional to an integral of type

$$\int \exp\left(\pm \frac{eaF \cos \theta}{kT}\right) 2\pi \sin \theta \, d\theta$$

and thus to

$$\sinh \frac{(eaF/kT)}{eaF/kT}.$$

If a quasi-equilibrium is set up, the capture time being unaffected by the field, it is clear that the electron concentration is increased by this factor, which therefore determines the drift velocity (cf. § 6.5.1).

Again for strong enough fields the electron will escape from the trap without thermal activation (Mott 1971, Marshall, Fisher, and Owen 1974).

If the traps are charged when empty (for instance compensated donors or the charged dangling bonds proposed for chalcogenides in Chapters 6 and 9), we expect Poole–Frenkel behaviour. This was introduced by Frenkel (1938) following earlier work by Poole (1916, 1917). A field F lowers the ionization energy of a centre by $\beta F^{1/2}$, where β is given by

$$\beta = 2e^{3/2}/\kappa^{1/2}. \tag{3.44}$$

κ is probably the high-frequency dielectric constant, though there is some uncertainty about this. Thus the release probability will be increased by the factor (Ieda et al. 1971)

$$\left(\frac{\beta F^{1/2}}{kT}\right)^{-1} \sinh\frac{\beta F^{1/2}}{kT}. \tag{3.45}$$

The electron concentration and thus the drift mobility will therefore be increased by this factor. These equations are changed to some extent when the treatment of Onsager is used instead of that of Poole–Frenkel (Pai 1975, Mott and Street 1977). This is discussed in Chapters 6 and 9.

Eqn (3.45) assumes that there is no tunnelling through the top of the barrier. At a depth ΔE below the top, the tunnelling factor is easily seen to be proportional to $\exp(-\alpha \Delta E)$, where ΔE depends on the field. For tunnelling to occur, α must be less than $1/kT$, so tunnelling is expected to set in suddenly as F increases or T decreases. The effect of tunnelling for strong fields has been treated in detail by Hill (1971) and applied to observations on SiO.

As pointed out by Simmons (1967) and Mark and Hartman (1968), for trap-limited mobility in which all the traps are normally full one expects for strong fields ($\beta F^{1/2} \gg kT$) a drift velocity proportional to

$$(\tfrac{1}{2}\beta F^{1/2}/kT).$$

The factor $\frac{1}{2}$ occurs because, while (3.45) gives the rate of escape from the centre, the number of recombination processes is proportional to n^2, where n is the free concentration.

4

THE FERMI GLASS AND THE
ANDERSON TRANSITION

4.1. Introduction
4.2. Metal–insulator transitions in crystals
    4.2.1. Band-crossing transitions
    4.2.2. Hubbard bands and the Mott transition
    4.2.3. Wigner crystallization
    4.2.4. Effect of disorder on Mott transitions
    4.2.5. Effect of correlation and distortion on Anderson transitions
4.3. Doped semiconductors
    4.3.1. Impurity conduction; direct currents
    4.3.2. Impurity conduction; alternating currents
    4.3.3. Metal–insulator transitions in doped semiconductors
    4.3.4. Metal–insulator transitions in metal–rare-gas systems
    4.3.5. Metal–insulator transitions in compensated semiconductors
4.4. Anderson transition in a pseudogap; magnesium–bismuth alloys
4.5. Anderson transition of type II; amorphous films of Fe–Ge
4.6. Two-dimensional conduction in an inversion layer
4.7. Fermi glasses where lattice distortion is important
    4.7.1. Impurity conduction in nickel oxide
    4.7.2. Conduction in glasses containing transition-metal ions
    4.7.3. Lanthanum–strontium vanadate ($La_{1-x}Sr_xVO_3$)
    4.7.4. Pyrolytic carbons
    4.7.5. Cerium sulphide
    4.7.6. Metal–insulator transitions in tungsten bronzes
4.8. Impurity conduction in magnetic semiconductors
4.9. Granular metal films
4.10. Polycrystalline aggregates

4.1. Introduction

A 'FERMI GLASS' is a material, solid or liquid, in which the density of states $N(E_F)$ for electrons with energies at the zero-temperature Fermi level is not zero, but in which the disorder is great enough to ensure that electrons in these states are localized in the Anderson sense (§ 2.3). The disorder can be due to the random distribution of two constituents in a crystalline alloy, or to the non-crystalline structure of a liquid or amorphous solid. A transition of Anderson type from non-metal to metal is said to occur when some parameter x, for example composition, stress, magnetic, or electric field, changes in such a way that states at E_F become non-localized (extended). The theory of the change in the conductivity σ

under these conditions has been described in § 2.8. In the usual case (transition of type I), where a mobility edge E_C above E_F exists for the Fermi glass, the transition occurs when x is varied in such a way that $E_C - E_F$ changes sign. We denote this quantity by ΔE. If ΔE is positive, conduction is activated, owing either to excitation of electrons to E_C or at low temperatures to variable-range hopping; when ΔE is negative, conduction is metallic, σ tending to a finite value as T tends to zero. In another form of Anderson transition (type II), the states in a half-full band (not strongly correlated) may all be localized and, as x is varied, non-localized states may appear first at E_F. In such a case the transition is from hopping to metallic behaviour, with no range of temperature for which conduction is by excitation to a mobility edge.

Many amorphous semiconductors have a finite density of localized states at the Fermi level and are in this sense Fermi glasses, showing variable-range hopping as the main transport mechanism at low temperatures (Chapter 6). This chapter, however, describes only systems which show an Anderson transition and which are solids; liquids are considered in the next chapter. Perhaps the main interest of such systems is that they give the clearest proof of the existence of Anderson localization, of the mobility edge, and of the minimum metallic conductivity; it is owing to their experimental investigation that these concepts can be applied with confidence to the problems of a non-degenerate gas in the conduction bands of amorphous semiconductors.

In the last chapter it has been shown that when an electron is in a localized state it will always deform the surroundings; this must be included in any description of a Fermi glass. Systems in which this is not important are in general those in which the hydrogen radius $\hbar^2 \kappa / m e^2$ is large compared with the lattice parameter, or in which for some other reason the radius $1/\alpha$ of localized states is large. Typical of these are donor states in (crystalline) silicon and germanium which are responsible for impurity conduction and in other semiconductors for which κ is large, and also two-dimensional conduction at a Si/SiO_2 interface (§ 4.6). As impurity conduction has been extensively investigated, we start the chapter with a discussion of this. Correlation, namely the Coulomb interaction between electrons and holes, plays a major role in impurity conduction, however, as indeed it does in most phenomena where metallic behaviour occurs in a narrow band; correlation can of course lead to a metal–insulator transition (the Mott transition). In the next section, therefore, we describe the effects of correlation, showing when it is important and when it is not. We then describe impurity conduction, some alloys such as amorphous Mg–Bi, and conduction at a Si/SiO_2 interface. Later sections will describe materials where distortion (polaron formation) is important, such as cerium sulphide and $La_{1-x}Sr_xVO_3$; a more complicated case is presented by VO_x with x

near to unity, where both correlation and polaron formation appear to play a role.

Figs. 4.1 and 4.2 collect together some of the effects at an Anderson transition which are predicted by the considerations of Chapter 2. Fig. 4.1

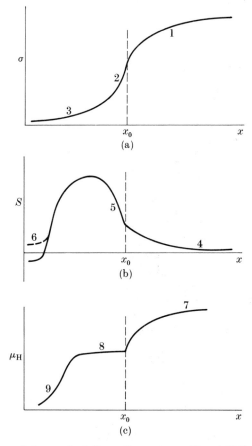

Fig. 4.1. Behaviour of the conductivity σ, thermopower S, and Hall mobility μ_H at an Anderson transition, plotted against a parameter x such that $E_F - E_C$ changes sign at x_0. The curve is at a constant temperature. The numbers refer to the following behaviour: 1, $\sigma =$ const. g^2; 2, $\sigma = \sigma_{min} \exp(-\varepsilon/kT)$; 3, hopping conduction; 4, metallic, $S =$ const. d ln g^2/dE; 5, $S = (k/e)(\varepsilon/kT + \text{const.})$; 6, hopping; the sign depends on that of d ln g/dE; 7, $\mu_H \propto g$; 8, $\mu_H =$ const., 9, hopping regime, μ_H due to excited electrons at the mobility edge.

shows the behaviour of the conductivity, thermopower, and Hall mobility as a function of a parameter x, for instance composition, which determines the ratio B/V_0 of § 2.3. In these figures x denotes this ratio. Localization occurs when $x = x_0$, where (for three dimensions) $x_0 \sim \frac{1}{2}$. Fig. 4.2 shows the plots as a function of $1/T$ for a Fermi glass, with a transition of type I.

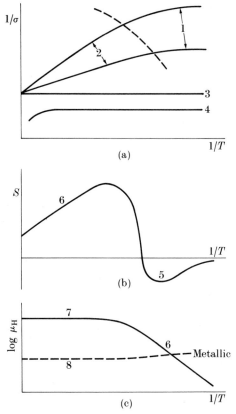

Fig. 4.2. Log(resistivity), thermopower, and Hall mobility at an Anderson transition as functions of $1/T$. In the top two curves of (a) E_F lies below E_C: 1 (to the left of broken curve), $\sigma = \text{const. } \exp(-B/T^{1/4})$, or $T^{1/3}$ in two dimensions; 2, $\sigma = \text{const. } \exp(-\varepsilon/kT)$; at 3, $E_F = E_C$ and $\sigma = \sigma_{\min}$; 4, E_F lies above E_C. (b) The thermopower when E_F lies below E_C; 5, the hopping regime; 6, the range of T when $S = (k/e)(\varepsilon/kT + 1)$. (c) The full curve shows μ_H when E_F lies below E_C: in (6) μ_H tends to zero as $T \to 0$ (the hopping regime); in (7) $\mu_H = \text{const.}(ea^2/\hbar)$; the broken curve (8) shows the metallic behaviour when E_F lies above E_C.

4.2. Metal–insulator transition in crystals†

4.2.1. Band-crossing transitions

In crystals the simplest form of metal–insulator transition is from a semi-metal, a material for which the conduction and valence bands overlap, to a non-metal, where they do not. The two situations are shown in Fig. 4.3. In the divalent metals ytterbium, barium, strontium, and calcium, all of which are cubic, pressure diminishes the overlap between the lowest band, with two electron states per atom, and the conduction band. When the two

† This section summarizes the treatment given by Mott (1974c).

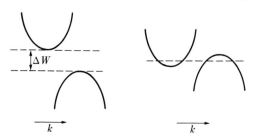

Fig. 4.3. Band-crossing transition; energy as a function of k; (a) non-metal, (b) metal.

bands cease to overlap, a metal–insulator transition is expected, and has been observed in ytterbium (McWhan, Rice, and Schmidt 1969, Jullien and Jerome 1971); discussions are given by Mott and Zinamon (1970) and Mott (1974c). Another example is bismuth, where the two overlapping bands separate on alloying with about 7 per cent of Sb (for references see a theoretical study by Martin and Lerner (1972)). If there were no inter-actions between electrons, these transitions would be continuous; the gap ΔW would go uniformly to zero, at which point an infinitely small number of electrons and holes would appear and then increase.

As first pointed out by Mott (1949, 1956), interaction between electrons and holes invalidates this conclusion, and at a transition of this kind there must be a discontinuous change in the number of current carriers; the argument was that a small number of electrons and holes would form pairs (excitons), and the system would not be conducting. For metallic conduc-tion there must be enough free carriers to screen the Coulomb field between carriers of opposite sign, so that excitons are not formed, and this was shown to lead to the formula

$$n^{1/3} a_{\mathrm{H}} \simeq 0 \cdot 25 \qquad (4.1)$$

with $a_{\mathrm{H}} = \hbar^2 \kappa / m^* e^2$, where κ is the background dielectric constant and $m^* = m_{\mathrm{e}} m_{\mathrm{h}} / (m_{\mathrm{e}} + m_{\mathrm{h}})$. Knox (1963) pointed out that before two bands begin to overlap, if the energy $(-m^* e^4 / 2\hbar^2 \kappa^2)$ of an exciton is (numeric-ally) greater than the band gap ΔW, excitons will form before ΔW vanishes. The literature on what happens to these excitons and the condi-tions under which they 'crystallize' is large (Keldysh and Kopaev 1965, Halperin and Rice 1968, Kohn 1967, further references in Mott 1974c). While the so-called excitonic phase is not ruled out, it now seems likely that a variation of the energy interval between the bands will *necessarily* lead to a discontinuous change in the density of electrons and holes from zero to the value at which the 'electron–hole gas' has its minimum energy. The properties of the electron–hole gas are known mainly from experi-ments on electron–hole droplets in strongly illuminated crystalline germanium. In these materials (Rogachev 1968, Benoit à la Guillaume and

Voos 1973, Benoit à la Guillaume, Voos and Salvan 1972, Hensel, Phillips, and Rice 1973)[†] the electrons and holes condense into droplets, giving a photoluminescence line of frequency that does not depend on the intensity, unless the excitation is so strong that the droplets expand to fill the whole specimen (Nakamura and Morigaki 1974). The theoretical problem is whether the electron–hole gas has a lower energy than a condensation of excitons. For isotropic energy surfaces and neglecting correlation, an estimate is as follows. The energy of an exciton is $-e^4 m^*/2\hbar^2 \kappa^2$, where

$$m^* = \frac{m_e m_h}{m_e + m_h}.$$

The kinetic energy of the electron gas together with that of the hole gas, there being n particles of each kind per unit volume, is

$$\frac{3}{5} \frac{\hbar^2 n^{2/3}}{2m^*}$$

and their electrostatic potential energy, assuming a uniform distribution of charge, is

$$-1\cdot1 \frac{e^2 n^{1/3}}{\kappa}.$$

The minimum comes when $n^{1/3} a_H \simeq 0\cdot1$, giving an energy $0\cdot1\, e^4 m^*/2\hbar^2 \kappa^2$, much less stable than for the free excitons. However, when the surfaces are anisotropic (Brinkman and Rice 1973) or when correlation is included (Vashishta, Battacharyya, and Singwi 1973), the reverse is usually the case. The electron–hole gas is then more stable and electron–hole droplets are formed if carriers are introduced either under illumination or through double injection (Marello et al. 1973). From this the conclusion is drawn that, just before two bands overlap and as soon as it is energetically favourable to form a droplet, the droplet will spread and fill the whole of the specimen. There is thus a discontinuous change in the number of carriers from zero to a value n given by

$$n^{1/3} a_H = \text{const.}$$

where $a_H = \hbar^2 \kappa/m^* e^2$ and κ is a background dielectric constant. The constant in this equation is difficult to calculate and will depend on the band form, but comparison with the observations on crystalline germanium gives a value of about $0\cdot3$. At the same time a discontinuity in the energy gap is expected; while at the transition no energy is required to form a droplet, an energy equal to the condensation energy per pair for a droplet is required to form a free electron and hole.

† A review is given by Hensel, Phillips and Thomas (1977).

The magnitudes of these discontinuities depend essentially on the background dielectric constant. For an overlap between a conduction and a valence band, where the optical excitation across the gap is allowed, this will probably be large, so the discontinuity in n will be small. In fact no discontinuity has been observed in ytterbium under pressure. A large discontinuity is most likely for the overlap between the two Hubbard bands considered in the next section, or between two d bands, because the oscillator strength for optical transitions between them is small.

An important consequence of the prediction of a discontinuity is that, if the free energy is plotted against volume or composition (x) in an alloy of composition P_xQ_{1-x}, a kink in the curve must occur as in Fig. 4.4. This

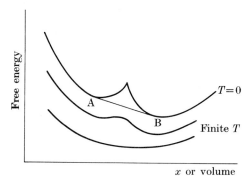

Fig. 4.4. Free energy against composition or volume for a system showing a metal–insulator transition. The top curve is for $T = 0$, the lower curves for higher temperatures.

means that at low temperatures there *must* be a discontinuous change in volume under pressure, from A to B in Fig. 4.4, and in alloy systems there must be a range of composition in which the alloy is unstable so that two phases will separate on annealing. The discontinuous change at low temperatures of the density of current carriers with continuously varying composition can thus only be observed in quenched unstable alloys, and at the time of writing we have no firm evidence of its occurrence except possibly (in the case of Hubbard bands) for lead–argon films (see § 4.3.4).

4.2.2. Hubbard bands and the Mott transition

For an array of one-electron centres with cubic structure, there are two possibilities. The first is that the crystal is a metal; the second is that it is an antiferromagnetic (or ferromagnetic) insulator. As first pointed out by Slater (1951), an antiferromagnetic lattice can split the conduction band, leading to a full and an empty band and insulating behaviour; these are now usually called Hubbard bands; the insulating behaviour does not depend on antiferromagnetic order, and continues above the Néel temperature. It is to be expected also for amorphous antiferromagnets, either at low temperatures where each moment is fixed in position or above the Néel temperature (see § 4.2.4).

In this section, then, we outline the properties of a crystalline array of one-electron centres, each described by an atomic wavefunction $\phi(r)$ behaving as $\exp(-r/a_H)$ for large r, at a distance a from each other sufficiently large for the tight-binding approximation to be useful (§ 2.3). If the number of electrons per atom deviates from an integral value, the model should always predict metallic behaviour, but with one or any integral number of electrons per atom this is not so, and the system is insulating. The most convenient description is in terms of the Hubbard intra-atomic energy U (Hubbard 1964), defined by

$$U = \iint (e^2/\kappa r_{12})|\phi(r_1)|^2|\phi(r_2)|^2 \, d^3x_1 \, d^3x_2. \tag{4.2}$$

For hydrogen-like wavefunctions this has been calculated and is (Schiff 1955)

$$U = 5e^2/8\kappa a_H. \tag{4.3}$$

If $\mathscr{I}$ is the ionization potential of each atom and $\mathscr{E}$ its electron affinity, then, assuming that the functions ϕ are not changed by the addition of an extra electron,

$$U = \mathscr{I} - \mathscr{E}. \tag{4.4}$$

The properties of such a system are the following.

(a) When the distance a between the centres is large so that the overlap energy integral I (§ 2.3) is small, the system is expected to be anti-ferromagnetic, with an energy below that of the ferromagnetic state (Anderson 1956) equal to

$$-2zI^2/U = -B^2/2zU. \tag{4.5}$$

The Néel temperature T_N will be such that kT_N is of this order.

(b) An extra electron placed on one of the atoms is able to move with a k vector just as in a normal band. It is said to have an energy in 'the upper Hubbard band'. It will polarize the spins on surrounding atoms antiparallel to itself, or parallel if the atomic orbitals are degenerate, forming a 'spin polaron'. Its bandwidth is probably not very different from B calculated without correlation.

(c) A hole formed by taking an electron away from one atom has similar properties, being able to move with a wavevector k and having a range of energies also not very different from B. One speaks of electrons in the singly occupied states as being in this 'lower Hubbard band', though it is preferable to think of a band of holes.

(d) The two bands will overlap when a is small enough, and consequently B great enough, to ensure that

$$B \geqslant U. \qquad (4.6)$$

A metal–insulator transition then occurs, sometimes known as the 'Mott transition'. It is similar to a band-crossing transition, being (if we neglect the arguments for a discontinuous change in the number of current carriers) from an antiferromagnetic insulator to an antiferromagnetic metal. However, as the overlap increases, the number n of free carriers increasing, the moments on the atoms and the Néel temperature will tend to zero and disappear when $n = 1/2z$ (Mott 1974c), as shown in Fig. 4.5; there

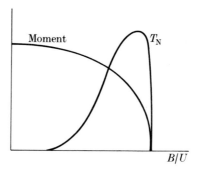

Fig. 4.5. Moment and Néel temperature at the transition between antiferromagnetic and normal metal, plotted against B/U.

may thus be *two* transitions, though the range of the antiferromagnetic metal may be absent. After the moments have disappeared, the electron gas is highly correlated, as first described by Brinkman and Rice (1970). This means that only a small proportion ζ of the sites are doubly occupied at any one moment, or unoccupied, as shown in Fig. 4.6. The spins of the electrons on the other sites are no longer arranged antiferromagnetically, but resonate quantum mechanically between their two positions. The band is no longer split into two Hubbard bands; instead there is a large enhancement by $1/2\zeta$ of the effective mass, leading to large values of the Pauli susceptibility and electronic specific heat, observed for instance in some metallic transition-metal oxides (Brinkman and Rice 1973).

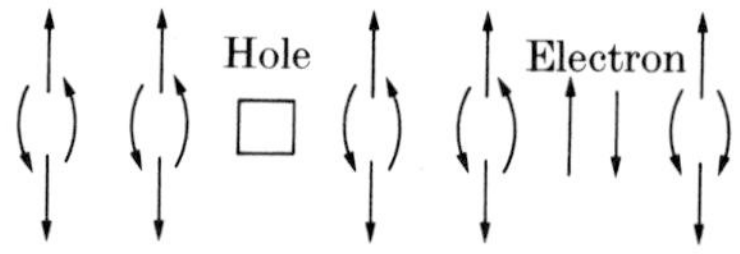

Fig. 4.6. Resonating moments in a highly correlated electron gas.

However, although there is no splitting, it is often convenient to draw the two overlapping Hubbard bands as in Fig. 4.7, each representing the density of states of 'spin polarons' (Mott 1972e, 1974c). This should be a valid procedure for calculations of conductivity, Hall effect, and thermopower, for instance in disordered systems.

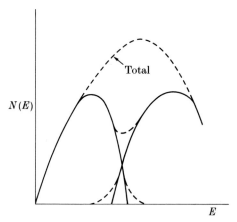

Fig. 4.7. Overlapping Hubbard bands. The density of states for current carriers is plotted against E, the dotted lines being for a non-crystalline system. 'Total' denotes the total density of states when all spin directions are included.

Just as for band-crossing transitions, a discontinuity in n is expected at the Mott transition; the discontinuity should be larger for the following reason. An optical transition across the gap involves transferring an electron from one atom to a neighbour, so the corresponding oscillator strengths and the consequent background dielectric constant are not likely to be large. Thus the kink in Fig. 4.4 will be important. Two phase regions are in fact observed in many systems. It appears that the change of n in, for instance, V_2O_3 under pressure, is so large ($>1/2z$ per atom) that the antiferromagnetic metallic region does not occur. However, Gautier *et al.* (1975) find for the system $NiS_{2-x}Se_x$ that there is a range of composition which is metallic and antiferromagnetic; their phase diagram is shown in Fig. 4.8.

The condition (4.6) for the transition can be evaluated for hydrogen-like wavefunctions, since U (eqn 4.3) and B are both known. Writing, using eqn (2.29) for values of αR near the transition

$$B \simeq 23z(e^2\alpha/\kappa)\exp(-\alpha R).$$

Because of the rapid variations of $\exp(-\alpha R)$, αR depends little on z; we have

$$\alpha R = \ln(36z) \simeq 5\cdot8$$

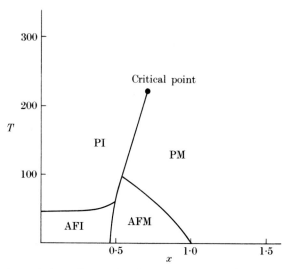

Fig. 4.8. Experimental phase diagram for the system $NiS_{2-x}Se$ (Gautier *et al.* 1975). AFI denotes antiferromagnetic insulator, PI the same when paramagnetic above the Néel point, AFM antiferromagnetic metal, and PM the same above the Néel point.

if z, the co-ordination number, is 6. Thus

$$n^{1/3} a_H \simeq 0 \cdot 2. \qquad (4.7)$$

This equation is used in § 4.3.2 for the discussion of the metal–insulator transition in doped semiconductors. It does *not* take account of any discontinuity in the number of carriers.

Eqn (4.7) is derived for hydrogen-like wavefunctions. It is pointed out particularly by Catterall and Edwards (1975) in their discussions of frozen solutions of alkali metals in hexamethylphosphoramide and more generally by Edwards and Sienko (1978) that κ cancels out from the ratio B/U, except in so far as it affects a_H. If a_H can be estimated independently, for instance from n.m.r. data, (4.7) should be valid for functions that are by no means hydrogen-like. Edwards and Sienko find that (4.7), with the constant equal to $0 \cdot 26$, is valid for the metal–insulator transition in a very wide range of semiconductors, in which n ranges from $\sim 10^{22}$ to $10^{14}\,\mathrm{cm}^{-3}$ (doped InSb).

In applying (4.6) to the transition in oxides, since $U - B$ is the energy necessary to form two carriers, U must be the screened value U_{scr} for which the lattice relaxes round each charge. U_{scr} may be much smaller (by a factor ~ 10) than the value of U in (4.5) which determines the Néel point (cf. Mott 1974c).

4.2.3. Wigner crystallization

The discussion of the Mott transition in the last paragraph is for a tight-binding band with one electron per atom. Similar arguments can be applied to any integral number, but for a non-integral number of electrons the system will always be metallic, unless compensating fixed charges are present to ensure charge neutrality. If so, they can trap the carriers in one or other Hubbard band, producing an extrinsic semiconductor (such as NiO doped with lithium, § 4.7.1).

However, in a partly filled band, or indeed in a free-electron gas, at a much smaller value of $n^{1/3}a_H$ a 'crystallization' to a non-conducting form was predicted many years ago by Wigner (1938). Various forms of electron crystallization have been observed, as for instance in Fe_3O_4, but an electron 'crystal' with a periodicity determined by n as Wigner predicted has yet to be observed with certainty (cf. Care and March 1975). The possibility has little influence on the processes described in this book, except perhaps those of § 4.6 (see for instance Kawaji and Wakabayashi 1977).

4.2.4. Effect of disorder on Mott transitions

The predictions of the last section are as follows.

(i) For a series of quenched alloys, if the composition goes through the critical value, there should be discontinuity in n, the number of carriers, from zero to a finite value.

(ii) There should be a range of composition for which the system is unstable, giving a separation of two phases if the system is annealed.

Metal–ammonia systems such as Na_xNH_3 show the well-known solubility gap (Thompson 1976, Mott 1975b) which we believe to be an example of this behaviour. Caesium vapour at high temperatures and pressures shows a critical point that we ascribe to the same cause. These are discussed by Mott (1974c). However, if the disorder is great enough, it is clear that the discontinuity is suppressed, an Anderson transition occurring first as the parameter favouring localization increases. This is shown by the fact that doped semiconductors such as Si:P show a transition of Anderson type with no discontinuity. A qualitative theory showing when a discontinuity is to be expected and when it is not has been given by Mott (1978a).

A second question of interest is the following. If an Anderson transition is observed, is the metallic electron gas highly correlated? If so, a pseudo-gap as in Fig. 4.7 is expected, with the following consequences.

(a) R_H is given by $0.7/necg$ and will be larger than the classical value (for a definition of $g = N(E_F)/N(E)_{free}$ see eqn (2.22)).

(b) The thermopower should change sign (and become positive) if the specimen is compensated.

(c) Anderson localization will be facilitated and occur when $g \simeq \frac{1}{3}$.
It now seems likely that this is not so for doped Si and Ge; this will be discussed further in §§ 4.3.3.

On the non-metallic side of the transition, crystalline materials are expected to be antiferromagnets. In the disordered case, also, each localized state should have a moment. Moreover, at low temperatures each moment should have a fixed orientation; the material is an amorphous antiferromagnet. It is believed that such materials have a sharp Néel temperature above which they become completely random, but this is not quite certain.† Below this temperature the moments should give a large broadening of any n.m.r. line; above it they should not. Also above it the term $(k/e)\ln 2$ (§ 2.13) should appear in the thermopower.

4.2.5. Effect of correlation and distortion on Anderson transitions

If correlation is strong, we may start our discussion with two slightly overlapping Hubbard bands, as in Fig. 4.7, and ask whether states at the Fermi energy are localized or not in the Anderson sense; unless there is a discontinuity in the number of current carriers, the transition takes place when E_C and E_F coincide.

However, if correlation is weak or if the number of carriers in the band considered is non-integral, we can start from the one-electron localized wavefunctions and ask what is the effect of correlation on them. We define a quantity U_A as the Hubbard intra-state interaction $\langle e^2/\kappa r_{12}\rangle$ for one of these Anderson localized states; its magnitude has been estimated (Mott 1974c), and it will certainly be much less than the corresponding value for an atom and tend to zero with α. However, it *must* ensure that states at E_F are singly occupied; there will in fact be a demarcation energy E_F' separating singly from doubly occupied states (Fig. 4.9). The interval $E_F - E_F'$ is U_A. As localization becomes stronger and stronger, E_F' will sink until all states are singly occupied. Here again, as in the last section, there must exist a Néel temperature above which the moments of the singly occupied states fluctuate between random directions. A Fermi glass must in principle always have a Néel temperature.

Another point of great importance is that, as shown in Chapter 3, there is always some displacement of the surrounding atoms when an electron is added to or removed from a localized state. As first pointed out by

---

†The experimental evidence comes from 'spin glasses', such as Fe dissolved in Au, where the interaction between moments is long range and fluctuating. For this alloy there is a sharp cusp (at T_t) in the susceptibility plotted against T (Cannella, Mydosh, and Budnick 1971), but the specific heat shows no sharp change. A theory in which it is supposed that the moments become free above T_t is given by Edwards and Anderson (1975). The subject is reviewed by Fischer (1977). At the time of writing it is not entirely clear whether amorphous systems have a true Néel point, below which at each site there is a fixed moment, or whether in a fairly narrow range of T the time scale for fluctuations of moments becomes very large.

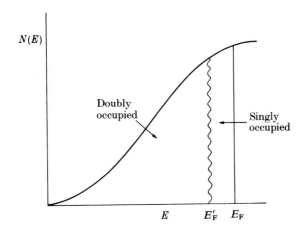

Fig. 4.9. Showing singly and doubly occupied states for a Fermi glass.

Anderson (1975), this may lead to a situation in which *all* states are doubly occupied or empty. Street and Mott (1975) and Mott, Davis, and Street (1975) use the concept for dangling bonds in chalcogenide glasses, supposing that an unoccupied dangling bond forms a bond with a lone-pair orbital on a neighbouring Se or Te atom, and the distortion energy is such that as many unoccupied bonds are formed as possible so that no unpaired spins remain. This concept is discussed further in Chapters 6 and 9.

4.3. Doped semiconductors

4.3.1. Impurity conduction; direct currents

The phenomenon now known as impurity conduction was first observed by Hung and Gleissmann (1950) as a new conduction mechanism predominant at low temperatures in doped and compensated crystalline germanium and silicon. Fig. 4.10 shows some typical results, due to Fritzsche and Cuevas (1960). It will be seen that at low temperatures the conductivity behaves like

$$\sigma = \sigma_3 \exp(-\varepsilon_3/kT) \tag{4.8}$$

and that σ_3 depends strongly on the concentration of donors (N_D), except at high concentrations where ε_3 disappears and 'metallic' behaviour occurs. It is now believed that impurity conduction with an activation energy ε_3 can only occur (except for concentrations near the metal–insulator transition where the two Hubbard bands overlap) when the material is compensated. This was first pointed out by Conwell (1956) and Mott (1956) and confirmed by the experimental work of Fritzsche (1958, 1959, 1960). Prior to 1960, measurements in silicon and germanium were made

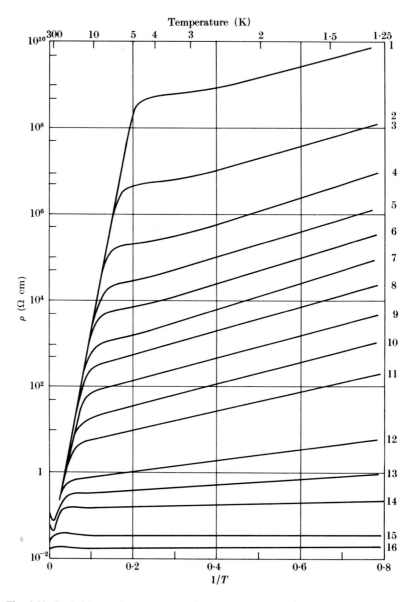

Fig. 4.10. Resistivity ρ of p-type germanium with compensation $K(=N_{\mathrm{D}}/N_{\mathrm{A}})=0.4$ (Fritzsche and Cuevas 1960). The concentrations of acceptors are as follows (in cm^{-3}): (1) 7.5×10^{14}; (2) 1.4×10^{15}; (3) 1.5×10^{15}; (4) 2.7×10^{15}; (5) 3.6×10^{15}; (6) 4.9×10^{15}; (7) 7.2×10^{15}; (8) 9.0×10^{15}; (9) 1.4×10^{16}; (10) 2.4×10^{16}; (11) 3.5×10^{16}; (12) 7.3×10^{16}; (13) 1.0×10^{17}; (14) 1.5×10^{17}; (16) 1.35×10^{18}.

with samples doped with suitable impurities, but Fritzsche and Cuevas (1960) introduced impurities by slow-neutron bombardment, producing transmutation of the germanium atoms; in this way it was possible to vary the concentration of donors, keeping the compensation K constant, which greatly facilitated investigation of the separate effects of N_D and K.

Impurity conduction is thought to take place in an impurity band separated from the conduction band and split by the Hubbard U, as in Fig. 4.11. States in the lower Hubbard band are Anderson localized owing to

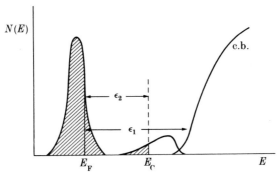

Fig. 4.11. Impurity band split by the Hubbard U, with light compensation; c.b. denotes the conduction band.

the random fields of the charged minority carriers. The compensation is defined (for an n-type semiconductor) by

$$K = N_A/N_D;$$

compensation is represented in Fig. 4.11 by a position of the Fermi energy such that the lower Hubbard band is not full. If K is small, ε_3 represents the energy required to remove a 'hole' (an empty donor) from the nearest negatively charged acceptor, and in fact an early paper (Mott 1956) represented ε_3 in this way; the thermopower S should then behave as $(k/e)(\varepsilon_3/kT + \text{const.})$. However, if K is not small, Anderson localization obtains, ε_3 is a hopping energy, and S should increase with T. This was first recognized by Twose (1959), developed in detail by Miller and Abrahams (1960), and reviewed by Mott and Twose (1961). In this analysis, the electron is supposed to jump to the nearest available empty site; variable-range hopping has been observed in certain cases (§ 4.3.3), but not as yet in lightly doped silicon and germanium.

The calculation of the conductivity consists of four parts.

(a) The calculation of the probability p per unit time that an electron jumps from one centre to another at a distance R, and with energy higher by W_D. This is the problem considered in Chapter 3, and we write it

$$p = p_0 \exp(-2\alpha R - W_D/kT). \tag{4.9}$$

In the calculation $1/\alpha$ is the radius of the centres, assumed spherically symmetrical and about 50 Å, and R typically is of order 300 Å; p_0 could have a wide range of values and may be of order $10^{12}\,\text{s}^{-1}$.

(b) An averaging over all energies W_D and all hopping processes† to give the observed hopping energy ε_3. In the work quoted, it is supposed that $\exp(-2\alpha R)$ is so small and sensitive to R that the electron always jumps to its nearest neighbour. For hopping to nearest neighbours Miller and Abrahams found for n-type material and for small values of the compensation K

$$\varepsilon_3 = \frac{e^2}{\kappa}\left(\frac{1}{R_D} - \frac{1\cdot35}{R_A}\right) \tag{4.10}$$

with a more complicated formula when this is not so. Here R_D and R_A are the averaged distances between donor and acceptor sites respectively. Since R_D is also the average donor–acceptor separation, the first term is simply the energy required to separate the carrier from the nearest charged acceptor. Calculations for all values of K were carried out and the plot of ε_3 against K is as in Fig. 4.12. For small K also ε_3 is proportional to $1/R_D$

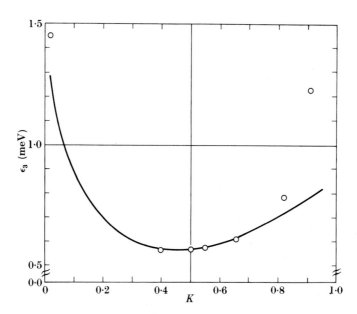

Fig. 4.12. Activation energy ε_3 as a function of compensation calculated from eqn (4.10) with $N_D = 2\cdot66 \times 10^{15}\,\text{cm}^{-3}$. The circles are experimental points. (From Mott and Twose 1961.)

† If K is small, the quantity ε calculated by Miller and Abrahams, or that deduced from observations, is not a mean of the individual hops; this is only so for reasonably large values of K.

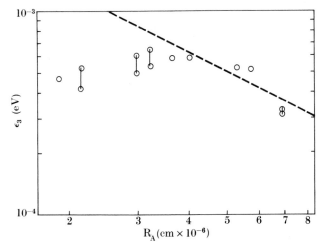

Fig. 4.13. The activation energy ε_3 for impurity conduction for the samples shown in Fig. 4.10, plotted against the average impurity separation R_A. The broken curve represents the results of eqn (4.10). (From Mott and Twose 1961.)

(i.e. $N_D^{1/3}$). According to Fig. 4.13 (from Mott and Twose 1961) this is so for low concentrations, but ε_3 drops for high concentrations. The drop may be due to the approach to the Anderson transition (at which ε_3 must vanish) or to the increasing dielectric constant caused by a high concentration of centres.† The former was suggested by Shklovskii and Shlimak (1972). Knotek (1977), however, proposes that W_D in (4.9) decreases because the kind of many-electron hops discussed in Chapter 2 (see eqn 2.66) can occur with higher probability.

(c) The next step in the calculation is to obtain the conductivity from the jump probability (4.9). This is carried out as in § 2.7, giving

$$\sigma = e^2 N(E_F) p R_D^2.$$

(d) Finally, since the factor $\exp(-2\alpha R)$ can vary enormously from one jump to another, the conductivity depends on an evaluation of the effect of a network of widely varying impedance elements. Pollak (1972) has criticized the way in which this was carried out by Miller and Abrahams; he finds that the average behaves‡ as $\exp(-2{\cdot}4\alpha R)$, a result near to that of Twose (1959) and differing from that of Miller and Abrahams ($\exp(-\text{const.}\ R^{3/2})$). Another treatment is due to Shklovskii (1973a) who

---

† Castellan and Seitz (1951) and Mott and Davis (1968) suggested that this may be the origin of the drop in ε_1 shown, for instance, in Fig. 4.17. However, it seems not to be shown for highly compensated samples (Fig. 4.10) where the Fermi energy is pinned, and may then be due in lightly compensated samples to a shift in the (unpinned) Fermi energy.

‡ Shklovskii (1973b) finds the same term (2.4); Seager and Pike (1974) find 2.8.

also considers a different percolation path from Miller and Abrahams and obtains a result in good agreement with experiment.

There have been few measurements of the thermopower for impurity conduction. For compensation K less than $\frac{1}{2}$ an n-type impurity band should give p-type thermopower (compare § 2.13). Fig. 4.14 shows results of Geballe and Hull (1955) for n- and p-type silicon, for concentrations just on the insulator side of the transition. A similar effect is observed in NiO (§ 4.5).

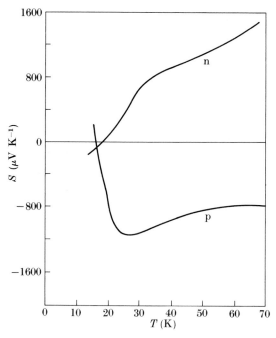

Fig. 4.14. Low-temperature values of the thermoelectric power S for samples of silicon containing 10^{18} cm^{-3} donors or acceptors, showing the reversal in the sign at low T. (From Geballe and Hull 1955.)

Some discussion of the Hall effect for hopping processes is given in § 3.9, together with the suggestion that it is small. A Hall effect for hopping has not been observed with certainty in impurity conduction, it being likely that any observed effects (e.g. Fritzsche and Cuevas 1960) are due to carriers excited to a mobility edge.

Observations and theories on magnetoresistance due to shrinking of the orbits are given by Mikoshiba and Gonda (1962), Mikoshiba (1962), and Knotek (1977).

At sufficiently low temperatures, some kind of variable-range hopping is always to be expected, the term W_D in (4.9) becoming smaller if transitions to distant states occur. We expect therefore a 'Miller and Abrahams' range with a constant activation energy ε_3, and a low-temperature range showing variable-range hopping $\sigma \propto \exp(-B/T^{1/4})$. Some of the evidence for this behaviour is discussed in § 4.3.3. As mentioned above, Knotek (1977) gives a different explanation for the Miller–Abrahams range.

4.3.2. Impurity conduction; alternating currents

In Chapter 2 we deduce that $\sigma(\omega)$, and hence the absorption coefficient α, should vary as $\omega^2\{\ln(I_0/\hbar\omega)\}^4$ for small ω in the limit as $T \to 0$. As far as we know, experiments at frequencies and temperatures low enough to verify this formula have not been made. More detailed formulae, applicable at frequencies where the absorption coefficient passes through a maximum and at temperatures such that $\sigma(\omega)$ decreases with temperature, have been proposed by Cumming et al. (1964) and by Blinowski and Mycielski (1964). Experiments in which comparison is made with these formulae are due to Tanaka and Fan (1963) and to Milward and Neuringer (1965).

The effect here, that of direct optical transition, is quite different from the Debye-type loss due to thermally activated hopping discussed in § 2.15. The latter was first observed by Pollak and Geballe (1961)[†] in n-type silicon. Some of their results are reproduced in Fig. 4.15. The theory given by these authors and by Pollak in subsequent papers (Pollak 1964) is similar to that of § 2.15 except that, for low compensation, the assumption of a random distribution of energies is certainly not valid, and this increases the complexity of the calculation. Nevertheless, the main results follow, namely that if σ is the observed conductivity

$$\sigma_{\text{a.c.}} = \sigma_{\text{d.c.}} + A\omega^s$$

with s close to $0 \cdot 8$ and A little dependent on T. As in § 2.15, the loss process is of the Debye type, and the main contribution comes from pairs of centres for which the difference W_D in the energy levels is of order kT, and the distance R between them such that

$$\nu_{\text{ph}} \exp(-2\alpha R) \sim \omega.$$

The frequency dependence predicted by the theory is such that

$$\sigma(\omega) \sim \omega\left(\ln \frac{\nu_{\text{ph}}}{\omega}\right)^4.$$

The dependence on $\omega^{0\cdot8}$ is verified, as may easily be seen, if $\nu_{\text{ph}} \sim 10^{12}\,\text{s}^{-1}$, as it appears to be in silicon and germanium. However, as is shown in Chapter 3 the quantity ν_{ph} is extremely sensitive to the parameters

---

[†] For more recent theoretical work, see Lyo and Holstein (1973).

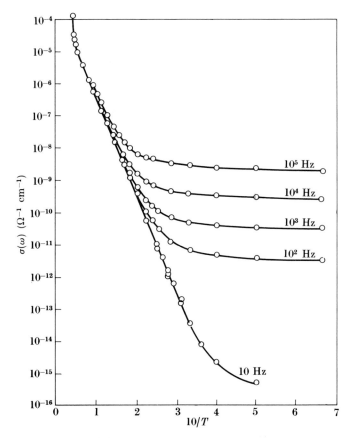

Fig. 4.15. a.c. conductivity $\sigma(\omega)$ of n-type silicon with low compensation. (From Pollak and Geballe 1961.)

involved, and, if there is strong polarization round the centre, ν_{ph} will contain the factor $\exp(-W_H/\tfrac{1}{4}\hbar\omega)$ at low temperatures.

As regards the temperature dependence, the treatment of § 2.15 supposes that compensation is not small and that kT is small compared with the bandwidth. Thus a fraction proportional to kT of the available electrons can take part in the hopping, and they can move into a fraction of empty centres also proportional to kT. Since kT also appears in the denominator of the expression for σ, because of the Einstein relationship, $\sigma \propto kT$. In the work of Pollak and Geballe, however, K was small and the 'holes' were non-degenerate, so one of the factors kT falls out and σ is independent† of T.

---

† At very low temperatures a drop of σ with T, so that $\sigma \propto T^2$, is to be expected (Pollak 1964).

Golin (1963) made measurements on p-type germanium, and his work differed from that of Pollak and Geballe in that the compensation K was large (0·4). In this case $\sigma(\omega)$ should be proportional to T. Some results are shown in Fig. 4.16, showing a dependence on T.

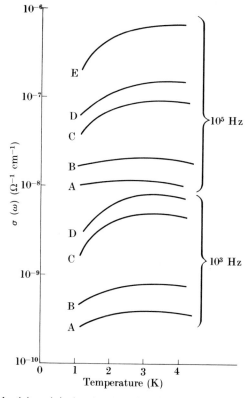

Fig. 4.16. a.c. conductivity $\sigma(\omega)$ of various samples of p-type germanium with high compensation ($K = 0\cdot4$) plotted against temperature for frequencies 10^3 and 10^5 Hz (Golin 1963). A dependence on T is shown (contrast Fig. 4.15).

4.3.3. Metal–insulator transitions in doped semiconductors

Another form of conduction was first identified by Fritzsche (1958), Fritzsche and Cuevas (1960), and Davis and Compton (1965). In this the electron is excited from a donor level to an *already occupied* donor,† and moves from one to another. In other words, it is excited to the mobility edge in the upper Hubbard band. The excitation energy ε_2 is thus $U - B$, it being assumed that both bands have the same width B. The authors quoted above, by analysing the conductivity–temperature curves, were able to

† The idea of negatively charged donors was probably first presented by Ansel'm (1953).

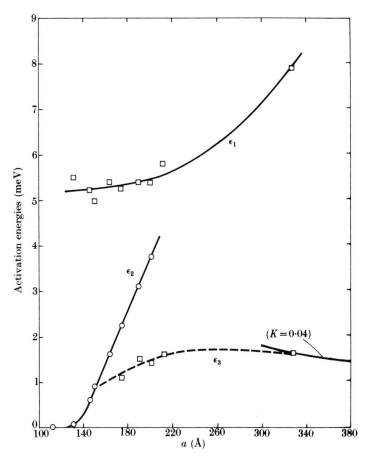

Fig. 4.17. Variation of the activation energies ε_1, ε_2, and ε_3 with distance a between donors for n-type germanium (Davis and Compton 1965). For ε_3 the calculations of Miller and Abrahams for $K = 0.04$ are shown (full curve).

separate the quantity ε_2 from the energy ε_1 required to ionize the centre. Their results for n-type germanium are shown in Fig. 4.17. In principle, conduction with the activation energy ε_2 could be by hopping between Anderson-localized states in the upper band or by excitation to a mobility edge E_C also in the upper Hubbard band. The experiments of D'Altroy and Fan (1956) on the dependence on frequency in a.c. conduction have been interpreted by Pollak (1964) as showing that this form of conduction is not by hopping. If it is not, ε_2 may be written

$$\varepsilon_2 = E_C - E_F \tag{4.10}$$

where E_C is the mobility edge in an upper Hubbard band.

In Fig. 4.11 we represent the density of states in the upper Hubbard band. Since this band is normally empty, its form should be calculated without correlation for a random assembly of wells in each of which the electron is described by an s wavefunction. Such calculations have been carried out by Gaspard and Cyrot–Lackmann (1973) and by Cyrot–Lackmann and Gaspard (1974). A feature is the long tail on the low-energy side, shown in Fig. 2.17.

The experiments of Norton (1976a, b) have identified the position and radius of the upper Hubbard band in Si:P in the non-metallic range of concentrations $(10^{14}-10^{17} \text{ cm}^{-3})$. Electrons are excited into the band, where at these concentrations they are metastable at low temperatures (2 K), because states in the tail are Anderson localized. They are then photoexcited into the conduction band. The depth of the state below the conduction-band edge in an isolated doubly occupied centre is 1.7 meV, contrasted with 40 meV when singly occupied. At a concentration of $3 \times 10^{15} \text{ cm}^{-3}$, the energy needed to excite them begins to increase, corresponding to the 'tail' of Figs. 2.17 and 4.11. Thus the radius of the doubly occupied state must be about 0.7×10^{-5} cm. The 'tail' lowers the energy of the trapped electrons to near 10 meV.

Yashihiro, Tokumoto, and Yamanouchi (1974) have obtained somewhat similar information from the absorption of submillimetre radiation in n-type Ge.

Fig. 4.17 shows that in n-type Ge there is no discontinuity in ε_2, which decreases linearly with change in concentration (except very near the transition, see below). This shows clearly that the transition is of Anderson type, and occurs when $E_F - E_C$ changes sign; this is so whether the specimen is compensated or not (see Fig. 4.10). Whether the electron gas is still highly correlated, or whether the two Hubbard bands have merged, is an important question that we shall discuss later in this section. We first review the evidence for an Anderson transition.

If the electron gas is not highly correlated, or if the Hubbard bands overlap, then we should expect charge transport by excitation to E_C at high temperatures and variable-range hopping at low temperatures. This may occur from 3×10^{18} down to 10^{18} cm^{-3} in Si:P. This effect seems to be shown by the results of Toyotomi (1973) on the Hall mobility of Si:P just on the insulator side of the transition shown in Fig. 4.18. It has been proposed (Mott, Pepper et al. 1975, Kamimura and Mott 1976) that this is to be interpreted as being due to electrons excited to the mobility edge, so that Friedman's formula (2.88) is valid at high T, but at low T, when most of the current is due to hopping at E_F which gives no appreciable Hall voltage, μ_H drops. Similar results are obtained by Vul (1974) for impurity conduction in diamond. Vul et al. (1976) have also measured the apparent free-electron density $(1/R_H)$ at 1.65 K in highly compensated GaAs as a

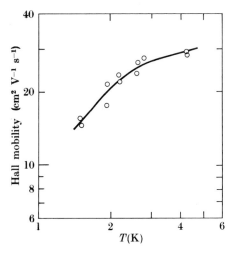

Fig. 4.18. Hall mobility in a specimen of Si:P just on the non-metal side of the metal–insulator transition. (From Toyotomi 1974.)

function of magnetic field, which shrinks the orbits. At high fields, when a rapid drop sets in at the non-metal side of the Anderson transition, a constant value of μ_H is observed, agreeing qualitatively with Friedman's prediction. However, it is not clear that Hall measurements such as those of Davis and Compton (1965) can be fitted to the model described here. Both for impurity bands and in the inversion layer (§ 4.6) the measured value of R_H often appears to give the total number of electrons, while σ measures the number excited to E_C (cf. Adkins 1978). This is not understood.

As regards the other features of variable-range hopping, some experimental evidence has been reviewed by Mott (1972e, 1974c). As an example in Fig. 4.19 we show more recent results due to Wallis (1973) for two concentrations of boron in Si:B. An early observation of it in doped germanium is that of Shlimak and Nikulin (1972).

At the metal–insulator transition in Si:P and similar materials, although there is no discontinuous change in n (or ε_2), a discontinuity has been observed in the Knight shift by Sasaki, Ikehata, and Kobayashi (1973), and by Ikehata, Sasaki, and Kobayashi (1975). A theoretical discussion is given by Mott (1974b). It is supposed that when states are localized at E_F, a magnetic field produces a large internal field in a small volume, proportional to H, in which spins are reversed. Thus most nuclei are not subjected to an additional internal field.

In the range of concentration when the two bands overlap, the phenomenon of e.s.r.-enhanced conduction has been observed by Morigaki and Onda (1972, 1974), which we believe is to be interpreted as

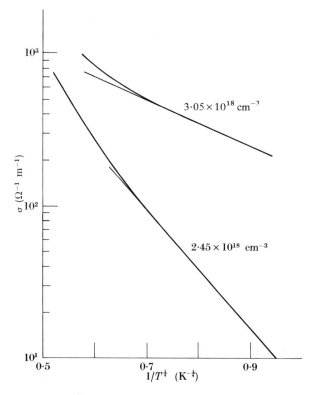

Fig. 4.19. Plot against $1/T^{1/4}$ of the conductivity of two specimens of Si:P. (From Wallis, unpublished.)

variable-range hopping with energy received from the radio-frequency radiation (Mott 1972*e*, 1974*c*; Kamimura and Mott 1976).

We now ask whether the electron gas is highly correlated at the transition. Fig. 4.20 shows the Hall coefficient R_H of Si:P measured by Yamanouchi, Mizuguchi, and Sasaki (1967). The drop in R_H below the value $1/nec$ as soon as states become localized is to be expected because the current is probably due to electrons excited to E_C. Particularly striking, however, is the fact that there is only a small deviation following Friedman's equation (2.89) from the classical equation $R_H = 1/nec$ in the metallic region. This has been confirmed for Ge:As by Wallis (1973). In some earlier publications it was assumed that the gas *must* be near the Mott transition and thus highly correlated, so that the only possible conclusion was that Friedman's equation is not applicable when a drop in $N(E)$ is due to overlap between Hubbard bands. It is, however, proposed by Mott (1976*a*, 1977*a*) that the Friedman equation can be used, that in

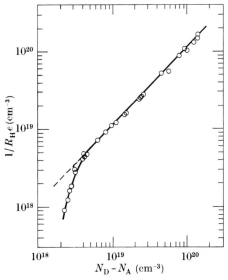

Fig. 4.20. The Hall electron density plotted against $N_D - N_A$ for varying compositions of Si:P at 4·2 K. (From Yamanouchi *et al.* 1967.)

Si:P the gas is not highly correlated, and that the transition should be given by a formula for Anderson localization for lateral (non-diagonal) disorder (eqn 2.39). The reason proposed for a small Friedman effect is that, for a random array of centres, Anderson localization would set in for a small degree of disorder (see § 2.3).

A further complication is that, according to Economou and Antoniou (1977), Anderson localization cannot appear in the *centre* of a band with only off-diagonal disorder. If this is correct, localization must *either* be due to charge fluctuation or to the effect of the Hubbard U. However, Debney's (1977) calculations (p. 22), using a more realistic model, give $n^{1/3}a_H \sim 0·35$. Probably this corresponds to the transition, and (4·7) to the concentration at which $N(E_F)$ becomes finite.

Further evidence that the Hubbard bands have not separated at the transition is provided by the measurements of Allen (1973) and of Mole (1978) on the thermopower of various compensated samples of Ge:As; no positive values were observed, as would be expected if there were a pseudogap in $N(E)$, with $d \ln N/dE$ negative.

Various authors (Mikoshiba 1968, Quirt and Marko 1973, and Sasaki 1976) have discussed the metal–insulator transition in terms of a model in which metallic and non-metallic regions coexist. Fig. 4.21, due to Sasaki, is a computer simulation of the type of distribution that random positions entail, for the concentration at which the metal–insulator transition occurs. In spite of the great fluctuations of density shown, the point of view of this

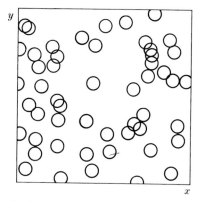

Fig. 4.21. Random distribution of impurities. (From Sasaki, private communication.)

book is that a treatment of this kind is only valid if the non-metallic regions are wide enough to prevent tunnelling, which is not likely to be the case here. The strongest evidence that localized and metallic regions do not coexist at the same energy is provided by the work of Geschwind *et al.* (1976) on spin-flip Raman scattering in n-type CdS. In the metallic region the linewidth of the scattered Raman radiation shows a broadening due to a Doppler shift; the broadened and unbroadened lines do not coexist at the same concentration of donors.

Isolated metal atoms such as that shown in Fig. 4.21 may well have a moment flipping from one orientation to the other by the Kondo mechanism. Toyozawa (1962) was the first to suggest that such moments might be responsible for the negative magnetoresistance shown by 'metallic' Si:P near the transition (see also Mott 1974c). Positive magnetoresistance is due to shrinkage of orbits. Both Ue and Maekawa (1971) and Quirt and Marko (1973) find that the e.s.r. magnetic susceptibility increases in metallic Si:P as the concentration of donors decreases towards the transition. This has been interpreted as a property of the highly correlated electron gas (Mott 1972e, 1974c, Berggren 1974). However, if the gas is not highly correlated, the extra susceptibility may be due to moments on nearly isolated atoms as in Fig. 4.21. On the other hand, Brown and Holcomb (1974) report n.m.r. data in which all carriers appear to participate in a single system. A satisfactory theory has not as yet been given.

Since there is no discontinuity in $N(E)$ at an Anderson transition (cf. § 2.4), no discontinuity is expected in the electronic specific heat (γT). This has been investigated down to 0·5 K in Si:P by Kobayashi *et al.* (1977) and there is no observable discontinuity at the transition at $3·2 \times 10^{18}\ \mathrm{cm}^{-3}$, though at $5 \times 10^{17}\ \mathrm{cm}^{-3}$ γ has fallen by an order of magnitude, suggesting that there is then only a weak overlap between Hubbard bands. Just on the metallic side of the transition there is a small (30 per cent) enhancement of

γ above the free-electron value, suggesting that the electron gas is not very highly correlated.

As the concentration of donors increases, the metallic impurity band will merge with the conduction band (Mott and Twose 1961, Matsubara and Toyozawa 1961). According to the latter authors, in Si:P this should occur at a concentration of 3×10^{19} cm^{-3}, while experiment (Alexander and Holcomb 1968) gives about 2×10^{19} cm^{-3}. For concentrations where merging has occurred, the conductivity should be treated by the methods of § 5.2.

An interesting point arises on whether a formula of the type

$$\sigma_{min} \simeq 0 \cdot 026 e^2 / \hbar a$$

is valid for a metallic impurity band in an *amorphous* semiconductor, if this is of the kind that can be doped. Here a is the distance between centres. A possible example is the amorphous film of Cd_2SnO_4 analysed by Nozik (1972), in which the donors are thought to be oxygen vacancies and which show metallic behaviour for carrier densities in the range 4×10^{18}–$1 \cdot 2 \times 10^{20}$ cm^{-3}. The conductivity, which is independent of temperature, is as follows; the value when $L \sim a$ but $N(E)$ is not eroded by disorder is also shown in the last column.

Concentration (cm^{-3})	$\sigma(\Omega^{-1}$ cm$^{-1})$	$\sigma_{min} = 0 \cdot 05 \, e^2/\hbar a$ (calc. with $a = n^{-1/3}$)	$0 \cdot 3 \, e^2/\hbar a$
$4 \cdot 3 \times 10^{18}$	15·6	17	—
$6 \cdot 2 \times 10^{19}$	100	—	—
$1 \cdot 2 \times 10^{20}$	385	—	400

We conclude that a should indeed be the distance between centres in amorphous as well as crystalline materials, and that for the higher concentrations $L \sim a$.

4.3.4. Metal–insulator transition in metal–rare-gas systems

Films of argon–copper and argon–lead have been prepared by condensation at liquid–helium temperatures. The results quoted here are due to Endo *et al.* (1973) following earlier work of Cate, Wright, and Cusack (1970) and Even and Jortner (1972). At helium temperatures these films show as a function of composition a discontinuous change in the resistivity by at least 10^7. Assuming that the films are homogeneous, this is perhaps the first example of a discontinuous Mott transition to be observed for an unstable series of alloys. Berggren, Martino, and Lindell (1974) show that the condition (4.7) is well satisfied. However, the conductivity at the transition ($\sim 30 \, \Omega^{-1}$ cm^{-1}) is perhaps an order smaller than the minimum metallic conductivity. This may mean that the assumption that the copper atoms are atomically dispersed is at fault, and that the transition is in fact due to a percolation path through macroscopic clusters of metal, as described in § 4.9. Quinn and Wright (1977) show that the lead–argon system shows a discontinuity in the conductivity at 50 per cent argon, and that for this system they found a single phase on both sides of the transition, suggesting that the predicted discontinuity does in fact occur.

Shanfield, Montano, and Barrett (1975) investigated xenon–iron films, and found that an Anderson transition occurred at 30 per cent iron. If both valencies of iron are present, this would mitigate against a discontinuity.

4.3.5. Metal–insulator transitions in compensated semiconductors

Experiments on doped *and compensated* semiconductors give one of the clearest examples of the Anderson transition. For substantial values of the compensation K, say around $0 \cdot 5$, the repulsion between electrons expressed through the Hubbard U (§ 4.3) does not essentially affect the conduction process, and we may make use of a model in which this interaction between the electrons is neglected. If N_A is not too great, therefore, the density of states will appear as in Fig. 4.11, an impurity band (the lower Hubbard band) appearing below the conduction band and being partially filled. The width of the impurity band would be determined by eqn (2.21) in the absence of disorder, but strong disorder is introduced both because of the random positions of the atoms in space and the random potential due to the charged minority centres. If K is increased, the Fermi energy E_F is shifted downwards, and at the same time the random potential becomes greater and E_C, the mobility edge, may be shifted upwards. If $E_F - E_C$ changes sign, an Anderson transition should be observed.

A transition that we now identify in this way was first observed by Fritzsche and Lark-Horovitz (1959) in p-type germanium with $N_A = 2 \cdot 5 \times 10^{17}$ cm^3 and values of K between zero and $0 \cdot 8$; these results are reproduced in Fig. 4.22. The results of Davis and Compton (1965) on n-type Ge with $N_D = 1 \cdot 7 \times 10^{-17}$ cm^{-3} are similar. The interpretation as an Anderson transition is due to Mott and Davis (1968). The observed minimum metallic conductivity is about $0 \cdot 5 \times 10 \, \Omega^{-1}$ cm^{-1}. With the distance between centres of 2×10^{-6} cm, eqn (2.55) gives just this value.

Detailed work by Allen and Adkins (1972) and Allen, Wallis, and Adkins (1974) was undertaken at temperatures down to $0 \cdot 1$ K on heavily doped and compensated germanium, with two objects in view: (a) to see whether $T^{1/4}$ behaviour continues down to lowest temperatures; (b) to see whether a minimum metallic conductivity exists, and to determine its magnitude. Fig. 4.23 shows some results from Allen *et al.* There is no sign down to these temperatures of any flattening of the curves at low temperatures, as would be expected if classical percolation theory could be applied

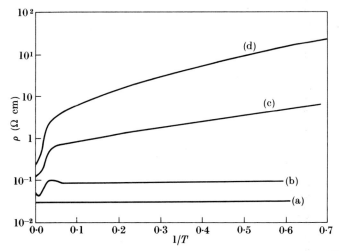

Fig. 4.22. Resistivity of heavily doped p-type germanium for $N_A = 2 \cdot 5 \times 10^{17}$ cm^{-3} and varying values of K, namely (a) 0·0, (b) 0·33, (c) 0·7, (d) 0·8. (From Fritzsche and Lark-Horovitz 1959.)

near the mobility edge. The minimum metallic conductivity is $\sim 15 \, \Omega^{-1}$ cm^{-1}, about twice the value deduced from (2.55). For a value of the compensation K equal to 0·3, the transition occurs at a donor concentration of $4 \cdot 5 \times 10^{17}$ cm^{-3}, at which without compensation the system would be metallic.

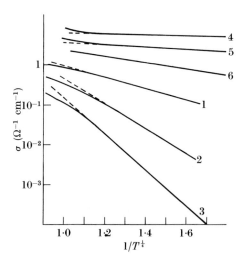

Fig. 4.23. Conductivity of some samples of Sb-doped Ge plotted against $1/T^{1/4}$. Value of the compensation: 1 and 2, 11 per cent; 3, 45 per cent; 4, 5, 6, 25 per cent. (From Allen *et al.* 1974.)

Allen and Adkins also discuss the term outside the exponential in the hopping formula (4.9), and also the high-temperature region, in which $\ln \rho$ is proportional to $1/T$, which they interpret as excitation to a mobility edge.

Important results were obtained by Ferre, Dubois, and Biskubski (1975) on the metal–insulator transition induced by a magnetic field in n-type InSb with donor concentrations near 10^{14} cm^{-3}. Their results, for conduction due to electrons excited to the mobility edge, are shown in Fig. 4.24.

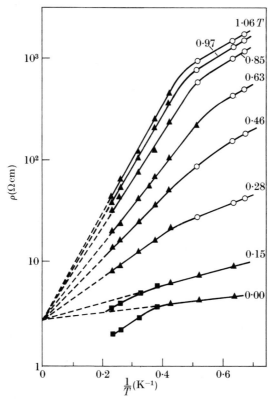

Fig. 4.24. Variation of resistivity of samples of n-type InSb in various magnetic fields. The triangles denote ε_2 conduction, the circles hopping (ε_3), and the squares ε_1. (From Ferre *et al.* 1975.)

Since the number of centres is not varied, $\sigma_{\min}$ is constant and all the plots of $\ln \sigma$ against $1/T$ extrapolate to the same point as for two-dimensional problems (§ 4.6). The value of $\sigma_{\min}$ obtained corresponds to $0.05 \, e^2/\hbar a$. Pepper (1976), in discussing these results, points out that a is ten times greater than in Ge:Sb, and that experimental values of $\sigma_{\min}$ have been observed over three orders of magnitude.

Earlier results also showing that in the ε_2 regime a magnetic field changed the activation energy but not the pre-exponential are those of Yamanouchi (1965) for n-type germanium and Gershenzon *et al.* (1973) in n-type InSb. Fritzsche (1962) finds that uniaxial compression changes ε_3 but not the pre-exponential in n-type Ge. In all such work it is supposed that the magnetic field shrinks the orbit, lowering the mobility edge E_C, but since $e^2/\hbar a$ does not contain the orbit radius, $\sigma_{\min}$ is not changed.

Shrinking the orbit will also decrease the hopping probability. This was treated by Mikoshiba (1962), who also discusses an effect due to a phase change in the wavefunctions.

4.4. Anderson transition in a pseudogap; magnesium–bismuth films

In § 2.11 it has been shown that, in amorphous materials, when a conduction and valence band overlap, the density of states should show a minimum or 'pseudogap', and that, if the density of states is small enough, Anderson localization is to be expected at the Fermi energy. An Anderson transition should therefore be observed in two alternative circumstances.

(a) The depth of the pseudogap is altered, as a consequence of changing composition, density, or structure. An example is liquid mercury at high temperatures and probably the liquid tellurium alloys, both described in the next chapter. Examples from disordered solids are not known as yet, except for overlapping Hubbard bands or broadened Landau levels (§ 4.6).

(b) The Fermi energy is shifted by a change in the composition which varies the number of electrons available. An example of an overlap between a conduction and a valence band is provided by the experiments of Ferrier and Herrell (1969) and Sik and Ferrier (1974) on amorphous films of the composition $Mg_{3-x}Bi_{2+x}$. Here a density of states as shown in Fig. 4.25 was assumed, with the additional assumption that a change in x will shift the Fermi energy without an important change in the density of states. Thus the conductivity is a minimum at $x = 0$, which is what is observed (Fig. 4.26). This behaviour is in sharp contrast to that of the

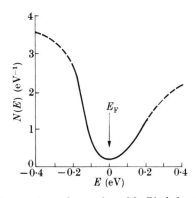

Fig. 4.25. Density of states per atom of amorphous Mg–Bi, deduced from the experimental results of Ferrier and Herrell (1969).

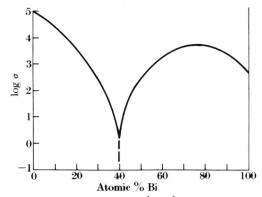

Fig. 4.26. Logarithm of the conductivity (in $\Omega^{-1}\,cm^{-1}$) as a function of composition for amorphous evaporated films of Mg–Bi. (From Ferrier and Herrell 1969.)

amorphous chalcogenides, where electrons are taken up in bonds, and changing the composition will not shift the Fermi energy from the minimum. Their results show also a large negative temperature coefficient for values of $|x|$ differing by about 0·2 from zero, for which the conductivity lies below $\sim 1000\ \Omega^{-1}\,cm^{-1}$, indicating either hopping or excitation to a mobility edge (Fig. 4.27). The results of Ferrier and Herrell on the ther-

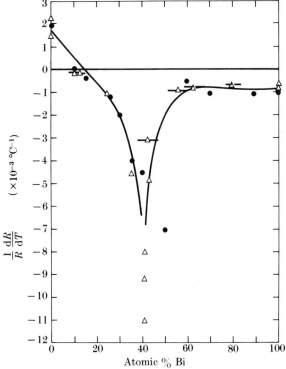

Fig. 4.27. Temperature coefficient of the resistivity of amorphous Mg–Bi. (From Ferrier and Herrell 1969.)

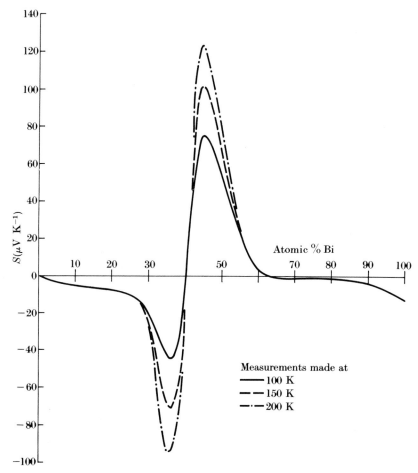

Fig. 4.28. Thermopower of amorphous Mg–Bi. (From Ferrier and Herrell 1969.)

mopower S are reproduced in Fig. 4.28; since S increases with temperature, this indicates hopping. Sik and Ferrier have plotted $\log \sigma$ against $1/T^{1/4}$ (Fig. 4.29) and obtained $T^{1/4}$ behaviour at low temperatures as expected.

The rigid-band model of Ferrier and co-workers has been criticized by Sutton (1975) on the basis of optical measurements. Sutton finds that Mg_3Bi_2 is a semiconductor with a gap of $0\cdot15$ eV, and excess Mg or Bi form states in the gap, as for a crystalline material.

4.5. Anderson transition of type II; amorphous films of Fe–Ge

Amorphous films of $Ge_{1-x}Fe_x$ have been investigated by Daver, Massenet, and Chakraverty (1974) and Massenet, Daver and Geneste (1974). The

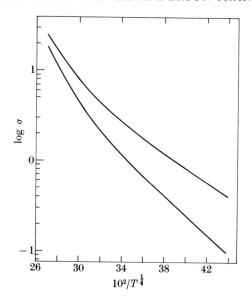

Fig. 4.29. Plot of log(conductivity) against $1/T^{1/4}$ for two specimens of Mg–Bi. (From Sik and Ferrier 1974.)

films were prepared by evaporation from an alloy ingot onto a cooled substrate. The Anderson transition with increasing iron content is shown in Fig. 4.30, with a minimum metallic conductivity of $\sim 200\ \Omega^{-1}\,\text{cm}^{-1}$. Massenet *et al.* have plotted the resistivity against $1/T^{1/4}$, obtaining good agreement for those showing activated behaviour except for one specimen (Fig. 4.31). The absence of evidence of excitation to a mobility edge is at first sight surprising. This must mean that E_F lies at a maximum in the density of states. The most obvious explanation is that the conditions in the band due to 'dangling bonds' are well on the metallic side of the Mott transition $(B \gg U)$, so that no overlapping Hubbard bands need be considered, and we observe here a property of a half-full band with localization purely due to disorder (§ 4.1). A further point of interest of this sytem is that spontaneous magnetization appears at about 20 per cent of iron and increases with Fe content. We attribute this to RKKY interaction between the iron moments, which will fall off as $\exp(-r/L)$ where L is the mean free path (de Gennes 1960) and therefore have a very short range and will be of ferromagnetic sign.

Similar results are shown by amorphous $FeSi_2$ for different states of annealing (Sharma, Theiner, and Geserich 1974). Also Kishimoto *et al.* (1976) find behaviour of this kind for the amorphous Si–Au system with unexplained deviations from $1/T^{1/4}$ behaviour at low temperatures.

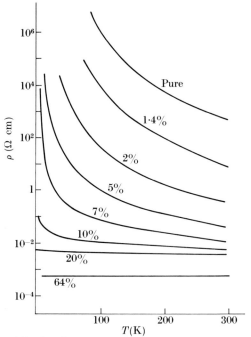

Fig. 4.30. Resistivity of Fe–Ge alloys versus temperature for different Fe concentrations as shown. (From Daver *et al.* 1974.)

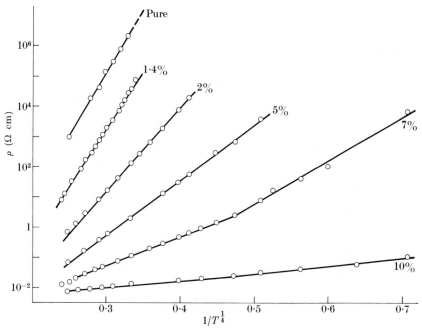

Fig. 4.31. Resistivity of Fe–Ge alloys versus $1/T^{1/4}$. (From Massenet *et al.* 1974.)

4.6. Two-dimensional conduction in an inversion layer

This section describes investigations of the current in the inversion layer at a Si–SiO$_2$ interface. Application of a voltage across the oxide can produce a situation in which the majority carrier is the minority carrier in the bulk semiconductor; the potential at the interface is illustrated in Fig. 4.32. As

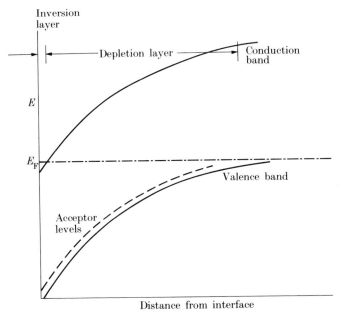

Fig. 4.32. Potential at interface of Si/SiO$_2$. (From Mott *et al.* 1975.)

first pointed out by Schrieffer (1957), under such conditions the energy of the electron perpendicular to the surface is quantized; at low temperatures a degenerate two-dimensional gas is formed, in which the wavefunction perpendicular to the surface consists of a single half-wave. Extensive investigations (Fowler *et al.* 1966, Stern 1972, 1974*a*, *b*, Dorda 1973) have confirmed that a two-dimensional gas of this kind can form and that higher subbands exist in which ψ in the perpendicular direction consists of two or more half-waves. These include Shubnikov–de Haas oscillations and photoexcitation to the upper subbands.

The resistivity in the surface layer was early interpreted as being due to diffuse scattering by surface roughness (Schrieffer 1955). A random potential may be due to such roughness or to charges on the oxide (Fang and Fowler 1968). It was suggested by Mott (1973*b*) and by Stern (1974, 1976) that the random potential might lead to Anderson localization at the Fermi level of a degenerate gas, and that an Anderson transition could be observed by changing the voltage across the layer and thus the position of the Fermi energy relative to the mobility edge.† This technique has the following advantages.

(a) The transition can be observed with a single specimen by changing the voltage.

(b) The low-temperature hopping conduction should vary as $\sigma \propto \exp(-B/T^{1/3})$, and it has proved possible to show that the index is nearer to $\frac{1}{3}$ than to $\frac{1}{4}$ (see below).

(c) The minimum metallic conductance σ_{min} is of the form const. $e^2/\hbar$, and the quantity a_E of eqn (2.56) is not involved.

According to Licciardello and Thouless (1975) the constant is 0·1 and does not depend on the form of the potential, within the kind of variation discussed by them, though Pepper (1977a) finds that long-range fluctuations give a larger value of σ_{min}.

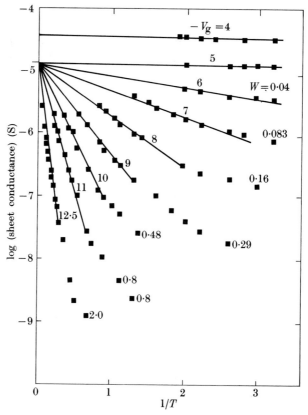

Fig. 4.33. Conductance of N-channel MOS device for various values of gate voltage V_g; W (in meV) is the activation energy $E_C - E_F$. (From Mott, Pepper *et al.* 1975.)

† The thickness of the specimen can also change the position of E_C (Soonpaa 1976, Pepper 1977b).

(d) σ_{min} can be determined either from the value of the gate voltage for which the plot of $\log \sigma$ against $1/T$ becomes flat, or from the extrapolation of those curves in the activated region to $1/T = 0$.

Extensive experiments have been carried out at low temperatures with MOS and MNOS transistors by Pepper and co-workers,† and a full account of this work is given by Mott et al. (1975), who compare the form of the 'Anderson transition' observed with that in three-dimensional systems. These authors obtain the theoretical value of σ_{min} to within a factor of 2. Tsui and Allen (1974, 1975) obtained values up to 10 times larger. Pepper (1977) has found it possible to reproduce the results of Tsui and Allen by the use of a substrate bias, which pushes the maximum of ψ further away from the interface. This is ascribed to the long-range nature of the fluctuations acting on the electron in the latter case; Pepper gives arguments to show that an increased value is then to be expected.

Fig. 4.33 shows the results of Pepper et al. for an N-channel MOS device in the region where conduction is due to excitation to a mobility edge and Fig. 4.34 that of the conductance in the low-temperature region plotted

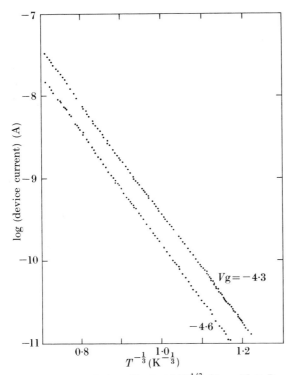

Fig. 4.34. Device current at low T, plotted against $1/T^{1/3}$. (From Mott, Pepper et al. 1975.)

†Pepper, Pollitt, and Adkins 1974a,b, Pepper, Pollitt, Adkins, and Oakley 1974.

against $1/T^{1/3}$. Mott, Pepper *et al.* (1975) find that, writing

$$\sigma = A \exp(-B/T^m)$$

$m = 0\cdot 32 \pm 0\cdot 02$. Since the theoretical value of B is

$$\{3\alpha^2/N(E_F)k\}^{1/3}$$

then, taking the (slowly varying) values of $N(E_F)$ observed, the variation of α with gate voltage could be deduced from the low-temperature plot and $E_C - E_F$ from that at high temperatures. The relation between them

$$\alpha = \text{const.}(E_C - E_F)^s$$

with $s = 0\cdot 73 \pm 0\cdot 08$, was in good agreement with the theoretical value of Abram (1973), discussed in § 2.3, as long as $E_C - E_F$ was not too small. For small values, however, the index s approaches unity, as shown in Fig. 4.35,

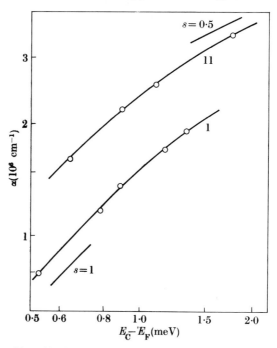

Fig. 4.35. α plotted logarithmically against $E_C - E_F$ for Si/SiO$_2$ interface for two specimens. Slopes with $s = 1$ and $s = 0\cdot 5$ are also shown. (From Pollitt 1976.)

taken from Pollitt (1976). This seems to be in agreement (Mott 1976*c*) with the considerations of § 2.7, rather than with the calculation of Abram. As shown there, we believe that a value of s less than unity is not compatible with the observation of a minimum metallic conductivity.

In the investigations of the energy dependence of α, $N(E)$ was determined from the rate of change of activation energy, $E_C - E_F$, when conduction was by excitation to the mobility edge. If n is the carrier concentration, such that

$$N(E) = \mathrm{d}n/\mathrm{d}W$$

it was found that, if the number of localized states was less than $\sim 3 \times 10^{11} \, \mathrm{cm}^{-2}$, then deep in the tail $N(E)$ decreased exponentially with energy (Pollitt 1976) and at the mobility edge $N(E)$ was close to half the free carrier value. However, if the number of localized states was greater than $\sim 3 \times 10^{11} \, \mathrm{cm}^{-2}$, the value of $N(E)$ derived became greater than the appropriate free carrier value (Pepper, Pollitt, and Adkins 1974a, b). Clearly this cannot represent the real behaviour of $N(E)$, and a possible explanation is that E_C rises as the localized states are populated. This effect was found in both n and p inversion layers and may be due to an increase in the random potential as the concentration of localized carriers increases.

The nature of the localization in the inversion layer implies that the random field arises from positive and negative charges within the SiO_2 which are in the form of pairs. The application of a substrate bias which pulls carriers away from the interface initially increases the number of localized states; then, as the carriers are pulled further from the interface, it increases the value of σ_{min}. Both effects are consistent with an increase in the range of the potential fluctuations (Pepper 1977).

A different form of localization is found when a high magnetic field ($\geqslant 50$ kG) is applied normal to the interface. The effect of the field is to split the density of states into Landau levels; owing to the disorder at the Si–SiO$_2$ interface, states in the tails of the Landau levels are localized. An Anderson transition can be observed in two ways: either, through a change in bias, sweeping the Fermi level through the Landau levels and observing activated conduction in the tails and metallic conduction where $N(E)$ is high, or by increasing the magnetic field when E_F lies in a minimum of $N(E)$. In this latter case, at low magnetic fields conduction is metallic and becomes activated as the magnetic field increases and $N(E_F)$ falls. Clear evidence of excitation to a mobility edge is obtained, with a common intercept in plots of $\log \sigma$ against $1/T$ (Nicholas, Stradling, and Tidey 1977). This is another example of a band-crossing Anderson transition like that discussed in § 4.4. Aoki and Kamimura (1977) have predicted that, when the mobility edge is in the tail of Landau levels, σ_{min} is reduced below $0 \cdot 1 \, e^2/\hbar$; this reduction in σ_{min} was observed experimentally by Nicholas et $al.$ (1977).

4.7. Fermi glasses where the lattice distortion is important

The theory of impurity conduction developed so far neglects the distortion of the crystal round the impurity centres. This is likely to be a good approximation only if the radii of the centres are large compared with the lattice parameter; otherwise the distortion of the lattice must be taken into account, and the polaron hopping energy W_H of Chapter 3 becomes important, whether the material is ionic or not.

4.7.1. Impurity conduction in nickel oxide

One of the clearest examples of the importance of W_H is that of impurity conduction in nickel oxide, observed by Bosman and Crevecoeur (1966) and by Springthorpe, Austin, and Smith (1965). Some results are shown in Fig. 4.36. Conductivity is due to lithium doping, the acceptor centres being

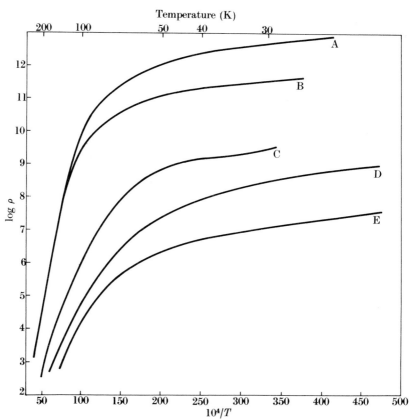

Fig. 4.36. Impurity conduction in crystalline NiO; logarithm of resistivity (in Ω cm) as a function of $1/T$. The values of x in the formula $Li_xNi_{1-x}O$ were as follows: A, 0·002; B, 0·003; C, 0·018; D, 0·026; E, 0·032.

lithium (Li^+), replacing Ni^{2+} in the lattice. To preserve electrical neutrality a neighbouring Ni ion has the charge of Ni^{3+}. Donors of unknown nature are present, so some of these centres acquire an electron that can hop from one to another. The activation energy drops from a value of 0·2–0·4 eV at high temperatures to ~0·004 eV or less at 10 K. The larger value is ascribed to the polaron term (W_H) of Chapter 3, which as we have seen should drop away to zero as $T/\Theta \rightarrow 0$. At low temperatures we should expect to find Miller and Abrahams' term ε_3, which can be estimated to be about 0·03 eV. The measured activation energies are much lower. A discussion of the reason for this is given by Austin and Mott (1969); it is possible that variable-range hopping is the explanation and if so the activation energy will always tend to zero.

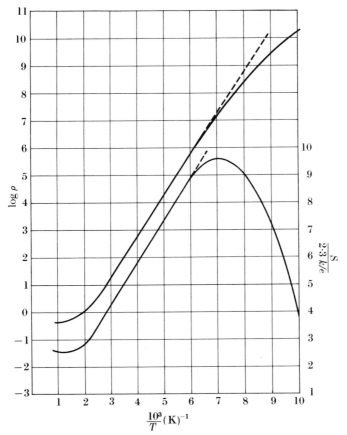

Fig. 4.37. Logarithm of resistivity (Ω cm) and thermopower ($S/2\cdot3(k/e)$) of Ni doped with 0·088 per cent Li$_2$O. (From Bosman and Crevecoeur 1966.)

Fig. 4.37 shows the thermopower,† which changes sign when impurity conduction carries most of the current. The reason for this is the same as for non-polar materials, as set out in § 4.3.1.

† See, however, § 3.10 for more recent results for single crystals.

4.7.2. Conduction in glasses containing transition-metal ions

Many glasses containing transition-metal ions, for instance vanadium or iron, are semiconductors. It is generally recognized that the conductivity in such glasses is due to the presence of ions of more than one valency, for instance V^{4+} and V^{5+} or Fe^{2+} and Fe^{3+}; an electron can pass from one ion to another, the process being similar to impurity conduction in nickel oxide as described in the last section. There is, however, one important difference: in a crystal the different sites (Li^+, Ni^{3+}) on which an electron may be located are crystallographically identical, differing only in their proximity to a charge minority centre that introduces the Miller–Abrahams spread of energies which determines W_D; in a glass however, the environments of the ions with the two valencies may or may not be different; there might be in principle a situation similar to that of Fig. 2.15, in which the configuration of atoms ensures during cooling of the glass that the electron on V^{4+} can form a bond. The evidence reviewed below shows that this does not happen in vanadium phosphate glasses; it probably does in the glasses containing Cu^+ and Cu^{2+} investigated by Drake and Scanlan (1970), for which a conducting and a non-conducting state can be formed depending on the rate of cooling.

For glasses in which the occupied and unoccupied sites are identical we may write for the conductivity[†]

$$\sigma = c(1-c)\frac{e^2 \nu_{el}}{RkT}\exp\left(-2\alpha R - \frac{W}{kT}\right). \tag{4.11}$$

Here R is the distance between the ions, c and $1-c$ are the proportions of V^{4+} and V^{5+}, α is as defined elsewhere (eqn 3.38) and

$$W = W_H + \tfrac{1}{2}W_D.$$

Fig. 4.38 shows some results for vanadium glasses,[‡] and it will be seen that W_H, of order 0·4 eV, drops towards zero at low T, as it does also for impurity conduction in NiO (Fig. 4.36). Also W_D must be small. There is other evidence for this (Kennedy and Mackenzie 1967), namely that vanadium glasses above about 200 K obey well the Heikes equation for the thermopower (§ 2.13)

$$S = \frac{k}{e}\ln\left(\frac{c}{1-c}\right).$$

This implies that W_D is less than kT (eqn 2.80), and we deduce that the sites must be very nearly identical.

[†] The term $\exp(-2\alpha R)$ only occurs if the transition is non-adiabatic and should be absent for small values of R (cf. § 3.5).

[‡] From R. M. Brown, Ph.D. Thesis, University of Illinois.

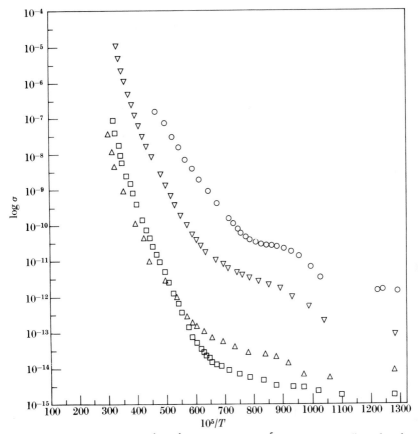

Fig. 4.38. Log conductivity (in Ω^{-1} cm^{-1}) as a function of $10^5/T$ of some vanadium phosphate glasses. Mole ratios of V_2O_5:P_2O_5 are as follows: △ 1:3, □ 1:1, ▽ 6:1, ○ 7:1. (From Schmid 1968.)

This small value of the disorder hopping energy is, however, contradicted by an analysis by Greaves (1973),[†] who finds a $1/T^{1/4}$ behaviour at low temperatures and interprets it as variable-range hopping; he deduces that the disorder energy separating adjacent sites is 0·4 eV. The matter therefore remains open. A value of W_D of this order for W_D due to random charges on V^{4+}, V^{5+} seems reasonable, but if so it is difficult to understand the thermopower.

Austin and Garbett (1973) show a temperature dependence of the thermopower below 200 K, *S increasing* at low T; it is not clear how this is to be explained, though it might relate to a small range of energies W_D, so that conduction is of Miller–Abrahams type. If so, as shown in § 2.13, S

[†] This article gives references to many experimental papers not cited here.

should *rise* with decreasing T until variable-range hopping sets in. These authors discuss high-field conduction, following an equation of the form $\sigma \propto \sinh(eRF/kT)$; values of R observed are larger than expected.

4.7.3. Lanthanum–strontium vanadate $(La_{1-x}Sr_xVO_3)$

The results of Dougier and Casalot (1970) and of Dougier (1975) on conductivity in crystals of this material are reproduced in Fig. 4.39. A clear case of an Anderson transition is shown, with a minimum metallic conductivity of $\sim 1000\ \Omega^{-1}\,cm^{-1}$, the disorder being supposed to be due to the random positions of the ions La^{3+} and Sr^{2+}. The results show that this is sufficient to localize the wavefunctions at the Fermi level in the vanadium 3d band when $x < 0.2$. The material is antiferromagnetic with $T_N \sim 100\ K$;

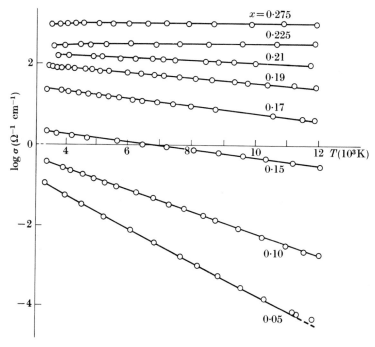

Fig. 4.39. Log(conductivity) of $La_{1-x}Sr_xVO_3$ as a function of $1/T$ for different values of x. (From Dougier 1975.)

with $x = 0$ we should expect the vanadium ions to be in the state V^{3+}, and the addition of strontium must produce holes in the lower Hubbard band (§ 4.4), so positive thermopower S is expected. This is shown in Fig. 4.40 for various values of x. For $x < 0.1$, S increases with decreasing temperature, suggesting excitation to a mobility edge; for $x = 0.15$ the reverse is the case, suggesting hopping.

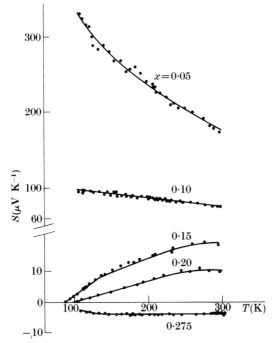

Fig. 4.40. Thermopower S of $La_{1-x}Sr_xVO_3$ as a function of T. (From Dougier 1975.)

The very small values of the activation energies ε, which we interpret in the higher-temperature range as $\varepsilon = E_C - E_F$, show that the d band is narrow, and this may be a material in which the carriers are appropriately described as a degenerate gas of small polarons (§ 3.11). At lower temperatures, however, Sayer *et al.* (1975) find $1/T^{1/4}$ behaviour. The hopping activation energy ε_3 in the composition range before this occurs is, according to both groups of authors, proportional to $(c - c_0)^{1.8}$, which may be because α behaves like $(c - c_0)^{0.6}$ (§ 2.4) so that the number of states available without tunnelling varies as $1/(c - c_0)^{1.8}$.

Very similar results are observed for $La_{1-x}Sr_xMnO$ (van Santen and Jonker 1950). These results are discussed by Methfessel and Mattis (1968). These materials are metallic ($\sigma \sim 10^{-2}\ \Omega^{-1}\ cm^{-1}$) and also ferromagnetic for x between 0·25 and 0·35. For $x = 0·5$ the conductivity drops to about $0·5\ \Omega^{-1}\ cm^{-1}$, depending little on T; we suppose that between 35 and 50 per cent or so the random field *just* produces Anderson localization, giving variable-range hopping.

4.7.4. Pyrolytic carbons

Bücker (1973) has investigated a series of semiconducting samples obtained by the thermal degradation of phenol formaldehyde resins at

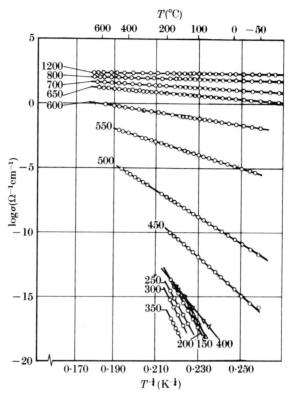

Fig. 4.41. Conductivity of pyrolytic carbons after annealing at the temperatures given by the numbers on the curves. (From Bücker 1973.)

different temperatures. His results are shown, plotted against $1/T^{1/4}$, in Fig. 4.41. The minimum metallic conductivity appears to be between 10^2 and $10^3 \, \Omega^{-1} \, cm^{-1}$.

4.7.5. Cerium sulphide

An early example of a phenomenon identified as an Anderson transition arose from the work of Cutler and Leavy (1964) on non-stoichiometric Ce_2S_3, which was interpreted in this way by Cutler and Mott (1969). This compound has the structure of Ce_3S_4, the excess sulphur being introduced as randomly situated cerium vacancies. The conduction band is thus highly disordered, there being a random distribution of centres which *repel* the electron. Changing the concentration to $Ce_{2-2x}S_{3-3x}$ introduces electrons into the conduction band, the charge being compensated by a small increase in the number of Ce vacancies. The number of electrons can thus be varied without any significant change in the potential. It is thus an ideal

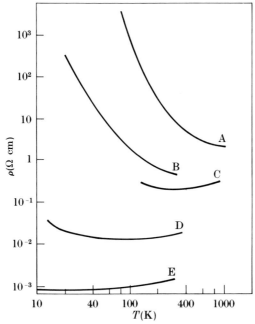

Fig. 4.42. Resistivity ρ of cerium sulphide of various compositions as a function of temperature. The electron concentrations (in 10^{18} cm^{-3}) are as follows: A; 0·5; B, 5·2; C, 14; D, 83; E, 1420. (From Cutler and Leavy 1964.)

system for the observation of the transition. Some results of Cutler and Leavy are shown in Fig. 4.42; the minimum metallic conductivity seems to be in the expected range of 10^2–10^3 Ω^{-1} cm^{-1}. Fig. 4.43 shows the thermopower S for two of the specimens marked in Fig. 4.42; the increase with T suggests that conduction is by hopping of electrons at E_F rather than by transport at the mobility edge. Cutler (unpublished) reports $T^{1/4}$ behaviour at low T. Data for the Hall coefficient are given by Cutler and Leavy. They show a Hall mobility with, very approximately, an activation energy of one-third of that for the conductivity. This could suggest that the carriers form polarons and that the main part of the hopping energy is of polaron type (W_H); the $T^{1/4}$ behaviour for low T could perhaps be due to the decrease towards zero expected for W_H. If this is correct, we should have to assume that as soon as Anderson localization occurs so does polaron formation, and that W_H due to the latter is much larger than the hopping energy due to disorder. This is perhaps not likely, and it appears more plausible to suppose that the Hall effect is due to electrons excited to a mobility edge, as in As_2Te_3 according to the interpretation of Nagels, Callaerts, and Denayer (1974; cf. Chapter 6).

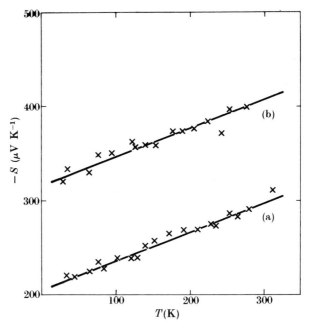

Fig. 4.43. Typical plots of thermopower S of two specimens of cerium sulphide as a function of temperature. (From Cutler and Mott 1969.)

4.7.6. Metal–insulator transitions in tungsten bronzes

The bronzes with composition $M_x WO_3$, where M is an alkali metal, show metallic behaviour for values of x above ~0·2 with a conductivity of about $500\ \Omega^{-1}\ cm^{-1}$ near the transition (Lightsey 1973) rising to 10^4–$10^5\ \Omega^{-1}\ cm^{-1}$ for high values of x. Below the transition one or more phase changes occur (for details see Hagenmuller 1971). The material is then a semiconductor; for small values of x, semiconducting behaviour is observed with a small binding energy of the electron to the sodium ion (0·04 eV). The absence of a Knight shift for the sodium nucleus (Fromhold and Narath 1964) suggests that the electron in the WO_3 d band overlaps the Na nucleus only slightly.

To understand this small activation energy we need to look at the properties of WO_3 which have been extensively investigated (Crowder and Sienko 1963). Fig. 4.44 shows the resistivity as a function of $1/T$. At low T there is a transformation to a structure of low symmetry, showing an activation energy for conduction of 0·4 eV (probably due to deep compensated donors); at 250 K there is a transformation to a state where the activation energy of conduction is ~0·02 eV, suggesting that the carriers are heavy polarons with a binding energy to the metal ion equal to $e^2/\kappa a$ (Chapter 3) and that κ (a static dielectric constant) has become very large

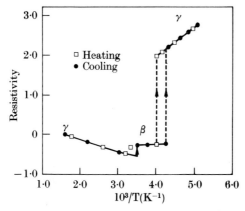

Fig. 4.44. log(resistivity) in Ω cm of WO_3 as a function of temperature.

(~ 100). For higher temperatures the donors are probably all ionized, and scattering by phonons causes the resistivity to rise.

For values of x for which the material is metallic, however, there is no sign of any heavy mass, and we suppose that the low-temperature behaviour of Fig. 4.44 must be assumed. For these concentrations the evidence of the electronic specific heat (Vest, Griffel, and Smith 1958, Zumsteg 1976) and of the magnetic susceptibility (Greiner, Shanks, and Wallace 1962, Zumsteg 1976) agree in determining $m_{eff}/m \simeq 1\cdot6$, but these quantities vary with x rather than $x^{1/3}$, as we should expect for a parabolic band form. This is believed to be a consequence of disorder (Tunstall 1975). This value of m_{eff} is in accord with the interpretation of the metal–insulator transition as a Mott transition (Mackintosh 1973) using eqn (4.7) and $m = m_e$. There seems no evidence that in the metallic phase the carriers are heavy, or form a 'degenerate gas of small polarons'.

With effective masses of order m_e, explanations of the metal–insulator transition based on classical percolation theory (Fuchs 1965, Lightsey 1973, Webman *et al.* 1975) are, we believe, unlikely to be correct. We think that, as x drops to $0\cdot2$, an Anderson transition is approached but is hidden by the change of structure. It can, however, be observed in *poly-crystalline* $Na_xWO_{3-y}F_y$ as shown by the results of Doumerc (1974) illustrated in Fig. 4.45. The fluorine ion F^-, taking the place of O^{2-}, adds an additional electron. A more detailed discussion of this system, and of the metal–insulator transition in some other bronzes, is given by Mott (1977*a*). The resistivity of polycrystalline systems is probably mainly due to grain-boundary regions.

If Anderson localization in the tungsten d band due to the random fields of the sodium ions is the correct description of the transition, and if $m_{eff} \sim m_e$, the random potential due to the Na^+ and F^- ions must be of the

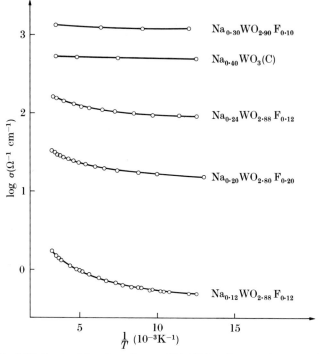

Fig. 4.45. Conductivity of specimens of $Na_xWO_{3-y}F_y$ plotted against $1/T$.

order 1 eV, which is not consistent with a high background dielectric constant. This is additional evidence that this disappears in the metallic phase.

The results of Lightsey, Lilienfeld, and Holcomb (1976) reproduced in Fig. 4.46 demonstrate that the Hall coefficient as x approaches 0.2 shows the anomaly predicted by Friedman (1971, § 2.14). Since a drop in $1/R_H$ by ~ 3 is observed, at the lowest metallic concentration the specimens must be near an Anderson transition.

4.7.7. *Vanadium monoxide* (VO_x)

Vanadium monoxide has the simple cubic structure. As normally prepared it contains a very high concentration (~ 15 per cent) of metal and oxygen vacancies, of which the ratio can be varied. It is not antiferromagnetic and must be considered a 'metal', or for some values of x a Fermi glass, the random field of the vacant sites being enough to cause Anderson localization at the Fermi energy. The resistivity plotted against $10^3/T$ for various values of x is reproduced in Fig. 4.47 showing a typical Anderson transition with $\sigma_{min} \simeq 2000 \ \Omega^{-1} \ cm^{-1}$. Fig. 4.48 shows ρ/ρ_{300} plotted on a

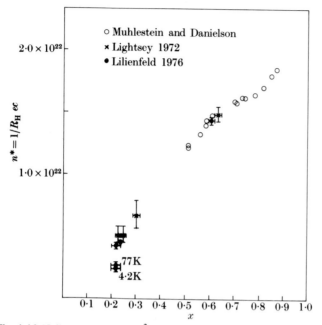

Fig. 4.46. Hall coefficient in cm^{-3} of Na$_x$WO$_3$ in the metallic range of x.

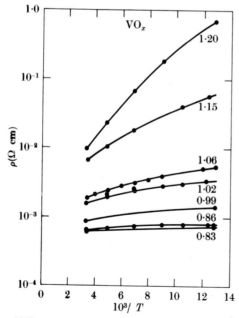

Fig. 4.47. Resistivity of VO$_x$ for various values of x plotted against $10^3/T$. (From Banus and Reed 1970.)

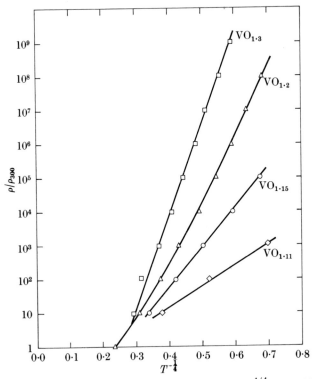

Fig. 4.48. Plot of ρ/ρ_{300} for VO_x on logarithmic scale against $1/T^{1/4}$. (From Mott 1971.)

logarithmic scale against $1/T^{1/4}$. The interpretation of this behaviour as an Anderson transition is due to Mott (1971), and the experimental material is that of Banus and Reed (1970), and of Goodenough (1972).

It is not known why small values of x are less effective in producing localization than large ones. The fact that 15 per cent of either kind of vacancy produces localization at all suggests that the vanadium d band is very narrow, and perhaps the carriers must be considered as a degenerate gas of small polarons.

Fig. 4.49 shows the thermopower S at room temperature as a function of x. The change of sign at $x = 0$ suggests strongly that two overlapping Hubbard bands are involved, and that this concept can be used for the thermopower of a highly correlated gas even when there is no antiferromagnetic order.

4.8. Impurity conduction in magnetic semiconductors

Rare-earth oxides containing excess metal can show impurity conduction, the donors being oxygen vacancies occupied by electrons and compen-

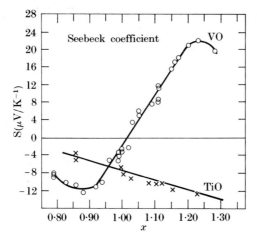

Fig. 4.49. Seebeck coefficient (denoted by S) at room temperature of VO_x and TiO_x as a function of x. S changes sign for x slightly greater than 1 in VO_x, suggesting that $\sigma(E)$ shows a minimum at E_F when $x \approx 1$. (From Mott 1971.)

sation probably being due to metal vacancies. Fig. 4.50 shows the results of von Molnar (1970) on EuO with excess europium. $1/T^{1/4}$ behaviour at low temperatures is well shown; an interesting conjecture is that hopping is due to interaction with spin waves rather than with phonons. The sudden increase in the resistivity at low temperatures is due to the following reason. Below these temperatures Hund's rule coupling between the

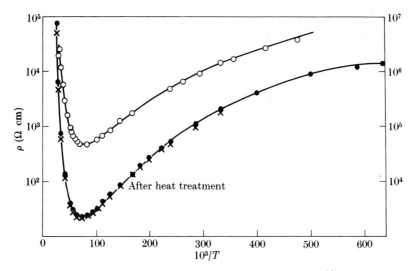

Fig. 4.50. Resistivity ρ of EuO doped with excess Eu, showing the $T^{-1/4}$ behaviour at low temperatures (von Molnar 1970). The scale on the right is for the upper curve.

trapped electron and the surrounding moments leads to their ferro-magnetic orientation in its neighbourhood. This lowers the free energy of the trapped electron, making it much more difficult to hop to an unoc-cupied vacancy.

A similar phenomenon occurs in EuS lightly doped with Gd or La, which provides an extra electron; this is shown in Fig. 4.51. Here at low tempera-tures conduction is metallic; the concentration of carriers is on the metallic side of the Mott transition, but near the Curie temperature the resistivity

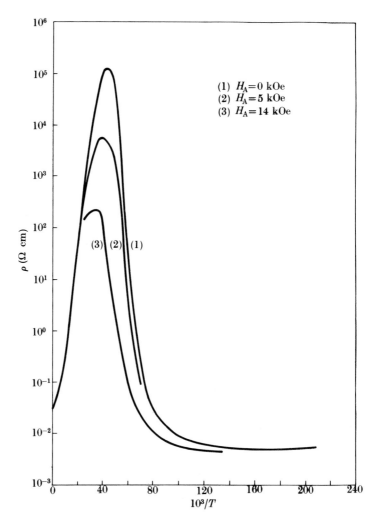

Fig. 4.51. Dependence on temperature of resistivity ρ of $Eu_{0.95}La_{0.05}S$ for various magnetic fields H_A. (From Methfessel and Mattis 1968.)

increases by seven orders of magnitude. So large an increase cannot possibly be due to spin scattering. According to the interpretation of Kasuya and Yanase (1968), the approach to the Curie temperature relocalizes the electrons on the impurity centres; there they orient the 4f spins on their nearest neighbours, but it would cost them a lot of free energy to move into the conduction band, since above the Curie point the entropy of disorder would prevent their forming a spin polaron. Current is thus carried by impurity conduction, the electrons hopping from one centre to another. The large activation energy arises because the electron has to jump from a site where it has had time to polarize its surroundings to an empty site where the spins are randomly oriented. This activation energy is destroyed if the moments are lined up by a magnetic field, and these materials therefore show an extremely large negative magnetoresistance.

Another magnetic semiconductor which shows $T^{1/4}$ behaviour is Cr_2S_x with x below 1·5, so that sulphur vacancies in the rhombohedral Cr_2S_x structures result. The results of Sugiura and Masuda (1973) on single crystals perpendicular to the c axis show an anomaly at the Néel temperature. These authors report that the resistivity parallel to the c axis is $\sim 10^4$ larger. If as they suggest we have here impurity conduction between electrons in vacant S sites, the rate of decay $\exp(-\alpha R)$ of the wavefunction must be highly anisotropic.

These results lead us to the concept of the 'spin polaron', namely an electron in a conduction band which orients spins parallel or antiparallel to itself. Some discussion is given by Mott (1974c). For the purpose of this book the important point is that a spin polaron is heavier than a free electron. The discussion relates to a carrier at zero temperature in an antiferromagnetic material such as NiO. We introduce the interactions J_1 between carrier and moments and J_2 between the moments themselves, and suppose that $J_1 > J_2$. Then let us assume that the carrier orients all moments within a radius R parallel to its own; it is then located in a 'box' of radius R, and, as in our discussions of polarons in Chapter 3, its kinetic energy is $\hbar^2\pi^2/2mR^2$. The carrier with its oriented cluster of spins is called a *spin polaron* and its total energy is

$$\frac{\hbar^2\pi^2}{2mR^2} + \frac{4\pi}{3}\frac{R^3}{a^3}J_2 - J_1. \tag{4.12}$$

Minimizing with respect to R we obtain

$$R^5 = \hbar^2\pi a^3/4mJ_2 \tag{4.13}$$

and the total energy is thus

$$\frac{5\hbar^2\pi^2}{6m}\left(\frac{4mJ_2}{\hbar^2\pi a^3}\right)^{2/5} - J_1. \tag{4.14}$$

Only if this quantity is negative will a spin polaron be formed in which the moments are fully oriented parallel to that of the carrier; otherwise there will be some smaller effect.

Following de Gennes (1960) we do not suppose that the moments in the spin polaron are all parallel up to a radius R, but rather that θ, the inclination to the spin direction in the absence of the carrier, tends gradually to zero as r/R exceeds unity. We can estimate the effective mass by computing the transfer integral when the polaron moves through one atomic distance. The spins will contribute a term proportional to

$$\prod \cos \theta_{r,r+1} \tag{4.15}$$

where $\theta_{r,r+1}$ is the change in the orientation of the spin when the carrier moves through one atomic distance. We may expect that $\theta_{r,r+1} \sim a/R$, so eqn (4.15) becomes

$$\prod (1 - a^2/2R^2)^{(R/a)^2}$$

and will thus vary for large R as

$$\text{const. } e^{-\gamma R/a}$$

where γ is some constant of order unity that we have not attempted to calculate. The effect of the moments on the effective mass m_p of a large spin polaron may be considerable, as we see by equating this with $\hbar^2/m_p a^2$. A large spin polaron will thus have effective mass considerably above that of the free carrier, but we do not think that spin–polaron formation can give rise to hopping motion in a perfect lattice.

Above the Curie or Néel temperature, a spin polaron will move by a diffusive mechanism. A tentative description is as follows. A moment on the periphery of the polaron will reverse in a time τ (the relaxation time for a spin wave). Each time it does so, the polaron can be thought to diffuse a distance $(a/R)^3 R$, and the diffusion coefficient is thus

$$D \simeq \tfrac{1}{6} a^6/R^4 \tau.$$

By Einstein's relation, we may write

$$\mu \simeq \tfrac{1}{6} e a^6/R^4 \tau k T. \tag{4.16}$$

The mobility thus decreases rapidly with increasing polaron radius.

Another treatment, in which the motion is assumed similar to that of domain-wall movement, has been given by Kasuya, Yanase, and Takeda (1970).

The important consequence of the heavy nature of the spin polaron is that Anderson localization occurs more easily. Evidence for this is provided by the work of Torrance, Shafer and McGuire (1972) on Eu-rich

EuO and that of Penney *et al.* (1973) on $Gd_{3-x}v_xS_4$ (v stands for a vacancy). They differ from cerium sulphide discussed in § 4.7.5 in that the Gd^{3+} ion carries a moment. Considering the latter case, in the degenerate electron gas each electron will form its spin polaron from these, so that owing to the enhanced mass Anderson localization by the field of the random vacancies will occur more easily. If the moments are lined up by a magnetic field, the spin polaron cannot form; the mass of the carrier drops and an·Anderson transition can occur. Fig. 4.52 shows this for a specimen

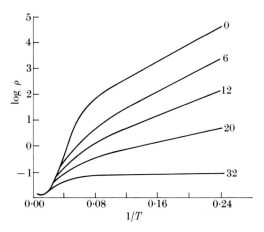

Fig. 4.52. Resistivity in Ω cm of a specimen of $Gd_{3-x}v_xS_4$ for magnetic fields shown in kOe (Penney *et al.* 1973).

in which the ratio of electrons to Gd ions is $0·88 \times 10^{-2}$. Since there are $1·8 \times 10^{22}$ Gd ions cm^{-3}, the mean distance a between electrons is $0·8 \times 10^{-7}$ cm. If σ_{min} is $0·026\, e^2/\hbar a$, this would give

$$\sigma_{min} \approx 70\, \Omega^{-1}\, cm^{-1}$$

in reasonably good agreement with the observations.

4.9. Granular metal films

Films of metallic particles encapsulated in an insulator have been investigated by many authors. Neugebauer and Webb (1962), Hill (1969), and Neugebauer (1970) realized that, if the particles are not in contact, an activation energy ε_2 is necessary to produce a positively and negatively charged particle so that the charge can tunnel from one particle to another. The analogy with the quantity ε_2 to produce two charged donors in (say) Si:P is close. If, however, the particles are in contact, so that percolation paths exist through the material, metallic conduction occurs and σ tends to a finite value as $T \to 0$.

Abeles, Ping Sheng et al. (1975) have investigated films prepared by co-sputtering metals (Ni, Pt, Au) and insulators (SiO$_2$, Al$_2$O$_3$), where the volume fraction of metal varies from 1 to 0·05 and the size of the metallic particles is below 100 Å. In the activated regime they found that the conductivity follows the law

$$\sigma = \text{const. } \exp(-A/T^{1/2}). \tag{4.17}$$

Their explanation is as follows. The energy ε_2 needed to form an electron–hole pair is of order $e^2/\kappa d$, where d is the diameter of a grain. If the tunnelling distance is comparable with the grain size, as shown by electron micrographs, then the product of the tunnelling probability and the number of carriers is proportional to

$$\exp\left(-2\alpha d - \frac{e^2}{\kappa d . kT}\right) \tag{4.18}$$

where $\alpha = (2mH)^{1/2}/\hbar$, H being the height of the barrier. Thus preferred channels exist for a value of d such that (4.18) is a maximum, when

$$d = \left(\frac{e^2}{2\kappa\alpha kT}\right)^{1/2}. \tag{4.19}$$

Inserting (4.19) in (4.18), a temperature dependence of the form (4.17) follows. Abeles, Pinch, and Gittleman (1975), in an investigation of W-AlO$_3$ films, find that the conductivity in the metallic region varies as $\sigma \propto (x - x_c)^p$, where x is the volume fraction of metal, $x_c = 0·47 \pm 0·05$ and $p = 1·9 \pm 0·2$. This is in good agreement with the expectations of percolation theory (§ 2.9).

Abeles et al. make the point that the energy levels of electrons in these very small particles are separated by ~ 10 meV, so that at low temperatures at any rate the motion of an electron from a charged to a neutral particle must be a hopping process, the electron exchanging energy with a phonon, though the small activation energy in the mobility is not considered. There is therefore a close analogy (except for the averaging process) with electrical conduction in uncompensated Si:P in the ε_2 region, where an electron is excited from one P atom to another, forming P$^-$, with an activation energy $U - \frac{1}{2}(B_1 + B_2)$. B_1 and B_2 are bandwidths (or mobility-gap widths) for the motion of the electron (P$^-$) and hole (P$^+$).

We now consider what happens when the proportion of metal increases; this is shown in Fig. 4.53 for nickel in SiO$_2$. It will be noted that the temperature coefficient of resistance changes sign when x (the proportion of metal) is about 0·6 as is to be expected from percolation theory. They also point out (Abeles and Ping Sheng 1974) that the conductivity at this point corresponds to $\sigma_{\min}$ for $a \sim 50$ Å. It seems to us likely that we have

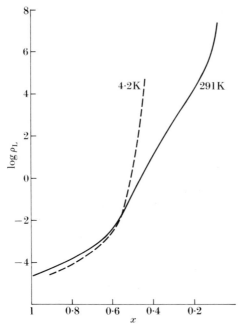

Fig. 4.53. Low-field resistivity ρ_L in Ω cm as a function of volume fraction x of Ni in Ni–SiO$_2$ sputtered films. The resistivities were measured in the plane of the film. The full and broken curves are smoothed values of the experimental data at 291 K and 4·2 K. (From Abeles and Ping Shen 1974.)

here a situation like that discussed in § 2.9; the concentration for the metal–insulator transition is given by percolation theory, but none the less the transition is of Anderson type, just as for uncompensated Si:P. If so, for concentrations x just below that for the metal–insulator transition, the $T^{1/2}$ behaviour should go over to $T^{1/4}$ hopping at low enough temperatures.

This behaviour seems also confirmed by the results of Abeles, Pinch, and Gittleman (1975) on dispersed W in Al$_2$O$_3$, who find that the temperature coefficient of resistance changes sign when $\sigma \sim 50\ \Omega^{-1}\ \mathrm{cm}^{-1}$, independent of the state of annealing. Writing $\sigma \propto (x - x_c)^p$, where x_c is the volume fraction of metal, p is $1\cdot9 \pm 0\cdot2$ and the transition occurs when $x_c = 0\cdot47 \pm 0\cdot05$.

The existence of $\sigma_{\min}$ for dispersed metallic particles as well as point centres is supported by results due to Dynes *et al.* (1978) on ultra-thin metal films condensed on a cold substrate. The films consist of islands connected by tunnel barriers. A minimum unactivated conductivity of $0\cdot12\ e^2/\hbar$ is found. We conjecture therefore that, provided kT is less than the energy between states in each island, the considerations relating to the Anderson transition and minimum metallic conduction can be applied.

4.10. Polycrystalline aggregates

A class of material in which σ may perhaps drop below the theoretical value of σ_{min} is provided by polycrystalline aggregates of, for instance, metallic transition-metal oxides in which each grain is covered with a fully oxidized layer; it is well known that the single crystals can show values of the conductivity ~ 100 times higher than sintered specimens. The resistance is therefore presumably due to the oxidized layer, through which the electrons can tunnel. The dependence on temperature would be weak, according to Simmons (1971) proportional to

$$1 + 6 \times 10^{-7} \, T^2$$

for typical barrier heights. One would expect the resistance to be very sensitive to oxide thickness but it may well be that the mechanism of oxidation involves tunnelling through the oxide (Cabrera and Mott 1948/49, Fehlner and Mott 1970) so that opaque layers cannot grow.

It should be pointed out, however, that if the tunnelling factor is small, states near the Fermi energy should be quantized just as in the last section, and conduction will be activated; truly metallic conduction, tending to a finite value as $T \to 0$, should not fall below $0 \cdot 1 \, e^2 / \hbar a$, a being the diameter of the crystals.

5

LIQUID METALS AND SEMIMETALS

5.1. Introduction
5.2. Scattering of electrons by a random distribution of centres; degenerate semiconductors
5.3. Resistivity of liquid metals; Ziman's theory
5.4. Resistivity of liquid alloys
5.5. Thermoelectric power of liquid metals
5.6. Hall effect in liquid metals
5.7. Density of states
5.8. Change of Knight shift on melting
5.9. X-ray emission spectra and photoemission
5.10. Liquid transition and rare-earth metals
5.11. Amorphous metals and grain boundaries
5.12. Injected electrons in liquid rare gases
5.13. Liquid semimetals and semiconductors
5.14. Liquid systems in which the depth of the pseudogap changes with volume
    5.14.1 Mercury
    5.14.2. Caesium
5.15. Liquid alloys with pseudogaps in which the composition is varied.
5.16. Liquid systems in which the depth of the pseudogap changes with temperature
    5.16.1. Tellurium
    5.16.2. Some liquid alloys of tellurium

5.1. Introduction

THE EXPLANATION given by Ziman (1961) of the electrical properties of liquid metals was one of the first and most important steps in the development of a theory of non-crystalline materials. The theory and its application to experiment have been extensively reviewed, both in published proceedings of international conferences† and in monographs and reviews (March 1968, 1977, Faber 1972), and the experimental material has also been reviewed by Busch and Güntherodt (1974). It is a weak-scattering theory, in which the scattering by each atom is considered small and the mean free path L is consequently large ($kL \gg 1$). A Fermi surface in k space can therefore be defined, which by symmetry must be spherical.

---

† For instance *The Properties of Liquid Metals, Second International Conference, Tokyo, Sept. 1972* (ed. S. Takeuchi), Taylor and Francis Ltd, London, 1973, and *Third International Conference, Bristol, July 1976, Liquid Metals* 3, Institute of Physics, London, 1977.

The pair distribution function of the liquid plays an essential role; otherwise the theory is equally applicable to amorphous metals or to problems such as the resistivity of heavily doped semiconductors, which for comparison is discussed in the next section.

This chapter, after an outline of Ziman's theory and the related problem of the drift mobility of electrons in liquid rare gases, will take as its main theme the behaviour of liquids in which the scattering is strong, kL is not large and Ziman's theory may no longer be applicable. There are in principle two ways in which this problem could be approached. The first would be to start from perturbation theory and go to higher orders of approximation. To use perturbation theory, we have to suppose that the scattering by each atom is fairly small. Of course the effect of an atom on the wavefunctions of a conduction electron in a metal is never small. In sodium for instance the wavefunctions will have two spherical nodal surfaces round each atom, as in a 3s atomic wavefunction. The possibility of treating the resistance of liquid metals by perturbation theory depends on the use of pseudopotentials, or model potentials in which the condition that ψ must be orthogonal to the wavefunctions of the inner shells, which leads to the spherical nodes, is replaced by the addition of a repulsive core to the potential. This can be chosen to give correctly the energies of the atomic states, or the phase shifts. If the phase shifts are small, perturbation theory can be used. If not, the phase shifts could be used in principle to give the scattering by each atom, but numerical results would then depend on the application of multiple-scattering theory. There is extensive theoretical work (Lax 1951, 1952, Phariseau and Ziman 1963, Beeby 1964, Ballentine 1965, Rubio 1969); more recently Khanna and Jain (1975) have studied liquid mercury using a t-matrix formalism. We do not, however, know of any analysis of the effect of multiple scattering as a correction to the Ziman theory. The other method, which we use in the later sections of this chapter, is to start from the situation when the mean free path has its minimum possible value $(L \sim a)$ so that the phase of the wavefunction varies in a random way from atom to atom. The appropriate theory has been developed in Chapters 2 and 4. The disadvantage of this treatment is that we have no way of calculating what happens in the intermediate region when kL is greater than unity but still not large; for this problem we have to depend on experiment.

For liquids (in contrast to amorphous solids) certain new problems arise, particularly because the structure of the liquid (the pair distribution function, or the density of states) depends strongly on temperature, and also because of the possibility of long-range fluctuations in density or structure, particularly near critical points. These will be discussed in this chapter.

In much theoretical work on liquid metals, as also in the theory of the resistance of disordered alloys, the change in the energy of the electron when it is scattered is neglected. It is not necessary to do this, as the work of Greene and Kohn (1965; see § 5.3) on liquid sodium has shown, and in alloys this neglect may lead to errors (see § 3.2). In this chapter we shall make this approximation, which probably leads to very small errors in liquid metals.

5.2. Scattering of electrons by a random distribution of centres; degenerate semiconductors

A degenerate electron gas with scattering by a random distribution of centres is the simplest example of the theory of this chapter; this model could apply in principle to a degenerate gas of electrons scattered by n-type centres in a highly doped semiconductor. We can ascribe to each centre a differential cross-section for scattering $I(\theta)$; $I(\theta)\,d\omega$ is the effective area for scattering by a centre through an angle θ into a solid angle $d\omega$. For the conductivity σ we can write

$$\sigma = ne^2\tau/m. \tag{5.1}$$

Here n is the number of electrons per unit volume, and the time of relaxation τ is given by

$$\frac{1}{\tau} = N_c v \int_0^\pi I(\theta)(1-\cos\theta)2\pi \sin\theta \, d\theta \tag{5.2}$$

where N_c is the number of centres per unit volume and v is the velocity of electrons at the Fermi energy. This formula was used in calculations of the resistance of disordered alloys in Nordheim's paper of 1931; it was derived by Edwards (1958) from the Kubo–Greenwood formula (compare § 2.4).

If the potential energy $V(r)$ of an electron in the field of a centre is small, $I(\theta)$ can be obtained from the Born approximation by writing

$$I(\theta) = |f(\theta)|^2$$

where

$$f(\theta) = \frac{m}{2\pi\hbar^2}\int V(r)\exp\{i(\mathbf{q}.\mathbf{r})\}\,d^3x$$

where as before $\mathbf{q} = \mathbf{k}' - \mathbf{k}$ and $\mathbf{k}$ and $\mathbf{k}'$ are the wavevectors before and after a collision. This can be evaluated to give

$$f(\theta) = \frac{2m}{\hbar^2}\int_0^\infty V(r)\frac{\sin\{2kr\sin(\tfrac{1}{2}\theta)\}r^2\,dr}{2kr\sin(\tfrac{1}{2}\theta)}. \tag{5.3}$$

Nordheim's (1931) application of these concepts to disordered alloys $A_x B_{1-x}$, where x is not small, is interesting as perhaps the first treatment of the resistance of a random system. If V_A and V_B are potentials of the two

atoms, the mean potential is $xV_A + (1-x)V_B$ and the deviations at A and B
are $(1-x)(V_A - V_B)$ and $x(V_A - V_B)$. Thus the scattered intensity and $1/L$
are proportional to $|V_A - V_B|^2$ and to

$$x(1-x)^2 + (1-x)x^2 = x(1-x)$$

(cf. Mott and Jones 1936, p. 289).

In the application of these formulae to a highly doped semiconductor,
the potential energy, which must be added to the lattice potential, should
be that of a screened Coulomb field, and may be written

$$V(r) = -(e^2/\kappa r)\exp(-\gamma r). \tag{5.4}$$

Here κ is a background dielectric constant. This is a true potential and not
a pseudopotential, and, since the potential is not strong enough to produce
nodes in the wavefunction, it is likely that the Born approximation may be
fairly good. The theory has been extensively reviewed (Mott and Twose
1961, Katz 1965, Mott 1967, § 7.8, Krieger and Strauss 1968, Meeks and
Krieger 1969, Krieger and Meeks 1973). For doped silicon and germanium
the results can be summarized as follows:

(a) For high carrier concentrations, using the formula

$$\sigma = \frac{Se^2L}{12\pi^3\hbar} \tag{5.5}$$

where S is the area of the Fermi surface and deducing the mean free path
L from comparison with experiment, one finds that L is of the order of
the distance between centres. Therefore the scattering is not weak, and
the use of the Born approximation (5.3) is suspect. It leads to values of L
about twice those observed.

(b) For lower values of n the observed conductivity decreases more
rapidly with n than the calculations predict, and near the metal–insulator
transition (Chapter 4) the discrepancy may amount to 10, the apparent
mean free path dropping below the distance between centres.

Krieger and Meeks (1973) find that agreement with the observations can
be obtained by taking into account the multi-valley nature of the conduction band and a revised form of the screening. However, if the distance
between centres is greater than the calculated mean free path, it is doubtful
if this method is valid. The discussion of Chapter 4, which shows that the
metal–insulator transition is of Anderson type, makes it likely that this
discrepancy is to be explained by the presence of a pseudogap between two
Hubbard bands, so that a factor g^2 should be introduced in (5.5), as in eqn
(2.52).

Katz, Koenig, and Lopez (1965) report a T^2 term in the electrical
resistance of n-type degenerate germanium, which they ascribe to electron–electron scattering (Baber 1937; for a review see Ziman 1960, Mott

1974c; also Hartman 1969 for measurements of the effect in certain semimetals, e.g. Bi). The T^2 term depends on the many-valley nature of the conduction band and disappears when the energies of the valleys are separated by stressing the crystal.

In compound semiconductors with high static dielectric constants the potential (5.4) might be expected to be very weak and the mobility consequently high. Allgaier and Scanlon (1958) and Allgaier and Houston (1962) in their work on PbS, PbSe, and PbTe find mobilities as high as $8 \times 10^5 \mathrm{~cm^2~V^{-1}~s^{-1}}$.

5.3. Resistivity of liquid metals; Ziman's theory

When the scattering centres are the atoms of a liquid metal (or an amorphous metal film), the atoms are not distributed completely at random; the amplitude scattered by two atoms at a vector distance $\mathbf{R}$ from each other is

$$\{1 + \exp(\mathrm{i}\mathbf{q} \cdot \mathbf{R})\} f(\theta) \tag{5.6}$$

where $\mathbf{q}$ as before is $\mathbf{k} - \mathbf{k}'$. Thus if we neglect multiple scattering the conductivity is given by eqn (5.5) where

$$\frac{1}{L} = \frac{1}{v\tau} = N \int S(\mathbf{q})(1 - \cos \theta) I(\theta) 2\pi \sin \theta \, \mathrm{d}\theta. \tag{5.7}$$

Here N is the number of atoms per cm^3 and $S(q)$ is the structure factor, given by

$$S(q) = N^{-1} \int \{1 + \exp(\mathrm{i}\mathbf{q} \cdot \mathbf{R})\}^2 P(R) \, \mathrm{d}^3 X.$$

$P(R)$ is here the pair distribution function, $P(R) \, \mathrm{d}^3 X$ being the probability that another atom is in the volume $\mathrm{d}^3 X$ at a distance R from the given atom. Using the Born approximation for $f(\theta)$ we can write for the resistivity ρ, following Faber and Ziman (1965),

$$\rho = \frac{3\pi}{\hbar e^2 v_F^2 \Omega} \int_0^{2k_F} \frac{|v(q)|^2 S(q) q^3 \, \mathrm{d}q}{4k_F^4} \tag{5.8}$$

where

$$v(q) = \int V(r) \, \mathrm{e}^{\mathrm{i}(\mathbf{q} \cdot \mathbf{r})} \, \mathrm{d}^3 x / \Omega$$

and the integral is over the volume Ω.

Fig. 5.1 shows schematically the behaviour of $S(q)$ and $v(q)$. The possibility of applying perturbation theory depends on the fact that $v(q)$ is small in the region where $S(q)$ is large.

The theory of the scattering of the electrons in a liquid metal is thus identical with that used for the scattering of X-rays or neutrons in a liquid.

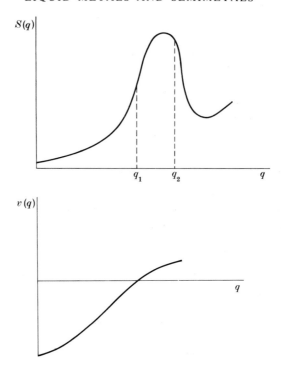

Fig. 5.1. The structure factor $S(q)$ and pseudopotential $v(q)$ for a liquid metal; q_1 and q_2 show the values of $2k_F$ for monovalent and divalent metals.

The first suggestion that the resistivity of liquid metals could be calculated in this way was made by Krishnan and Bhatia (1945) and Bhatia and Krishnan (1948). Ziman (1961) put forward the idea again,[†] using for $V(r)$ the newly discovered atomic pseudopotential, and made a detailed comparison with experiment. One of the most successful applications of Ziman's theory is to the temperature dependence of the resistivity, also discussed by Bhatia and Krishnan. This is large and positive for mono-valent metals, and small and negative for divalent metals. The explanation in terms of the observed behaviour of the structure factor $S(q)$ is as follows.

Fig. 5.2, deduced by North, Enderby, and Egelstaff (1968) from their neutron-scattering measurements, shows $S(q)$ for liquid lead at various temperatures. It will be seen that for monovalent metals the resistivity is determined by the left-hand side of the peak, and indeed since $v(q)$ has a zero near $q = 2k_F$ (Fig. 5.1), probably well below the maximum in $S(q)$. It is observed[‡] that the resistivity of monovalent liquid metals at constant

[†] See also Baym (1964).
[‡] See, for example, Lien and Sivertsen (1969), who observe for Na and K a linear dependence on T at constant volume over a temperature range of about 200 K.

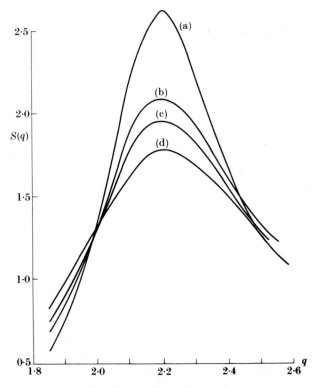

Fig. 5.2. The function $S(q)$ for liquid lead determined from neutron scattering at different temperatures: (a) 340°C; (b) 600°C; (c) 780°C; (d) 1100°C. (From North *et al.* 1968.)

volume is proportional to the absolute temperature; this suggests that $S(q)$ is also proportional to T over the range in which $|v(q)|^2$ is significant. For very low q, the structure factor $S(q)$ will be given by the Ornstein–Zernike formula

$$S(q) = kT/\beta\Omega_0 \qquad (5.9)$$

where β is the bulk modulus and Ω_0 the atomic volume. This represents the contribution from macroscopic fluctuations of density and will be true for liquids or solids. But the formula should not be true near $q = 2k_F$, even for monovalent metals, and Fig. 5.2 shows that it is not. There has been a good deal of controversy as to whether, with experimental values of $S(q)$, the linear dependence of ρ on T can be explained. Greenfield (1966) and Wiser and Greenfield (1966) deduce for liquid sodium from observed values of $S(q)$ that the calculated value of $d(\ln\rho)/dT$ is half the observed value. We think that the observed linear dependence of ρ on T must depend on the fact that for real monovalent metals $|v(q)|^2$ vanishes near $q = 2k_F$ (see Fig. 5.3).

Table 5.1 (partly from Cusack 1963) shows some values of the mean free paths in metals deduced from the free-electron formula

$$\sigma = ne^2 L/\hbar k_F, \qquad k_F = mv_F/\hbar = (3\pi^2 n)^{1/3}$$

where n is taken to be the number of valence electrons per unit volume and m is taken to be the free-electron mass. It will be observed that $k_F L \gg 1$ for most of the normal metals, though not for tellurium and some liquid alloys. To most of the liquid metals, therefore, the Ziman theory should be applicable. It should, however, be mentioned that the use of the Born approximation is not beyond criticism; $v(q)$ is necessarily equal to $\frac{2}{3}E_F$ at $q = 0$, which is not small. L is large only because $S(q)$ is small for small q, as we shall see below.

TABLE 5.1

	Li	Na	Cu	Zn	Hg	Pb	Bi	Te	PbTe	HgTe
Valence	1	1	1	2	2	4	5	6	5	4
$L(\text{Å})$	45	157	34	13	7	6	4	0·9	0·5	0·3
$k_F(\text{Å}^{-1})$	1·1	0·89	1·33	1·56	1·34	1·54	1·63	1·60	1·69	1·65
$E_F(\text{eV})$	4·6	3·0	6·6	9·2	6·9	9·0	10·0	9·7	10·9	10·3

The difficulty in using the theory to obtain numerical values of the conductivity σ derives from uncertainties in both $S(q)$ and (particularly) $v(q)$, as well as doubts about the validity of the Born approximation. Fig. 5.3 shows some values of $v(q)$ calculated by Animalu and Heine (1965). All curves show a zero in the neighbourhood of the maximum of $S(q)$ and this makes the conductivity very sensitive to the position of $v(q)$, as is shown particularly by the calculations of Ashcroft and Lekner (1966), who use various forms of $v(q)$.

The zero in $v(q)$ means that the scattered intensity vanishes at a certain angle θ, the scattered amplitude due to s-type and p-type phase shifts being of the form $A + B \cos \theta$. In the Born approximation, A and B are real, but not if exact phase shifts are taken, and the success of the theory in obtaining a fair approximation to the resistivity suggests therefore that the Born approximation is sufficient and the phase shifts really are small. On the other hand, the phase shifts η_l for a single atom should satisfy the Friedel sum rule

$$\frac{2}{\pi} \sum (2l+1)\eta_l = z$$

where z is the valency, an equation that can be used to derive the relationship (if all η_l are treated as small)

$$v(0) = \tfrac{2}{3}E_F$$

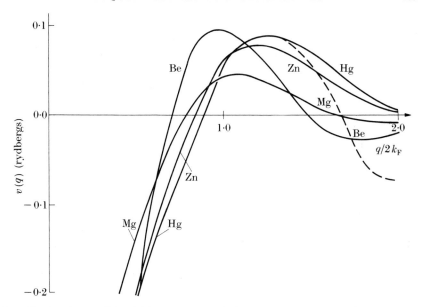

Fig. 5.3. Pseudopotentials $v(q)$ for certain metals (Animalu and Heine 1965). The broken curve for Hg is due to Evans (1970).

(Harrison 1966, Heine 1970). Heine (1970) points out that it would not be correct to take exact phase shifts, calculate $I(\theta)$, and put the result in eqn (5.7); the scattering at small angles is never small because of the Friedel rule, and multiple-scattering theory would give comparable corrections. Perturbation theory works because $S(q)$ is small for small q, and $v(q)$ small for large q.

The same considerations enable a distinction to be made between monovalent and polyvalent metals. In monovalent metals, $S(q)$ is small nearly up to $q = 2k_F$, so the resistance is small compared with that which would be produced if the atoms were distributed at random. In polyvalent metals the two quantities are comparable.

The attempts to obtain detailed agreement between theory and experiment are reviewed by Faber (1972). We may mention particularly the calculations of the resistivity of crystalline and liquid sodium carried out by Greene and Kohn (1965), and by Hasegawa (1964). Both authors find that both in the solid and liquid the calculated resistance is about one-half of that observed. Darby and March (1964), however, obtain fair agreement for the solid by taking into account the variation of elastic constants with volume, which is large.

A success for the theory has been the calculation of the conductivity for liquid rubidium at high temperatures and low pressures. Block *et al.* (1976)

have determined the structure factor up to $1400°C$ and $200\,\text{bar}$, and
Pfeifer, Freyland, and Hensel (1976) have determined the density,
conductivity, and thermopower. With a density change from $1\cdot42$ to
$0.98\,\text{g cm}^{-3}$ the conductivity drops from $3\cdot1$ to $0\cdot5\times10^4\,\Omega^{-1}\text{cm}^{-1}$, and
using the observed structure factor the calculated values agree very well
indeed.

5.4. Resistivity of liquid alloys

Ziman's theory was extended to liquid alloys by Faber and Ziman (1965;
see also Faber 1967, 1972). In order to give a complete description, for a
binary alloy the three separate pair correlation functions $S_{11}(q)$, $S_{22}(q)$,
and $S_{12}(q)$ are needed, giving the Fourier transforms of the probability that
an atom of type 1 or 2 is at a given distance from another atom of its own
kind or of the opposite kind. The analysis of Faber and Ziman is based on
the assumption that these quantities are identical, and for many alloy
systems the assumption gives a good description of the observations. The
theory gives for a concentration c of one component a resistivity $\rho =
\rho' \pm \rho''$, where

$$\rho' = (3\pi N/\hbar e^2\Omega v_F^2)\langle(1-c)S|v_1(q)|^2 + cS|v_2(q)|^2\rangle$$

$$\rho'' = (3\pi N/\hbar e^2\Omega v_F^2)\langle c(1-c)(1-S)|v_1(q) - v_2(q)|^2\rangle.$$

The angular brackets denote an average heavily weighted towards large
values of $q(q \lesssim 2k_F)$. It will be seen that the second term is likely to be
small for polyvalent metals because $S \sim 1$ over the important range of
scattering. Thus ρ' makes the major contribution, and the scattered
intensities from atoms 1 and 2 must be added together. Fig. 5.4 shows that
this is so for Pb–Sn. On the other hand, for monovalent metals $S \ll 1$, and
ρ'' makes the major contribution. The interference between waves scattered by different atoms is important, just as in crystalline alloys, and
curves such as those shown for Ag–Au and Na–K are obtained.

A theoretical account of the curve reproduced for Cu–Sn needs a
determination of the three separate partial structure factors, which can be
obtained from a combination of X-ray and neutron diffraction (Enderby,
North, and Egelstaff 1966). Enderby and Howe (1968) have shown that
good agreement with experiment can be obtained when this is done.

Mercury alloys (amalgams) normally show the behaviour shown in Fig.
5.5, the resistivity dropping sharply with concentration; the evidence was
summarized by Mott (1966), though his suggested explanation is not correct, a revised description being due to Mott (1973b). This is based on the
proposal by Evans (1970) that, because of the proximity of the full 5d band
to the Fermi surface, there is an abnormally large d phase shift and that the
value of this is very sensitive to energy. The result is that, as shown in Fig.
5.5, $v(q)$ for mercury is negative for $q/2k_F$ near the value 2, so that for

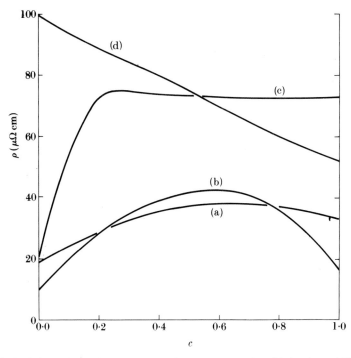

Fig. 5.4. Resistivity of liquid alloys as a function of concentration: (a) Ag–Au at 1200°C; (b) Na–K at 100°C; (c) Cu–Sn at 1200°C; (d) Pb–Sn at 400°C. (From Faber and Ziman 1965.)

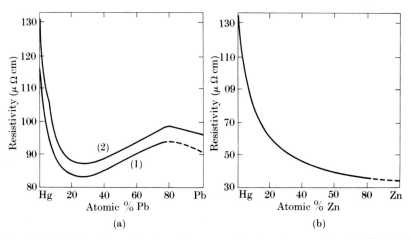

Fig. 5.5. Resistivity of some liquid amalgams, as a function of composition: (a) Hg–Zn; (b) Hg–Pb. (From Adams and Kravitz 1961.)

large angles the scattered amplitudes from Hg and from the alloyed atoms have opposite signs and interfere destructively.

There appears to be a weak pseudogap in liquid mercury ($g \approx 0.7$); this follows from calculations due to Ballentine and Chan (1973), and also from photoemission work due to Oelhafen (1975, 1976). Since the Hall effect indicates two electrons per atom, we suppose that $L \geqslant a$ and the Edwards cancellation theorem (eqn 5.16) is applicable. (See § 5.14.1.)

5.5. Thermoelectric power of liquid metals

As shown in § 2.13, for materials in which the conductivity is determined by electrons with energies near E_F, we may use the 'metallic' formula for the thermoelectric power, which we write in the form

$$S = \frac{\pi^2}{3}\frac{k}{e}\frac{kT}{E_F}\xi$$

where

$$\xi = \{d(\ln \sigma)/d(\ln E)\}_{E=E_F}.$$

Ziman (1961) and Bradley et al. (1962) applied this formula to liquid metals using equation (5.8) for σ; they found

$$\xi = 3 - 2\eta \tag{5.10}$$

where

$$\eta = \frac{\{|v(q)|^2 S(q)\}_{q=2k_F}}{\langle v^2 S(q) \rangle}$$

and the angular brackets denote as before an average over q that is defined in their papers. For most metals, comparison with experiment gives values of η close to unity, so the thermopower is negative (n type). This is because the scattering amplitude $v(q)$ for most liquid metals has a zero near $2k_F$; in other words, the probability of scattering through an angle 180° is small and decreases with E. Eqn (5.10) cannot give a value of ξ greater than 3; for mercury the experimental value of ξ is 5, and Bradley et al. and subsequent workers (Faber 1971) ascribe this to a breakdown of the assumption that the scattering potential $v(q)$ is independent of energy; Evans (1970) ascribes this to the d phase shift.

5.6. Hall effect of liquid metals

For most liquid metals, measurements of the Hall coefficient R_H, when interpreted by the use of the formula

$$R_H = 1/nec \tag{5.11}$$

give a value of n, the number of electrons per unit volume, equal to the actual number assuming all electrons in the outer shell are free. The values quoted by Faber (1972) show that this is so, and deviations† occur only for metals for which $kL \sim 1$, so that the conductivity falls below $3000 \, \Omega^{-1} \, cm^{-1}$. Exceptions are certain alloys of liquid mercury (e.g. Hg–Cd, where R_H falls on alloying, (Faber, loc. cit. p. 510)), liquid transition and rare-earth metals where R_H is, surprisingly, sometimes *positive* (§ 5.10), and expanded fluid mercury (§ 5.14.1).

5.7. Density of states

In second-order perturbation theory, the energy of an electron with wavenumber k is

$$E = E_k + v(0) + \frac{\Omega}{8\pi^3} \int \frac{|v(q)|^2 S(q) \, d^3q}{E_k - E_{k+q}} \tag{5.12}$$

where

$$E_k = \hbar^2 k^2 / 2m.$$

The density of states for a liquid or amorphous material can be evaluated using the formula (cf. eqn 2.1)

$$N(E) = \frac{4\pi k^2}{8\pi^3 (dE/dk)}$$

(Faber 1967). These formulae are not exact, because one cannot treat the changes in E that result from the term $|v(q)|^2$ without at the same time treating the scattering. Edwards (1961, 1962) and Faber (1971) treat the two together.

If E is known as a function of k, the density of states can be calculated, and in this way a number of authors have discussed the changes in the density of states due to the last term in eqn (5.12) (for some references see Ballentine 1977, Faber 1972).

A point of particular interest is the production of a pseudogap (a minimum in the density of states). This is to be expected if $v(q)$ in eqn (5.12) is large at $q = 2k_F$; the same condition would lead in crystals to a large band gap near the Fermi energy (as in γ-brasses for example) and in the liquid to high resistivity. Fig. 5.6 shows results for the resistivity of liquid Sn–Ag due to Halder and Wagner (1967); the composition at the maximum, where $\rho \sim 10^{-4} \, \Omega \, cm$, whence $L \sim 6 \, \text{Å}$, corresponds to an electron–atom ratio of 1.6, about that for γ-brass formation. Here the use of the Ziman–Faber formulae with appropriate pseudopotentials gives good agreement with

† Ballentine (1977) discusses some deviations due to spin–orbit coupling.

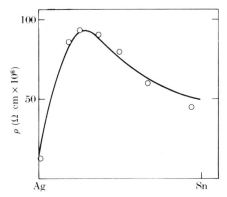

Fig. 5.6. Experimental points and theoretical curve for the resistivity of liquid silver–tin alloys (Halder and Wagner 1967). (From Faber 1972, p. 519.)

experiment, as the solid curve shows. This may be fortuitous (Faber 1972, p. 519) or may be due to the cancellation of the factor g^2 (eqn 5.16).

On the experimental side for normal metals the evidence is that the density of states changes little on melting, as the next three sections show.

5.8. Change of Knight shift on melting

Table 5.2 shows the change in the Knight shift K on melting for a number of metals. For some metals there is little change. Since K is proportional to the density of states $N(E_F)$ at the Fermi surface, this indicates that $N(E_F)$ does not change much in these metals. This does not necessarily mean that $N(E_F)$ has the free-electron value; the effect of $v(0)$ is the same in the liquid as in the solid. But it does mean that either the effect of band structure for all these solids is small, or that the term $S(q)|v(q)|^2$ gives the same change in $N(E_F)$ as the corresponding term in the crystal. In view of the calculations of Ballentine and others, showing that for the liquid this term is small, it seems likely that for *these particular metals* there is little band-structure effect on $N(E_F)$ in the crystal. This problem has been reviewed by Ziman (1967). It would be particularly interesting to make the

TABLE 5.2

Change of Knight shift K on melting

Metal	Li	Na	Rb	Cs	Cu	Cd	Hg	In	Sn	Bi	Te
Liquid K_L	0·026	0·116	0·662	1·46	0·25	0·8	2·45	0·75	0·73	1·40	0·38
$\dfrac{10^2(K_L - K_S)}{K_L}$	0	+2	+1	−2	+5	24	0	−1	−3	80	100

From Faber 1972.

comparison for a metal where there is likely to be a large change (e.g. Be, Ca).

Ziman (1967) has plotted the observed Knight shift of a large number of liquid metals against the free-electron density of states and shows that the values lie closely about a straight line; he argues from this that there is little deviation of the density of states from the free-electron value.

We shall discuss in § 5.16 some other cases where the density of states certainly falls below the free-electron value, and increases with T; this effect is reflected in the Knight shift.

5.9. X-ray emission spectra and photoemission

Soft X-ray emission bands frequently show a structure. Until recently it was thought that these were due to band-structure effects and some may be. Thus the persistence of the structure shown by the L_{III} band in Al into the liquid phase (Fig. 5.7), as observed by Catterall and Trotter (1963), was

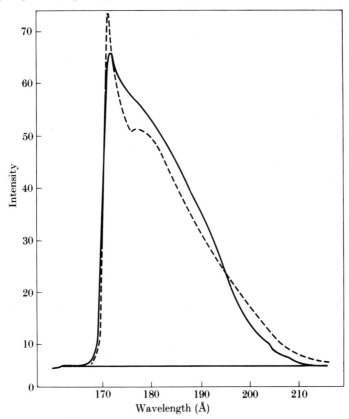

Fig. 5.7. X-ray emission bands of solid (broken curve) and liquid aluminium. (From Catterall and Trotter 1963.)

cited as evidence that the term $|v(q)|^2$ produces about the same result in the liquid as in the crystalline state. However, the work of Roulet, Gayoret, and Nozières (1969) now makes it clear that a peak at the Fermi limit can be caused by a many-body effect, which should be the same in the liquid as in the solid. Also Peterson and Kunz (1975) find the 'spike' at the Fermi limit in the $L_{2,3}$ emission in sodium virtually unchanged on melting. In view of these developments it has not yet proved possible to use X-ray spectra to make firm predictions about the change of $N(E)$ on melting.

Owing to this effect in X-ray emission, photoemission can give a more dependable test for the change of the density of states on melting. A review is due to Spicer (1973). Data for gold due to Eastman (1971) show little change on melting in emitted intensities for $h\nu = 40\cdot8$ eV and $h\nu = 26\cdot9$ eV, though for lower energies there is some loss of structure.

5.10. Liquid transition and rare-earth metals[†]

Liquid transition metals have the following properties.

(1) The conductivity is usually about $10^{-4}\ \Omega^{-1}\ cm^{-1}$ and depends little on temperature, and the change on melting is abnormally small.

(2) The Hall coefficient does not satisfy the normal formula and is often positive, as in Fe, Co, and Mn (by extrapolation from results on alloys with germanium), and also in liquid Ce and La.

There has been some controversy about the interpretation of their electrical properties. Evans, Greenwood, and Lloyd (1971) and Dreirach *et al.* (1972) argue that the full description involves ascribing a mean free path only to the s electrons, which resonate in the d shells; Mott (1972*d*) argues that, as in the solid (cf. Coles and Taylor 1962), different mean free paths L_d and L_s must be introduced for d and s electrons, that $L_d \sim a$ and that L_s is determined by s–d transitions.

Enderby and Dupree (1977) present measurements of the thermopower of liquid Fe, Co, and Ni which show that a proportionality between $1/\tau$ and calculations of $N(E)$ for the d band (Keller *et al.* 1974) can account for the results, thus supporting the model of Mott (1972*d*). Further data on liquid Cu–Ni is due to Dupree *et al.* (1977). The occurrence of a positive Hall coefficient has not been explained. Busch and Gunther̈odt (1974) find that $N(E_F)$ for solid iron is decreasing at E_F according to current calculations, and postulate that the same may be true in the liquid and that the positive term may have its origin here. Our hesitation in accepting this is because, if the current carriers have their Fermi energy at a value where $N(E)$ is decreasing, we should always expect $L \sim a$, which would first of all give a resistivity higher than observed and secondly, according to the arguments of § 2.14, a negative value of R_H is expected if the Hall coefficient is determined by three-site coincidences.

[†] For a review see Guntherödt and Kunzi (1973) and Busch and Guntherödt (1974).

Guntherödt, Hauser, and Künzi (1974, 1977) have reviewed the behaviour of liquid rare-earth metals. The behaviour of dysprosium is shown in Fig. 5.8. It shows the same general behaviour as that of the transition metals. However, s–f transitions are not normally to be expected. The conductivity of the liquid, $5 \times 10^3 \ \Omega^{-1} \text{cm}^{-1}$, is about what one would expect when $L \sim a$, if disorder produced no diminution of $N(E)$, so that $g \sim 1$. This is perhaps why there is so little change in resistivity on melting; thermal vibrations before melting have produced as short a mean free path as is possible. An interesting result is that liquid europium has a negative temperature coefficient of resistance (TCR). The resistivity is high, of order $250 \ \mu\Omega$ cm, and we believe the negative TCR to be due to the effect discussed for amorphous alloys in the next section.

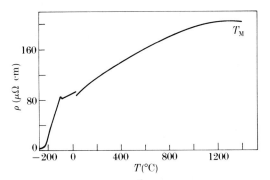

Fig. 5.8. Electrical resistivity in Ω cm $\times 10^6$ of dysprosium as a function of T. (From Guntherödt *et al.* 1974.)

There has been a considerable literature on the calculation of the density of states in liquid transition metals, in which a d state is hybridized with the s–p states. References are given in articles by Chang *et al.* (1975) and by Keller, Fritz, and Garritz (1974).

5.11. Amorphous metals and grain boundaries

The structure of amorphous alloy films, such as liquid-quenched Fe–P–C and Pd–Si alloys, electrodeposited Ni–P, and vapour-quenched Cu–Mg and Ag–Cu, has been reviewed by Wagner (1969). His general conclusion is that the order is somewhat greater than in the liquid, but the breadth of the diffraction lines is not dissimilar (see also Wright 1977).

According to Mader *et al.* (1963) and Mader (1965), amorphous alloys can normally be deposited on a cold substrate if there is a difference in the atomic radii of more than 10 per cent; such films are stable up to $\sim 0.3 \ T_M$, where T_M is the melting point. Fig. 5.9 shows the same results for Cu + 50% Ag evaporated onto a substrate at 80 K. The results show that the resistivity of the film is about half that of the liquid.

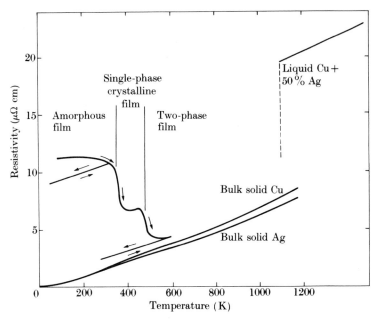

Fig. 5.9. Resistivity of amorphous films of Cu + 50 per cent Ag evaporated at 80 K, showing reversible and irreversible behaviour on annealing. (From Mader 1965.)

As a contrast to the comparatively low resistivity of amorphous films, it is worth mentioning that Andrews, West, and Robeson (1969) have measured the grain-boundary contribution to the resistivity in Cu and Al and find $\sim 3 \times 10^{-12} \, \Omega \, cm^2$; Kasen (1970) finds $1 \cdot 35 \times 10^{-12} \, \Omega \, cm^2$ for aluminium. Assuming the width of a grain boundary to be 3×10^{-8} cm, this corresponds to a resistivity of $\sim 10^{-4} \, \Omega \, cm$, which is about 10 times the resistivity of liquid copper at the melting point. Grain boundaries therefore behave as if they are much more disordered than the liquid.

The extensive work of Duwez and co-workers (for a review see Duwez 1976) on splat-cooled alloys containing a transition metal (e.g. $Pd_{0 \cdot 8}Si_{0 \cdot 2}$) show resistivities of order $1-2 \times 10^{-4} \, \Omega \, cm$. We think that s–d transitions are responsible for the mean free path. Mooij (1973) and Tangonan (1975) point out that, for alloys of this type, $d\rho/dT$ becomes negative if $\rho > 170 \, \mu\Omega \, cm$. The condition $L \sim a$ must be satisfied in this case, so phonon scattering cannot increase the resistivity; then $\rho \propto (1 - AT^2)$, where A depends on dN_d/dE, $N_d(E)$ being the density of states in the d band (Mott and Jones 1936, p. 270, Coles and Taylor 1962). A convenient expression for a nearly full d band with degeneracy temperature T_0 is $A = \pi^2/6T_0^2$. An alternative theory in which phonons affect $N(E)$ is due to Brouers and Brauwers (1975).

The mechanical and magnetic properties of these 'metallic glasses' have attracted much attention (see for instance Gilman 1975), but will not be reviewed here. These glasses are most stable for compositions near the liquid eutectic, and it is believed (see for instance Tauc and Nagel 1976) that this occurs at a given electron–atom ratio and that they are stabilized because E_F lies in a pseudogap, as in the Jones (1934) theory of alloys obeying the Hume-Rothery rules (cf. Mott and Jones 1936, Chapter V). The theory by which the position of the gap is calculated in terms of $S(q)$ should be as in § 5.7. Presumably the gap is not deep enough for the mean free path to have its minimum value $(L \simeq a)$ or for σ to drop below $\sim \frac{1}{3}e^2/\hbar a$; otherwise values of the conductivity below those quoted above would be observed.

5.12. Injected electrons in liquid rare gases[†]

A number of investigations have been made of the transport of electrons in solid and liquid rare gases. Thus Miller, Howe, and Spear (1968) have produced carriers by pulses of 40 keV electrons and measured the drift mobility. Similar work has been carried out by Halpern *et al.* (1967) and by Schnyders, Rice, and Meyer (1966). The drift velocities for solid and liquid krypton at 113 K are shown in Fig. 5.10. It will be seen that at low fields

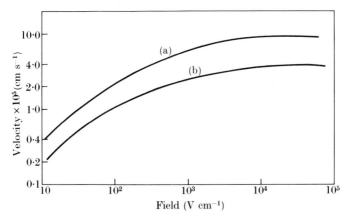

Fig. 5.10. Drift velocities of electrons in (a) solid and (b) liquid krypton as a function of field (Miller *et al.* 1968). The values of the thickness are from 185 to 585 μm.

the mobility in the liquid is high (2000 cm^2 V^{-1} s^{-1}). This corresponds to a long mean free path of several hundred atoms. There is clearly no significant trapping by localized states. We have suggested in § 2.9 that localized states do not occur when the wavefunctions are s like at the bottom of a band.[‡]

[†] A review of the experimental material is given by Gallagher (1975).
[‡] This was first proposed by Jortner *et al.* (1965).

For the liquid as for the solid the concept of a *deformation potential* can be used to calculate the mobility (cf. Chapter 3). If E_0 is the change in potential at the bottom of the conduction band for unit expansion, we should expect for the mean free path L

$$\frac{1}{L} = N \int S(q) l^2 (1 - \cos \theta) 2\pi \sin \theta \, d\theta \qquad (5.13)$$

where $l = \Omega_0 m E_0 / 2\pi\hbar^2$; the quantity l is called the scattering length. For $S(q)$ we can take the Ornstein–Zernike formula (5.9), so that $1/L = 4\pi l^2 kT/\beta$, where β is the bulk modulus. The mobility μ is $e\tau/m$ $(= eL/\surd(mkT))$, which can be written

$$\mu = \frac{e\beta}{4\pi l^2 kT} \frac{1}{\surd(mKT)}. \qquad (5.14)$$

A striking feature of the results shown in Fig. 5.10 is that the mobilities for liquid and solid are in the ratio of the bulk moduli β for the two states, so l and m appear to be the same in liquid and solid. This is further evidence that, for a band built up mainly from s orbitals, there is little change in the band structure on melting.

The effective mass in the solid is about $0.5\ m_e$ for solid argon, so in liquid argon we are very far from the nearly-free-electron approximation. If we started from eqn (5.12) in a calculation of $N(E)$, we would have to assume a large dependence of $v(0)$ on E. The high value of the mobility depends on the smallness of $S(q)$. The scattering length is comparable with a. Discussions of the detailed behaviour of l have been given by Lekner (1967, 1968); an interesting result is that l goes through zero as the gas expands because the deformation potential changes sign. This leads to a maximum in the mobility. The observations of Kimura and Freeman (1974) on the mobility of electrons in xenon are reproduced in Fig. 5.11 and show the effect well, as does the work of Miller *et al.* (1968) on argon. Near the critical point μ drops to $21\ \text{cm}^2\ \text{V}^{-1}\ \text{s}^{-1}$, presumably because of the small bulk modulus, or according to Lekner and Bishop (1972) because the long-range fluctuations produce a mobility edge, and localization is possible.

In liquid rare gases there is strong evidence for Wannier exciton states (see Raz and Jortner 1970, 1973, Asaf and Steinberger 1974). Since the bottom of the conduction band is so little distorted this is not surprising. In situations where a substantial range of localized states exists at the extremities of the bands, it is doubtful whether exciton line spectra can be observed (§ 6.7.1).

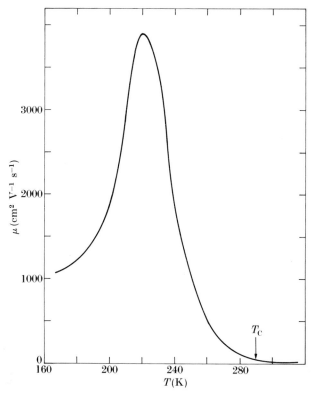

Fig. 5.11. Mobility of electrons in liquid xenon as a function of temperature. The field was 16 V cm^{-1}. (From Kimura and Freeman 1974.)

5.13. Liquid semimetals and semiconductors

The review by Ioffe and Regel (1960) and the books by Gubanov (1963) and by Glazov, Chizhevskaya, and Glagoleva (1969) were the first to treat liquid semiconductors in any detail.† In our view there is no essential difference in the theories necessary to treat electrical conduction in liquid and in solid non-crystalline semiconducting materials, but liquids are more complicated because the arrangement of atoms, and therefore the structure factor $S(q)$ and the density of states, can and do change with temperature, while in amorphous solids below the glass transition temperature the change due to lattice vibrations is probably much smaller. In non-crystalline solids we normally suppose that a resistance which decreases with increasing temperature means either thermally activated hopping or excitation to a mobility edge. For liquids this is not necessarily so. As we have seen, in

† A recent review is that by Cutler (1977).

most liquid metals the change in $S(q)$ is responsible for the dependence of resistivity on temperature, which may have either sign; this may be so too for semimetals and even semiconductors.

In liquids and amorphous materials a semimetal is one in which the Fermi energy lies in a pseudogap (a minimum in $N(E)$), a semiconductor one in which it lies in a gap, or at any rate in a range of E where $N(E)$ is small and states are localized. In this section we discuss divalent metals at low densities (mercury), monovalent metals (caesium), and liquids such as tellurium and its alloys where a pseudogap between a valence and conduction band depends on the structure. The relevant theory has been described in § 2.11 and § 4.1; we summarize here the different situations which it leads us to expect.

Regime I. The mean free path L is large ($k_FL \gg 1$), $\sigma > e^2/3\hbar a$ (3000–4500 Ω^{-1} cm^{-1} depending on the atomic volume a^3) and the Ziman theory is applicable. The conductivity is given by

$$\sigma = \frac{S_F e^2 L g^2}{12\pi^3 \hbar} \tag{5.15}$$

where k_F is the wavevector at the Fermi energy E_F,

$$g = \frac{N(E_F)}{N(E_F)_{free}}$$

and S_F is the Fermi surface area ($4\pi k_F^2$). If L_{Ziman} is the value of L calculated from perturbation theory (eqn 5.7), the Edwards (1962) cancellation theorem gives (cf. eqn (2.24))

$$L = L_{Ziman}/g^2 \tag{5.16}$$

so eqn (5.15) reduces to

$$\sigma = \frac{S_F e^2 L_{Ziman}}{12\pi^3 \hbar}.$$

It will be noted that the effective mass does not appear in these equations. The Hall coefficient R_H should be given by the classical expression $R_H = 1/nec$, and the Hall mobility is the same as the conductivity mobility

$$\mu_H = e\tau/m_{eff} = eL\hbar/k_F.$$

Regime II (conductivity between ~3000 and ~300 Ω^{-1} cm^{-1}). The scattering is so strong that $L \sim a$, and consequently

$$\sigma = \frac{S_F e^2 a g^2}{12\pi^3 \hbar}. \tag{5.17}$$

The cancellation theorem no longer applies because L does *not* depend on

g. k_F is not, however, a good quantum number; if we write $k_F = \pi/a$ (for a divalent metal), $\sigma = \frac{1}{3}e^2g^2/\hbar a$. For the Hall coefficient Friedman (§ 2.14) gives

$$R_H \simeq 0.7/neg \qquad (5.18)$$

and for the Hall mobility

$$\mu_H \simeq 0.1 \, ea^2 g/\hbar. \qquad (5.19)$$

μ_H thus decreases with increasing depth of the pseudogap. The thermopower will be given by

$$S = \frac{\pi^2}{3} \frac{k^2 T}{e} \frac{2d(\ln g)}{dE}. \qquad (5.20)$$

The Knight shift is proportional to $N(E_F)$ and thus to g. An approximate relationship $\sigma \simeq \{N(E_F)\}^2$ is expected, where $N(E_F)$ can be deduced from the Pauli paramagnetism or Knight shift. If instead of a we take a_E (§ 2.11), however, and write

$$1/a_E^3 = g/a^3$$

then a proportionality between σ and $\{N(E)\}^{5/3}$ may be more appropriate (see Fig. 5.23).

Regime III. If g falls below the value ~ 0.3 (cf. § 2.11), states at the Fermi energy will become Anderson localized. The liquid will behave as a semiconductor, the conductivity being given by

$$\sigma = \sigma_{min} \exp(-\varepsilon/kT)$$

where ε, the energy required to excite an electron to the mobility edge, is given by $\varepsilon = E_C - E_F$. The thermopower is

$$S = (k/e)(\varepsilon/kT + A). \qquad (5.21)$$

The theory of § 2.13 predicts $A = 1$, but negative values are sometimes found, for a reason that is not clear. The Hall mobility is given by Friedman's formula (§ 2.14)

$$\mu_H = \frac{1}{18} \frac{2\pi ea^2}{\hbar} \left(\frac{B}{V_0}\right)_{crit} \simeq \frac{ea^2}{6\hbar} \qquad (5.22)$$

and the conductivity mobility by

$$\mu_C = \frac{\sigma_{min}}{eN(E_C)kT}$$

it being assumed that transport by electrons at E_F makes a negligible contribution.

Regime IV. As ε increases, there arises the possibility that transport by carriers at E_F gives the predominant term in the conductivity; in solids this regime can always be reached by lowering the temperature and the transport mechanism is by hopping. In liquids there is another possibility, namely that a carrier in a localized state drifts like a heavy ion, carrying the configuration that traps it with it. This is especially probable if it can distort its surroundings and 'dig its own hole', as is certainly the case for solvated electrons in metal–ammonia (Thompson 1976).

Regime V. When the density of states at the Fermi energy vanishes, the main contribution to the conductivity must be due to electrons or holes or both excited to a mobility edge; the properties are then just the same as in regime III.

One result of this analysis is that only for values of the conductivity greater than $\sim 3000 \, \Omega^{-1} \, cm^{-1}$ (normal liquid metals) can we expect values of μ_H above $0 \cdot 1 \, cm^2 \, V^{-1} \, s^{-1}$. That this is so is shown in Fig. 5.12, due to Allgaier (1970).

Of particular interest are liquid systems in which it is possible to go from one regime to another by changing the composition, temperature, or density. If the volume of a divalent metal can be increased, the conduction

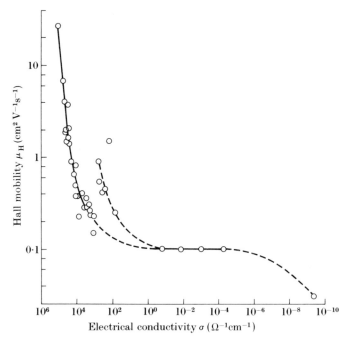

Fig. 5.12. Hall mobilities of a number of liquids plotted against conductivities. (From Allgaier 1970.)

and valence bands can be separated, leading to a pseudogap (a minimum in $N(E)$) and ultimately to a gap. This is possible in mercury at high temperatures, for which in the next section we shall show how all the above regimes can be identified. The same is true for fluid caesium, liquid salts such as $KI_{1-x}K_x$ and metal–ammonia solutions; for these, two Hubbard bands separate and at low temperatures a two-phase region forms, as described in the last chapter. Mercury and caesium are described next; for the application of these ideas to metal–ammonia see Mott (1974c, 1976e), Thompson (1976).

5.14. Liquid systems in which the depth of the pseudogap changes with volume

5.14.1. Mercury

Liquid mercury is anomalous in many ways. The mean free path deduced from (5.15) is 7 Å (Cusack 1963) and the Hall coefficient obeys the equation $R = 1/nec$ quite accurately with n given by two electrons per atom. The anomalous drop in the resistivity on alloying illustrated in Fig. 5.5 was first discussed by Mott (1966) in terms of a pseudogap which fills up, but this explanation proved wrong and Evans et al. (1970) showed that it could be explained by the pseudopotential illustrated in Fig. 5.3. This behaviour is due to a large d contribution to the scattering cross-section. The same result is obtained by Khanna and Jain (1975) who use a t-matrix formulation. This means that the electron waves scattered by the alloying element have the opposite phase to those scattered by the mercury atoms, and interfere destructively. There is some evidence for a pseudogap in liquid mercury with $g \approx 0.7$, which is reviewed by Mott (1972b). This is not supposed to affect the electrical properties, the Edwards cancellation theorem (5.16) being applicable. The normal value of the Hall coefficient can be taken as experimental evidence that the Friedman formula (5.22) does *not* apply even for small values of the mean free path L as long as $L > a$. Additional evidence that any pseudogap does not affect electrical properties is provided by the observation of Choyke, Vosko, and O'Keefe (1971) that the marked peak in the imaginary part of the high-frequency dielectric constant of solid mercury at $h\nu = 2$ eV completely disappears in the liquid, where the conductivity $\sigma(\omega)$ obeys the Drude formula.

On the other hand, the measurements of Hensel and Franck (1966, 1968), Hensel (1970), Schmutzler and Hensel (1972), and Even and Jortner (1972, 1973) on the electrical conductivity, thermopower, and Hall coefficient of expanded fluid mercury at low densities consequent on high temperatures give ample evidence for the formation of a pseudogap as the liquid expands. Fig. 5.13 shows the conductivity as a function of volume. If g is deduced from the observed conductivity σ using (5.17) when σ falls

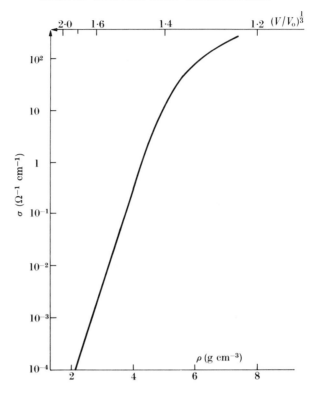

Fig. 5.13. Specific conductivity σ of mercury at 1550°C as a function of density (Hensel and Franck 1968); V is the volume and V_0 the molar volume.

below $3000\ \Omega^{-1}\ cm^{-1}$ (the value for $L \sim a$), then the observed Hall coefficient is given well by Friedman's formula (5.18) down to $g \approx 0.3$ ($\sigma = 300\ \Omega^{-1}\ cm^{-1}$), where Anderson localization occurs. This is shown in Fig. 5.14. The liquid then enters regime III, and conduction is due to electrons excited to the mobility edge; this results in a constant value of μ_H, given roughly by Friedman's formula (5.22). Both the drop and the subsequent constant value are observed by Even and Jortner and shown in Fig. 5.14. This explanation of their results, as regards the constant value of μ_H, was proposed by Mott (1975c).

That the fluid (near the critical point) goes into regime IV, conduction being by electrons at E_F and near the bottom of a pseudogap, is indicated by the measurements of the thermopower by Duckers and Ross (1973) which falls precipitately to a low value.

For lower densities conduction is again at E_C (true semiconducting behaviour, regime V). In this regime, since we expect $\sigma =$

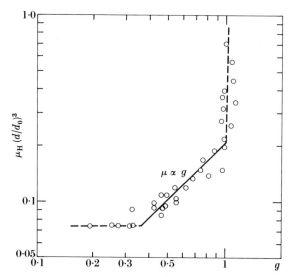

Fig. 5.14. Normalized Hall mobility (in $cm^2\,V^{-1}\,s^{-1}$) of expanded fluid mercury (Even and Jortner 1972). The quantity g is deduced from the observed conductivity for $g < \frac{1}{3}$, and for $g > \frac{1}{3}$ by extrapolation.

$\sigma_{min} \exp(-E/kT)$ and $S = (k/e)\{E/kT + A\}$ with $A = 1$ for charge transport at a mobility edge, it follows that

$$\log \sigma = \log \sigma_{min} - \frac{|S|(e/k) - A}{2 \cdot 3}. \qquad (5.23)$$

A plot of $\log \sigma$ against S is shown in Fig. 5.15(a). The slope is exactly as predicted. However, extrapolation to $eS/k = 1$ gives a rather small value of $\sigma_{min}(\sim 20\,\Omega^{-1}\,cm^{-1})$. Results very similar to these are found for liquid Se in the range up to 1700°C and 100 kbar (Hoshino, Schmutzler, and Hensel 1976, Hensel 1976); these are reproduced in Fig. 5.15(b). Here σ_{min} appears to be about $10\,\Omega^{-1}\,cm^{-1}$. However (see § 2.13), we expect $S = (k/e)\,2\ln 2$ at an Anderson transition where $\sigma = \sigma_{min}$ and Fig. 5.15 for Hg gives $\sigma = 400$ at this value of S; if σ_{min} remained constant, the behaviour shown by the broken curve would be expected. In Se there is a much smaller change.

These small values of σ_{min} may mean that the carriers in these expanded liquids are able to form 'polarons' by bonding together two or more atoms, but with an activation energy in the mobility which is small. If so, the constant μ_H shown in Fig. 5.14 is an example rather of the Friedman–Holstein behaviour (§ 3.9) with small W_H. When conduction becomes metallic, so that carriers compete for Hg atoms, polarons would no longer form. Similar behaviour for liquid Se–Te is described in Chapter 10.

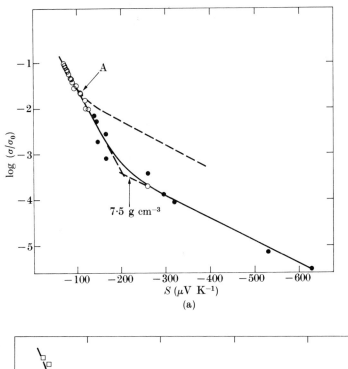

(a)

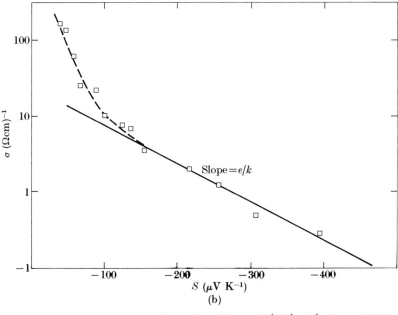

(b)

Fig. 5.15. (a) Logarithm of conductivity σ/σ_0 with $\sigma_0 = 10^4 \, \Omega^{-1} \, \text{cm}^{-1}$ of expanded fluid mercury *versus* thermopower S. (From Schmutzler and Hensel 1972.) (b) Similar results for liquid selenium.

Devillers (1974) has pointed out that the electrical gap deduced from the work of Schmutzler and Hensel and of Duckers and Ross (1973) lies well below the optical gap (Hensel 1970) and that a 'theoretical' gap calculated by Devillers and Ross (1975) lies between them. They give an explanation in terms of density fluctuations. We think that for these low densities there may be important band tails due to fluctuations and that the mobility gap will be substantially greater than the optical gap, as Devillers (1974) proposes. Band-theoretical calculations of the optical gap by Overhof, Uchtmann, and Hensel (1975) seem to confirm this model, as do those of Yonozawa et al. (1977).

Measurements of the Knight shift K by El-Hanany and Warren (1975) in the range of density for which a metallic pseudogap is expected do not confirm its presence; the reason for this is not understood. K drops rapidly when σ falls below $\sim 200 \ \Omega^{-1} \ \text{cm}^{-1}$.

Popielawski (1972) and Popielawski and Gryko (1977) have treated the low-density limit (density $\sim 2 \ \text{g/cm}^3$) by assuming that each Hg atom lowers the conduction band by the mean of the electron's potential energy in its neighbourhood, and that weak scattering theory can be applied as in § 5.13. Whether this is so or not depends on whether $kL \gg 1$.

In the description given here, apart from any effect of long-range fluctuations, the metal–insulator transition in mercury is of Anderson type, occurring where $\sigma \simeq 300 \ \Omega^{-1} \ \text{cm}^{-1}$. At the critical point the conductivity is very much lower ($\sim 10^{-3} \ \Omega^{-1} \ \text{cm}^{-1}$). There is no suggestion of a two-phase region in the neighbourhood of the metal–insulator transition, which according to the consideration of Chapter 4 must always occur in a crystal. The reason suggested by Mott (1974c) is that the background dielectric constant is large, so the discontinuity represented by eqn (4.7) is small. Therefore, as in Si:P, Anderson localization sets in before the discontinuity due to band overlap can occur.

Cohen and Jortner (1973, 1974a,b) have given a quite different explanation of the metal–insulator transition in mercury. Their description is the same as ours for values of σ above $\sim 300 \ \Omega^{-1} \ \text{cm}^{-1}$, where the kink in the plot of μ_H is observed (Fig. 5.14); below this they postulate an 'inhomogeneous regime', in which conduction is still essentially metallic but is along classical percolation channels.

5.14.2. Caesium

Experiments on this material at low densities have been carried out by Freyland, Pfeifer, and Hensel (1974). The most striking difference from mercury is that, at the critical point, the conductivity ($\sim 10^3 \ \Omega^{-1} \ \text{cm}^{-1}$) is six orders of magnitude higher than that for mercury at its critical point. The critical point for caesium is, we believe, that for the transition between a metallic and non-metallic liquid as described in § 4.3, and the factor g is due to two overlapping Hubbard bands. The two-phase region occurs, in contrast to mercury, because for two Hubbard bands the background dielectric constant is small.

Freyland *et al.* (1974) have investigated the thermopower of liquid caesium in the semiconducting range, and have plotted σ logarithmically against S. Here extrapolation to $eS/k = 1$ (as in Fig. 5.15) and putting in the value of σ when $S = (k/e)2\ln 2$ gives the *same* value of $\sigma_{min}(\sim 300\ \Omega^{-1}\ cm^{-1})$.

5.15. Liquid alloys with pseudogaps in which the composition is varied

The first alloy system of this kind to be investigated was $Mg_{3-x}Bi_{2+x}$ (Ilschner and Wagner 1958). Their results showed a sharp maximum in the resistivity at $x = 0$. The work on solid amorphous alloys at these compositions (§ 4.2.5) was undertaken as a consequence of this work to separate the effects of thermal activation of carriers from that of structural changes. The solid alloys were treated by a rigid-band model, the Fermi energy

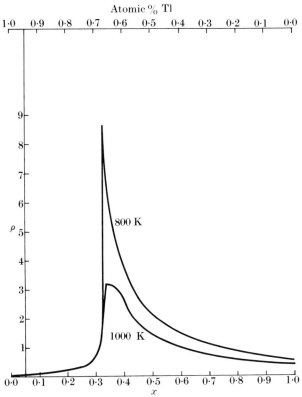

Fig. 5.16. Resistivity ρ in $\Omega\ cm \times 10^{-3}$ of liquid Te_xTl_{1-x}. (From Cutler and Mallon 1965.)

moving through the pseudogap when the composition varied and the thermopower changing sign.

Another example is the liquid system $Te_{1-x}Tl_{2+x}$ on which detailed work has been undertaken by various authors, including Cutler and Mallon (1965), Cutler and Field (1968), Enderby and Simmons (1970), Regel *et al.* (1970), Cutler (1971), Donally and Cutler (1968, 1972), and Cutler (1974). Fig. 5.16 shows the resistivity of these alloys as a function of composition, and Fig. 5.17 shows the thermopower. The sharpness of the

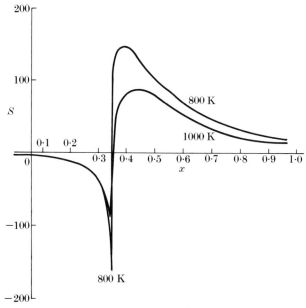

Fig. 5.17. Thermopower in $\mu V\,K^{-1}$ of liquid $Te_x Tl_{1-x}$.

transition on the thallium-rich side is not compatible with a broad pseudo-gap. Cutler (1971) has given thermodynamical evidence to support the hypothesis that molecules of the composition $TeTl_2$ can form and partially dissociate as the temperature is raised; an excess of either constituent can be dissolved in the liquid formed from these molecules.

We interpret the properties of such materials, therefore, in the following way. The Te^{2-} ion in the molecule (e.g. $TeTl_2$) forms a closed shell. Thus the wavefunctions near the bottom of the conduction band are s like on the Te atoms. Therefore the density of states in the conduction band should be parabolic, and there will not be an extensive range of localized states. The situation is similar to that of the conduction band in liquid argon (§ 5.13). The resistances of the thallium-rich alloys do not depend on temperature, and are within the metallic range of values ($\sim 10^{-3}\,\Omega$ cm).

One supposes therefore that an excess of Tl above the composition $TaTl_2$ gives rise to a degenerate electron gas in this conduction band. Cutler and Field (1968) have shown that the addition of other metals leads to the same result, and also that the metal atom gives up all its valence electrons to the degenerate gas. We suggest that the reason may be that the 'molecule' $TeTl_2$ has a dipole, and that when there is an excess of Tl the metal ion is solvated. This would provide the energy required to raise the energies of the electrons to the conduction band of the liquid formed from the molecules $TeTl_2$. The field round the excess metal ions would thus be weak, and the degenerate electron gas should have properties similar to those of electrons in concentrated metal–ammonia solutions, where again the NH_3 molecules have dipoles. An unexplained fact is that measurements of the Hall coefficient interpreted through the equation $R_H = 1/nec$ give a temperature-independent value of n ten times greater than that due to the excess of Tl atoms (Donally and Cutler 1972), these being supposed monovalent (Cutler 1974). The Hall mobility shows a discontinuity at the stoichiometric composition.

This behaviour is in sharp contrast with that on the tellurium-rich side. The Hall coefficient is then negative, in accordance with the general prin-

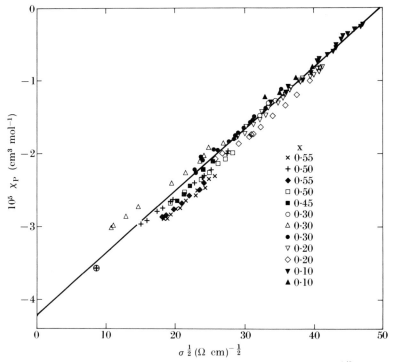

Fig. 5.18. Pauli susceptibility of $Te_{1-x}Tl_{2+x}$-liquid alloys plotted against $\sigma^{1/2}$ for various temperatures and compositions. (From Cutler 1977.)

ciple for such materials first discovered by Enderby and Walsh (1966); the thermopower is, however, p like and the conductivities are greater than $100\,\Omega^{-1}\,cm^{-1}$ and probably metallic. They show a Pauli-type paramagnetism χ_P which can be separated from the diamagnetic term, and over a wide range of composition and temperature $\sigma \propto \chi_P^2$ (Gardner and Cutler 1977, Fig. 5.18). Assuming that χ_P is proportional to $N(E)$ and σ is proportional to $\{N(E)\}^2$, this gives strong support to the pseudogap model with constant mean free path (eqn (5.17)). However, the Hall mobility μ_H does not obey Friedman's law (5.19); Donally and Cutler (1972) find it to be independent of T. This seems to be a general law for Te and its alloys, which is not fully understood. The drop in the thermopower S with increasing T must be interpreted by supposing that $d \ln N(E_F)/dE$ decreases with increasing T faster than $1/T$.

While the thermopower of liquid Te–Tl alloys changes sign at a fixed composition, as also does that of Pb–Te, this is not so for Ga–Te (shown in Fig. 5.19) or for the chalcogenide glasses in general. The two classes of

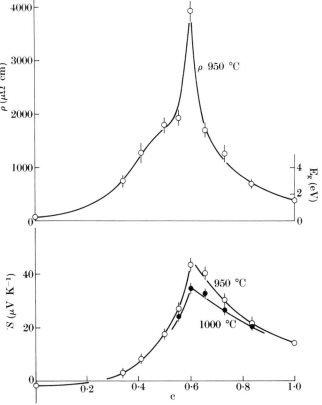

Fig. 5.19. Resistivity ρ and thermopower S of liquid Ga–Te plotted against concentration c of Te. (From Valiant and Faber 1974.)

liquids are discussed by Valiant and Faber (1974). For the chalcogenide glasses, in the liquid as in the amorphous solid state, all outer electrons are apparently taken up in bonds and the Fermi energy is pinned near mid-gap, probably by charged dangling bonds as explained in Chapters 9 and 10. As the temperature is raised, the gap (to be discussed in the next section) turns into a pseudo-gap, eventually giving metallic behaviour. For Ga–Te Tschirner et $al.$ (1976) find a strong maximum in R_H at 60 per cent Te, and μ_H about 10^{-1} cm^2 V^{-1} s^{-1}, which is about what we should expect from the Friedman formulae (§ 2.14).

Warren (1972a,b) and Brown, Moore, and Seymour (1972) have measured the Knight shift K and the nuclear relaxation as a function of x in $Ga_{1-x}Te_x$ and Tl_xTe_{1-x}, materials in the two different classes. For the Knight shift a deep minimum at $x = 0.66$ is very marked, rising sharply on the Ga-rich side.

5.16. Liquid systems in which the depth of the pseudogap changes with temperature

5.16.1. Tellurium

This liquid is a semimetal in regime II (p. 182). The conductivity is 1300 Ω^{-1} cm^{-1} at 675 K and rises with temperature to 2750 Ω^{-1} cm^{-1} at

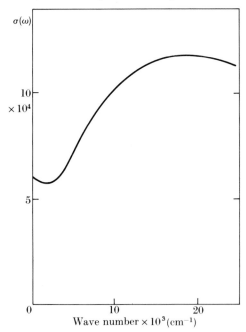

Fig. 5.20. The dependence on wavenumber ω/c of the conductivity $\sigma(\omega)$ (in Ω^{-1} cm^{-1}) of liquid tellurium. (From Hodgson 1963.)

1100 K (Warren 1972*a,b*). This is somewhat less than the value
($\sim$5000 Ω^{-1} cm^{-1}) that we should expect with $g = 1$ (no pseudogap) and six
electrons per atom, and so it seems probable that at the melting point the
liquid metal has a pseudogap in the density of states with $g^2 \sim 0\cdot4$ at the
Fermi energy, and that $g \rightarrow 1$ as the temperature rises. The optical properties shown in Fig. 5.20, due to Hodgson (1963), give evidence for a
pseudogap, showing low values of $\sigma(\omega)$ for small ω but larger values for
transitions across the gap.

Urbain and Übelacker (1966) have measured the magnetic susceptibility
of solid and liquid tellurium, their results being shown in Fig. 5.21. They

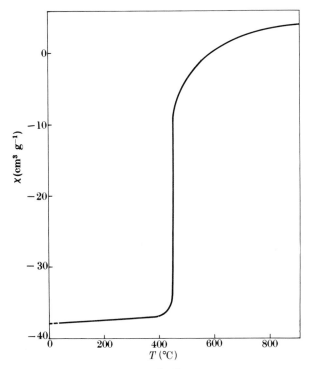

Fig. 5.21. Magnetic susceptibility χ in cm^3 g^{-1} of solid and liquid tellurium (Urbain and
Überlacker 1966). The free-electron value according to the Pauli theory is shown by the
horizontal line framing the top of the diagram.

suggest that, after the susceptibility of the ions has been subtracted, the
remaining susceptibility approaches the Pauli free-electron value for six
electrons per atom. We believe that this also may indicate that $N(E_F)$
gradually approaches the free-electron value ($g = 1$) as the temperature is
raised.

Cabane and Froidevaux (1969) and Warren (1972a,b) have measured the Knight shift; the results of the former authors are shown in Fig. 5.22(a). These confirm the values of g estimated above. Also the plot shows that σ is proportional to K^2 and thus to g^2 (i.e. to $N(E)^2$, as should be the case when $L \sim a$). In contrast, the data of Tièche and Zareba (1963) show that

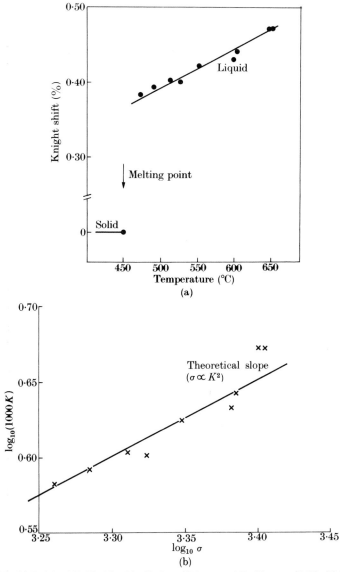

Fig. 5.22. (a) Knight shift K of liquid tellurium (Cabane and Froidevaux 1969). (b) Plot of log $(1000K)$ versus log σ, showing the linear relationship $\sigma \propto K^2$, where K is the Knight shift.

R_H is proportional to g^2 so that μ_H is approximately independent of g. This, as already stated, is a property of Te-rich Te–Tl, and as we shall see in Chapter 10 of $Te_{1-x}Se_x$. Various explanations have been attempted (Cohen and Jortner 1973); within the context of the models used here, it remains unexplained.

Cabane and Friedel (1971) show that there is a gradual transition from twofold to threefold co-ordinations as the temperature is raised; this is discussed in Chapter 10.

5.16.2. Some liquid alloys of tellurium

Andreev (1976) has surveyed the temperature dependence of the conductivity of a large number of these alloys, including PbTe, SnTe, GeT, In_2Te_3, InTe, and some similar alloys of Se. According to this author, they all show the behaviour sketched in Fig. 5.23. Here in regime 1 the gap $(E_F - E_C)$ is a

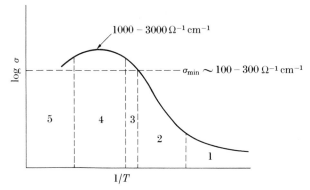

Fig. 5.23. Schematic behaviour of conductivity of semiconducting Te and Se alloys as a function of $1/T$. (From Andreev 1976.)

linear function of T, in regime 2 it is non-linear, rapidly disappearing at the Anderson transition, in regime 3 the conductivity σ is proportional to g^2 rising to the value for $g = 1$ in regime 4, while in regime 5 there is a drop in σ due to a decrease in density. For the flat part (regime 4) Andreev finds that, on comparing with the equation $\sigma = e^2 S_F L / 12\pi^3 \hbar$, $L \sim 0.4\,a$ gives the best fit assuming that *all* outer electrons contribute to S_F.

For In_2Te_3 and Ga_2Te_3 the investigations of Warren (1970a,b, 1972a,b) on the Knight shift provide the most convincing evidence that a pseudogap is formed as the temperature is lowered. If g is determined from the Knight shift, the conductivity is found to drop as g^2 until g reaches a value $g \sim 0.2$; thereafter it drops faster, indicating that states are localized at E_F and conduction is by excitation to E_C. An unusual enhancement of the nuclear relaxation rate above that expected from the Korringa relation gives evidence for localization.

It is of some interest to enquire whether σ varies as K^2 or $K^{5/3}$; the latter would be expected if the minimum mean free path is the distance a_E between localized states and $N(E) \propto 1/a_E^3$. Plots of the two types are shown in Fig. 5.24, and cannot as yet distinguish between the two possibilities.

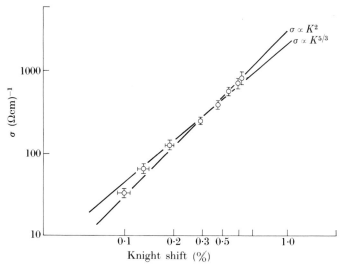

Fig. 5.24. Conductivity σ of Ga_2Te_3 plotted against different powers of theKnight shift K.

Warren and Brennert (1974) and Warren *et al.* (1974) have investigated the system $Ga_2(Se_x Te_{1-x})_3$. The Knight shift shows g decreasing with increasing x, as one might expect. When localization occurs, they give evidence that conduction is at E_C; however, they consider that E_C is not sharp but that conduction is over a range of some multiples of kT at E_C. They also show that in the crystal this system shows a range of solid state immiscibility for $0.5 < x < 0.9$ and that the conductivity of the liquid shows a structure here, indicating long-range fluctuations.

Liquid selenium and selenium–tellurium alloys (Perron 1971) are discussed in Chapter 10. Perron has also investigated the Hall mobility, and his results show that the Friedman relationship ($\mu_H \propto g$) is *not* obeyed, μ_H being roughly constant; in this respect these alloys behave like liquid tellurium and other tellurium alloys.

6

NON-CRYSTALLINE SEMICONDUCTORS

6.1. Introduction
6.2. Preparation and classification of materials
6.3. Methods of determining structure
6.4. Electrical properties of non-crystalline semiconductors
    6.4.1. Density of states
    6.4.2. The mobility edge
    6.4.3. Temperature dependence of the d.c. conductivity
    6.4.4. Behaviour in the liquid state
    6.4.5. a.c. conductivity
    6.4.6. Thermopower
    6.4.7. Hall effect
    6.4.8. Magnetoresistance
    6.4.9. Field effect
6.5. Drift mobility and photoconduction
    6.5.1. Drift mobility
    6.5.2. Photoconductivity
    6.5.3. Quantum efficiency
6.6. Conduction at a mobility edge *versus* hopping
6.7. Optical absorption
    6.7.1. Absorption edges and Urbach's rule
    6.7.2. Effect of externally applied fields
    6.7.3. Interband absorption
    6.7.4. Absorption at high energies
    6.7.5. Intraband absorption
    6.7.6. Photoluminescence
    6.7.7. Vibrational spectra. Density of phonon modes
6.8. Other measurements
    6.8.1. Specific heat and thermal conductivity
    6.8.2. Photoemission and density of states
    6.8.3. Electron spin resonance

6.1. Introduction

THE APPLICATION of the concepts developed in earlier chapters to non-crystalline semiconductors is one of the main purposes of this book. In this chapter we describe some of the experimental techniques that have been used to characterize these materials, the kinds of results obtained, and the models used to interpret them. More detailed descriptions of specific materials are given in the final chapters. We find that the models developed in previous chapters, namely those of localized states, mobility edges, states in the gap arising from structural defects, and variable-range hopping conduction, serve as a framework into which most of the observed data can be fitted.

6.2. Preparation and classification of materials

The two normal ways of preparing amorphous solids are (i) by condensation from the vapour as in thermal evaporation, sputtering, glow-discharge decomposition of a gas, or other methods of deposition, or (ii) by cooling from a melt.† The first methods produce thin films and the second bulk material. If a material can be prepared in the amorphous phase from a melt, it is generally also possible to prepare it by deposition. If a suitable thinning technique can be found, such as etching or 'blowing' the glass into thin-film form, it is frequently possible to bridge the 'thickness gap' between the two methods of preparation. However, there will inevitably be structural differences between the same material prepared by different methods, and these must be considered in any comparison of properties.

Materials that are obtained by cooling from the molten state are called glasses and generally have a smaller tendency to crystallize compared with those that can be prepared only by deposition. In certain cases this reluctance to crystallize makes it possible to heat the material through the softening range of temperature into the liquid state without any discontinuous change in properties. Such stable glasses may, however, undergo a second-order phase transition at the so-called glass transition (transformation) temperature T_g. This transition, which corresponds to the accessibility of new configurational energy states or degrees of freedom, marks the onset of softening and is accompanied by an increase in the heat capacity and thermal expansion coefficient (Fig. 6.1). Unlike the melting temperature T_m of the crystal, it is not a particularly well-defined temperature and depends on the rate of cooling or heating. Less stable glasses, for instance those with compositions near the border of a glass-forming region in a multicomponent system (see Fig. 6.3), may, on heating slowly, undergo phase separation and crystallization. Further heating causes melting, and certain properties of the liquid state, such as the temperature dependence of conductivity, are then often similar to those in the disordered solid at temperatures below crystallization. In order to prepare such glasses, rather faster quenching techniques are needed to avoid devitrification. In the case of As_2Te_3, for example, 'splat cooling' is used to prepare the glass. In contrast, the crystallization process in As_2Se_3, a stable glass, is so slow as to allow preparation by cooling the melt at a very low rate. As an example of a material intermediate between these two extremes, we show in Fig. 6.2 results of differential thermal analysis on

---

† Other methods include electrolytic decomposition from solution, and prolonged irradiation of crystalline material with high-energy particles such as neutrons or ions. For a general review, see Owen (1973).

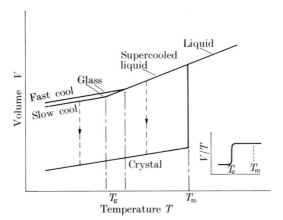

Fig. 6.1. Schematic variation of volume *versus* temperature for a glass-forming material. A glass is formed on cooling the liquid state below the glass-transition temperature T_g, but the value of the latter as well as the volume (and structure) of the glass depend on the cooling rate. Crystallization can occur from the supercooled liquid or the glass (vertical arrowed lines). The crystal, which generally has a higher density than, but a similar expansion coefficient to, the glass, melts at T_m.

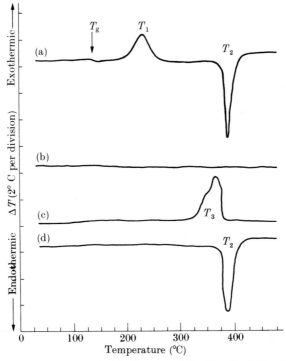

Fig. 6.2. Differential thermal analysis (DTA) traces of $Ge_{16}Te_{82}Sb_2$ (a) heating $25°C\ min^{-1}$; (b) fast cooling; (c) slow cooling; (d) heating. (From Fritzsche and Ovshinsky 1970.)

$Ge_{16}Te_{82}Sb_2$ reported by Fritzsche and Ovshinsky (1970). Heating the glass at 25°C min^{-1} as shown in curve (a) produces a small step at T_g, the glass transition temperature, followed by an exothermic crystallization peak at T_1 and an endothermic melting peak at T_2. Fast cooling (greater than 50°C min^{-1}) (curve (b)) shows no reaction, and the high-temperature disordered state is frozen in. Slow cooling (less than 10°C min^{-1}) shows (curve(c)) an exothermic transformation at T_3 below which the material is partially crystallized; subsequent heating of this material (curve (d)) shows the T_2 endotherm only.

Fig. 6.3 illustrates the glass-forming regions in a few ternary systems. For other examples and more detailed information on the preparation, stability, and physical properties of glasses the reader is referred to Rawson (1967), Turnbull (1969), and Owen (1973).

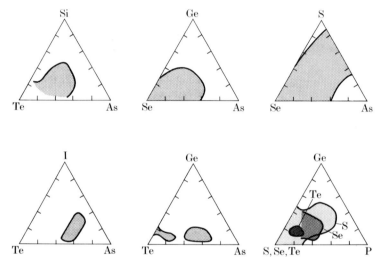

Fig. 6.3. Approximate glass-forming regions in several ternary systems (Hilton and Brau 1963, Haisty and Krebs 1969b, Flaschen et al. 1959, Pearson et al. 1962, Krebs and Fischer 1971, Krebs 1969; see also Hilton 1970).

The stability or otherwise of glasses can frequently be understood from structural considerations. Thus, for example, selenium in the hexagonal crystalline state is composed of helical chains stacked parallel to each other (Chapter 10). The binding in the chains is covalent and strong; between the chains it is weak, perhaps of van der Waals type. In the liquid state the chains can be considered as randomly oriented. Fast cooling of the melt does not allow time for reorientation of these chains before the viscosity becomes too high for this to occur, so the glassy state is formed. Addition of Te to the melt is believed to lead to shorter chains, the Se–Te bond being weaker than the Se–Se bond, and devitrification on cooling in easier. In contrast, As tends to act as a cross-linking or branching additive, retarding the reorientation necessary for crystallization.

Generally speaking, most amorphous solids that can be prepared by cooling from the melt are insulators or wide-gap semiconductors in which the energy gap is greater than about 1 eV. Examples include Se, As_2Se_3, and similar chalcogenide compounds or multicomponent systems, $CdGeAs_2$, and the common borosilicate glasses. Exceptions are 'glassy metals' (§ 5.12), which can be prepared from the melt by rapid cooling.

Materials such as Te, Ge, Si, B and InSb, which cannot be produced in the glassy state by normal quenching from the melt, can be obtained in an amorphous form by deposition. In the case of Ge and Si it has been established that the co-ordination numbers in the liquid lie between 6 and 8 compared with 4 in the solid, and a different co-ordination in the liquid and solid is the normal rule for this class of materials.† The band gaps are generally smaller than for stable glasses, less than about 1 eV; there are, however, exceptions, e.g. dielectric oxide films. For these substances great care is necessary to avoid unwanted crystallization, and frequently the properties of such films are found to be very sensitive to the conditions of deposition and to subsequent annealing treatments. Keeping the substrate temperature low inhibits crystallization, and many materials have to be deposited at liquid-nitrogen temperatures or below in order to obtain an amorphous state.‡

At the present time, classification of amorphous semiconductors is probably best achieved by division into groups of materials having the same short-range structural co-ordination. A general rule, first stressed by Ioffe and Regel (1960), which applies to many amorphous semiconductors, is the preservation of the first co-ordination number of the corresponding crystal.§ For compositions for which there is no crystalline phase the rule is, of course, meaningless, but it does suggest that, in the As–Se system for example, Se-like structural units might be expected to be dominant at low concentrations of As, while at higher concentrations there may be a tendency to favour As_2Se_3-like units; at all compositions the natural co-ordination of each element is likely to be preserved.

This scheme of classification, based on co-ordination number, is shown for some representative materials in Table 6.1 and has been used to some extent to divide the experimental results into those described in Chapters 7, 8, 9, and 10. Notable omissions from Table 6.1 include the common silicate, borate, and phosphate glasses containing metal ions, transition-metal oxide glasses (§ 4.7.2), and many multicomponent glasses containing

† Tellurium, discussed in Chapter 10, is an exception and there are others.
‡ For detailed information on the preparation of thin films the reader should consult the references cited in later chapters on individual materials, and the books by Chopra (1969) and Holland (1963).
§ There are exceptions such as Ge(S, Se, Te) and As_2Te_3.

elements mentioned in Table 6.1 but not in stoichiometric proportions. The freedom to depart from stoichiometric proportions is one of the most important features of semiconducting glasses.

TABLE 6.1

Classification of non-crystalline semiconductors according to their nearest-neighbour co-ordinations

Nearest-neighbour co-ordination	Type (column of periodic table)	Examples
2	VI	Se, Te
3	V	As, Sb, P
3 and 4	IV	C
4	IV	Ge, Si
4	IV–IV	SiC
4	III–V	(Ga, In) (P, As, Sb)
4	III–VI	(Ga, In) (S, Se, Te)
4	III–IV–V	$(Cd, Zn) (Si, Ge, Sn)_x (As, P)_2$
6	III	B
3–2	V–VI	$(As, Sb, Bi)_2 (S, Se, Te)_3$
4–2	IV–VI	SiO_2, (Ge, Sn, Pb)(S, Se, Te)

Another scheme of classification might be to group materials according to whether their amorphous structure is best described by continuous random networks, by assemblies of distorted layers, molecules, or polymeric units, or by close packing of atoms, etc. The problem here is that insufficient structural data exist to classify many materials properly; for instance amorphous As has been modelled both by a continuous random network and by a distorted layer structure; amorphous Se is possibly a mixture of molecules (closed rings) and polymeric chains. Yet another scheme (Fritzsche 1973) would be to isolate those materials whose uppermost valence band is made up of non-bonding (lone-pair) orbitals, e.g. those containing a large amount of a two-fold co-ordinated chalcogenide element S, Se, Te, or O.

6.3. Methods of determining structure

An amorphous solid is one in which three-dimensional periodicity is absent.† The arrangement of atoms, however, will not be entirely random as in a gas. The binding forces between atoms are very similar to those in the crystal and, although *long*-range order is absent, *short*-range order will generally be present.‡

† This definition makes the terms disordered, non-crystalline, amorphous, glassy, and vitreous, synonymous. However, the last two terms are generally restricted to non-crystalline solids prepared by quenching from the liquid state.

‡ What is meant here by short-range order is a reasonably well-defined bond length, bond angle, and co-ordination number. Compared with a crystal, short-range structure is frequently distorted (because of variations in the bond angle for example) and for certain properties, such as charge transport, this can be as significant as the absence of long-range order.

A complete theoretical description of the properties of an amorphous solid would need a full knowledge of the structure. Even within the restraints imposed by forces between individual atoms and the tendency towards short-range order, there is an infinite number of allowed structures for any amorphous material. However, in view of the importance of short-range order in determining many physical properties, it is valuable to determine the nature and extent of this as far as possible. Techniques used for structural investigations include electron, X-ray, and neutron diffraction, EXAFS (extended X-ray absorption fine structure), and infrared and Raman spectroscopy, as well as several less direct methods.

Diffraction patterns from an amorphous solid should consist of broad haloes or rings, vanishing rapidly with increasing angle, without any evidence of spots or sharp rings which would indicate some degree of crystallinity. From the angular dependence of the scattered intensity $I(s)$, one determines a diffraction function $F(s)$ given by

$$F(s) = s\left\{\frac{I(s)}{f^2(s)} - 1\right\}$$

where $s = 4\pi \sin \theta/\lambda$ and $f(s)$ is the atomic scattering factor. The radial distribution function $J(r)$ is defined in terms of the atomic density $\rho(r)$ at distance r from any chosen atom:

$$J(r) = 4\pi r^2 \rho(r) = 4\pi r^2 \rho_0 + rG(r)$$

where ρ_0 is the average atomic density and

$$G(r) = \frac{2}{\pi} \int_0^\infty F(s) \sin sr \, ds.$$

The radial distribution function (RDF) is therefore the Fourier transform of the scattered intensity and gives a one-dimensional description of the atomic distribution. Appropriately normalized, the RDF displays a number of peaks at certain values of r giving the average separation between nearest neighbours, next nearest neighbours, and so on (see Fig. 6.4). The width of a peak, after corrections for thermal broadening, describes the radial fluctuation in the corresponding interatomic distance, and the area below each peak is equal to the number of atoms contained in the respective co-ordination shell. In the case of germanium shown here, the first two co-ordination numbers are 4 and 12 as in the crystal and the average first interatomic distance is, within experimental error (about ± 1 per cent), unchanged. The second-neighbour distance, however, shows a distribution about the crystalline value owing to bond-angle distortions. The curve is distributed about a parabola representing the RDF of a hypothetical amorphous solid of the same density ρ_0 but with matter uniformly distributed in space. The decreasing amplitude of the oscillations with r is a consequence of the lack of long-range order.

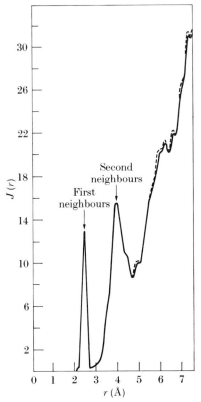

Fig. 6.4. RDF of amorphous germanium. The curves were derived from electron diffraction
measurements on two similar samples. (From Graczyk and Chaudhari 1973*a*.)

There are several problems involved in obtaining a reliable RDF, not the
least of which is concerned with termination errors in the Fourier integral
which can introduce spurious ripples into $J(r)$. There are various pro-
cedures for eliminating these (Grigorovici 1973, 1974) but considerable
care is necessary. In amorphous solids containing more than one type of
atom the situation is more complicated and, for a complete study, partial
interference functions are required. These can sometimes be derived by
using electron diffraction in conjunction with X-ray diffraction, since the
atomic scattering factor f depends on the atomic number Z as $Z^{1/3}$ for the
former and as Z^2 for the latter. However, inelastically scattered electrons
have to be eliminated and other corrections are necessary.

Neutron diffraction, particularly when used on a series of samples with
different isotopic compositions, is a powerful technique for structural studies. It has been used extensively on liquids (Chapter 5) but not, at present,
to any great extent on amorphous solids (see, however, Wright 1974,

Leadbetter *et al.* 1977, Betts *et al.* 1972, Bellisent and Tourard, 1976, 1977). In some cases this is because large samples are not available.

EXAFS is a technique for studying structural arrangements which involves measurement of the energy dependence of the absorption coefficient beyond an X-ray absorption edge. Electrons, photoejected from a deep electronic state (e.g. the K-shell) of a particular atom, are back-scattered by neighbouring atoms, interfere with the outgoing wave, and modulate the absorption coefficient in a way that depends on the phase of the returning wave (see reviews by Stern 1976, Pendry and Gurman 1977, Sayers 1977). Although the data can be Fourier transformed to yield an RDF, analysis is not as straightforward as for X-ray or electron diffraction, owing particularly to the difficulty of treating multiple scattering events. However, the technique has a particular advantage as far as alloys are concerned, because each type of atom has its own absorption edge and associated fine structure. Furthermore, use of synchrotron radiation sources, having over four orders of magnitude more intensity than conventional X-ray tubes, has dramatically reduced the time required to obtain spectra. Several papers on the theory of EXAFS have been published (Stern, Sayers, and Lytle 1975, Stern 1974, Ashley and Doniach 1975, Lee and Pendry 1975, Hayes, Sen, and Hunter 1976, Gurman and Pendry 1976, Lee and Beni 1977). A typical spectrum for crystalline selenium and Fourier transformed data for crystalline and amorphous selenium are shown in Fig. 6.5. The power of the technique to produce accurate information about the first co-ordination shell is seen in these data which enable one to deduce a decrease in the Se–Se distance of 0·024 Å and a reduced amplitude of vibration in the amorphous form. Experiments on GeO_2 (Sayers, Stern, and Lytle 1975), GeSe, $GeSe_2$, and As_2Se_3

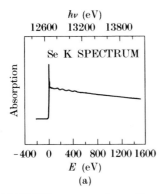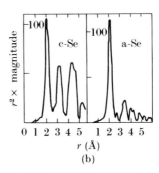

Fig. 6.5. (a) EXAFS spectrum of crystalline selenium. The energy is measured relative to the binding energy of the K-shell electron (12 658 eV). (b) Plots of r^2 times the magnitude of the Fourier transform of EXAFS data for crystalline and amorphous selenium. (From Sayers 1977.)

(Sayers, Lytle, and Stern 1974), and As_2Te_3 (Pettifer, McMillan, and Gurman 1977) have provided information on local co-ordinations around the individual atoms in these alloys in both the crystalline and amorphous forms. The co-ordination number of Cu in amorphous $As_{1-x}Se_x$ alloys has been investigated by Hunter, Bienenstock, and Hayes (1977a,b) and of As in doped amorphous Si by Hayes, Knights, and Mikkelsen (1977).

Infrared and Raman spectra reveal vibrational modes at frequencies governed by atomic masses, force constants, and the geometrical arrangements of atoms. A comparison of results from crystalline and amorphous materials enables recognition of modes associated with local molecular groupings of atoms, e.g. the AsX_3 unit in As chalcogenides (Lucovsky 1974). For tetrahedrally co-ordinated amorphous semiconductors the most obvious feature of the infrared and Raman spectra is the breakdown of selection rules operative in the crystal (see § 6.7.7) leading to structure determined to a large extent by the phonon density of states. This can be calculated for various structural models and compared with experiment.

Other methods that provide information on the structure of amorphous materials include nuclear magnetic resonance (n.m.r.), differential thermal analysis (DTA), and measurement of viscosity. High-resolution electron microscopy is useful in detecting ordered coherently diffracting regions in a material, and in the case of Ge it fuelled a controversy as to whether this material is best described as a continuous random network or as an assembly of small ordered regions (microcrystallites). Defects in the basic structure of an amorphous material, such as crystallization, segregation, or phase separation, and large voids are fairly easily detectable by electron microscopy; however, the presence of small voids (say <100 Å) is generally inferred from density and small-angle X-ray or electron-scattering measurements (see Chapter 7).

Structural models of amorphous solids or liquids can be constructed by hand or with the aid of a computer. The scattering properties of such a model can then be calculated and its RDF derived in order to compare with experiment. Although good replication of the experimental RDF is a necessary requirement of any model, it should be noted that it can rarely be sufficient, as RDFs are relatively insensitive to different topologies (see for example § 7.1.2). In the case of tetrahedral elemental materials only the first and second co-ordinations are well defined by the RDF, and in other materials, especially those containing more than one kind of atom, even this is not the case. Other requirements of a model are that it should have the correct density and be capable of explaining the metastability of the material, its heat of crystallization, and effects of annealing, as well as its behaviour under pressure or uniaxial stress. The co-ordinates of models can be refined to reproduce the correct distribution of nearest-neighbour bond lengths and angles; they can also be energetically relaxed using appropriate force constants for bond stretching and bond bending. These co-ordinates can be used to calculate the vibrational properties and hence the infrared or Raman spectra, or the electronic density of states, all of which can be compared with experiment. The most successful attempts at model building to date have been for four- or three-fold co-ordinated elemental materials (Ge, Si, As) and for SiO_2, with other simple covalently bonded binary systems following closely behind. Later chapters will contain more detailed structural information on specific materials. The reader is also referred to reviews by Grigorovici (1973, 1974), Moss (1974), and Wright and Leadbetter (1976).

6.4. Electrical properties of non-crystalline semiconductors

In this section we discuss the density of electron states, mobility edges, and a variety of electrical measurements on non-crystalline semiconductors. Generally speaking the experimental techniques are the same as or similar to those employed for crystalline materials; however, the models used in this book for interpreting data are somewhat different. We shall summarize points of view put forward in earlier chapters and develop further some of the concepts related to electrical properties.

Most of our discussion will be confined to materials that are microscopically homogeneous. We believe that this is normally the case for glasses cooled from the melt (excepting SiO_2 containing Na_2O in a certain composition range) and for thin films deposited under certain conditions. However, in many materials local variations in density and composition undoubtedly occur and these will produce, amongst other things, spatial variations or so-called 'elastic fluctuations' in the band gap (Fig. 6.6(a)). In addition fluctuations of an 'electrostatic' nature (Fig. 6.6(b)) may also be

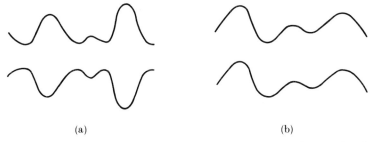

(a)   (b)

Fig. 6.6. Possible forms of spatial fluctuation in the band edges of amorphous semiconductors: (a) elastic; (b) electrostatic.

present if there are spatial variations in charge density. For example many deposited films of Ge and Si are known to contain voids (Chapter 7) and if any of these remain charged (after reconstruction of unsatisfied orbitals) such potential fluctuations will result. If the scale of these fluctuations in space is such that tunnelling between neighbouring wells can be neglected, then, for electrical transport by carriers in conduction and valence bands, a classical percolation treatment, rather than one based on mobility edges, may be more appropriate. The concept of variable-range hopping by electrons with energies at the Fermi energy E_F must remain valid at low temperatures if $N(E_F)$ is finite, and so parts of this chapter may be applied even to such inhomogeneous films. For glow-discharge-deposited Si and Ge (believed to be void free) and for the majority of chalcogenide films and glasses, we think that any long-range potential fluctuations are small in magnitude and indeed most data can be interpreted without any need to consider them.

6.4.1. Density of states

The density of electron states, as we showed in § 2.1, remains a valid concept for non-crystalline as for crystalline materials and its form can be determined by various experimental methods. We suggested in § 2.10 that the states lie in bands separated by energy gaps just as in crystals, and that, as for crystals, states in the gap are a consequence of defects. Furthermore, as long as the short-range order present in the crystalline phase is essentially unchanged (i.e. similar bond lengths, bond angles, and local co-ordination) the gross features of the crystalline density of states are preserved, though there are differences. For the most part, however, it is the similarities between crystals and non-crystals which were, in the past, the most surprising feature so long as the band structure of crystals was thought to depend essentially on long-range order. Recently, however, some new (and some old) techniques for calculating densities of states, that do not rely on lattice periodicity, have been developed which emphasize the importance of short-range structure. These will be discussed in later chapters, where it will be shown that they are capable of reproducing the gross band structures of some non-crystalline materials. It has not yet been possible, however, to calculate, except for certain idealized potentials, (§ 2.12), the form of the density near band edges, which frequently determines transport properties. However, the proportionality between $N(E)$ and $E^{1/2}$ obtained for crystals is not to be expected; for some purposes we shall assume that $N(E) \propto E^n$, where n is to be determined empirically.

The question of 'states in the gap', whether of intrinsic or extrinsic nature, is of considerable importance. In an early paper, which considerably influenced the development of the subject, Cohen, Fritzsche, and Ovshinsky (1969) supposed that the non-crystalline structure would lead to over-lapping band tails of localized states as in Fig. 6.7(a). Those derived from the conduction band would be neutral when empty and those from the

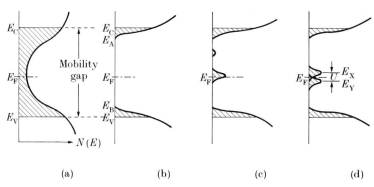

Fig. 6.7. Various forms proposed for the density of states in amorphous semiconductors. Localized states are shown shaded. (a) Overlapping conduction and valence band tails as proposed by Cohen *et al.* (1969), the CFO model; (b) a real gap in the density of states, suggested here as being appropriate for a continuous random network without defects; (c) the same as (b) but with a partially compensated band of defect levels; (d) the same as (b) but with overlapping bands of donor (E_Y) and acceptor (E_X) levels arising from the same defect. (The model for chalcogenides, which involves two-electron energy levels, is not shown in this diagram.)

valence band neutral when full. In the overlap region they would be charged, leading to centres with unpaired spins. Such overlapping states would pin the Fermi energy. The other principal feature of this model was the existence of 'mobility edges' at energies in the band tails. These are identified with the critical energies separating localized from extended states introduced earlier (Mott 1966), so that the model is sometimes called the Mott–CFO model. The difference between the energies of the mobility edges in the valence and conduction bands is called the 'mobility gap'. Although there is considerable evidence for the concept of mobility edges (see Chapter 4, § 6.4.2, and following sections), the proposal of overlapping tails is now considered unlikely to apply to amorphous semiconductors and insulators that are transparent in the visible or infrared. We believe that an ideal amorphous semiconductor in which all bonds are saturated and in which there are no long-range fluctuations should have a density of states as in Fig. 6.7(b) with a true band gap. *Deep* tails should arise only from *gross* density or bond-angle fluctuations. However, strongly overlapping tails as envisaged by Cohen *et al.* (1969) probably exist in some liquids such as expanded fluid mercury which undergo a metal–insulator transition on changing the volume or temperature (§ 5.15.1).

Real non-crystalline materials as we have seen, however, are thought to contain imperfections, such as impurities, or dangling bonds at point defects or microvoids as outlined in § 2.10, and these, just as in crystals, may lead to levels within the band gap. In evaporated films of Ge and Si and some of their alloys, the conductivity, particularly at low temperatures, is due to hopping conduction between such defect states by electrons with energies near the Fermi level. The density of such states depends on the conditions of deposition. In chalcogenides, however, though defect states can be inferred from many types of measurement, they do not normally carry the d.c. current at temperatures where this is measurable. In these materials transport occurs in the valence or conduction band.

None the less, for chalcogenides and for many other amorphous semi-conductors, there is a particularly interesting feature which may be said to be their most surprising and indeed characteristic property. This is that the Fermi level is located near mid-gap and appears to be pinned† there over a wide temperature range. This is not a universal property; indeed Le Comber and Spear (1976) and Spear and Le Comber (1976), by glow-discharge decomposition of SiH_4 with PH_3 or BH_3, have prepared speci-mens that are heavily doped, the Fermi energy being shifted towards and even near to the conduction or valence band (see Chapter 7). In glasses and perhaps evaporated or sputtered films of silicon and germanium, impurities do not normally have this effect, apparently because of the strong tendency to saturate bonds (§ 2.10). None the less amorphous semiconductors are not normally 'intrinsic', in the sense that valence and conduction bands control the Fermi energy. The evidence is summarized for instance by Fritzsche (1973, 1974); the continuation of a straight plot of log σ *versus* $1/T$ down to low temperatures, in addition to thermopower and field-effect measurements, all point to a Fermi energy pinned in some way near mid-gap. For chalcogenides the evidence is discussed further in Chapter 9.

Alternative suggestions to the CFO model for states in the gap are shown in Fig. 6.7(c) and (d). In Fig. 6.7(c) a band of deep acceptors is partially occupied by electrons originating from a weaker band of donors. The role of donors and acceptors can of course be reversed. This simple model, proposed by Davis and Mott (1970), was based on several experi-mental results which implied a finite density of states at E_F. As long as the total density of states in the gap is not large, the model allowed optical transparency without the need for assumptions about the magnitude of the matrix elements; however, no explanation was offered as to why the controlling states should lie near mid-gap. Mott (1972*b*) suggested that if the states arose from a defect centre, e.g. a dangling bond, then they could act both as deep donors (E_Y) *and* acceptors (E_X), single and double occupancy conditions leading to two bands separated by an appropriate correlation energy or Hubbard U. This is illustrated in Fig. 6.7(d).

Even without additional compensating centres, on this model E_F should lie between the two bands if they do not overlap or be pinned within them if they do. If they overlap strongly, as would be expected if the density of centres is high, then the model becomes essentially indistinguishable from

† For a definition of pinning see § 2.9.

that illustrated in (c); if they are well separated, then the model approaches that suggested (albeit for different reasons) by Marshall and Owen (1971) and discussed also by Roberts (1973).

Since the above models were proposed, many experimental data have emerged providing a clearer picture of the density distribution of states in the gaps of amorphous semiconductors. In particular we mention here the field-effect results of Spear (1974b) on amorphous Si prepared by glow-discharge decomposition of SiH_4 (see Chapter 7). For this material it appears that the Fermi level is located near a minimum between two maxima in the density of states. Spear proposed that the centres responsible may arise from *pairs* of dangling bonds at defects similar in nature to the divacancy in the crystal (Watkins and Corbett 1965). The lower (E_Y) and upper (E_X) levels associated with this defect correspond to bonding and anti-bonding states and are thus separated by more than the Hubbard U. The model is discussed further in Chapter 7. In evaporated Si and Ge the density of states is too high to allow an analysis of its distribution by field effect. Studies of hopping conduction (Chapter 7) as a function of temperature (Pollak *et al.* 1973, Knotek 1974, 1975b) have been interpreted as implying a *maximum* in the density of states at E_F; a similar deduction can be made in Ge-Fe alloys. Rather different conclusions have been reached by Beyer and Stuke (1975a) from conductivity and thermopower measurements as a function of annealing (Chapter 7). The situation in evaporated Si and Ge is complicated by the presence of voids as well as point defects such as divacancies. By analogy with dangling bonds on cleaved crystalline surfaces (Kaplan *et al.* 1975, Lemke and Haneman 1975), we may expect a fairly complete pairing of such bonds on the internal surfaces of voids, but the resulting states may lie at energies different from those associated with isolated divacancies. However, in amorphous materials, voids with an odd number of dangling bonds are possible, so one spin may remain unpaired. In addition, as mentioned before, if any voids remain charged, long-range fluctuations of potential are expected.

A model for the location of the Fermi level near mid-gap in amorphous semiconductors, based on such potential fluctuations, has been suggested by Fritzsche (1971) and other authors who propose fluctuations of an 'electrostatic' nature of the form illustrated in Fig. 6.6(b) but of such an extent that minima in the conduction band edge lie below maxima in the valence band. The mathematical consequences of potential fluctuations have been little explored, but one prediction is that for conduction in the bands the activation energy for electrical conductivity should be greater than that for the thermopower (§ 2.9), and another is that, since electrons and holes move in different percolation channels, band-to-band photoluminescence is not expected.

If we adopt for our model of gap states in amorphous Ge and Si one in which two bands of localized levels arising from defects overlap, giving a finite density of states at the Fermi level, an e.s.r. signal and Curie-type paramagnetism are expected, as well as variable-range hopping as the predominant mechanism of current transport at low temperatures. The same would follow from the model of Fig. 6.7(c). These phenomena are indeed observed in evaporated amorphous Si and Ge (Chapter 7) and e.s.r. in glow-discharge-deposited Si if the temperature of deposition is not too high (Stuke 1977). For *pure* annealed chalcogenide glasses or films, however, none of these phenomena are normally observed† (unless one first illuminates at low temperatures, see § 9.6). However, there is plenty of evidence, to be reviewed later, for gap states and for pinning of the Fermi level near mid-gap in chalcogenides. This could in principle be explained by a similar model to that proposed for Ge and Si but in which the two defect bands do not overlap to any observable extent. Such a model fails, however, to explain much other evidence, namely luminescence at a photon energy of approximately half the gap, a Pollak–Geballe type of a.c. conductivity (§ 2.19), and the observed temperature dependence of the field effect. This evidence will be presented later; sufficient to say here that a rather different model is required for gap states in chalcogenides (and probably amorphous As and similar low-co-ordination materials as well). We describe here a model due to Street and Mott (1975) and Mott, Davis, and Street (1975), already mentioned briefly in §§ 2.10 and 3.5.

The first attempt to reconcile the absence of e.s.r. and paramagnetism with an essentially pinned Fermi energy was due to Anderson (1975, 1976); he supposed that the number of orbitals or bonding situations available to the electrons in a glass is greater than the number of electron pairs, and that all orbitals are doubly occupied or empty. This is because a doubly occupied orbital represents a bond, which contracts, and the energy given up is then assumed to be more than sufficient to compensate for the (Hubbard) repulsion between the two electrons. Anderson showed that this 'negative Hubbard U' model fixes the Fermi energy without introducing single-occupied paramagnetic centres. A variant of this concept for chalcogenides, which we believe is able to explain more experimental data,‡ is that the extra orbitals are at specific defects, namely dangling bonds which would be neutral if occupied by one electron. It is supposed that the unoccupied dangling bond forms a covalent bond with the lone-pair orbital on a neighbouring chalcogen atom and so has a large binding energy (like a V_K centre, see § 3.3). If we denote by D^+, D^0, and D^- the dangling bonds

---

† See, however, results on sputtered As_2Te_3 due to Hauser and Hutton (1976) discussed in § 9.4.

‡ For a comparison of Anderson's model with that presented here see Mott (1977d).

with zero, one and two electrons, the reaction

$$2D^0 \rightarrow D^+ + D^-$$

is assumed to be exothermic. Thus all dangling bonds are positively or negatively charged; the Fermi energy is determined by the energies of the states D^+ and D^- and there are no free spins. If so, any broadening owing to disorder must be small so that the bands of localized states do not overlap. Charge transport by carriers in these gap states involves simultaneous hopping by two electrons carrying with them the local distortion or polarization cloud (Chapter 3) and is not normally observed, except perhaps in a.c. conduction (§ 6.4.5).

With this model the 'Franck–Condon' density of states may vanish in the gap so that the material is transparent; we may write $N(E_F) = 0$. In contrast, the 'true' density of states which we shall call $N_0(E_F)$ may be quite large.

Further discussion of this model and of its consequences are discussed later in this chapter and in Chapter 9.

6.4.2. The mobility edge

The concept of a mobility edge separating localized from extended states has been introduced in Chapter 2 and is a feature of all the models illustrated in Figure 6.7. It is defined as the energy separating states that are localized from those that are extended. We believe that an energy of this kind always exists for conduction and valence bands in non-crystalline systems, though for some, such as the conduction band in liquid argon (§ 5.12) and in SiO_2 (§ 9.15), it is very near the band edge. For systems known as 'Fermi glasses' in which the Fermi energy of a degenerate electron gas can be moved through this critical energy, very clear evidence for a mobility edge has been obtained (Chapter 4). For amorphous semiconductors, because of the pinning of E_F discussed in the last section, it has not been possible to explore mobility edges directly in this way, although in *doped* glow-discharge-deposited Si one must be close to achieving this. There is, however, much evidence for their existence based on electrical measurements, particularly conductivity, thermopower, Hall effect, and drift mobility, and it will be reviewed in this chapter. As to the nature of the disorder which produces mobility edges we shall suppose that it is *intrinsic*, i.e. arising from the *presence of short-range disorder* rather than from random fields due to point defects. For elements such as Si local variations in bond angle are likely to be the principal source, but in some materials a spread in bond lengths or alternatively a variation in density or composition *over distances of a few bond lengths* may contribute. Estimates of the Anderson disorder energy V_0 (§ 2.3) from such sources are necessarily crude, but perhaps a value of order 1 eV is not unreasonable, leading to a range of localized states at band edges of $\sim 0 \cdot 1 – 0 \cdot 3$ eV in many amorphous materials (§ 2.10), and less in some others.

In this section we consider the nature of the conduction processes both above and below a mobility edge under non-degenerate conditions, i.e. when E_F is many multiples of kT below E_A (Fig. 6.7(b)) and the carrier concentrations at E_A and E_C are determined by Boltzmann statistics. The carriers may be thermally or optically excited across the gap or else injected from an electrode; in the latter two cases the equilibrium Fermi level is to be replaced by a pseudo-Fermi level.

(i) *Hopping conduction between E_C and E_A (variable-range hopping in a tail of localized states).* We assume for illustration (Fig. 6.8) that the form

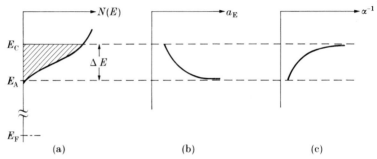

Fig. 6.8. (a) The density of states $N(E)$ at the edge of a conduction band with a range ΔE of localized states; (b) variation with energy of the average separation a_E between localized states; (c) variation with energy of the decay length α^{-1} of the localized states. At the mobility edge, E_C, $\alpha^{-1} \to \infty$.

of the density of states near the bottom of a conduction band is of the form $N(E) = C(E - E_A)^n$. One can define a separation a_E in space between localized states (§ 2.6) by

$$a_E^3 = 1/N(E)V_0$$

where V_0 is a parameter characterizing the disorder. Averaged over the whole conduction band (if all the states were localized), $\langle a_E \rangle$ could equal the average distance a between atoms; in the tail $a_E > a$. The spatial extent of the localized states will also vary with energy. At E_C, $\alpha^{-1} \to \infty$ and the states become extended; below E_C and close to E_C, α^{-1} falls as $(E_C - E)^{2/3}$ (§ 2.9). Deeper in the tail we assume that α^{-1} can be regarded as constant, although this is clearly an approximation. Taking particular forms for the energy dependence of a_E and α^{-1}, Weiser et al. (1974) determined at what energy the current would be carried, assuming that hopping occurred between nearest-neighbour states. We believe, however, that fixed-range nearest-neighbour hopping occurs only under conditions where kT is comparable with the total energy range of localized states. If $kT \sim \Delta E$, then extended-state conduction E_C would certainly dominate. Instead, following Grant and Davis (1974), we consider that hopping is of variable-range type, and proceed to determine around what energy this occurs and to calculate the temperature dependence in the hopping range, the hopping energy, and the mobility. The analysis is as follows.

First we determine the energy E_m at which the number of carriers is a maximum. For the form of the density of states chosen this is easily shown to be nkT above E_A. The density of states at E_m is

$$N(E_m) = C(nkT)^n$$
$$= N(E_C)\left(\frac{nkT}{\Delta E}\right)^n$$

where $N(E_C)$ is the density of states at the mobility edge. We now write the hopping probability around E_m (under the self-consistent assumption that the energy range of states involved is small) as

$$\nu = \nu_{ph} \exp\left(-2\alpha r - \frac{w}{kT}\right) \tag{6.1}$$

where r and w are the variable (with T) average hopping length and energy and ν_{ph} is a quantity related to the phonon frequency to be discussed below.

The hopping energy w is now written as $\frac{3}{4}\pi r^3 N(E_m)$, following the procedure of § 2.7 for Fermi-level hopping in a constant density of states. With $N(E_m)$ given above, the hopping probability can be maximized to yield the following expressions for r and w;

$$r = \frac{9(\Delta E)^n}{8\pi\alpha N(E_C)n^n(kT)^{n+1}}$$
$$w = \frac{3}{4}\left(\frac{8}{9}\right)^{3/4}\left\{\frac{(kT)^{3-n}(\Delta E)^n\alpha^3}{\pi n^n N(E_C)}\right\}^{1/4} . \tag{6.2}$$

The hopping probability will then vary with T as

$$\nu = \nu_{ph} \exp(-BT^{-(n+1)/4})$$

where

$$B = 3\left(\frac{8}{9}\right)^{3/4}\left\{\frac{(\Delta E)^n\alpha^3}{\pi N(E_C)n^n k^{n+1}}\right\}^{1/4} .$$

We estimate the average hopping length and energy using some reasonable parameters. For a linear density of tail states ($n = 1$), $E = 0\cdot3$ eV, $\alpha^{-1} = 10$ Å, and $N(E_C) = 2 \times 10^{21}$ cm^3 eV^{-1}, the values at $T = 300$ K are

$$r = 30\cdot5 \text{ Å} \qquad w = 0\cdot05 \text{ eV}.$$

The calculated values, particularly that of r, are rather insensitive to the parameters chosen, because of the exponent $\frac{1}{4}$. We note that r is considerably larger than a_E (~ 18 Å at kT above E_A, taking $V_0 = 1$ eV) and that w is $\sim 2kT$.

We now relate the hopping probability to the mobility and finally to the conductivity. As in § 2.7 we write

$$\mu_{hop} = eD/kT \quad \text{and} \quad D = \tfrac{1}{6}\nu r^2. \tag{6.3}$$

Thus

$$\mu_{hop} = \tfrac{1}{6}e\nu r^2/kT \tag{6.4}$$

where r is given by eqn (6.4). The mobility should then have a temperature dependence of the form

$$\mu_{hop} = C\nu_{ph}T^{-(n+3)/2}\exp(-BT^{-(n+1)/4}). \tag{6.5}$$

The temperature dependence of the pre-exponential term is not complete, however, because there may be a temperature dependence in ν_{ph}. If ν_{ph} was to be given by a formula of the Miller–Abraham type then this would contain a factor $r^2 w$, i.e. a temperature dependence of $T^{-(3n-1)/4}$. As was discussed in § 3.5, however, the Miller–Abrahams formula for ν_{ph} is unlikely to be appropriate for this situation, i.e. when the radius of the localized state is larger than the appropriate phonon wavelength.[†]

The uncertainty in the magnitude and temperature dependence of the pre-exponent (Chapter 3), as well as certain percolation aspects to the current paths, makes it difficult to estimate the magnitude of the mobility expected for hopping conduction in a band tail. We may, however, deduce that, even ignoring any temperature dependence of the pre-exponent, the temperature dependence of the mobility does not exhibit a simple activation energy but has, on an Arrhenius plot, a slope which decreases with falling temperature, such that at a given value of T the tangential slope is equal to $n + 1$ times the true hopping energy (Grant and Davis 1974). An analysis of band-edge hopping conduction by Butcher (1976a,b), based on percolation theory, results in a temperature dependence for μ_{hop} of a form similar to eqn (6.5) but with the power of T raised to $(n + 1)/(n + 4)$ instead of $(n + 1)/4$ as derived here.

The conductivity associated with band-tail hopping may be written as $ne\mu_{hop}$, where n is the number of electrons in the tail, given by

$$n = \int_{E_A}^{\infty} N(E)\exp\left(-\frac{E - E_F}{kT}\right)dE \tag{6.6}$$

[†] Emin (1974) has argued that for hopping energies comparable with or greater than the maximum phonon energy, ν_{ph} should be negligibly small. We discuss this in § 3.5.1, and find that a reasonable estimate for ν_{ph} may be $\omega/(\alpha a)^3$, where a is the distance between atoms.

which yields a term containing $\exp\{-(E_A - E_F)/kT\}$. This activated temperature dependence will normally dominate over that occurring in the mobility.

(ii) *Conduction in extended states above* E_C. As the temperature is raised, the conduction process described in (i) will give maximum current at an increasing value of the energy in the band tail. However, the current will never be carried in localized states close to E_C if, as we assume, there is a sharp jump in the mobility there. Rather, at some critical temperature, or more accurately over some temperature range, there will be a transition to conduction by carriers in extended states.

Here the mobility is given by

$$\mu_{\text{ext}} = eD/kT \qquad D = \tfrac{1}{6}\nu_{\text{el}}a^2 \qquad (6.7)$$

where ν_{el} is an electronic frequency $(\sim\hbar/ma^2)$ and a is now the distance in which phase coherence is lost. Thus at $T = 300$ K

$$\mu_{\text{ext}} \sim \frac{1}{6}\left(\frac{e}{kT}\right)\frac{\hbar}{m} \sim 6 \text{ cm}^2 \text{ V}^{-1}\text{ s}^{-1}. \qquad (6.8)$$

Alternatively one can write

$$\mu_{\text{ext}} = \frac{\sigma_{\min}}{eN(E_C)kT} \qquad (6.9)$$

where $\sigma_{\min}$ is the minimum metallic conductivity (§ 2.6) given by

$$\sigma_{\min} = 0\cdot026\, e^2/\hbar a \qquad (6.10)$$

for a co-ordination number of 6 and for vertical disorder. For lateral disorder the value is perhaps three times greater (§ 2.6). Eqn (6.10) then yields $\sigma_{\min} \sim 150\ \Omega^{-1}\text{ cm}^{-1}$ for a value of $a = 4$ Å. In general $\sigma_{\min}$ may perhaps lie between 100 and 600 $\Omega^{-1}\text{ cm}^{-1}$ in most materials, depending on the value of a. For $\sigma_{\min} = 150\ \Omega^{-1}\text{ cm}^{-1}$ and $N(E_C) = 2 \times 10^{21}\text{ cm}^{-3}\text{ eV}^{-1}$, eqn (6.9) yields $\mu_{\text{ext}} \sim 20\text{ cm}^2\text{ V}^{-1}\text{ s}^{-1}$ at $T = 300$ K.

Consideration of the thermopower, the Hall effect, and the drift mobility for charge transport above and below E_C will be made in later sections and summarized in § 6.6.

6.4.3. *Temperature dependence of the d.c. conductivity*

With the models described above for the density of states and mobility edges in an amorphous semiconductor, there are three mechanisms of conduction which we may expect to find in appropriate ranges of temperature.

(a) Transport by carriers excited beyond the mobility edges into non-localized (extended) states at E_C or E_V. The conductivity is (for electrons)

$$\sigma = \sigma_{min} \exp\left(-\frac{E_C - E_F}{kT}\right)$$

where σ_{min} has been discussed in Chapter 2 and the previous section. A plot of $\ln \sigma$ versus $1/T$ will yield a straight line if $E_C - E_F$ is a linear function of T over the temperature range measured.† If so, we write

$$E_C - E_F = E(0) - \gamma T \tag{6.11}$$

and the slope of such a plot will be $E(0)/k$, and the intercept on the σ axis will be $\sigma_{min} \exp(\gamma/k)$.

(b) Transport by carriers excited into localized states at the band edges and hopping at energies close to E_A or E_B. For this process, assuming again conduction by electrons,

$$\sigma = \sigma_1 \exp\left\{-\frac{E_A - E_F + w_1}{kT}\right\} \tag{6.12}$$

where w_1 is the activation energy for hopping. As discussed in the previous section, w_1 should decrease with decreasing temperature on account of the variable-range nature of the hopping transport. However, as the principal temperature dependence is through the carrier activation term, an approximately linear dependence of $\ln \sigma$ versus $1/T$ is again expected.

An estimate of σ_1 is not easy to make but it is expected to be several decades smaller than σ_{min}, partly because of a lower effective density of states near E_A compared with E_C and also because of a lower mobility as discussed in § 6.4.2. In addition the experimental slope and intercept of $\ln \sigma$ versus $1/T$ at $1/T = 0$ will be affected by any temperature dependence of $E_A - E_F$ analogous to that affecting the slope and intercept associated with process (a).

(c) If the density of states at E_F is finite, then there will be a contribution from carriers with energies near E_F which can hop between localized states by the process analogous to impurity conduction in heavily doped crystalline semiconductors. We may write for this contribution

$$\sigma = \sigma_2 \exp(-w_2/kT) \tag{6.13}$$

where $\sigma_2 \lesssim \sigma_1$ and w_2 is the hopping energy, of the order of half the width of the band of states if the form of the density of states is as shown in Fig. 6.7(c). At temperatures such that kT is less than the bandwidth, or if $N(E)$

---

† Even if E_F is pinned with respect to the band edges, a temperature dependence of the activation energy is expected because of the temperature variation of the band gap.

is as shown in Fig. 6.7(a) or (d), hopping will not be between nearest neighbours and variable-range hopping of the form

$$\sigma = \sigma_2' \exp(-B/T^{1/4}) \qquad (6.14)$$

with $B = 2\{\alpha^3/kN(E_F)\}^{1/4}$ is to be expected, at a temperature sufficiently low for $N(E_F)$ to be considered constant over an energy range $\sim kT$. This result was derived in § 2.7, where it was stated that different analyses yield rather different values of the numerical constant B, varying from 1·70 to 2·05 (see also Pike and Seager 1974). The pre-exponential terms σ_2 and σ_2' are, as for σ_1, not easy to evaluate mainly because of uncertainties in the term ν_{ph} (see Chapter 3).

The total conductivity for all processes is obtained as an integral over all available energy states. Thus for states above E_F

$$\sigma = \int \sigma(E)\, dE$$

where (§ 2.2)

$$\sigma(E) = eN(E)\mu(E)kT\frac{df(E)}{dE}$$

and $f(E)$ is the Fermi–Dirac function. Fig. 6.9 shows schematically the variation of $N(E)$, $\mu(E)$, $f(E)$ and $df(E)/dE$ with E for states above E_F,

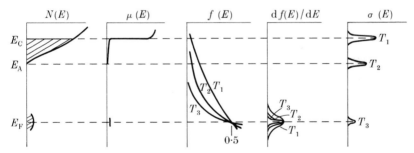

Fig. 6.9. Illustration of the effect of temperature on the mode of conduction: $T_1 > T_2 > T_3$.

and the manner in which $\sigma(E)$ may vary with temperature. This is also illustrated on a plot of $\ln \sigma$ versus $1/T$ in Fig. 6.10. If the density of defect states at E_F is high, then process (b) may not be dominant in any temperature range and a direct transition from (a) to (c) will result.

Davis and Mott (1970), expressing the conductivity in the high-temperature range for amorphous semiconductors investigated at that time in the form $C \exp(-E/kT)$, plotted C against E; they found values of C clustering round $10^3\ \Omega^{-1}\ cm^{-1}$ for most materials, which could well correspond to the quantity $\sigma_{min} \exp(\gamma/k)$. The value of the pre-exponential

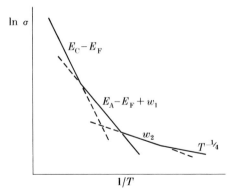

Fig. 6.10. Illustration of the temperature dependence of conductivity expected on the model of Fig. 6.9. The activation energies associated with various processes described in the text are indicated.

factor has sometimes been used as a test for conduction at a mobility edge, but in view of uncertainties in the term $\exp(\gamma/k)$, we do not feel this to be very reliable. In § 6.6 it will be shown how combined conductivity, thermopower, drift mobility, and Hall-effect measurements as a function of temperature may lead to a fairly unambiguous interpretation.

6.4.4. Behaviour in the liquid state

In this section we discuss the behaviour of the conductivity of some non-crystalline semiconductors in the liquid state. Two classes of materials can be distinguished as follows.

(i) Those that cannot be prepared by quenching from the melt, e.g. Ge, Si, and III–V compounds. For these the liquid state is metallic, the conductivity exhibiting a sudden increase on melting after prior crystallization. The co-ordination number in the liquid state is different from that in the solid. Fig. 6.11 (from Ioffe and Regel 1960) shows the conductivity and density in a few such materials of this type.

(ii) For amorphous semiconductors that can be obtained by quenching from the melt there are two types: stable glasses that do not crystallize even when heated very slowly, and glasses that do not crystallize on fast heating but do when heated slowly. There is no sharp distinction between the two, but the large variation in crystallization times found in practice makes the distinction a useful one. For both types, however, the retention of semiconducting properties in the liquid state seems general. Fig. 6.12 (from Male 1970) gives a few examples. The slope of the plot of $\ln \sigma$ *versus* $1/T$ is frequently higher in the liquid than in the solid. As the slope gives the activation energy for conduction extrapolated to $T = 0$, the indication here is that the gap decreases with T faster in the liquid state and the decrease is

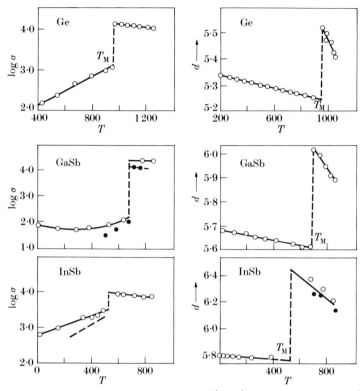

Fig. 6.11. Abrupt changes in conductivity σ (in $\Omega^{-1}\,cm^{-1}$) and density d (in $g\,cm^{-3}$) in some non-glass-forming materials on melting. (From Ioffe and Regel 1960.)

not linear; *the gap is actually smaller in the liquid* and may close to zero at a sufficiently high temperature. Some indication of this is seen in the curve for As_2Te_3, which turns over at a value of $\sigma \sim 10^3\,\Omega^{-1}\,cm^{-1}$. Other examples of this gradual change towards metallic properties are seen in similar curves for liquid alloys of Te–Se (Chapter 10). It is likely that in all cases there is a gradual increase in co-ordination number as the liquid is heated, as is described for various materials in Chapter 5. Numerous examples of this kind of transition to a metallic state for ternary glasses have been given by Haisty and Krebs (1969a,b).

6.4.5. a.c. conductivity

As we have seen in § 6.4.3, there are three mechanisms of charge transport that can contribute to a direct current in amorphous semiconductors. They can all contribute to the a.c. conductivity as follows.

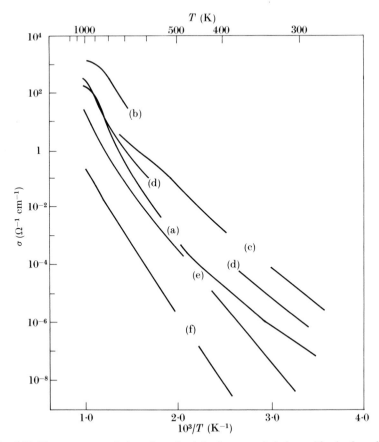

Fig. 6.12. Temperature variation of conductivity for several chalcogenides in the solid and liquid states: (a) $As_{30}Te_{48}Si_{12}Ge_{10}$; (b) As_2Te_3; (c) $As_2S_3Tl_2Te$; (d) As_2SeTe_2; (e) As_2Se_2Te; (f) As_2Se_3. (From Male 1970.)

(a) Transport by carriers excited to the extended states near E_C or E_V. For these we might expect that $\sigma(\omega)$ would be given by a formula of the Drude type,

$$\sigma(\omega) = \frac{\sigma(0)}{1 + \omega^2\tau^2} \qquad (6.15)$$

The time of relaxation τ will, however, be very short ($\sim 10^{-15}$ s) and a decrease in $\sigma(\omega)$ as ω^{-2} (i.e. free carrier intraband absorption) is not expected until a frequency $\sim 10^{15}$ Hz is reached. This corresponds to an energy quantum lying above the fundamental optical absorption edge in materials of interest here. In any case, as we have seen in § 2.10, the Drude formula is hardly applicable for such small values of τ. It is, however, in excellent agreement with experiment for most liquid metals, but not for liquid Te (§ 5.15). Even when τ is large, deviations from the Drude formula are expected if the density of states varies with energy over a range $\hbar/\tau$. In § 6.7.4 we calculate the contribution to $\sigma(\omega)$ due to free carriers that is expected in amorphous semiconductors. It is sufficient to state here that in the electrical range of frequencies (up to, say, 10^7 Hz) no frequency dependence of the conductivity associated with carriers in extended states is expected.

(b) Transport by carriers excited into the localized states at the edges of the valence or conduction band. No complete theoretical treatments of $\sigma(\omega)$ for hopping under conditions of non-degenerate statistics are known, but we might expect a similar dependence on frequency to that derived under degenerate conditions (see (c) below), and thus as $\omega\{\ln(\nu_{ph}/\omega)\}^4$. This varies approximately as ω^s, where $s < 1$ when $\omega < \nu_{ph}$. In order to estimate the frequency at which such an increase is expected, a comparison with the magnitude of the d.c. hopping conduction (§ 6.4.2) would be required. The temperature dependence of this component of the a.c. conductivity should be the same as that of the carrier concentration at the band edge, so that for the conduction band it should increase as $\exp\{-(E_A - E_F)/kT\}$.

(c) Hopping transport by electrons with energies near the Fermi level, provided $N(E_F)$ is finite. There have been several theoretical treatments of $\sigma(\omega)$ for this mode of conduction (see the review by Pollak 1976). $\sigma(\omega)$ should increase with frequency in a manner similar to that for process (b). However, the exponential dependence on the temperature will be absent, and $\sigma(\omega)$ should be proportional to T if kT is small compared with the energy range over which $N(E_F)$ may be taken as constant, and independent of T if kT is larger than the width of some well-defined defect band in which E_F lies. An analysis of observed results has been given by Davis and Mott (1970) using the formula given by Austin and Mott (1969) and reviewed in § 2.15, namely

$$\sigma(\omega) = \tfrac{1}{3}\pi e^2 kT\{N(E_F)\}^2 \alpha^{-5}\omega\left\{\ln\left(\frac{\nu_{ph}}{\omega}\right)\right\}^4. \tag{6.16}$$

Assumptions involved in this formula have been discussed by Pollak (1971a,b, 1976) and by Butcher (1974a,b): the main ones are that hopping is assumed to be between independent *pairs* of centres, i.e. multiple hopping can be neglected, and also that there is no correlation between the hop energy and the hop distance. In Pollak's formula our factor $\pi/3$ is replaced by $\pi^3/96$. Butcher and Hayden (1977a) obtained a similar formula again with a slightly different numerical factor $3\cdot 66\pi^2/6$.

The frequency dependence predicted by eqn (6.16) can be written as $\sigma(\omega) \propto \omega^s$, where s is a weak function of frequency if $\omega \ll \nu_{ph}$. A plot of

$\ln \sigma(\omega)$ *versus* $\ln \omega$ is therefore approximately linear with slope s given by

$$s = \frac{d[\ln\{\omega \, \ln^4(\nu_{ph}/\omega)\}]}{d(\ln \omega)}$$

$$= 1 - \frac{4}{\ln(\nu_{ph}/\omega)}.$$

A plot of $\omega \, \ln^4(\nu_{ph}/\omega)$ *versus* ω for various values of ν_{ph} is shown in Fig. 6.13(a) and of s *versus* ν_{ph}/ω (on a heavily contracted scale) in Fig. 6.13(b).

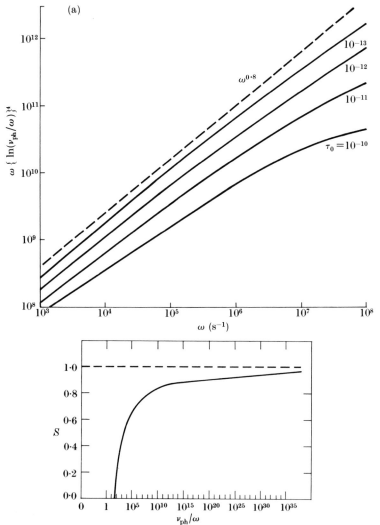

Fig. 6.13. (a) Plot of $\omega \, \ln^4(\nu_{ph}/\omega)$ against ω for various values of $\nu_{ph} = \tau_0^{-1}$. (b) Plot of $s = 1 - \{4/\ln(\nu_{ph}/\omega)\}$ against ν_{ph}/ω.

At a particular frequency, for instance $\omega = 10^4\,\text{s}^{-1}$, s varies from 0·4 to 0·8 for ν_{ph} in the range 10^7–10^{13} Hz and values of s outside this range are therefore unlikely.

Using a more general analysis of Debye-type hopping conduction, which can be applied to both variable-range hopping and random potential barrier hopping as well as hybrid cases, Butcher and Morys (1973, 1974) have derived formulae for both the real (σ_1) and imaginary (σ_2) parts of the conductivity, the leading terms of which are

$$\sigma_1(\omega)/\sigma_1(\infty) = 1{\cdot}84\ \omega\tau_0\ \log_{10}^4(\omega\tau_0)^{-1}$$

$$\sigma_2(\omega)/\sigma_1(\infty) = -0{\cdot}359\ \omega\tau_0\ \log_{10}^4(\omega\tau_0)^{-1}.$$

(6.17)

Here $\tau_0 = 1/\nu_{\text{ph}}$ and the expressions have been normalized by dividing by $\sigma_1(\infty)$. Fig. 6.14 (dotted curves) shows these results compared with those

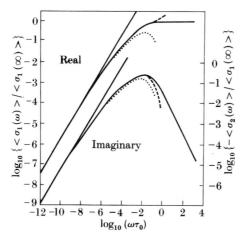

Fig. 6.14. Comparison of exact results (solid curves) with asymptotic (dotted curves) and complete (broken curves) expansions for variable-range hopping. The straight lines have slope 0·82. (From Butcher and Morys 1974.)

obtained from fuller expressions (broken curves) and also exact numerical results (solid curves) (Butcher and Morys 1973). The ratio $\sigma_2(\omega)/\sigma_1(\omega)$ obtained from eqns (6.17) is

$$\sigma_2(\omega)/\sigma_1(\omega) = 0{\cdot}293\ \log_{10}(\omega\tau_0)^{-1}.$$

The corresponding ratio for random potential barrier hopping is five times higher, offering the possibility of distinguishing between the two processes experimentally (see also below).

An experimental determination of ν_{ph} would be of interest. A low value is expected if polarization is large at the localized sites and furthermore it should vary with T when $T > \Theta/2$ (§ 3.6). According to the pair-hopping mechanism, a low ν_{ph} should manifest itself by a low value of s (see Fig. 6.13(b)).

Formulae based on the assumption of pair hopping yield $\sigma(\omega) = 0$ at $\omega = 0$. Other treatments of a.c. hopping conduction include the d.c. limit (see Pollak 1976 for references). It will be sufficient here to note that the d.c. conductivity due to hopping is finite (if $T \neq 0$, process (c) above) and a smooth transition to a frequency-independent conductivity is expected at sufficiently low frequencies. A condition for $\sigma(\omega)$ to exceed $\sigma(0)$ is that the a.c. hopping length should be less than the d.c. hopping length. For variable-range hopping, the latter is given by eqn (2.62), namely

$$r_0 = \tfrac{3}{4} \{3/2\pi\alpha N(E_F)kT\}^{1/4} \tag{6.18}$$

and the a.c. hopping length by eqn (2.96)

$$r_\omega = \frac{1}{2\alpha} \ln\left(\frac{\nu_{ph}}{\omega}\right). \tag{6.19}$$

These two lengths are shown in Fig. 6.15 as a function of $N(E_F)$, T, and ν_{ph}/ω, taking $\alpha^{-1} = 8$ Å. Thus at $T = 100$ K and $\nu_{ph} = 10^{12}$ Hz, a transition

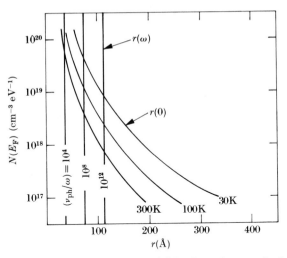

Fig. 6.15. Variation of hopping lengths $r(0)$ and $r(\omega)$ for d.c. and a.c. conduction as a function of $N(E_F)$, ν_{ph}/ω, and temperature.

from d.c. to a.c. hopping is expected at $\omega \sim 2 \cdot 2 \times 10^{-3}\,\text{s}^{-1}$ for $N(E_F) = 10^{18}\,\text{cm}^{-3}\,\text{eV}^{-1}$ and $\sim 5 \cdot 7 \times 10^3\,\text{s}^{-1}$ for $N(E_F) = 10^{19}\,\text{cm}^{-3}\,\text{eV}^{-1}$. For higher densities of states than the latter, a frequency-dependent conductivity requires unreasonable values of the parameters.

In Fig. 6.16 we illustrate schematically the frequency dependence of the conductivity expected for the three conduction mechanisms outlined

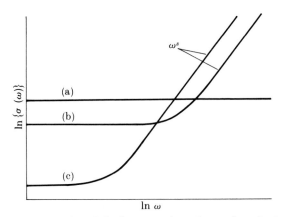

Fig. 6.16. Schematic illustration of the frequency dependence of conductivity for the three modes of conduction described in the text.

above. For the two hopping processes (b) and (c), $\sigma(\omega)$ increases as ω^s. In practice the mechanism giving the highest conductivity at a particular temperature is observed; in the situation shown this would mean that hopping at the band edge would not be observed at any frequency. Conversely, if the onset of the ω^s behaviour for process (b) occurred at a lower frequency, then this would operate to the exclusion of process (c). A distinction between the a.c. conductivity processes (b) and (c) can be made by observing their temperature dependencies.

Measurements of a.c. conductivity in the frequency range 10–10^5 Hz can conveniently be made using bridge techniques with a sensitive null detector. Experimental difficulties arise when the dielectric loss tan δ is small, and the sample geometry is chosen to make the loss as large as possible. A Q-meter can be used up to $\sim 10^7$ Hz, the upper limit for the frequency if electronic circuits are used. For higher frequencies, microwave techniques (slotted line up to about 10^{10} Hz, waveguide to 10^{12} Hz) can be employed. The real and imaginary parts of the conductivity are normally determined by regarding the sample as a resistor and capacitor in parallel. From these measurements, the dielectric constant, the loss, the refractive index, and the optical absorption constant can also be deduced. It is frequently useful to determine the frequency dependence of these parameters when attempting to decide on mechanisms and models.

Data on a.c. conductivity for several amorphous semiconductors will be presented in later chapters. In many materials a variation of $\sigma(\omega)$ proportional to ω^s and with a weak temperature dependence is observed at frequencies sufficiently high or temperatures sufficiently low that any d.c. processes are overtaken. In several chalcogenides (Rockstad 1970, 1971, Ivkin and Kolomiets 1970, Owen 1967, Owen and Robertson 1970,

Lakatos and Abkowitz 1971, Crevecoeur and de Wit 1971, Kočka, Triska, and Stourac 1976, Kočka 1976) s is found to be $\sim0\cdot8$–$1\cdot0$ and densities of states at the Fermi level ranging from 10^{18} cm^{-3} eV^{-1} up to 10^{21} cm^{-3} eV^{-1} have been inferred from application of eqn (6.16). However, as emphasized by Fritzsche (1973) and others, the values of $N(E_F)$ obtained in this way are often up to two orders of magnitude higher than values deduced by other methods. It appears that only in a few cases is pair hopping the likely mechanism. Perhaps the best examples are glow-discharge-deposited silicon and germanium for which Abkowitz, Le Comber, and Spear (1976) deduced, from eqn. (6.16), values of $N(E_F)$ in agreement with estimates using the field-effect technique. However, even in these data (see Chapter 7) the value of s ($\sim0\cdot95$) seems too high (see Fig. 6.13(b)), requiring values of $\nu_{ph} \sim 10^{30}$ s^{-1}!

Other authors have also questioned whether observation of a frequency-dependent conductivity varying as ω^s necessarily implies hopping conduction (Fritzsche 1973, Pollak and Pike 1972, Jonscher 1975). It seems clear that it does not, as several alternative mechanisms are possible. Quite generally what is required is a system with an extremely wide distribution of relaxation times τ (although as emphasized by Pollak (1976) the *form* of the distribution is important to obtain s close to $0\cdot8$). The mechanism is therefore not restricted to processes which also give d.c. conduction or to electronic processes at all.

An example of a non-electronic mechanism is provided by a model originally devised to explain the linear variation with temperature of the specific heat observed in many glasses at low temperatures (see § 6.8.1). The central hypothesis of the model (Anderson, Halperin, and Varma 1972, Phillips 1972) is that there should be a certain number of atoms, or groups of atoms, that can sit with only slightly differing energies in two equilibrium positions. The barrier separating the two minima in energy should be sufficiently great to prohibit resonant tunnelling, but small enough so that thermal equilibrium can occur during the time span of the experiment (say 10^{-10} s $< t < 10^3$ s). It is also required that the pairs of minima under question are (accidentally) degenerate to within a few kT. The probability distribution of the energy difference between all pairs will of course cover a much wider range; what is necessary is that this distribution be *smooth* on the scale of kT. Pollak and Pike (1972) have discussed the possibility that movement of atoms between the two sites can account for the a.c. conductivity behaviour, assuming of course that the transition involves a net motion of charge.

In chalcogenides, interpretation of the ω^s behaviour, with $s \sim 0\cdot8$, in terms of pair hopping poses problems. The fact that the values of $N(E_F)$ deduced appear in many cases to be too high has already been mentioned. Furthermore, on our model for gap states in chalcogenides, outlined in

§ 6.4.1 and discussed further in Chapter 9, it is supposed that electrons are trapped in pairs at defect sites, the Coulomb repulsive energy between the electrons being outweighed by an energy gain associated with lattice distortion. Tunnelling of these bipolarons might therefore be considered a possible mechanism for a.c. conductivity. However, Phillips (1976) has argued that the activation energy associated with bipolaron tunnelling is so large (~half the band gap) that the transition rate would be cut down to exceedingly low values except perhaps for very close pairs of sites (Mott and Street 1977) (see § 9.3). Even then a low effective value of ν_{ph} and a correspondingly low value of s are expected.

Elliott (1977, 1978) has proposed a model for the mechanism responsible for a.c. conductivity in chalcogenides which overcomes most of the problems associated with interpretation in terms of the Austin–Mott formula. It is based on the model of charged defect centres outlined in § 6.4.1 and in Chapter 9. The two electrons in a D^- site are assumed to transfer to a D^+ site by hopping over rather than by tunnelling through the potential barrier separating them. The required broad spectrum of hopping times arises from a quasi-continuous distribution of barrier heights derived from overlapping Coulomb potentials on close pairs. On this model the a.c. conductivity for $N/2$ pairs of sites is

$$\sigma(\omega) = \frac{\pi^2 N^2 \kappa}{24} \left(\frac{8e^2}{\kappa W_M} \right)^6 \frac{\omega^s}{\nu_{ph}^{s-1}} \tag{6.20}$$

where κ is the dielectric constant and W_M is the barrier height separating distant pairs. The exponent s is related to W_M by

$$1 - s = 6kT/W_M.$$

W_M can be approximately equated to the band gap of the material and hence s takes, at room temperature, values between ~0·81 and 0·95 for chalcogenides with band gaps lying between 0·8 and 2·3 eV. Furthermore s is temperature dependent in the manner commonly observed. Values of N deduced from data on arsenic chalcogenides and Se lie between $1·8 \times 10^{18}$ and $2·2 \times 10^{19}$ cm^{-3}.

A similar theory derived for the case of single polarons (rather than bipolarons) hopping over a barrier, has been applied to data on amorphous arsenic (Elliott and Davis 1977) in which it is known that there exists a high concentration of D^0 centres (see Chapter 8).

Collating earlier data, Davis and Mott (1970) noted a correlation in several materials between the magnitude of the a.c. conductivity (at a given frequency) and the d.c. conductivity or the band gap (see inset to Fig. 6.17). This correlation, not understood at the time, is explained quite naturally on Elliott's theory. Other things being equal, $\sigma(\omega)$ is expected to vary inversely as the sixth power of the band gap. According to eqn (6.20), a plot of $\log[\sigma(\omega)\kappa^5(\omega\tau)^\beta/N^2]$, where $\tau = \nu_{ph}^{-1}$ and $\beta = 1 - s$, *versus* $\log W_M$ should have a slope of -6, which is approximately as observed (Fig. 6.17) when W_M is taken to be the band gap B.

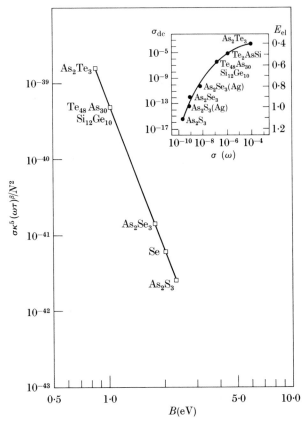

Fig. 6.17. Plot of $\ln\{\sigma(\omega)\kappa^5(\omega\tau)^\beta/N^2\}$ at $\omega = 10^6\,\text{s}^{-1}$ *versus* $\ln B$ for various chalcogenide glasses. B is the band gap. According to theory this line should have a slope of 6; it has a slope of 6·3. The inset shows the correlation between $\sigma(\omega)$ and the activation energy for d.c. conduction ($E_{el} \sim B/2$) (Davis and Mott 1970). (From Elliott 1978.)

The densities of states deduced from a.c. conductivity data on evaporated films of germanium, assuming pair hopping to be responsible, have values so high that they violate the conditions for applicability of eqn (6.16). This is discussed in § 7.1.4, where it is suggested that the presence of voids may be responsible for the a.c. conductivity. For materials like evaporated germanium, in which the d.c. conductivity is associated with variable-range hopping conduction, observation, as the frequency is increased, of the d.c. to a.c. transition for the *same set of states* is possible in principle. Arizumi,

Yoshida, and Safi (1974) and Arizumi *et al.* (1974), amongst others, have studied this transition and fitted their results to a theory of electronic hopping by Scher and Lax (1973), an analysis which includes the d.c. limit.

At frequencies ~ 1 MHz and higher, it is commonly found that $\sigma(\omega) \propto \omega^s$ with $s > 1$ and often $s \sim 2$. Although this can sometimes be an experimental artefact related to electrode resistance (Street, Davies, and Yoffe 1971), it has been observed (e.g. Owen and Robertson 1970) by a slotted-line technique for which no contact to the sample is required. Other authors have reported power dependences larger than expected on the hopping model (see Pollak 1971a,b). It is possible that direct photon absorption (§ 2.5) becomes dominant at high frequencies in some cases. However, Pollak estimates that this mechanism fails to account for the observed magnitude of $\sigma(\omega)$ by several powers of 10.

Austin and Garbett (1971) suggest a quite different mechanism for the high-frequency conductivity (above 1 MHz). In their model it arises from a long low-energy tail to phonon absorption processes. These are normally confined to the near and far infrared (say 10^{12}–10^{14} Hz) in crystalline materials and are associated with transverse optical modes at $k = 0$. Unless there are more than three atoms per unit cell, optical absorption by a single acoustic phonon is forbidden (Zallen 1968). In an amorphous material relaxation of the selection rules may allow such absorption, which, with frequency, would follow the acoustic phonon density of states. In materials with a high degree of ionicity or strong electron–phonon coupling, such a process is observed as 'ultrasonic loss'. Amrhein and Mueller (1968) have also suggested that, when the wavelength of the electromagnetic wave is greater than the mean free path, a temperature-insensitive absorption results from interaction with acoustic phonons.

Austin and Garbett's suggestion is reinforced by the magnitude of the a.c. conductivity, which can be converted to an absorption constant by the relationship

$$\sigma(\omega) = n_0 \alpha / 377$$

The units here are $\Omega^{-1}\,\mathrm{cm}^{-1}$ for σ and cm^{-1} for α; n_0 is the refractive index. This will perhaps be made clearer in Fig. 6.18(a), in which experimental measurements of optical absorption and a.c. conductivity for amorphous As_2Se_3 are displayed over a wide frequency range in the same diagram, with $n_0 = 4$. For convenience, the frequency scale is shown in terms of several commonly used units. The a.c. conductivity data are from Ivkin and Kolomiets (1970), the phonon absorption data from Austin and Garbett (1971), and the fundamental optical absorption edge data from Edmond (1966). All results refer to room temperature. Data in the large frequency gap between 10^8 and 10^{12} Hz have been obtained for a variety of chalcogenides by Taylor, Bishop, and Mitchell (1970) and by Strom and

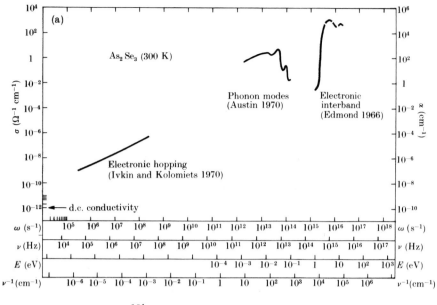

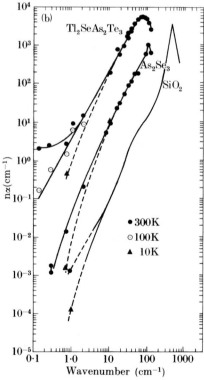

Fig. 6.18. (a) a.c. conductivity and optical absorption over a broad frequency range in amorphous As_2Se_3; (b) Frequency dependence of refractive index times optical absorption in various glasses. (From Strom and Taylor 1977.)

Taylor (1974, 1977) using microwave cavity perturbation techniques in the range 10^9–10^{11} Hz and infrared absorption above 10^{11} Hz. A few of their results are shown in Fig. 6.18(b). For As_2S_3 and As_2Se_3 the absorption was found to be independent of temperature, while for $Tl_2SeAs_2Te_3$ this was not the case and for this material the points shown represent a temperature-independent component that dominates below 200 K. The dependence of $n_0\alpha$ on frequency is very close to ω^2 for As_2Se_3 and As_2S_3 and to $\omega^{1.8}$ for $Tl_2SeAs_2Te_3$ from 4×10^9 Hz up to the first vibrational peak.

The ω^2 dependence of the far-infrared absorption is not in fact unique to chalcogenides but is present in several silicate and oxide glasses as well as in lucite and polystyrene. It appears reasonable to ascribe it to disorder-allowed one-phonon acoustic mode absorption (Strom, Schafer, and Taylor 1977).

6.4.6. Thermopower

In § 2.13 we derived formulae appropriate to the thermoelectric power. We now discuss further those that are relevant to amorphous and liquid semiconductors and consider some experimental data.

The thermoelectric power or Seebeck coefficient is measured by $\Delta V / \Delta T$, where ΔV is the voltage developed between two points of the material maintained at a small temperature difference ΔT. For an n-type *crystalline* semiconductor, it is given by

$$S = -\frac{k}{e}\left(\frac{E_C - E_F}{kT} + A\right) \tag{6.21}$$

where E_C is the energy of the conduction-band edge and AkT is the average energy of the transported electrons measured with respect to E_C. The value of A depends on the nature of the scattering process and is normally a constant between 2 and 4. If the current is carried by holes, the sign of S is reversed and $E_C - E_F$ is replaced by $E_F - E_V$. For ambipolar conduction the thermopower associated with each carrier is weighted according to the contribution each makes to the total current.

In amorphous semiconductors we do not expect any major modification to these formulae, mainly because the transport term A makes only a small contribution to S when $E_C - E_F$ or $E_F - E_V \gg kT$. For current carried in extended states, A is expected to be equal to unity (§ 2.9.3) and E_C or E_V refers to the appropriate mobility edge. For current carried in localized states at the band edges, A will again be small and E_C and E_V are to be replaced by E_A and E_B respectively. If there are several parallel mechanisms, a weighted mean must be taken, as for crystalline semiconductors. The sign of S is therefore a reliable indicator of whether the material is n type or p type, in contrast to the Hall effect which is not (§ 6.4.7).

At the temperature corresponding to a transition from hopping at E_A to transport in extended states at E_C, there will be a change in the slope of the curve of S *versus* $1/T$. Because the intercept at $1/T$ remains virtually unchanged, S has to pass through a transitional region between two curves. This can be spread over a fairly wide temperature range within which the slope of S has little significance (see § 6.6).

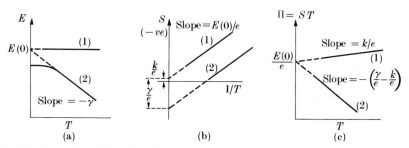

Fig. 6.19. Illustration of the effect of a temperature-dependent energy gap on the slope and intercepts of plots of S against $1/T$ and Π against T: (a) $E = E(0) - \gamma T$; (b) $S = -k/e\{E(0)/kT - \gamma/k + 1\}$; (c) $\Pi = (E(0)/e) - (\gamma/e - k/e)T$. $\gamma = -dE/dT$ and is zero in curves (1).

Measurements of S as a function of temperature provide perhaps the most direct way of determining the temperature coefficient γ of the activation energy for conduction, a quantity of importance as was shown in § 6.4.3. This is illustrated in Fig. 6.19. As before, we assume that over a limited temperature range (for n-type material)

$$E = E_C - E_F = E(0) - \gamma T \qquad (6.22)$$

giving for S (with $A = 1$)†

$$S = -\frac{k}{e}\left\{\frac{E(0)}{kT} - \frac{\gamma}{k} + 1\right\}. \qquad (6.23)$$

A plot of S against $1/T$ has a slope that yields $E(0)$ (as does a plot of $\ln \sigma$ against $1/T$, § 6.4.3), and the intercept on the S axis at $1/T = 0$ yields γ. Alternatively a plot of the Peltier coefficient $\Pi = ST$ against T yields γ from its slope.

Measurements of the values of the thermoelectric power in amorphous chalcogenide semiconductors have shown them to be p type in the majority of cases reported. Detailed measurements as a function of temperature are rather difficult because of problems associated with measuring small voltages across high-resistivity material. Data have been obtained by Owen and Robertson (1970), Seager, Emin, and Quinn (1973), Callaerts *et al.*

† See Butcher and Friedman (1977) and § 2.13.

(1970), Grant *et al.* (1974), and Seager and Quinn (1975). Some of these results will be discussed in § 6.6 and in Chapter 9.

Measurements by Edmond (1966) in several liquid chalcogenides are shown in Fig. 6.20b. Comparison of the slopes of these lines with those obtained from conductivity data (Fig. 6.20a) is made in Table 6.2. The agreement is not as good as expected but it *may* be within experimental error. It should be pointed out that if an activation energy in the mobility (such as expected for hopping conduction) was contributing to the slope of the plot of $\ln \sigma$ *versus* $1/T$, then this would worsen the discrepancy. There is of course the possibility of mixed conduction processes being involved and Moustakas and Weiser (1975) have interpreted data on this basis. In the case of As_2Se_3 (Hurst and Davis 1974, cf. Chapter 9) and liquid Se (Chapter 10), the activation energies for conductivity and thermopower have been found to agree.

TABLE 6.2

Activation energies for conduction in various liquid chalcogenides as determined from the data of Fig. 6.20.

	$E_s(0)$ (eV)	$E_\sigma(0)$ (eV)	$\gamma(\text{eV K}^{-1})$
As_2Se_3	1·21	1·06	$1\cdot00\times10^{-3}$
As_2Se_2Te	1·04	0·84	$9\cdot80\times10^{-4}$
As_2SeTe_2	0·95	0·69	$1\cdot06\times10^{-3}$
As_2Te_3	(0·77)	(0·56)	$1\cdot01\times10^{-3}$
$As_2Se_3.Tl_2Te$	0·50		$4\cdot05\times10^{-4}$

$E_s(0)$ is the slope of the plot of S against $1/T$, and $E_\sigma(0)$ is the slope of the plot of $\ln \sigma$ against $1/T$. γ is the temperature coefficient of E_s as inferred from the intercepts on the S axis at $1/T = 0$ of Fig. 6.20(b).

Also shown in Table 6.2 is the temperature coefficient of E_F-E_V, determined from the extrapolated intercept of the plot of S against $1/T$ on the S axis at $1/T = 0$, under the assumption that $A = 1$. Apart from the liquid containing Tl, the values of γ are all close to 10^{-3} eV K^{-1}. The value for As_2Se_3 may be compared (see also Chapter 9) with the temperature coefficient of the optical gap, $\beta = 1\cdot65 \times 10^{-3}$ eV K^{-1}. Although E_F-E_V is approximately half the optical gap, it is not necessarily true that $\beta = 2\gamma$. These high-temperature coefficients yield very large values for $\exp(\gamma/k)$, which are consistent with the large values of the intercept on the σ axis of the curves of $\ln \sigma$ against $1/T$ shown in Fig. 6.20(a). As mentioned in § 6.4.4, these high values also explain why the slopes of plots of $\ln \sigma$ against $1/T$ yield higher activation energies in the liquid than in the solid amorphous state.

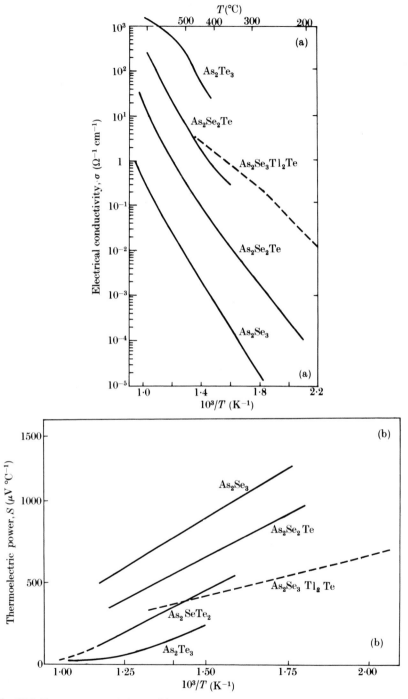

Fig. 6.20. Temperature variation of (a) conductivity and (b) thermoelectric power in various liquid chalcogenides. (From Edmond 1966.)

The temperature at which $S = k/e$ corresponds to that at which $E_F - E_V$ becomes zero. For As_2Te_3 this is, from Fig. 6.20(b), about 770 K (497°C), at which point the conductivity is observed to be about 200 Ω^{-1} cm^{-1} (Fig. 6.20(a)). As the temperature is increased further, the conductivity becomes less dependent on the temperature and appears to saturate at about 2×10^3 Ω^{-1} cm^{-1}. It is tempting to assume that the mobility gap also closes to zero at the same time as $E_F - E_V$ vanishes; however, there is always the possibility that E_F is moving rapidly towards E_V until the semiconductor becomes degenerate p type. Similar curves of σ and S for Se–Te liquid alloys have been obtained by Perron (1967) and are discussed in Chapter 10.

Analysis of thermopower data becomes less simple when S is small ($\leq k/e$, i.e. 86 μV K^{-1}). If it is small not because of ambipolar conduction but because $E_F - E_V \sim kT$, the situation then approaches that in a metal, with the current carried by electrons with energies within a few multiples of kT of the Fermi level. In this case the appropriate formula (§ 2.9.3) is

$$S = \frac{\pi^2}{3} \frac{k^2 T}{e} \left(\frac{\partial \ln \sigma}{\partial E} \right)_{E = E_F}. \tag{6.24}$$

If the current is carried by electrons in extended states near E_F and $L \sim a$, σ is proportional to $\{N(E)\}^2$. Therefore the sign of S will depend on whether the density of states in the vicinity of E_F increases or decreases with energy. There is no indication of S becoming negative in Edmond's data for As_2Te_3.

If states at E_F are localized so that conduction is by variable-range hopping, eqn (6.24) was proposed by one of us (Mott 1967), and is correct only in so far as it predicts a value increasing with T. More recent developments are reviewed in § 2.13; different investigators find that S increases as $T^{1/2}$ or $T^{1/4}$, and that it is proportional to d ln $N(E)/dE$ for $E = E_F$.

In amorphous Si and Ge, small values of the thermopower are observed at *low* temperatures. This is because in these materials the density of states at the Fermi energy is large (not because of a closing pseudogap but because of the presence of defects) and conduction occurs there by hopping according to process (c) described in § 6.4.3. The analysis of § 2.13 should be appropriate in these cases also and it will be used to interpret data presented in Chapter 7. At higher temperatures, when conduction is in one of the bands, E_σ is generally found to be larger than E_S, which can be interpreted in terms of conduction at a band edge and a temperature-activated mobility. However, in glow-discharge-deposited germanium (Jones, Spear, and Le Comber 1976) and silicon (Jones, Le Comber, and Spear 1977) $E_\sigma = E_S$ under certain conditions; presumably conduction then occurs beyond a mobility edge in extended states (see Chapter 7).

6.4.7. Hall effect

In crystalline semiconductors, measurements of the Hall voltage are an important complement to those of the conductivity. Interpretation in terms of the microscopic mobility is straightforward for materials in which the mean free path of the carriers is sufficiently long for the Boltzmann equation to be appropriate and in which the energy-band structure near the band edges is fairly simple. For carriers with very short mean free path, however, as in an amorphous semiconductor, interpretation of observed Hall mobilities in terms of conventional theory leads to values of the mean free path much below the interatomic distance, and often a sign of carrier different from that obtained from the thermopower.

Our understanding of the Hall effect under these conditions depends on the work of Friedman and of Holstein reviewed in Chapters 2 and 3. We summarize our conclusions as follows.

(a) For charge transport at a mobility edge,

$$\mu_H = Cea_E^2/\hbar \tag{6.25}$$

where C, depending on the Anderson localization criterion, is of order $0 \cdot 1$. This is one or two orders smaller than the conductivity mobility and, unlike the conductivity mobility, is independent of T. If a_E is a few ångströms, it is of order $0 \cdot 1 \, \text{cm}^2 \, \text{V}^{-1} \, \text{s}^{-1}$. The sign of the Hall effect is negative both for electrons and for holes if the atomic orbitals are s like, but Emin (1977*a,b*) finds that, if they are antibonding orbitals and the wavefunctions lose phase memory between each orbital, then positive values can occur for electrons. The reversal of sign for both electrons and holes applies only when the number of sites involved in producing the Hall effect is odd. This is normally expected to be the minimum number required—namely three, but if the number of sites is even then the Hall effect has the normal sign.

(b) For hopping between Anderson-localized states, μ_H appears to be negligibly small. There is no real theoretical proof of this (cf. § 3.9) and our conclusion depends on a variety of observations.

(c) For polaron hopping in crystals

$$\mu_H = \text{const.} \; T^{-3/2} \exp(-\tfrac{1}{3} W_H/kT). \tag{6.26}$$

In non-crystalline materials we conjecture that μ_H drops sharply below this value if $W_D \gg kT$, so that preferred percolation paths occur.

Early experimental results due to Male for chalcogenides are shown in Fig. 6.21(a) agreeing with Friedman's formula for conduction at a mobility edge and giving negative R_H in contrast to the positive values of the thermopower. In amorphous germanium Clark (1967) also found a negative Hall effect and estimated a Hall mobility at room temperature of about

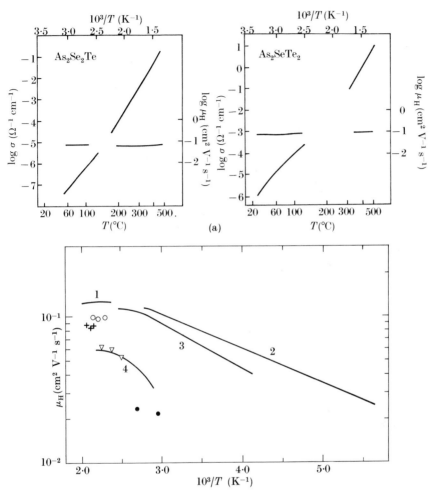

Fig. 6.21. (a) Temperature variation of conductivity and Hall mobility in two chalcogenide glasses in the solid and liquid states (Male 1967). (b) Temperature variation of Hall mobility in various non-crystalline materials (from Mytilineou and Roilos 1978): 1, As_2Se_3; 2, As_2Te_3; 3, $As_2Se_{0.75}Te_{2.25}$; 4, $As_2Se_{2.25}Te_{0.75}$; ∇, $As_2Se_{2.625}Te_{0.375}$; +, $As_2Se_{2.5}S_{0.5}$; ●, As_2SeSTe; ○, $As_2Se_{2.94}Te_{0.06}$.

10^{-2} cm^{-1} V^{-1} s^{-1}. This is close to the experimental limit of measurement using conventional techniques. More recent observations are due to Roilos and Mytilineou (1974), Nagels *et al.* (1974), Seager, Emin, and Quinn (1973), and Mytilineou and Roilos (1978). Typical results are those of the latter authors shown in Fig. 6.21(b). Two quite different interpretations have been given of these results. While Emin and co-workers consider that the carriers (holes) in chalcogenides form small polarons and this behaviour is to be interpreted through eqn (6.26), Nagels *et al.* (1974) and, following them, Mott, Davis, and Street (1975) suppose that the flat part is to be interpreted in terms of Friedman's equation (6.25) and the drop at low temperature occurs because most of the current is carried by hopping, which is supposed to make no contribution to R_H. We return to a comparison of the two hypotheses in § 6.6.

Positive Hall coefficients have been observed in $CdGeAs_2$ by Callaerts *et al.* (1970), and Nagels *et al.* (1970), in amorphous arsenic by Mytilineou (see Chapter 8), and in n-type glow-discharge-deposited silicon by Le Comber, Jones, and Spear (1977) (see Chapter 7). A change in the sign of the Hall effect with temperature has been observed by Meimaris *et al.* (1977) in certain $CdGe_xAs_2$ glasses. Emin's (1977*a,b*) considerations of the sign of the Hall effect are probably applicable to these cases.

6.4.8. Magnetoresistance

In crystalline semiconductors measurements of the magnetoresistance, the fractional change $\Delta\rho/\rho$ of resistivity in a magnetic field, can, like the Hall effect, be used to determine carrier mobilities. Normally $\Delta\rho/\rho$ is positive and proportional to the square of the magnetic induction B. For hopping conduction in doped crystalline semiconductors the magnetoresistance can be either positive or negative (see Chapter 4).

Mell and Stuke (1970) have reported magnetoresistance measurements on several amorphous semiconductors using an alternating (1·5 Hz) magnetic field up to 25 kG. The magnetoresistance was found to be negative over a wide temperature range. Only at very low magnetic fields and low temperatures was a positive effect found for amorphous germanium (Chapter 7). More recently Mell has found a positive effect in amorphous films of InAs and InP. For most materials $\Delta\rho/\rho$ was found to be proportional to B^n, with n being close to unity at high temperatures and falling to a value of the order of one-half near room temperature. Figure 6.22 shows the temperature dependence of the magnetoresistance at 25 kG for a number of materials. Following Mell and Stuke (1970), we believe that this negative magnetoresistance is contributed by that part of the current that is carried by electrons hopping with energies near the Fermi level, the fall-off at high temperatures occurring because the current is then mainly due to electrons or holes in the extended states where $\Delta\rho/\rho$, like the Hall effect, is small.

In impurity conduction in doped crystalline semiconductors, a large positive magnetoresistance can occur owing to shrinking of the orbits. This is unlikely to be important for the much smaller orbits deep in the gap of non-crystalline materials. Movaghar and Schweitzer (1977) account for the effect in the following way. Hopping to states that already contains an electron is only possible for electrons with the opposite spin. In the

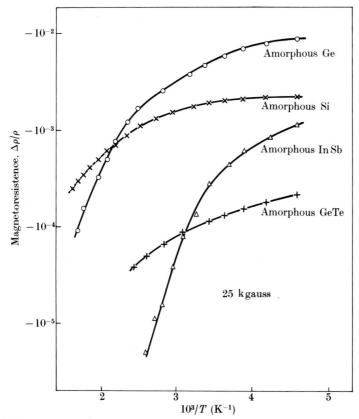

Fig. 6.22. Temperature variation of magnetoresistance in a few amorphous semiconductors. (From Mell and Stuke 1970.)

presence of a magnetic field, which affects the probability of a spin flip, the chance that an electron can find an empty site with the right spin is increased; a negative magnetoresistance is thus deduced.

6.4.9. Field effect

Measurements of the field effect, namely the increased conductivity in the surface layer when a field is applied to the surface of a conductor, provides a powerful method of determining the density of states in certain amorphous semiconductors. In the conventional field-effect geometry, a semiconductor sample forms one plate of a capacitor and the other is a metal electrode (the gate), separated from the sample by a thin sheet of insulating dielectric such as mylar or SiO_2 (Fig. 6.23(a)). A voltage on the gate induces onto the sample a charge which resides either in surface states at the interface between insulator and semiconductor or within a region

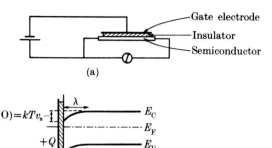

(a)

$$\Delta E(0)=kTv_{\mathrm{s}}-\overline{\overline{\underline{}}}$$

(b)

Fig. 6.23. (a) Experimental arrangement for field-effect measurements. (b) Variation of potential with distance x below the surface of the semiconductor. A charge $+Q$ on the gate electrode induces $-Q$ which, in the absence of surface states, is distributed throughout a space charge region λ either in gap states or in the bands.

beneath the semiconductor surface in which a space charge is found. The portion of the charge that is mobile within the semiconductor can be determined, given some independent knowledge about the mobility and the form of the space-charge region, by monitoring the transverse (source–drain) current. If the density of states at the Fermi level in the semiconductor is finite, then part of the charge may be immobilized there, and, aside from the difficulty of distinguishing this from charge immobilized in surface states, the possibility exists of determining the density of gap states in the bulk of the semiconductor. The analysis by which this can be done is as follows. The charge in the surface layer will lead to band bending, as shown in Fig. 6.23(b) for a negative induced charge. If ΔE is the decrease in the level of the conduction band at a distance x from the surface, then the change in the conductivity there is

$$\frac{\Delta\sigma(x)}{\sigma_0}=\frac{b}{1+b}\left\{\exp\left(\frac{\Delta E}{kT}\right)-1\right\}+\frac{1}{1+b}\left\{\exp\left(-\frac{\Delta E}{kT}\right)-1\right\} \qquad (6.27)$$

where b is the ratio of electron to hole current when $\Delta E = 0$. The relative change in conductance (ΔG) is obtained by integration over the sample thickness d. If V changes exponentially with distance with a screening length λ, then $\mathrm{d}V/V = \mathrm{d}x/\lambda$ and the integral becomes (Fritzsche 1973)

$$\frac{\Delta G}{G_0}=\frac{\lambda}{d}\left\{\frac{b}{1+b}F(-v_s)+\frac{1}{1+b}F(v_s)\right\} \qquad (6.28)$$

where

$$F(v_{\mathrm{s}})=\int_0^{v_s}\left(\frac{e^{-v}-1}{v}\right)\mathrm{d}v=\sum_{m=1}^{\infty}\frac{(-v_s)^m}{mm!}$$

and v_s is the reduced surface potential $\Delta E(0)/kT$. This function, shown in Fig. 6.24, is known as a field-effect curve and, with the assumptions under which it has been derived, can be compared with experiment to yield the charge immobilized at the Fermi level and hence $N(E_F)$.

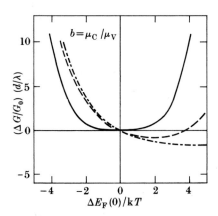

Fig. 6.24. Predicted change of conductance $\Delta G/G_0$ versus reduced surface potential v_s for various values of the ratio b of electron to hole current when $v_s = 0$: solid curve, $b = 1$; broken curve, $b = 0\cdot1$; chain curve, $b = 0\cdot01$. (From Kastner and Fritzsche 1970.)

To obtain λ and $\Delta E(0)$ in the case of constant $N(E)$ near the Fermi level we use Poisson's equation

$$\frac{d^2V}{dx^2} = \frac{4\pi e^2 \Delta E N(E_F)}{\kappa}$$

which gives an exponential decay of V and

$$\lambda^2 = \frac{\kappa}{4\pi e^2 N(E_F)}. \tag{6.29}$$

The relation between charge per unit area Q and $\Delta E(0)$ is

$$Q = \frac{1}{4\pi\kappa} \int \frac{d^2V}{dx^2} dx = \frac{1}{4\pi\kappa}\left(\frac{dV}{dx}\right)_{x=0}$$

and with $V = \Delta E(0)\, e^{-x/\lambda}$, this gives

$$4\pi\kappa Q = \Delta E(0)/\lambda. \tag{6.30}$$

If $\Delta E(0) \gg kT$, the leading term in $\Delta G/G$ will be proportional to $\exp\{\Delta E(0)/kT\}$. For small values of the gate voltage and for p-type conductors ($b = 0$) these equations give

$$\Delta G/G_0 = \frac{(\lambda^2 d)4\pi\kappa Q}{kT}$$

$$= \frac{Q}{N(E_F)dkT}.$$

(6.31)

In this case the activation energy for conduction is the same as for bulk conductors.

If the Fermi energy is displaced into a region in which $N(E)$ is not constant, a more elaborate numerical procedure is necessary, which has been used notably by Spear and Le Comber (1972) to determine $N(E)$ in the whole gap region in glow-discharge-deposited silicon.

More complete analyses, including the effects of surface states, have been given by Barbe (1971) and by Neudeck and Malhotra (1975). It is clear that if the surface state density is greater than about $10^{14}\,\mathrm{cm}^{-2}\,\mathrm{eV}^{-1}$, then determination of $N(E_F)$ by the field-effect technique is not possible, unless communication with the surface states is slower than with those at E_F, in which case modulation at a sufficiently high frequency may render them inactive. Although this is likely for 'slow' states residing within an oxide layer on the semiconductor surface or in the insulator, it is unlikely to be the case for true interface states.

If $N(E_F)$ is not finite, the screening length (cf. § 9.4.1) is temperature dependent, and the temperature dependence of the field-effect current depends both on the conductivity of the affected region and on its width. If there are no surface or gap states, all the induced charge must be in the conduction or valence band; the screening length is then not relevant.

If the mobility is not activated, neither is ΔG. Also if there are a set of levels at an energy ε from E_F, this takes practically all the surface charge, and the extra charge in the conduction band will vary with T as $\exp\{-(E-\varepsilon)/kT\}$ where $E = E_C - E_F$. A model of this kind has been used by Marshall and Owen and is discussed in § 9.4.1.

Measurements of the field effect in amorphous semiconductors include data on multi-component STAG glasses by Kastner and Fritzsche (1970), Egerton (1971), Levy, Green, and Gee (1974), and Marshall and Owen (1976), on evaporated Si and Ge by Malhotra and Neudeck (1974, 1975), on glow-discharge-deposited Si by Spear and Le Comber (1972), Spear (1974b), Madan, Le Comber, and Spear (1976), and on carbon by Adkins and Hamilton (1971). Anderson (1974) using CdSe thin-film transistor structures has deduced the energy distribution of states within the gaps of various insulators. Results of some of these experiments will be described in subsequent chapters. Pepper et al. (1974a,b) have used the field effect in MONS devices to explore the nature of localized states at band edges in crystalline Si. This work was described in Chapter 4. Field-effect measurements on chalcogenides are described in § 9.4.1.

6.5. Drift mobility and photoconduction

6.5.1 Drift mobility

The phenomena described so far in this chapter depend on 'states in the gap', only in so far as these determine the position of the Fermi energy E_F and, if $N(E_F)$ is finite, hopping conduction by electrons with energies near E_F. Measurements of drift mobility and still more of photoconduction, however, must often be interpreted in terms of deep traps and recombination centres not necessarily at E_F; this is why we consider them together.

Direct measurements of the drift mobility have proved very useful in determining the mechanism of charge transport of electrons or holes. Carriers are injected at one point of a sample and their transit time t_t to another point at a distance d is measured under the influence of an electric field F. The drift mobility is then defined by

$$\mu_D = d/Ft_t. \qquad (6.32)$$

In the experiments described below the materials have high resistivity, typically greater than $10^7\,\Omega$ cm, and the dielectric relaxation time $(10^{-12}\rho\kappa/4\pi)$ is large compared with t_t; the excess carriers are *not* screened by other carriers, as in experiments on the drift mobility of minority carriers in crystalline semiconductors (Shockley 1950). Both hole and electron transits can thus be observed. Fig. 6.25(a) illustrates a typical 'sandwich' arrangement in which the semiconductor film (perhaps 50 μm thick) is equipped with two blocking electrodes, one of which is semi-transparent if optical injection is used. The carriers can be electrically injected as a pulse (with duration less than t_t) from one of the electrodes, or created in pairs by a flash of strongly absorbed light or by an electron beam. If the injected charge is kept smaller than $C_s V$, where C_s is the

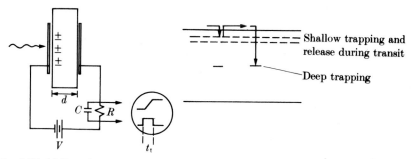

Fig. 6.25. (a) Experimental arrangement for drift mobility studies: $\mu = d/Ft_t = d^2/Vt_t$; (b) schematic representation of shallow and deep trapping processes.

capacity of the film, then the internal field remains essentially constant during transit. The time for the selected carrier to drift to the opposite electrode is determined by observing either the square-like current transient ($CR \ll t_t$) or the ramp-like charge (voltage transient ($CR \gg t_t$)).

The technique has been used by Spear to establish polaron hopping in sulphur (Chapter 3) and electron mobility in liquid rare gases (Chapter 5). In amorphous materials the main interest is in the effect of localized states (traps). In a crystalline material one could argue as follows. If there are N_t traps with discrete energy ε_t, below the conduction band and if μ_0 is the conductivity mobility, then the drift mobility μ_D is given by

$$\mu_D = \mu_0\left(\frac{n_0}{n_0 + n_t}\right)$$

where n_0 and n_t are the carrier densities in the band and in the traps respectively. Assuming a thermal equilibrium distribution between the free and trapped carriers during transit, this ratio can be expressed in terms of the trap parameters, so that

$$\mu_D = \mu_0\left\{1 + \frac{N_t}{N_c}\exp\left(\frac{\varepsilon_t}{kT}\right)\right\}^{-1}$$

which approximates to

$$\mu_D = \mu_0\left(\frac{N_c}{N_t}\right)\exp\left(-\frac{\varepsilon_t}{kT}\right) \tag{6.33}$$

except at high temperature where the probability of thermal release is high. Here N_c is the effective density of states at the band edge. These quantities are frequently not known, but $e\mu_0 N_c$ is the pre-exponential factor in the dark conductivity, and if conduction is in extended states this should be (§ 6.4.3) $\sigma_{min}\exp(\gamma/k)$. Thus if μ_D is plotted against $1/T$, the slope yields the trap depth, and, if μ_0 and N_c are known, the intercept yields the trap density. The formula can be generalized to situations in which there is a continuous distribution of traps or, in amorphous semiconductors, to a continuous range of localized states below E_C as will be shown below.

Normally for unambiguous interpretation it is important to ascertain that a well-defined transit time exists and that it scales correctly with field and sample thickness. Some spreading in the sheet of charge during transit owing to diffusion is inevitable, but more commonly a spectrum of arrival times results from a statistical spread in trapping and release time. The assumption that there is thermal equilibrium between electrons in the conduction band and the traps will not be valid if ε_t is too large. If the thermal release time is the reciprocal of

$$\nu_{ph}\exp\left(-\frac{\varepsilon_t}{kT}\right)$$

and t_p is the duration of the pulse, the condition that a trap will release electrons during the passage of the pulse is

$$\nu_{ph} t_p \exp\left(-\frac{\varepsilon_t}{kT}\right) > 1. \tag{6.34}$$

Deeper traps lead to a tail in the current pulse or to a lack of saturation in the charge collected (Fig. 6.26). Under these conditions one can still determine a time t'_t from which a drift mobility can be determined, but one is then probing only those traps with release times less than t'_t.

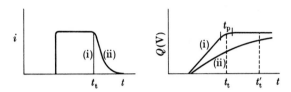

Fig. 6.26. Shape of current i and charge Q (or voltage V) transients: (i) case where the trap depth is not greater than that given by eqn (6.34); (ii) case where the release time of the traps is greater than the pulse time t_p.

For amorphous semiconductors we distinguish between 'intrinsic' traps (localized states) extending from a band edge at E_A to E_C and deep extrinsic traps due to defects. We make the qualitative distinction that hopping between the former is possible, while in measurements of the drift mobility hopping between the latter is negligible.† With this assumption, we can define, as in our discussion of conductivity, an *intrinsic* drift mobility

$$\mu_D^0 = \mu_{ext}\left(\frac{\Delta E}{kT}\right)^n \exp\left(-\frac{\Delta E}{kT}\right) + \mu_{hop} \exp\left(-\frac{w}{kT}\right). \tag{6.35}$$

Here μ_{ext} is the mobility $\sigma_{min}/N(E_C)e$ at the mobility edge, $\Delta E = E_C - E_A$ and n comes from the assumption that the form of the density of states is $N(E) \propto (E - E_A)^n$. The hopping energy is w and is expected to be a function of temperature as discussed in § 6.4.2. The condition that this is the measured drift mobility is that, during the duration of the pulse (t_p), *most of the electrons are within a range kT above E_A and not in lower states* (Spear and Le Comber 1972). This may be *either* because the density of states at a depth ε_t below E_A is too low, so that

$$\frac{N_t}{N_A} \exp\left(\frac{\varepsilon_t}{kT}\right) \leqslant 1 \tag{6.36}$$

† Hopping between deep states is, however, always possible and as we have seen is responsible for the $T^{1/4}$ behaviour (if the states are at E_F) observed in some materials.

(where N_A is the effective density of states at E_A), or because the release time is too long, so that trapped electrons contribute only to the tail of the pulse.

The number of amorphous semiconductors so far investigated by this technique is relatively small. As mentioned above, the material should have a high resistivity. In glow-discharge-deposited silicon, Le Comber and Spear (1970) and Le Comber, Madan, and Spear (1972) find a kink† in a plot of $\ln \mu_D$ against $1/T$ as predicted by eqn (6.35). The steady-state d.c. conductivity also shows a change of behaviour at the same temperature (200 K), supporting the suggestion that here a transition from conduction in delocalized (extended) states to conduction in localized states occurs. This work has given the strongest experimental support for the concept of a mobility edge, and is described in Chapter 7. Although deep states are undoubtedly present, the release time must be too long for equilibrium to be set up during the passage of the pulse. However, measurements on chalcogenides indicate that gap states *do* control the drift mobilities (see Chapter 9). In amorphous selenium well-defined transit times have been observed both for electrons and holes and the temperature dependence of the drift mobilities studied. These results are described and discussed in Chapter 10. Results of Hughes (1973, 1975, 1977) for electrons and holes in SiO_2 are described in Chapter 9; for electrons the drift mobility is not activated, suggesting that the energy difference between the band edge and the mobility edge is less than kT at the temperature of the experiment; for holes the mobility is activated, probably owing to polaron formation.

Drift-mobility studies as a function of pressure have been made in amorphous selenium by Dolezalek and Spear (1970). No change in the magnitude or activation energy of the drift mobility was found, providing evidence against hopping conduction at room temperature in this material. Similar measurements on As_2Se_3 by Pfister (1974), however, reveal a decrease in the transit time for hole transport.

Observations of transients under varying conditions of field, temperature, and thickness can be used to determine the carrier lifetime τ or the time for a carrier to become lost by a deep trapping event. The product $\mu_D \tau$ is the range per unit applied electric field, and is an important material parameter (in the commercial process of xerography for example). Measurement of the total charge collected under conditions where the range is much larger than the sample thickness enables a direct measurement of quantum efficiency to be made. Measurements on selenium as a function of the energy of the incident radiation will be discussed in § 6.5.3 and in Chapter 10.

In drift-mobility measurements on certain materials, e.g. As_2Se_3 and Se (at low temperatures), the current pulse shape departs from the ideal

† See Figs. 6.29, 7.39, and 7.40 and also Moore (1977).

rectangular shape. Fig. 6.27(a) illustrates data obtained on Se. As the temperature is lowered from 297 K to 123 K, it becomes increasingly difficult to identify the time t_t at which the bulk of the injected carriers reach the opposite electrode until eventually a smooth decrease of the current with time is observed. However, a plot of log i *versus* log t still yields a break point as shown (Fig. 6.27(b)).

For a sheet of carriers dispersed by Gaussian statistics during transit, the tail of the current pulse is expected to increase as $t_t^{1/2}$ (i.e. to become relatively steeper), and this is found to be so at high temperatures. At slightly lower temperatures, the fall off in i for times less than t_t could be

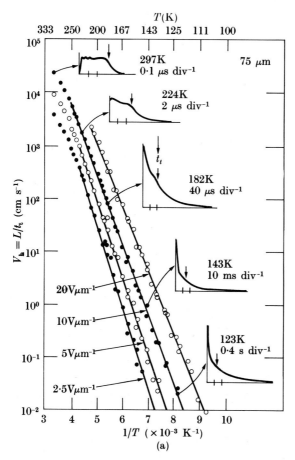

Fig. 6.27. (a) Temperature dependence of hole drift velocity in amorphous selenium at various values of electric field. The shape of the current transients at various temperatures are illustrated; the arrows indicate transit times. (b) (p. 252) Current transients plotted on a log–log basis for various values of temperature and electric field F. The scales have been normalized to illustrate the scaling with respect to the transit time. (From Pfister 1976.)

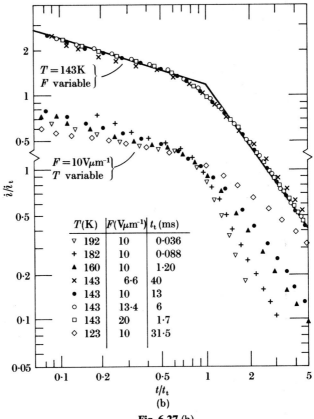

T(K)	F(Vμm^{-1})	t_t (ms)
▽ 192	10	0·036
+ 182	10	0·088
▲ 160	10	1·20
× 143	6·6	40
• 143	10	13
○ 143	13·4	6
□ 143	20	1·7
◇ 123	10	31·5

Fig. 6.27 (b)

explained by deep trapping events which cause carriers to be held back from the main pulse, possibly to contribute eventually to the tail. Such a model has been proposed by Marshall and Owen (1971). The most successful analysis of the pulse shapes over a wide range of temperatures and transit times, however, has been provided by Scher and Montroll (1975). Their theory is able to explain the experimental observations that at low temperatures the dispersion is proportional to t_t and the drift mobility μ_D appears to depend on sample thickness. The model also predicts a field dependence of μ_D.

The Scher and Montroll theory is formulated in terms of a model in which transport occurs by hopping between localized states; however, the same idea could be applied to transport at a mobility edge interrupted by multiple trapping, if there is some fluctuation in trap depth, or to the hybrid case of hopping conduction interrupted by trapping in deeper states. The principal idea is that a distribution in hopping times (which are assumed to arise from fluctuations in hopping distances rather than hopping energies) is so large that the maximum in the carrier concentration remains close to the generation region. The current transient then represents simply the time dependence of the arrival of carriers from the leading edge of the distribution. The dispersion of the carrier sheet and the mean displacement of the charge from the front surface both increase with time in the same manner, their ratio remaining constant. The probability for a carrier, after having arrived at a site at time $t = 0$, to hop to its next site is taken as $\psi(t) = t^{-(1+\alpha)}$, where $\alpha\,(0 < \alpha < 1)$ is a parameter depending on the hop distance, the hop energy, and the spatial extent of the localized state. This weak dependence with time, which is to be contrasted with a Gaussian probability $\exp(-t/\tau)$ that rapidly vanishes with increasing time, leads to a time dependence of current behaving as

$$
\begin{aligned}
i &\sim t^{-(1-\alpha)} \qquad t < t_t \\
i &\sim t^{-(1+\alpha)} \qquad t > t_t.
\end{aligned}
\tag{6.37}
$$

Here t_t is the transit time for the *leading* edge of the carrier distribution and shows clearly on a log plot of the transient current shape. The sum of the slopes on either side of t_t is equal to 2. The mean displacement of the carrier sheet depends on time as t^α and the transit time t_t for a sample of thickness L is thus proportional to $L^{1/\alpha}$.

With the further assumption that the asymmetry between forward and reverse hops increases linearly with field strength F, $t_t \sim F^{-1/\alpha}$ and the transit time scale as

$$
t_t \propto (L/F)^{1/\alpha}.
\tag{6.38}
$$

Since $\alpha < 1$, a drift mobility μ_D deduced from L/t_tF (eqn (6.32)) is predicted to have a superlinear dependence on field and to depend on the sample thickness.† Under conditions when the transit time is much longer than the individual event times, the non-linear effects disappear and the mobility again becomes well defined.

The results of Pfister (1976) shown in Fig. 6.27 are in good agreement with the theory of Scher and Montroll; in particular one should note the universality of the normalized current transients for a range of field strengths and the independence of the activation energy in the drift velocity as the shape of the transient changes for a fixed field. Between 140 K and 170 K, the sum of the slopes of the lines on either side of t_t is close to the theoretical value of 2. Above 170 K the dispersion becomes Gaussian; below 140 K the sum is smaller than 2 and it is argued that trapped charge causes pulse distortions, flattening the initial part of $i(t)$. These results

† The concept of a thickness-dependent drift mobility is not very satisfying and use of the term 'effective drift mobility' might be preferable.

show also the predicted superlinear dependence of t_t on sample thickness, the parameter α decreasing from $\sim 0\cdot75$ to $\sim 0\cdot51$ as the temperature is lowered. The field dependence of the mobility apparently contains a contribution additional to that predicted by theory. A model for the field dependence of drift mobilities, based on the Scher and Montroll formalism, has been given by Pfister (1977a,b).

A detailed study of time-dependent transport in amorphous As_2Se_3 has been made by Pfister and Scher (1977). These authors interpret the results in terms of a model in which the carriers hop through localized states but interact with discrete trapping levels of lower density. Alternative models in which transport proceeds via extended states but is trap controlled have been proposed by Marshall (1977) and by Silver and Cohen (1977) and Silver (1977). Pollak (1977), Schmidlin (1977), and Noolandi (1977) have discussed the formal equivalence of the stochastic transport theory of Scher and Montroll (1975) with that of a multiple trapping model. The character of transient conduction in both amorphous Se and As_2Se_3 has, at the time of writing, not been resolved in detail, in spite of the large amount of data available.

A review of non-Gaussian transient transport in disordered solids has been given by Pfister and Scher (1979).

6.5.2. Photoconductivity

A discussion of photoconductivity in an amorphous semiconductor involves two separate problems. When a quantum of radiation is absorbed, an electron and a hole are created. They attract each other, and particularly at low temperatures may recombine, having made no contribution to the current. The next section, in which experiments and models for quantum efficiency are described, deals with this aspect of the problem. In this section we treat situations in which G, the number per cubic centimetre per second of *free* electrons and holes produced by the radiation, can be equated to the number of quanta absorbed.

We define i_p as the excess current per unit volume produced by the radiation. For simplicity we suppose that this is mainly carried either by electrons or by holes, though the analysis can easily be generalized to the case when both contribute. Then two situations are of interest. In the first, electrons and holes are trapped and do not escape and they recombine from these traps, that is from states fairly near the Fermi level. In this case the photocurrent i_p is proportional to G (monomolecular behaviour). In certain cases i_p is proportional to the drift mobility μ_D. Thus in drift-mobility experiments the carriers form a quasi-equilibrium between states near the mobility edge, localized states at the band edge, and perhaps defect states. If recombination takes place from the lowest of these states, and is not temperature activated, the T dependence of i_p is that of μ_D.

In the second case the recombination rate of the electrons is proportional to the number of holes (trapped or otherwise). In such a case the plot of photocurrent against $1/T$ is as in Fig. 6.31, with i_p proportional to $G^{1/2}$ (bimolecular behaviour) when $i_p > i_d$, i_d being the dark current, but proportional to G when $i_p < i_d$.

We consider the first situation. Glow-discharge-deposited silicon provides an example. The spectral dependence of the photocurrent will be discussed in § 7.7. The density of gap states depends strongly on the temperature of deposition. Fig. 6.28 shows schematically how a band of

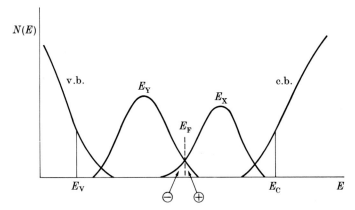

Fig. 6.28. Schematic illustration of overlapping acceptor (E_X) and donor (E_Y) gap states in amorphous silicon.

deep donors can overlap a band of acceptors. Charged centres occur when the bands overlap. According to Anderson and Spear (1977), these charged centres trap holes and electrons; holes are trapped first, and the photocurrent is highly sensitive to the number of *charged E_Y* states which trap electrons. Thus n-type doping (Chapter 7) removes these, and monomolecular recombination disappears, going over to bimolecular. The monomolecular current for a field F is given by

$$i_p = eF\mu_D G\tau$$

where μ_D is the drift mobility for electrons above E_A and τ is the lifetime. This should be so, whether or not a quasi-equilibrium with E_X is established, because, if it is, μ_D is reduced by $\exp(-\Delta E/kT)$, but so is $1/\tau$. Apart from this, $1/\tau$ will normally be little dependent on T, as discussed in Chapter 3.

Fig. 6.29 shows plots of i_p against $1/T$ for films deposited at different temperatures T_d, compared with μ_D measured directly. The similarity of the curves shows, with the kink at the same temperature, that μ_D is an 'intrinsic' property, depending little on the density of defects (which

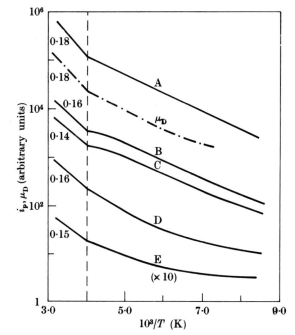

Fig. 6.29. Temperature dependence of photocurrent i_p and drift mobility μ_D in glow-discharge-deposited silicon. The numbers show the activation energy in eV. Curves A, B, and C refer to samples deposited at 500 K, D at 350 K, and E at 320 K. Curves A, D, and E were obtained with higher light intensities than curves B and C. The excitation energy was 2 eV. The current is monomolecular, i.e. proportional to G. (From Spear *et al.* 1974.)

determines τ) or the hydrogen concentration. However, τ is highly sensitive to T_d (Fig. 6.30).

For glow-discharge-deposited silicon with a low density of states in the gap, bimolecular behaviour can be observed at high intensities, particularly (as already remarked) if the charged E_Y centres are removed by n-type doping. Bimolecular behaviour is normal in chalcogenides. In both it is thought that the minority carriers (holes in Si, electrons in chalcogenides) are trapped first; τ does *not* depend on the density of traps. What happens is that the concentration of trapped holes builds up until the (equal) concentration of free electrons gives a recombination rate equal to G. In this case the photocurrent gain can greatly exceed unity.

The analysis for this case is as follows. The carriers (for instance electrons) form a quasi-equilibrium either in the conduction band itself or with shallow defect traps below it. If Δn is the excess density of these carriers due to the radiation, then the photocurrent i_p is given by

$$i_p = e\mu_D F \,\Delta n \qquad (6.39)$$

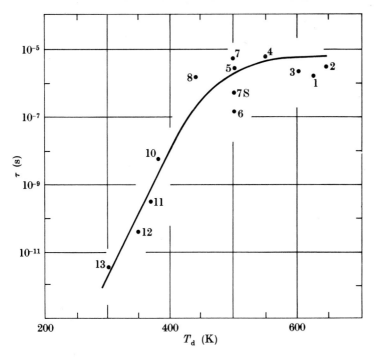

Fig. 6.30. Recombination lifetime of photogenerated carriers as a function of deposition temperature T_d for glow-discharge-deposited silicon. (From Spear 1974a.)

where F is the field and where μ_D is the drift mobility for this carrier, as described in § 6.5.1. The lifetime of one of these carriers in the states between which the quasi-equilibrium is maintained is denoted by τ; then Δn is given by

$$\Delta n = G\tau \qquad (6.40)$$

where G is the number of carrier pairs generated per unit time and per unit volume. After the radiation is cut off, Δn decays as

$$\frac{d \ln(\Delta n)}{dt} = -\frac{\Delta n}{\tau}. \qquad (6.41)$$

From the decay of the photocurrent τ can be determined, and experiments are described in § 9.7 in which this is done. τ is determined by recombination with the other carrier, either directly or after it has been trapped in a defect state. If i_p is great compared with the dark current i_d, i_p is proportional to Δn. Then eqn (6.40) shows that Δn and hence i_p are proportional to $G^{1/2}$ (the bimolecular law), and (6.41) gives a decay as $1/(t+\text{const.})$.

We next define a quantity b, such that bN is the chance per unit time that a majority carrier recombines when there are N minority carriers (trapped or otherwise) per unit volume. For a crystalline semiconductor we could write

$$b = vA$$

where v is the thermal velocity of the carrier and A is a capture cross-section. For electrons with energies above the mobility edge, this procedure is allowable; for electrons hopping at the band edge and holes trapped in centres it is preferable to introduce the chance p per unit time that an electron in a band-edge localized state overlapping the recombination centre recombines. Then

$$b = pa^3 \tag{6.42}$$

where a is the spatial extent of the state and

$$1/\tau = b(\Delta n + n_0) \tag{6.43}$$

where n_0 is the concentration of trapped or free minority carriers in the dark. We now see that the steady-state excess carrier density is given by

$$\Delta n = \frac{G}{b(\Delta n + n_0)} \tag{6.44}$$

from which the current may be found from (6.39). When the current is switched off, the rate of decay is given by (6.41) with Δn substituted from (6.44), which leads to

$$\frac{d \ln(\Delta n)}{dt} = -b\, \Delta n (\Delta n + n_0). \tag{6.45}$$

If $\Delta n \gg n_0$ (regime II as defined below)

$$1/\Delta n = bt + \text{const.} \tag{6.46}$$

characteristic of bimolecular decay, and is dependent on T only through the recombination mechanism (Chapter 3). If $\Delta n \ll n_0$ (regime I), however,

$$\Delta n = \text{const.} \exp(-bn_0 t) \tag{6.47}$$

with n_0 strongly dependent on T.

The variation with temperature of the photocurrent is similar for all chalcogenides investigated and is of the general form shown in Fig. 6.31. Following Simmons and Taylor (1974), we label the three parts of the curve regimes I, II, and III. In regime I the photocurrent is less than the dark current and increases with decreasing temperature; in regime II the photocurrent is larger than the dark current and decreases with decreasing

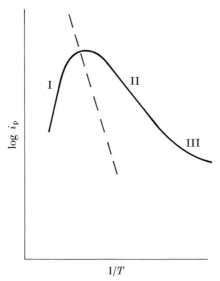

Fig. 6.31. Temperature dependence of photocurrent in chalcogenides. The broken line is the dark current (schematic).

temperature; in regime III the photocurrent does not fall so rapidly as in regime II and can approach a temperature-independent value. In regime I the drop in the photocurrent and in τ with increasing T occurs because Δn is less than the equilibrium number, the photocurrent i_p being less than the dark current i_d; recombination of, say, electrons is with thermally generated holes, the number of which increases with T. In this regime i_p is thus proportional to the light intensity and thus to G. In regime II $i_p \gg i_d$ and recombination is between electrons and holes, both of which are generated by the radiation; i_p is proportional to $G^{1/2}$, though Arnoldussen et al. (1974) have reported that in chalcogenides at low intensities it is proportional to G. In regime III $i_p \propto G$; this behaviour is discussed for chalcogenides in Chapter 9. Fig. 6.32 shows the transition between the two regimes I and II from results of Weiser et al. (1970).

All these regimes can be explained in principle with the model given above. In regime II we suppose that n_0 is negligible if recombination is with trapped holes (though this is not the same as saying that $i_p > i_d$). Then eqn (6.40) becomes

$$\Delta n = \surd(G/b) \tag{6.48}$$

and the photocurrent is proportional to $G^{1/2}$. The temperature dependence is that of $\mu_D b^{-1/2}$, and thus probably that of the drift mobility (see below). In regime I

$$\Delta n = G/bn_0 \tag{6.49}$$

and the temperature dependence is that of μ_D/n_0. In principle we have here, through the temperature dependence of n_0, a way of finding the position of any gap state that traps the minority carrier.

If minority carriers (holes) are trapped prior to recombination, the rate of recombination is independent of the trap density. As we have seen, however, in glow-discharge silicon at any rate, the model described above is only applicable for a low density of gap states. For a high density the monomolecular mechanism takes over, the electrons descending into the E_Y states before recombining. This is strikingly shown in glow-discharge-deposited amorphous silicon, where the drift mobility μ_D is known from direct experiment. Spear (1974b) has deduced from the photocurrent in regime II (using values of μ_D from his drift mobility experiments) the recombination lifetimes shown in Fig. 6.30, plotted there against the temperature T_d of deposition. Only for high T_d (low density of gap states), where τ is independent of this density, is bimolecular behaviour observed.

If a quasi-equilibrium is established with the E_X states, but recombination is from E_A, then μ_D is decreased by $\exp(-\Delta E/kT)$, but $1/\tau$ is decreased by $\exp(-\frac{1}{2}\Delta E/kT)$. This case seems to have been observed in doped a-Si by Anderson and Spear (1977).

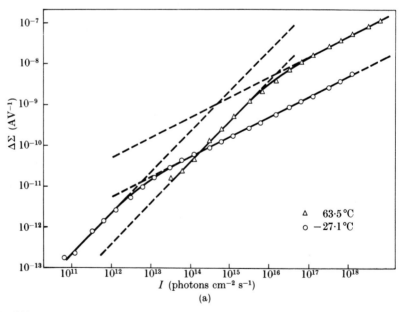

Fig. 6.32. Photoconductivity in amorphous $2As_2Te_3 . As_2Se_3$: (a) photoconductance $\Delta\Sigma$ against photon flux at two temperatures using He–Ne laser (6328 Å) excitation (b) (p. 261) photoconductance $\Delta\Sigma$ (linear and square root regimes of (a)) and dark conductance Σ against inverse temperatures. (From Weiser *et al.* 1970.)

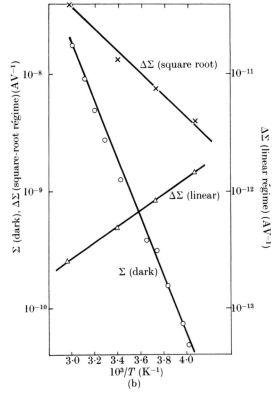

Fig. 6.32 (b)

For chalcogenides τ seems to show little temperature dependence in regime II. Some results for these due to Moustakas and Weiser (1975) and to Main (1974) are shown in Figs. 6.33 and 6.34.

We now consider the quantity b, particularly for the case of radiationless recombination, though recombination with radiation is of course possible, as in the photoluminescence observed in glow-discharge-deposited silicon at low temperatures (§ 6.7.6). An *essential* assumption for our analysis is that the majority carriers normally recombine from the lowest level of band-edge or discrete states involved in the quasi-equilibrium. If this is not so, and if they recombine from a level $\Delta\varepsilon$ higher up, b will contain a factor $\exp(-\Delta\varepsilon/kT)$; this is well shown in a detailed application of the Shockley-Read (1952) statistics due to Simmons and Taylor (1975).

We suppose then that a majority carrier combines with a trapped carrier from a discrete or band-ege localized state in its vicinity. Mott, Davis, and Street (1975) introduce two possibilities. The first is a low-temperature regime (i), in which the recombination always occurs before the majority

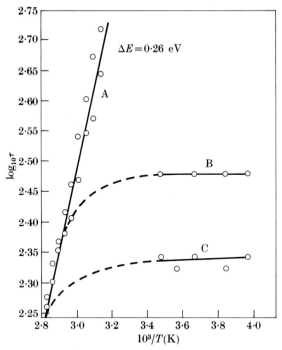

Fig. 6.33. Temperature dependence of recombination lifetime (ns) in amorphous As_2Te_3: A, 2×10^{18} photons $cm^{-2} s^{-1}$; B, 1.2×10^{19} photons $cm^{-2} s^{-1}$; C, 2.3×10^{19} photons $cm^{-2} s^{-1}$. (From Moustakas and Weiser 1975.)

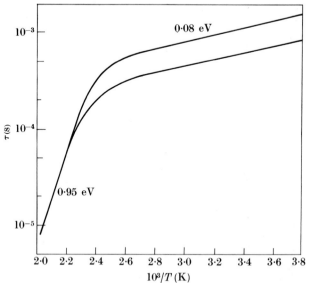

Fig. 6.34. Temperature dependence of recombination lifetime in amorphous As_2Se_3 for 10^{15} (upper curve) and 2×10^{15} photons $cm^{-2} s^{-1}$. (From Main 1974.)

carrier hops away from the recombination centre. If w is the activation energy for hopping and p_0 the probability of recombination per unit time, this occurs if

$$p_0 > \nu_{ph} \exp(-w/kT).$$

In this case

$$b = a^3 \nu_{ph} \exp(-w/kT). \tag{6.50}$$

In such a case, b is not independent of T. Since, in experiments available at the time of writing, b seems nearly independent of T, it is assumed that a high-temperature regime (ii) obtains and the carrier normally hops away before recombination. In this case

$$b = p_0 a^3. \tag{6.51}$$

Mott, Davis, and Street (1975) suppose also that charged traps capture the minority carrier in both silicon and chalcogenides, so that the centre, after capturing an electron or hole, is neutral. If it were charged, a^2 in (6.51) should be replaced by πr_e^2, where (Onsager 1938)

$$e^2/\kappa r_e \sim kT \tag{6.52}$$

and (6.51) should be multiplied by $\exp(U/kT)$, where

$$U \simeq e^2/\kappa a$$

to take account of the fact that the population of states for which the carriers are near to each other will be increased by Coulomb attraction.

With all these assumptions, then $b = p_0 a^3$, where p_0 is the radiative or non-radiative transition probability. If radiation is not emitted, then probably the multiphonon process described in § 3.5.2 must be involved, which is likely to be faster than radiative transitions only if n, the number of phonons of energy $\hbar\omega$ emitted, is less than about 8. Such a process is only independent of T if the factor $\{1 - \exp(-\hbar\omega/kT)\}^{-n}$ is nearly constant. If $\hbar\omega$ is $\sim 0 \cdot 1$ eV, this is likely to be so at room temperature.

For glow-discharge-deposited silicon in the bimolecular range, we follow Spear (1974b) in supposing that holes are captured by E_Y centres. If these are charged, they must be centres at the top of the E_Y band where they overlap the E_X band. At room temperature the subsequent recombination with an electron is not observed to be radiative, and Mott, Davis, and Street (1975) suggest that the centres involved may include a hydrogen atom, which gives a high value of $\hbar\omega$ and thus a high non-radiative transition probability (§ 3.5.2). However, at low temperatures this transition is observed in the photoluminescence spectrum, and a possibility (§ 6.7.6) is that these are transitions to centres without hydrogen.

For chalcogenides photoconductivity is interpreted in terms of the D^+, D^- centres introduced earlier and discussed in Chapter 9. Electrons are thought to be trapped by deep D^+ centres, holes establish a quasi-equilibrium with D^- centres, and recombination is a tunnelling process in which two D centres revert to D^+ and D^-.

We have already stated that for silicon this treatment is correct only for a low density of gap states; in this limit, the density of gap states does not appear in the equation for the photocurrent. For high densities in glow-discharge-deposited silicon i_p is no longer found to be proportional to $G^{1/2}$ but approaches G, as discussed earlier. For very high densities electrons and holes must drop down to the Fermi level by tunnelling from one state to another with emission of phonons. One might expect in such a case a 'demarcation level', namely an energy in the gap above which electrons normally escape into the conduction band and below which they normally drop down (see Rose 1963). Arnoldussen et al. (1974) have discussed photoconduction in chalcogenides in these terms and Weiser and Brodsky (1970) used it for a treatment of photoluminescence (for an alternative treatment based on more recent results see Chapter 9). In this kind of photoconduction the current should be proportional to G and to $\exp\{-(E_C - E_D)/kT\}$, where E_D is the demarcation level, which itself increases with decreasing temperature.

If the electrons can jump from one state in the gap to another, they must be able to contribute to the current after they have fallen below the demarcation level. This kind of conduction has been discussed by Fischer and Vornholz (1975) as an interpretation of their observations on evaporated amorphous germanium. They consider that if the ranges of energy of deep electron and hole traps overlap slightly, giving a finite value of $N(E_F)$, carriers will be quickly thermalized to E_F and recombine there slowly enough for the photocurrent to be mainly due to carriers there. An approximate behaviour of the photocurrent as $\exp(-B/T^{1/4})$ is observed.

We now discuss the spectral dependence of the current on exciting radiation. In contrast with most crystalline semiconductors, which exhibit a peak in the photoconductivity at a photon energy corresponding to the onset of interband electronic transitions, amorphous semiconductors with small gaps have a spectral response of photoconductivity that rises at approximately the same photon energy as the optical absorption edge and remains relatively constant at higher energies (Fig. 6.35). As the fall-off on the high-energy side of the edge in crystals is attributed to the increased role of surface recombination for carriers generated by strongly absorbed light, this observation presumably implies very similar rates of recombination at the surface and in the bulk. As we shall see in § 6.7, the optical absorption edge of nearly all amorphous semiconductors is far from sharp,

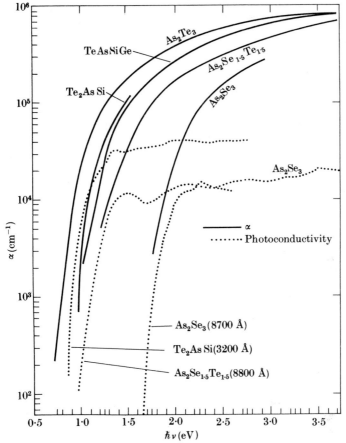

Fig. 6.35. Photoconductivity and optical absorption against photon energy in chalcogenides at room temperature. The thicknesses of the films in which photoconductivity was measured are given in parentheses. (From Rockstad 1970.)

and in fact is normally characterized by an absorption constant that rises exponentially with photon energy. These two features, together with a possible wavelength dependence of the quantum efficiency, make the determination of mobility gaps from photoconductivity data uncertain.

6.5.3. Quantum efficiency

We now discuss G, the rate of generation of carriers. We may write

$$G = \eta\left[\frac{I_0(1-R)\{1-\exp(-\alpha d)\}}{d}\right] \qquad (6.53)$$

where the quantity in square brackets is the number of photons absorbed per second in a sample of thickness d in the direction of the incident radiation, I_0 is the incident flux density, α the absorption coefficient, and R the reflectivity. The quantum efficiency for photogeneration is denoted by η. It is sometimes assumed to be unity for suitable excitation; however, it can be measured absolutely and for some materials (notably amorphous Se and to a lesser extent As_2Se_3) it is found to be dependent on photon energy, electric field, and temperature, as we shall now see.

If, in a photoconductivity experiment, the wavelength of the excitation is such that absorption occurs by one-electron interband transitions (i.e. energy of radiation quanta comparable with the fundamental absorption edge), we may assume that one carrier pair (electron and hole) is *created* per absorbed photon. However, if any of these pairs undergo recombination with each other before they separate (geminate recombination), then they cannot of course contribute to the photocurrent. The quantum efficiency is defined as the number of *free* electrons (or holes) created per absorbed photon.

The most straightforward technique for the determination of quantum efficiency involves the use of strongly absorbed, pulsed excitation in a sandwich cell of the type described in § 6.5.1 for the measurement of drift mobility. Sufficiently high electron fields are used to ensure that the drifting carriers (of one sign) do not suffer a range limitation by deep trapping events during their passage across the film. Another similar method is the xerographic discharge technique in which the rate of change of voltage at the free surface of a corona-charged film is measured under steady illumination. Extensive measurements of this kind have been made on amorphous Se by Tabak and Warter (1968), Pai and Ing (1968), and Pai and Enck (1975), and on amorphous As_2S_3 by Ing and Neyhart (unpublished). In both these materials the quantum efficiency is found to depend on photon energy, temperature, and electric field, reaching unity only for high values of one or more of these parameters. Fig. 6.36 (from Hartke and Regensburger 1965) compares the quantum efficiency and optical absorption edges in amorphous Se at room temperature. The quantum efficiency does not saturate at unity until a photon energy $\sim3\,\text{eV}$ is reached. Although it is not obvious where exactly to locate the optical absorption edge (this problem will be discussed in § 6.7), it is clear that there is considerable absorption at $2 \cdot 1\,\text{eV}$, at which energy the quantum efficiency has fallen to very small values. Further discussion of photogeneration in amorphous Se will be given in Chapter 10.

Main and Owen (1973) have determined the photon energy dependence of quantum efficiency in amorphous As_2Se_3 and As_2S_3 by a direct normalization of the photoconductivity in terms of photocurrent per absorbed photon. Their results for As_2S_3 (Fig. 6.37) show that the quantum efficiency falls with decreasing photon energy well above the optical absorption edge (taken to be where $\alpha = 5 \times 10^3\,\text{cm}^{-1}$), although the effect is

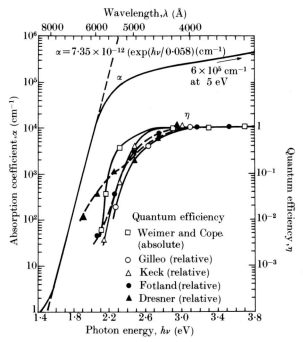

Fig. 6.36. Optical absorption edge and quantum efficiency in amorphous selenium at room temperature. (From Hartke and Regensburger 1965.)

not so marked as in Se. As_2Te_3 films do not exhibit the phenomenon, presumably because the parameters in this material, particularly the dielectric constant, allow for easier separation of the electron and hole.

A simple model for the temperature and wavelength dependence of quantum efficiency is as follows. In crystalline semiconductors, dissociation of an exciton or the excited state of a donor will not lead to photoconduction unless the carriers separate from each other, or the electron separates from the donor. An early example of the latter behaviour is provided by the work of Pohl (1937) on F centres in alkali halides, which shows a sharp drop in quantum efficiency below $-150°C$ in NaCl for example. The explanation of this (Gurney and Mott 1938, Mott and Gurney 1940) is as follows. If p_0 is the probability per unit time of radiative or multiphonon recombination of the exciton, W is the binding energy of the electron and hole so that the probability per unit time of escape into the conduction band is $\nu_{ph} \exp(-W/kT)$, the quantum efficiency η, should be given by

$$\eta = \frac{1}{1 + (p_0/\nu_{ph}) \exp(W/kT)}. \qquad (6.54)$$

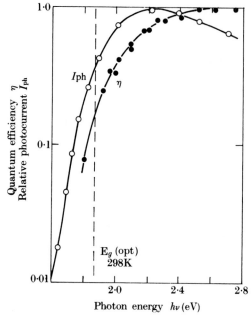

Fig. 6.37. Spectral dependence of normalized photocurrent and quantum efficiency for As$_2$Se$_2$ at room temperature. (From Main and Owen 1973.)

p_0/ν_{ph} is probably small ($\sim 10^{-4}$ for radiative transitions) so at low temperatures η is proportional to $\exp(-W/kT)$, saturating at near unity when $kT > W/\ln(\nu_{ph}/p_0)$.

If recombination is radiative, the number of quanta emitted per absorbed quantum will be $1 - \eta$.

In non-crystalline materials various theoretical discussions of the exciton have been given (Majerníková 1974a,b, Bonch-Bruevich and Iskra 1975; see also § 6.7.1). What is certain, however, is that bound electron–hole pairs can form under their mutual potential energy $-e^2/\kappa r$. Two models are possible. If the binding energy W of an exciton ($m^*e^4/2\hbar^2\kappa^2$ with $m^* = m_e m_h/(m_e + m_h)$) is greater than the range of localized states in at least one band, then we can probably use the concept of a Wannier exciton in its lowest state; the broadening by random fields is known to give an Urbach edge (§ 6.7.1). In the contrary case we may think of carriers in band-edge localized states, at a distance a as small as possible, bound with an energy $e^2/\kappa a$ (Pankove and Carlson 1976). A difference from the crystalline case is that, for electrons at the band edge, either $L \sim a$ or motion is by hopping and we may think of a diffusive motion by which the electron and hole approach each other. Davis (1970) and Knights and Davis (1974) have treated this in the following way, illustrated in Fig. 6.38.

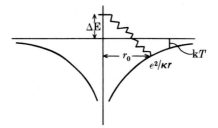

Fig. 6.38. Thermalization of electron–hole pair in their mutual Coulomb field. Starting with an excess kinetic energy ΔE, the carrier pair ends up with a binding energy $e^2/\kappa r_0$.

We suppose that an electron is excited to an energy ΔE above the bottom of the conduction band. If the electron loses one phonon quantum $\hbar\omega$ per time $1/\omega$ (cf. Chapter 3), it will diffuse a distance r_0 given by

$$r_0 = \left(\frac{\Delta E + e^2/\kappa r_0 D}{\hbar\omega^2}\right)^{1/2}$$

before it is thermalized; D is here the diffusion coefficient. If now r_0 is greater than the Onsager radius r_e defined by (6.52), the particles will escape. If not, they will normally come together and form an 'exciton'. In this there are two possibilities.

(a) Recombination normally then takes place. In this case the quantum efficiency is (cf. Mott 1977f), if $r_0 < r_e$,

$$\eta = \frac{1}{1 + \exp(E_0/kT)} \tag{6.55}$$

with

$$E_0 = \frac{e^2}{\kappa}\left(\frac{1}{r_e} - \frac{1}{r_0}\right).$$

(b) The electron and hole normally escape before recombination. In this case $\eta \sim 1$.

The latter case is of interest in connection with photoluminescence, in which case eqn (6.54) is valid with $1 - \eta$ for the efficiency.

Knights and Davis (1974) also describe the electric-field dependence of the quantum efficiency on a simple Poole–Frenkel model in which the effective barrier for escape is lowered by the field. Although the essence of the model is believed to be correct and fair agreement with experimental data for Se was obtained, a better analysis in terms of the Onsager theory of dissociation has been given by Pai and Enck (1975). This is discussed further by Mott (1977f) and in Chapter 10.

6.6. Conduction at a mobility edge *versus* hopping

With the exception of evaporated Ge and Si and some small-band-gap alloys, for which $T^{-1/4}$ behaviour and small values of the thermopower indicate conduction by electrons at the Fermi level, most amorphous semiconductors have conductivities and thermopowers that are temperature activated with an energy close to one-half of the optical gap. Whether conduction occurs by hopping in localized states at an extremity of the valence (or conduction) band or by charge transport in extended states beyond a mobility edge, depends on the form of the density of states, the range of energies of localized states, the increase in mobility at E_C, and the temperature. In this section we summarize the expected dependence on temperature of the d.c. conductivity σ, the thermopower S, the drift mobility μ_D, and the Hall mobility μ_H on passing through the transition from one mode of conduction to the other.

The model used to illustrate these forms of behaviour is shown on the right of Fig. 6.39 and the activation energies marked on the left assume no

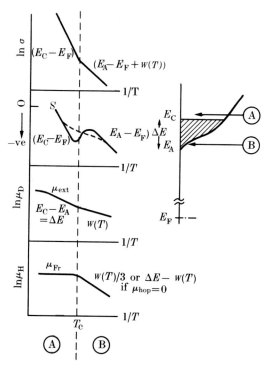

Fig. 6.39. Schematic representation of temperature dependence of d.c. conductivity σ, thermopower S, drift mobility μ_D, and Hall mobility μ_H for carriers in a band with a mobility edge as shown on the right.

temperature variation of E_C, E_A, or E_F relative to each other. Conduction above T_c is predominantly by carriers at E_C and below T_c at E_A. In some materials a more gradual transition than illustrated may occur if the change in mobility at E_C is not great.

In the temperature range marked A both the conductivity and the thermopower are activated with an energy $E_C - E_F$. The drift mobility, which is trap limited in this regime, is activated with an energy $E_C - E_A = \Delta E$, unless it is determined by discrete states below E_A (§ 6.5.1). It should saturate to μ_{ext} (which contains a T^{-1} dependence) at sufficiently high temperatures (see eqn (6.33)). The Hall mobility should be independent of temperature and have a value given by the Friedman expression (§ 6.4.7).

In the temperature range B the conductivity is activated with energy $E_A - E_F$ together with the contribution $w(T)$ due to hopping. A change of slope of magnitude $\Delta E - w(T)$ therefore occurs at T_c. If it is assumed that none of the hopping energy contributes to the thermopower, then the curve of S versus $1/T$ has a slope of $E_A - E_F$. However, because the thermopowers extrapolated to $1/T = 0$ are similar for processes A and B, the form of S in the transition region can take several forms depending on the sharpness of the transition in σ. The total thermopower for parallel conduction mechanisms is simply the sum of the individual thermopowers weighted according to the contribution each mechanism makes to the conductivity; the slope of S in the transition region has little meaning. The drift mobility in region B should exhibit a temperature-dependent hopping energy $w(T)$ and therefore the change in slope of the plot of μ_D versus $1/T$ at T_c is equal to $\Delta E - w(T)$ as in the conductivity.

The temperature dependence of the Hall mobility in region B is uncertain. If the main part of the hopping energy $w(T)$ is of polaron type, then an activation energy of $\frac{1}{3}w(T)$ is expected (§ 3.9). If, however, we assume that this is not the case, the contribution to the Hall effect is negligible and the temperature dependence of the Hall mobility will simply reflect the decreasing contribution to the Hall effect from carriers moving above E_C. As for the thermopower, the weighting factors for parallel mechanisms are just the relative contributions each makes to the conductivity. The activation energy found by plotting $\ln \mu_H$ versus $1/T$ in region B is therefore expected to be $\Delta E - w(T)$.

As an example of a gradual transition from process A to process B we show in Fig. 6.40 data due to Nagels et al. (1974) on p-type As_2Te_3 (+1% Si). The solid curves are theoretical curves based on the models for σ, S, and μ_H using the parameters given in the underline.

Results on glow-discharge-deposited silicon by Le Comber and Spear (1970), which display a much sharper transition, and on glow-discharge-deposited germanium (Jones et al. 1976) will be described in Chapter 7. Data for amorphous arsenic appear in Chapter 8.

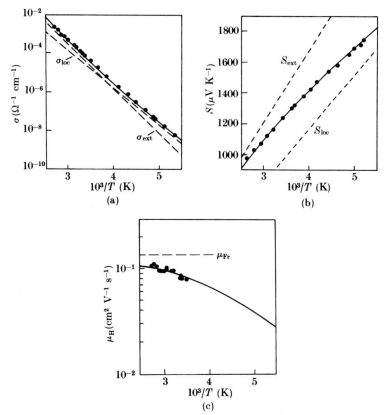

Fig. 6.40. (a) Conductivity, (b) thermopower, and (c) Hall mobility of As_2Te_3 with 1% Si, plotted from the results of Nagels *et al.* (1974). The solid curves are theoretical using the following parameters (conduction is by holes): $E_F - E_B = 0.35$ eV, $\Delta E = 0.11$ eV, $w = 0.03$ eV, $\sigma_0 = 3200 \ \Omega^{-2}$ cm^{-1}, $\sigma_1 = 90 \ \Omega^{-1}$ cm^{-1} (σ_0 and σ_1 are the intercepts on the $1/T = 0$ axis for conduction in extended and localized states respectively). (From Mott, Davis, and Street 1975.)

6.7. Optical absorption

In Fig. 6.18 the optical absorption in amorphous As_2Se_3 was shown over a very wide range of frequencies. This section is concerned with those absorption processes that occur at the higher end of the frequency spectrum, particularly those associated with interband electronic transitions. In §§ 6.7.1 and 6.7.2 we consider the fundamental absorption edge, which in the materials with which we are concerned lies between 0.3 and 2.5 eV, and the effects of externally applied electric fields on this. The absorption at slightly higher energies (associated with absorption coefficients $\alpha \gtrsim 10^4$ cm^{-1}), which may provide information on the combined density of states at the valence-band and conduction-band edges, is discussed in § 6.7.3. Interband absorption involving valence and conduction states lying deeper in their respective bands is the subject of § 6.7.4. The use of synchrotron radiation extends the range normally covered by ultraviolet spectroscopy (6–15 eV) out to several hundred electronvolts and enables the observation of absorption due to transitions from deep-lying atomic-like levels to the conduction band. X-ray and u.v. photoemission spectra will be discussed in a separate section (§ 6.8.2).

Reference to Fig. 6.18 shows a region of the spectrum in the range 10^{14}–10^{15} s^{-1} in which no measurements are shown for As$_2$Se$_3$. In this range intraband or free-carrier absorption could be detected, if it were not for impurity and structure-dependent absorption which frequently dominates the spectrum of materials on the low-energy side of the fundamental edge. There is, however, some evidence for intraband absorption in As$_2$Se$_3$ at high temperatures (liquid phase), when the d.c. conductivity is greater than $10^{-2}\,\Omega^{-1}\,\mathrm{cm}^{-1}$, and at room temperature in chalcogenides with a smaller band gap. Intraband absorption will be the subject of § 6.7.5. Absorption by phonons (and Raman scattering) will be discussed in § 6.7.6.

Perhaps the most important feature of optical absorption processes in amorphous semiconductors is that certain selection rules (particularly that of k conservation) which apply to optically induced transitions in crystalline materials are relaxed.

6.7.1. Absorption edges and Urbach's rule

Before discussing the optical absorption edges observed in amorphous semiconductors, we review briefly the types of edges that have been found in crystals and their interpretation.

Basically there are two types of optical transition that can occur at the fundamental edge of crystalline semiconductors, direct and indirect. Both involve the interaction of an electromagnetic wave with an electron in the valence band, which is raised across the fundamental gap to the conduction band. However, indirect transitions also involve simultaneous interaction with lattice vibrations. Thus the wavevector of the electron can change in the optical transition, the momentum change being taken or given up by phonons. (The radiation imparts negligible momentum to the electron.)

If exciton formation (electron–hole interaction) is neglected, the forms of the absorption coefficient α as a function of photon energy $\hbar\omega$ depend on the dependence on energy of $N(E)$ for the bands containing the initial and final states. For simple parabolic bands ($N(E) \propto E^{1/2}$) and for direct transitions

$$\alpha n_0 \hbar\omega \sim (\hbar\omega - E_0)^n$$

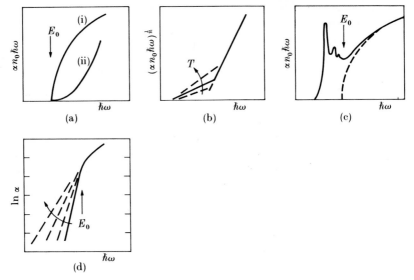

Fig. 6.41. Schematic illustration of different kinds of absorption edges in crystalline semiconductors: (a) (i) direct allowed, (ii) direct forbidden; (b) indirect, $n = 2$ allowed, $n = 3$ forbidden; (c) direct allowed with exciton formation; (d) exponential 'Urbach' edges.

where $n = \frac{1}{2}$ or $\frac{3}{2}$ depending on whether the transition is allowed or forbidden in the quantum-mechanical sense. E_0 is the optical gap and n_0 the refractive index. This type of absorption, shown schematically in Fig. 6.41, is independent of temperature apart from any variation in E_0. For indirect transitions

$$\alpha n_0 \hbar \omega = \frac{(\hbar\omega - E_0 + h\nu_{\mathrm{ph}})^n}{\exp(h\nu_{\mathrm{ph}}/kT) - 1} + \frac{(\hbar\omega - E_0 - h\nu_{\mathrm{ph}})^n}{1 - \exp(-h\nu_{\mathrm{ph}}/kT)}$$

The two terms here represent contributions from transitions involving phonon absorption and emission respectively, and have different coefficients of proportionality and temperature dependences. For allowed transitions $n = 2$ and for forbidden transitions $n = 3$. In each case multiple-phonon processes can occur, leading to additional pairs of terms.

In general both direct and indirect transitions can occur in a crystalline semiconductor. However, in materials in which the smallest gap is a direct one, indirect transitions, which are associated with smaller absorption coefficients, are not observed. An exception to this is the special case of vertical transitions involving absorption of a phonon of very small wavevector.

All the above types of optical transitions are modified when the electron–hole interaction is not ignored. The mutual attraction allows bound states of the electron and hole, namely excitons, to be formed with energy less than that of the free pair. For direct allowed transitions absorption into these exciton states is in the form of a hydrogenic spectrum below the continuum at E_0 (Fig. 6.41(c)). Moreover the continuum itself is modified in shape. For direct forbidden transitions the $n = 1$ line is missing and the continuum absorption, although increased above the edge, starts at $\alpha = 0$.

For allowed and forbidden indirect transitions inclusion of electron–hole interaction does not lead to a series of absorption lines, transitions into bound exciton states giving rise to a continuous absorption. However, additional terms similar to those given in the equation above will occur, corresponding to the transitions into bound exciton states. For these the power n is $\frac{1}{2}$ and $\frac{3}{2}$ for allowed and forbidden transitions respectively. For transitions into the continuum the corresponding power laws are $\frac{3}{2}$ and $\frac{5}{2}$.

Examples of most of the above relationships have been observed in crystalline semiconductors, and investigations on certain materials have yielded detailed information concerning the electronic structure at the band extrema.

A completely different type of optical absorption edge is observed in several materials, in particular the alkali halides, CdS, and trigonal selenium. It is an absorption constant that increases as the exponential of the photon energy. Thus, in contrast with 'direct allowed' edges, for example, which lead to a rather rapid rise in the absorption coefficient over several decades within a few tenths of an electronvolt near the energy gap, one observes in this case a gradual increase in the absorption extending over perhaps several electronvolts. This so-called Urbach edge (Urbach 1953, Dexter and Knox 1965) frequently obeys the empirical relationship

$$\alpha = \alpha_0 \exp\left\{ -\frac{\gamma'(E_0 - \hbar\omega)}{kT} \right\} \tag{6.56}$$

where γ' is a constant and T is the absolute temperature down to a critical value T_0 and equal to T_0 for lower temperatures. Thus the edge becomes broader as the temperature rises above T_0 (Fig. 6.41(d)). No single explanation has been accepted for this behaviour in crystalline materials, which is unfortunate in view of the fact that, as we shall see, it appears to be the type of edge most characteristic in amorphous semiconductors. A short review of the theoretical situation has been given by Hopfield (1968). We shall briefly summarize the various models that have been proposed.

(a) *Bound exciton interaction with lattice vibrations.* Toyozawa (1959a,b, 1964) first proposed that the normal Gaussian shape of an exciton line becomes exponential in its leading edge when quadratic terms in its interaction with phonons are considered. This theory seems to account for the exponential tails observed in alkali halides particularly well. The difficulty seems to be in justifying the fact that the quadratic terms outweigh the linear terms in the interaction. For other theoretical treatments based on the same ideas the reader should see Mahr (1963), Mahan (1966), and Keil (1966).

(b) *Electric-field broadening of the absorption edge.* In the presence of an electric field, the absorption coefficient associated with direct allowed transitions between parabolic bands is modified in the manner shown in Fig. 6.42. At photon energies below the onset of the field-free absorption

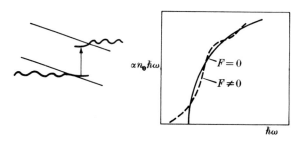

Fig. 6.42. Schematic illustration of the Franz–Keldysh effect and the effect on a direct allowed optical absorption edge.

an approximately exponential tail is introduced, and at higher energies the absorption coefficient oscillates. The 'red shift' at low energies is the Franz–Keldysh effect and is due to a finite probability for tunnelling of the band Bloch states into the energy gap. Clearly, if this explanation is invoked to explain Urbach behaviour, then the origin of the electric field has to be considered. In principle this could arise from charged impurity states (Redfield 1963). However, the temperature dependence of the slope must be taken into account in any proposed model. Dexter (1967) introduces this by postulating that the electric field arises from the vibrating atoms in the material. In ionic solids optical phonons are involved; in the more covalent materials instantaneous changes in electronic charge clouds have to be invoked. The magnitude of the effect depends on the electron–phonon interaction, which is strong in piezoelectric solids like CdS and trigonal Se.

(c) *Electric-field broadening of an exciton line.* Dow and Redfield (1970) treated the problem of absorption for direct excitonic transitions in a uniform electric field. A numerical study of the shape of the absorption showed that, in contrast with the Franz–Keldysh results, its variation with photon energy is accurately exponential. These authors therefore propose that the 'spectral' Urbach rule arises from an electric-field broadening of an exciton line. Fig. 6.43 shows the results of the calculations. The parameter f is the electric-field strength expressed as the ratio of the potential-energy drop due to the field across the radius of the exciton to the

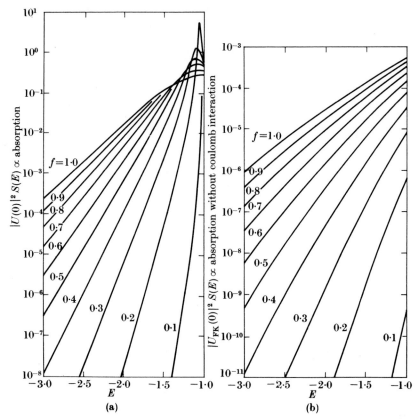

Fig. 6.43. Dow and Redfield's theory for optical absorption in the presence of an electric field: (a) exciton effects included; (b) exciton effects not included (Franz–Keldysh result). The parameter f is a measure of the strength of the electric field. The energy E is measured from the conduction-band edge in units of the binding energy of the unperturbed exciton. (From Dow and Redfield 1970.)

exciton binding energy. The Bohr radius a of the exciton ground state and the Rydberg constant R are given by

$$a = \frac{\hbar^2 \kappa}{m^* e^2} \qquad R = \frac{e^2}{2\kappa a} = \frac{m^* e^4}{2\kappa^2 \hbar^2}$$

where m^* is the reduced mass of the electron–hole pair. Thus

$$f = \frac{F|e|a}{R}$$

$$= 3 \cdot 89 \times 10^{-10} \left(\frac{m}{m^*}\right)^2 \kappa^3 F$$

where m is the free-electron mass and F is in $V\,cm^{-1}$. For $m^* = m$ and $\kappa = 6$, the line marked $f = 0{\cdot}6$ corresponds to a field $\sim 7 \times 10^6\,V\,cm^{-1}$. It is interesting that a classical criterion for direct field ionization of an exciton $(F_C = \kappa R^2/4e^3)$ corresponds to $f = 0{\cdot}125$. As seen from the curves, a discernible exciton peak remains for fields almost an order of magnitude larger than this. The influence of the electron–hole interaction on the shape of absorption curves can be seen by comparison with Fig. 6.43, in which the Coulomb interaction has been 'turned off', giving the normal Franz–Keldysh result

$$\alpha = \alpha_0 \exp\{-C(E_0 - \hbar\omega)^{3/2}\}.$$

If Dow and Redfield's model is a correct description of Urbach's rule, it is of course still necessary to consider the origin of the internal electric field and the temperature dependence of the slope. This point will be considered later.

As mentioned above, the fundamental absorption edge in most amorphous semiconductors follows an exponential law, i.e. $\ln \alpha$ is proportional to $\hbar\omega$. At the time of writing only certain thin films of Ge and Si and perhaps InSb stand out as notable exceptions. (Absorption edges in Ge and Si will be presented in Chapter 7.) However, because of the experimental difficulties associated with measuring small values of the absorption constant in thin films, it is not easy to test for exponential behaviour in those amorphous semiconductors that can be prepared only in this form. Verification of exponential behaviour is thus established with certainty only for those amorphous semiconductors that can be obtained as a glass by melt quenching, even though of course thin films of these materials are used for transmission measurements at high values of α. Fig. 6.44 shows the room-temperature absorption edges of a few amorphous semiconductors. Exponential behaviour is observed up to $\alpha = 10^2\,cm^{-1}$ in some materials, and up to $10^4\,cm^{-1}$ in others.

As discussed above there are several plausible explanations for the existence of an exponential absorption edge. In amorphous semiconductors there is an additional possibility which has been suggested by many authors, namely that it arises from electronic transitions between (localized) states in the band-edge tails, the density of which is assumed to fall off exponentially with energy (Tauc 1970a, Lanyon 1963). We think that this explanation is unlikely (see Davis and Mott 1970). The main evidence against such an interpretation is that, as shown in Fig. 6.44 and Table 6.3, slopes of the observed exponential absorption edges are very much the same in a variety of materials; it would seem unlikely that the state tailing is so similar. We shall, however, show below that there is an interesting correlation for elemental materials between the slope of the Urbach tail and the co-ordination number.

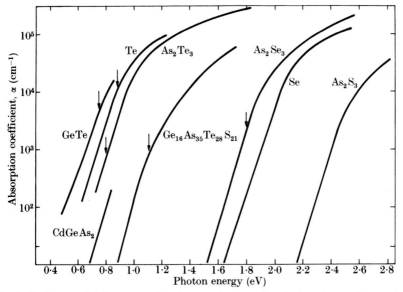

Fig. 6.44. Exponential absorption edges in amorphous semiconductors at room temperature (references as for Table 6.3). The arrows mark the value of $2E$ for those materials for which the electrical conductivity has been observed to obey the relation $\sigma = C \exp(-E/kT)$.

In view of the likely presence of strong internal fields in amorphous semiconductors and the accurately exponential behaviour predicted by Dow and Redfield's theory for field broadening of an exciton line, this model is obviously attractive as an explanation for the type of edge shown

TABLE 6.3

Approximate values of Γ for the amorphous semiconductors of Fig. 6.44 in the region where $\alpha = \alpha'_0 \exp(\Gamma \hbar \omega)$, and values of the photon energy $\hbar \omega$ corresponding to $\alpha = 10^2 \text{ cm}^{-1}$ ($T = 300$ K)

	$\Gamma(\text{eV}^{-1})$	$\gamma' = \Gamma kT$	$\hbar\omega(\alpha = 10^2 \text{ cm}^{-1})$ (eV)	Reference
GeTe	15	0·38	0·50	Bahl and Chopra (1969)
Te	18	0·47	0·61	Stuke (1970a)
As_2Te_3	19	0·49	0·70	Rockstad (1970)
$CdGeAs_2$	19	0·49	0·80	Červinka et al. (1970)
$Ge_{16}As_{35}Te_{28}Si_{21}$	22	0·57	0·99	Fagen et al. (1970)
As_2Se_3	20	0·52	1·64	Edmond (1966) (from Owen 1970)
Se	17	0·44	1·77	Hartke and Regensburger (1965)
As_2S_3	19	0·49	2·28	Kosek and Tauc (1970)

in Fig. 6.44. If the average of the observed slopes (Table 6.3) is taken as $17\,\text{eV}^{-1}$, the corresponding value of the parameter f in the figure is $\sim 0\cdot 5$ for $\kappa = 6$ and $\sim 1\cdot 0$ for $\kappa = 8$ (assuming $m^* = m$). Thus electric fields of strength 10^6–$10^7\,\text{V cm}^{-1}$ are required. Before proceeding further we should pay attention to the following questions.

(a) Do excitons exist in amorphous semiconductors?

(b) What possible sources are there of internal electric fields?

(c) If the fields are random and have a spatial variation in magnitude, do they average to give the same result as a constant uniform field?

With regard to (a), some discussion has already been given in § 6.5.3; both for chalcogenides and for amorphous silicon (§ 6.7.6) electron–hole pairs bound together are known to exist. However, sharp exciton lines are hardly to be expected, particularly for an excited state of the exciton or pair. Silicon dioxide, where the exciton binding energy is large, is an exception (Chapter 9), and in amorphous Mg–Bi alloys (Slowick and Brown 1972) exciton lines are observed when excitation is from core levels but not for excitation from the valence band. If the exciton binding energy is small compared with the range of localized states, then the Coulomb force between an electron and a hole should lead simply to an effective lowering in energy of the relevant states by $\sim e^2/\kappa a_E$ where a_E is the distance between those states. However, it should be noted that Dow and Redfield's theory predicts, for a given electric field, a *large* effect on the absorption edge when the exciton binding energy is *small*.

As regards the origin of internal electric fields in amorphous semiconductors we may consider longitudinal optical phonons (as Dexter (1967) did in crystalline materials), or static spatial fluctuations in potential arising from the lack of long-range order, variations in density, or charged defect centres. It is not easy to calculate the magnitude of the electric fields produced by these possible sources without knowledge of specific parameters (Tauc 1970b). It is relevant to note that experimental evidence has been obtained for exponential broadening of absorption edges created by charged impurities in crystals (Afromowitz and Redfield 1968). Also Olley (1973) has shown that bombardment of crystalline PbI_2 with helium ions quenches a sharp exciton line and induces an exponential leading edge (Fig. 6.45); similar behaviour has been observed in other materials.

With regard to (c), general calculations are difficult (Bonch-Bruevich 1970a,b, 1973, Bonch-Bruevich and Iskra 1975, Majerníková 1974a,b, Lukes and Somaratna 1970, Redfield 1963). Dow and Redfield (1972) have shown that the exponential tail derived for uniform fields survives the necessary averaging in the case of certain random distributions. However, Majerníková (1975) has pointed out that a Gaussian distribution of random fields yields an exponential absorption edge *without* the need to include exciton effects.

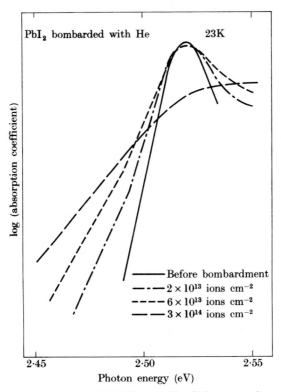

Fig. 6.45. Optical absorption edge of initially crystalline PbI_2 measured at various stages of bombardment with He ions. The excitonic feature broadens into an exponential tail. (From Olley 1973.)

For electrostatic potential fluctuations such that valence and conduction bands remain parallel (Fig. 6.6(b)), Dow and Hopfield (1972) calculate an optical absorption edge that is suppressed by the lack of spatial correlation between electrons and holes but the shape of which is not very different from the density of states averaged over many potential fluctuations.

Any model that is invoked to explain exponential absorption edges in amorphous semiconductors should consider the effect of temperature. Experimentally it is observed that there is only a slight change of slope with temperature below room temperature. In this case it may be preferable to refer to 'a spectral Urbach rule'. There is, as might be expected, a displacement towards lower photon energies as the temperature is raised, but this presumably is due to the temperature coefficient of a gap. Above room temperature (from data on a limited number of materials) the slope decreases as the temperature is raised. However, it appears necessary to reach the liquid state before the temperature dependence predicted by Urbach's empirical rule (eqn (6.56)) is obeyed. Thus, for amorphous semiconductors, T_0 is considerably higher than in crystalline materials obeying Urbach's rule. If the exciton model with electric-field broadening is appropriate, then the temperature dependence of the dielectric constant will certainly be involved, although probably not to the third power as in Dow and Redfield's theory, because F itself will probably be proportional to at least $1/\kappa$. If the field is considered to arise from longitudinal optical phonons, then it may be possible to account for the high value of T_0 by regarding the amorphous state as one that contains a large concentration of frozen-in phonons (see Chopra and Bahl 1972).

For As_2S_3, Street, Searle, Austin, and Sussman (1974) propose that the slope of the Urbach tail is determined by two contributions to broadening, one arising from static disorder and the other from phonons. They estimate the former, which is insensitive to temperature, to be about three times the latter.

In view of the uncertainties concerning the nature of the exponential absorption edges in amorphous semiconductors, we may ask whether it is possible to define an *optical gap*. In crystalline materials obeying Urbach's rule, a gap is normally determined by locating the *focal point* of the family of edges obtained as a function of temperature. Thus, allowing E_0 in eqn (6.56) to be a linear function of temperature, so that $E_0 = E_0(0) - \beta T$ we obtain

$$\alpha = \alpha_0 \exp\left[-\frac{\gamma'\{E_0(0) - \hbar\omega - \beta T\}}{kT} \right] \qquad (6.57)$$

The absorption coefficient is thus independent of T when $\hbar\omega = E_0(0)$ and this is the focal point. In trigonal selenium for example (Roberts, Tutihasi, and Keezer 1968) it lies at $\alpha = 10^5$ cm^{-1}. As the temperature dependence of Urbach's rule is not normally seen in solid amorphous semiconductors, this procedure cannot be used. In § 6.7.3 we shall show how the form of the absorption above the exponential edge may be extrapolated to give an optical gap. This energy frequently turns out to lie close to the 'knee' of the absorption edge, i.e. the energy where $\ln \alpha$ ceases to be linear with $\hbar\omega$. However, this is a very crude marker, and we shall see that it is expected to give values less than the mobility gap.

It has been pointed out by Stuke (1970a) that the mobility gap in many amorphous semiconductors corresponds to a photon energy at which the optical absorption coefficient has a value $\sim 10^4$ cm^{-1}. The assumption here is that the mobility gap is equal to twice the activation energy for electrical conduction, that is twice the slope of a plot of $\ln \sigma$ against $1/kT$, so that the conduction is truly intrinsic. However, if the position of the Fermi level is determined by states in the gap (§ 6.4.1), then twice the activation energy observed in electrical conduction is not a quantity of particular significance. In any case one should correct the slope of a curve plotting $\ln \sigma$ against

$1/kT$ for any temperature variation of the mobility gap in order to compare it with a room-temperature absorption edge. Applying this correction leads to the positioning of the mobility gap in many materials at an energy where the absorption coefficient is approximately 10^3 cm^{-1}.

The possible existence of sharp absorption edges in amorphous Ge reported by some workers (Chapter 7) demands special attention. There seems to be no reason why random fields should not occur in this material; the density of defect centres is probably larger than in the chalcogenides and small voids may be present even in samples deposited onto high-temperature substrates. The exciton binding energy in crystalline Ge is $\sim 2 \text{ meV}$, an exceedingly small value arising partly because of the small effective mass of the electron–hole pair $(m^* \sim 0 \cdot 33 \text{ m})$ but partly because of a large dielectric constant $(\kappa = 15 \cdot 8)$. Although it is not clear what value to take for the effective mass in amorphous Ge, it might be considered that the absence of an exponential edge similar to that observed in the chalcogenides arises from the higher dielectric constant in this material, which could make exciton effects negligible. However, there is some difficulty with this suggestion. From the curves of Fig. 6.43(a), the slope of the optical absorption edge can be found to be given roughly by the relation

$$\Gamma = \text{const.} \left\{ \frac{(m^*)^{2b-1}}{\kappa^{3-2b}} \right\} F^{-b} \qquad (6.58)$$

where b lies between $0 \cdot 18$ and unity. Thus, for the same magnitude of electric field F, the slope decreases with decreasing effective mass and increasing dielectric constant. To obtain a sharp edge necessitates the assumption of random electric fields considerably lower in magnitude than estimated for the chalcogenides. Alternatively, it could be that high fields do exist in the vicinity of defects, but their spatial extent is small (on account of the high dielectric constant); their effects on optical properties may thus be reduced in importance relative to the chalcogenides. Certain films of germanium, however, *do* exhibit exponential edges (Chapter 7), and in these cases the corresponding slope is one-half to one-third that commonly appearing for chalcogenides. This would seem to be more consistent with the model of field-broadened absorption edges discussed above.

An interesting correlation exists between the slope of the Urbach edge and valency N in elemental amorphous semiconductors. This is shown in Fig. 6.46. The values for germanium and silicon are taken from edges that exhibit exponential behaviour. A possible explanation of this trend is that materials having lower co-ordination (higher valency) more easily form ideal amorphous networks with fewer defects and voids and perhaps less departure from optimum covalent band angles and lengths than in

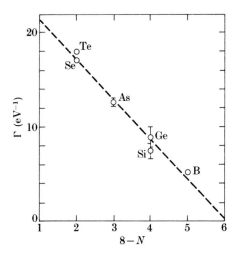

Fig. 6.46. Values of Γ for absorption edges of elemental materials, the absorption coefficient of which obeys $\alpha = \alpha_0' \exp(\hbar\omega/\Gamma)$. A higher value of Γ means a steeper edge. Data for Te from Stuke (1970*a,b*), for Se from Hartke and Regensburger (1965), for Ge from Connell *et al.* 1974, for Si from Loveland *et al.* (1973/74), and for B from Berezin *et al.* (1974). N is the valency.

those with higher co-ordination. All these features could reduce random microfields.

The effect of hydrostatic pressure on exponential absorption edges has been measured in only a few materials. In these a parallel shift to lower energies with increasing pressure has been found. Presumably this reflects the pressure coefficient $(\partial E_0/\partial P)_T$ of some gap E_0. It is interesting that in the materials studied the sign of this quantity is opposite to that expected from the sign of the temperature coefficient $(\partial E_0/\partial T)_P$ referred to above. Either an increase in temperature or an increase in pressure shifts the edge to lower energies. The coefficients are related by the thermodynamic relationship

$$\left(\frac{\partial E}{\partial T}\right)_P = \left(\frac{\partial E_0}{\partial T}\right)_V - \frac{\alpha_V}{K_s}\left(\frac{\partial E_0}{\partial P}\right)_T \tag{6.59}$$

where α_V is the volume expansivity $(\mathrm{d}V/\mathrm{d}T)_P/V$ and K_s is the compressibility $-(\mathrm{d}V/\mathrm{d}P)_T/V$. The second term is the contribution to $(\partial E_0/\partial T)_P$ due to dilation and, with the negative sign in front, makes a positive contribution. As $(\partial E_0/\partial T)_P$ is negative, $(\partial E_0/\partial T)_V$ must make a sizable negative contribution. However, it appears that this behaviour, not common in most crystalline materials for which $(\partial E_0/\partial T)_V$ is normally rather small and attributable to electron–phonon interaction (Fan 1951), is not a characteristic of the amorphous state. In fact a similar behaviour occurs in crystalline As_2Se_3 and similar materials.

6.7.2. Effects of externally applied fields

Application of external electric fields changes the optical properties of both crystalline and amorphous semiconductors. The effects are generally small, necessitating the use of modulation techniques. For crystals, electro-absorption and electroreflectance measurements have proved to be very useful in revealing fine structure not resolved by conventional spectroscopy (see Cardona 1969); for amorphous semiconductors the signals are broad but nevertheless interesting and can provide information concerning internal microfields.

For amorphous As_2S_3, Kolomiets et al. (1970) observed a shift of the optical edge towards lower energies. The displacement of the Urbach tail was found to be independent of photon energy and proportional to the square of the electric field; it amounted to 1 part in 10^5 at 10^5 V cm^{-1}. Kolomiets et al. (1970) interpreted these data in terms of the theory of Franz (1958) which is based on an exponential distribution of tail states and predicts a parallel shift of the exponential absorption edge in an electric field F according to the relation

$$E = \frac{e^2 h^2 \Gamma^2 F^2}{24 m^*} \tag{6.60}$$

where Γ is the slope of the edge. Interpreted in this way an effective mass ratio $m^*/m = 7{\cdot}5$ was computed. Similar results for amorphous As_2Se_3 (Kolomiets, Mazets, and Efendiev 1970) yield $m^*/m = 2{\cdot}9$ and for Se (Stuke and Weiser 1966, Drews 1966) $m^*/m = 4{\cdot}5$.

More detailed studies, which include the effect of sample preparation as well as temperature on the electroabsorption signal have been made on Se, Se–Te alloys, As_2S_3, and As_2Se_3 by Roberts, Keating, and Shelley (1974), on As_2S_3 by Street, Searle, Austin, and Sussmann (1974) and on Se and As–Se alloys by Sussmann, Austin, and Searle (1975). In some respects the data are conflicting, particularly concerning the exact value of the power describing the field dependence and also the variation with wavelength. For Se, Roberts et al. (1974) and Sussmann et al. (1975) agree that an electric field does not produce a parallel shift in the Urbach tail; in fact the edge becomes steeper.† Results from Sussmann et al. (1975) are shown in Fig. 6.47. In several types of film $\Delta\alpha/\alpha$ passes through a maximum close to the upper energy limit of the Urbach tail (marked by an arrow). The signal increases by about 50 per cent as the temperature is lowered from 300 K to 120 K.

† It may be worth mentioning that, at a given photon energy, the fractional change in transmitted intensity is proportional to the absolute change $\Delta\alpha$ in the absorption coefficient α. A value of $\Delta\alpha/\alpha$ that is independent of photon energy implies a parallel displacement of the optical absorption edge on a logarithmic scale.

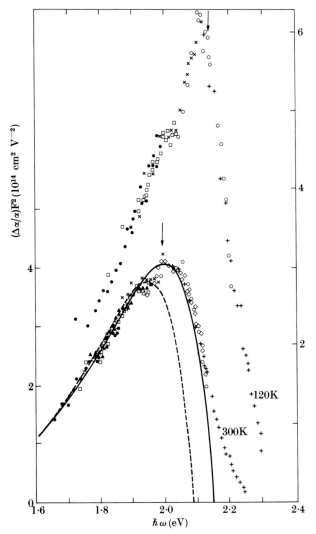

Fig. 6.47. Electroabsorption spectra of amorphous selenium at 120 K (right-hand scale) and 300 K (left-hand scale); □, ◇, ▲, blown films; ● hot-pressed films; ○, ×, +, evaporated films. The arrows mark the upper energy limit of the linear Urbach tail at each temperature. The full and broken lines are theoretical. (From Sussmann *et al.* 1975.)

A steepening of the absorption edge on application of an external field is *not* predicted by most theoretical treatments. The theory of Franz (1958), referred to above, predicts a constant value of $\Delta\alpha/\alpha$ at low photon energies, tailing to zero at higher frequencies. A theory due to Esser (1972),†

† See also more recent papers co-authored by Esser (1975). *Phys. Status Solidi B* **68**, 265; **71**, 63; **72**, 535.

based on the random microfield model of Bonch-Bruevich (1970), predicts a decrease of $\Delta\alpha/\alpha$ as the photon energy increases. Street, Searle, Austin, and Sussmann (1974) have calculated the electroabsorption expected for the field-broadened exciton model of Dow and Redfield (1972). For a Gaussian distribution of microfields, they find

$$\frac{\Delta\alpha}{\alpha} \propto \left(\frac{E_{exc} - \hbar\omega}{F_s^4}\right)^{2/3} F^2 \qquad (6.61)$$

where E_{exc} is the position of the ground-state exciton peak in the absence of microfields, F_s is a measure of the root-mean-square microfield, and F is the applied field. If the random fields are due to Coulomb centres, Street *et al.* (1974*a*,*b*,*c*) find

$$\Delta\alpha/\alpha \propto (E_{exc} - \hbar\omega)^{-2} \qquad (6.62)$$

which does give an increase in $\Delta\alpha/\alpha$ as $\hbar\omega$ increases; however, in this case, the edge is not strictly exponential, either with or without the applied field.

Although a microfield distribution composed of both Coulombic and Gaussian components could be postulated and a theoretical fit to the data obtained, it might be more meaningful to assume, following Sussmann *et al.* (1975), that the field-free breadth of the absorption edge is caused by two contributions, one of which is insensitive to an applied external field and another that is not. A physical basis for this model may be a non-isotropic distribution of microfields in systems having local anisotropy. Using appropriate parameters, Sussmann *et al.* (1975) obtain the theoretical fits shown by the solid lines in Fig. 6.47. An alternative model, proposed by Roberts *et al.* (1974), is based on the Stark effect associated with Se_8 molecules.

For As_2Se_3 and As_2S_3, the field-free slope of the exponential absorption edge is higher than for Se (see Table 6.2) and, although for the same applied field $\Delta\alpha$ is larger, $\Delta\alpha/\alpha$ varies less with photon energy in the region of the Urbach tail (Roberts *et al.* 1974; Sussmann *et al.* 1975). Nevertheless a similar model to that suggested for Se may still be appropriate.

Early electroreflectance measurements on amorphous Ge films (Piller *et al.* 1969) which purported to show a spin–orbit split-valence band, has subsequently been shown to be of doubtful validity owing to the presence of signals associated with interference fringes.

6.7.3. Interband absorption

In this section we discuss the form of the optical absorption edge expected in amorphous semiconductors in the absence of any electric-field or exciton effects. For materials exhibiting an exponential edge, the results should be appropriate at photon energies above the exponential tail. The following assumptions are also made.

(a) The matrix elements for the electronic transitions are constant over the range of photon energies of interest.

(b) The k-conservation selection rule is relaxed. This assumption is made in amorphous semiconductors because, near the band edges at least, $\Delta k \sim k$ and thus k is not a good quantum number. There is evidence (Berglund and Spicer 1964) for a relaxation of the k-conservation rule for some interband transitions even in certain crystalline materials. On an E–k diagram such transitions would be non-vertical. However, no phonon absorption or emission processes are invoked to conserve momentum and all the energy required is provided by the incident photons. Such transitions are termed non-direct as opposed to indirect.

The conductivity at frequency ω under these conditions is given by

$$\sigma(\omega) = \frac{2\pi e^2 \hbar^3 \Omega}{m^2} \int \frac{N(E)N(E+\hbar\omega)|D|^2 \, \mathrm{d}E}{\hbar\omega} \qquad (6.63)$$

where Ω is the volume of the specimen and D the matrix element of $\partial/\partial x$. This follows from the analysis of Chapter 2. The corresponding absorption coefficient is given by

$$\alpha = \frac{4\pi}{n_0 c} \sigma(\omega) \qquad (6.64)$$

where n_0 is the refractive index.

The matrix element D for transitions between states in different bands will be taken to be the same as that for transitions between extended states in the same band, without the factor m/m^* (§ 2.5), so that

$$D = \pi (a/\Omega)^{1/2} \qquad (6.65)$$

where a is the average lattice spacing. It should be pointed out that we are taking the matrix element to be the same whether or not either the initial or final state, or both, are localized. This is an assumption that needs justifying. The argument given by Davis and Mott (1970) is that the smaller value of D that may be expected when the wavefunctions are localized is compensated by the increased value of the normalization factor (see also Tauc and Menth 1972). However, when both initial and final states are localized the lack of spatial overlap between states may cut down D to small values.

Neglecting for the moment this last possibility, we find for interband transitions (ignoring the variation of n_0 with $\hbar\omega$)

$$\alpha(\omega) = \frac{8\pi^4 e^2 \hbar a}{n_0 c m^2} \int \frac{N_V(E)N_c(E+\hbar\omega)}{\hbar\omega} \, \mathrm{d}E \qquad (6.66)$$

where the integration is over all pairs of states in the valence and conduction bands separated by an energy $\hbar\omega$.

If the density of states at the bottom of the conduction band is represented by $N_c(E) = \text{const.}(E - E_A)^s$ and at the top of the valence band by $N_v(E) = \text{const.}(E_B - E)^p$, then, by making the substitution

$$y = \frac{E_A - \hbar\omega - E}{E_A - \hbar\omega - E_B}$$

and denoting the gap by $E_0 = E_A - E_B$ we find

$$\alpha(\omega) = \text{const.} \left\{ \int_0^1 (1-y)^p y^s \, dy \right\} \frac{(\hbar\omega - E_0)^{p+s+1}}{\hbar\omega}. \tag{6.67}$$

The quantity in braces is a known integral, namely

$$\frac{\Gamma(s+1)\Gamma(p+1)}{\Gamma(s+p+2)}$$

Without knowledge of the form of $N(E)$ at the band edges, it is speculative to take the calculation further. Under the assumption of parabolic bands (Tauc 1970a), $s = p = \frac{1}{2}$, leading to

$$\alpha(\omega) = \frac{\text{const.}(\hbar\omega - E_0)^2}{\hbar\omega}. \tag{6.68}$$

The absorption in many amorphous materials is observed to obey this relation above the exponential tails. A few examples are shown in Fig. 6.48. The constant E_0 can be used to define an optical gap, although it may represent an extrapolated rather than a real zero in the density of states.

The quadratic relation between $\alpha(\omega)\hbar\omega$ and $\hbar\omega$ given above has also been derived by Davis and Mott (1970) using different assumptions. Using the notation of Fig. 6.7 and assuming that the densities of states at the band edges are linear functions of the energy, that

$$N(E_C) = N(E_V) \qquad E_C - E_A = E_B - E_V = \Delta E$$

and that transitions in which both the initial and final states are localized can be neglected, we find from eqn (6.66)

$$\alpha(\omega) = \frac{(4\pi/n_0 c)\sigma_{\min}(\hbar\omega - E_0)^2}{\hbar\omega \, \Delta E}. \tag{6.69}$$

Here, E_0 is $E_A - E_V$ or $E_C - E_B$, whichever is the smaller, and

$$\sigma_{\min} = \frac{2\pi^3 e^2 \hbar^3 a}{m^2} \{N(E_C)\}^2$$

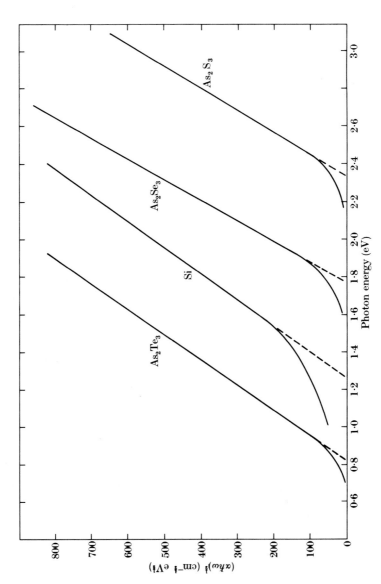

Fig. 6.48. Examples of absorption edges whose functional dependence on photon energy is given by $\alpha \hbar \omega = B(\hbar \omega - E_0)^2$. Data for As_2Te_3 from Rockstad 1970, Si from Brodsky *et al.* 1970, and As_2Se_3 and As_2S_3 from Felty and Myers (private communication). The slopes and intercepts for these materials (particularly silicon) vary somewhat according to the method of preparation.

is the quantity discussed in § 2.6 and in § 6.4.2. We can make a rough estimate of the magnitude of the absorption coefficient by taking $\Delta E \sim$ 0·2 eV, $n_0 = 4$, and $\sigma_{min} \sim 200\ \Omega^{-1}\ cm^{-1}$. Thus, if energies are in electron-volts,

$$\alpha(\omega) \sim \frac{10^5 (\hbar\omega - E_0)^2}{\hbar\omega}\ cm^{-1}. \qquad (6.70)$$

In Table 6.3 values of B and E_0 are given for a few amorphous semiconductors whose optical absorption obeys the relationship $\alpha\hbar\omega = B(\hbar\omega - E_0)^2$. In certain cases the above relationship holds over a much larger range of photon energies than is consistent with our estimate of ΔE and the agreement may therefore be fortuitous.

TABLE 6.3

Room-temperature experimental values of E_0 and B for a few amorphous semiconductors whose optical absorption coefficient α obeys the relation $\alpha\hbar\omega = B(\hbar\omega - E_0)^2$ in a range of photon energies above the exponential edge

	$E_0(eV)$	$B\ (cm^{-1}\ eV^{-1})$	Reference
GeTe	0·70	$2\cdot1\times10^5$	Tsu, Howard, and Esaki (1970)
As_2Te_3	0·83	$4\cdot7\times10^5$	Weiser and Brodsky (1970)
	0·82	$5\cdot4\times10^5$	Rockstad (1970)
Si	1·26	$5\cdot2\times10^5$	Brodsky, Title, Weiser, and Pettit (1970)
As_2Se_3	1·76	$8\cdot3\times10^5$	Felty and Myers (private communication)
As_2S_3	2·32	4×10^5	Felty and Myers (private communication)

There are some notable exceptions to the quadratic frequency dependence of the absorption coefficient, and it should not therefore be regarded as a characteristic phenomenon of amorphous semiconductors. The absorption coefficient in amorphous Se exhibits, above the exponential tail, a relation of the form (see Chapter 10)

$$\alpha(\omega)\hbar\omega = \text{const.}(\hbar\omega - E_0). \qquad (6.71)$$

Some speculations regarding this behaviour have been made by Davis and Mott (1970). Certain multicomponent materials have been found by Fagen (private communication) to have an absorption coefficient that obeys the relation

$$\alpha(\omega)\hbar\omega = \text{const.}(\hbar\omega - E_0)^3. \qquad (6.72)$$

These are shown in Fig. 6.49. This relationship can be derived by relaxing one of the assumptions used above in obtaining the quadratic relation, namely that transitions between localized states can be neglected, and assuming instead that they have the same matrix elements as all other transitions. In this case we find (Davis and Mott 1970)

$$\alpha(\omega) = \frac{(4\pi\sigma_0/3n_0c)(\hbar\omega - E_0)^3}{\hbar\omega(\Delta E)^2}. \qquad (6.73)$$

Here E_0 is equal to $E_A - E_B$, using again the notation of Fig. 6.7.

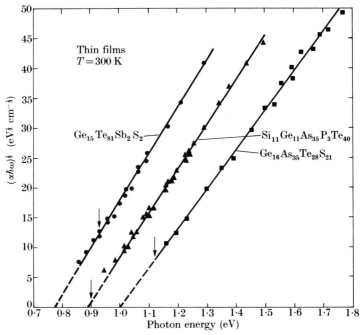

Fig. 6.49. Absorption edges in three multicomponent glasses which obey the relationship $\alpha\hbar\omega \sim (\hbar\omega - E_0)^3$. The arrows mark the values of $2E$ in the formula for the conductivity $\sigma = C \exp(-E/kT)$. (From Fagen, private communication.)

Amorphous germanium has been reported in several papers to have an exponential absorption edge followed at higher energies by the quadratic variation. Other measurements by Donovan, Spicer, and Bennett (1969) have revealed a sharp absorption edge, but no simple power law is capable of describing its spectral shape (see Chapter 7). Presumably in this case the density of states at the band edges is not given by a simple power law either.

In summary, it appears that the absorption edge of many amorphous semiconductors can be described by a simple power law, at least over a limited range of absorption coefficients, which enables an optical gap E_0 to be defined. However, without independent knowledge of the density of states and matrix elements as a function of energy, we can only speculate on whether this is a real gap in the density of states or some other characteristic energy related to the mobility gap.

The relation between the energy gaps in amorphous materials and those in the corresponding crystalline state (when this exists) is of interest. A general rule appears to be that, if the local atomic order is not appreciably altered in the amorphous phase, the gaps in the two states are not *appreciably* different. As expected this rule works best for 'tight-binding materials', i.e. those for which the band structure is determined mainly by nearest-neighbour overlap integrals.

Connell and Paul (1972) find little correspondence between the energy gaps (or their pressure coefficients) in crystalline and amorphous forms of Si, Ge, and III–V compounds. For GaP in particular they find a large discrepancy (0·42 eV amorphous, 2·22 eV crystalline). They suggest that this may be explained by the occurrence of 'wrong bonds' (Ga–Ga or P–P) due to the presence of five-fold rings in the amorphous structure. However, the Penn gap (§ 7.16) is not altered greatly. In crystalline Te the smallest gap occurs in a direction in the Brillouin zone which does not correspond to that along the chain structure. This suggests that interaction between chains is important, and it is not surprising to find a large change in the gap (∼0·7 eV amorphous, ∼0·25 eV crystalline); even if the chain structure is preserved to some degree, the interaction between the chains must change. In selenium the chain interaction is less and the gaps in the trigonal crystal and amorphous states are similar (∼2 eV) (see Fig. 10.6). Crystalline As is a semimetal; amorphous films of As are quite transparent below ∼1 eV; clearly a change of local structure suggests itself here (Chapter 8).

6.7.4. Absorption at high energies

Beyond the fundamental absorption edge both crystalline and amorphous semiconductors continue to absorb strongly. Measurements up to about 20 eV can be made most conveniently using an ultraviolet diffraction-grating reflectometer. The optical constants are then derived from a Kramers–Kronig analysis of the reflectivity.

In crystalline semiconductors, the interband absorption in this range is characterized by a succession of peaks related to structure in the density of states of both the valence and conduction bands. As an example, Fig. 6.50 shows (broken curve) the experimentally determined absorption spectrum of crystalline germanium above the edge (at 0·7 eV). Note that the imaginary part of the dielectric constant is related to α by the relation $\varepsilon_2 = n_0 c \alpha / \omega$. The structure shown here has been interpreted as arising principally from the transitions marked by arrows in Fig. 7.36 (Chapter 7), which is a theoretical calculation of the electronic band structure of crystalline Ge in two principal directions of the Brillouin zone. It should be stressed that these assignments are not unambiguous; in particular it is now considered that the transition X_4–X_1 is not truly representative of the large peak at 4·5 eV but that a much larger region of the zone is involved.

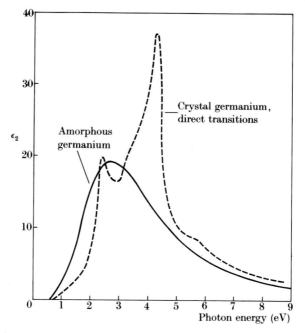

Fig. 6.50. Spectral dependence of the imaginary part of the dielectric constant ε_2 as determined by Kramers–Kronig analysis of reflectance data for amorphous (solid curve) and crystalline (broken curve) germanium. (From Spicer and Donovan 1970a.)

However, it serves to illustrate the principles of interpretation of optical absorption spectra of crystals in this range of photon energies. These are as follows.

(a) Vertical transitions (corresponding to no change in the electron wavevector) at critical points in the zone contribute strongly to the absorption. Critical points occur where the valence and conduction bands are parallel to k space because the joint or combined densities of states thus have maximum values.

(b) The joint density of states is not the only parameter of importance. Certain transitions are forbidden by symmetry requirements. Others are damped or enhanced by a matrix element, which may have considerable structure as a function of energy. For example, the 4·5 eV transition in Ge referred to above is Umklapp enhanced (Phillip 1966), (see Fig. 6.51). The 'optical density of states' can thus be considerably different from the actual density of states obtained by integrating all the various overlapping bands over the whole Brillouin zone.

Amorphous semiconductors show far less structure in their absorption spectra. The solid curve in Fig. 6.50 is the experimentally determined spectrum for amorphous Ge. Other materials show a similar lack of struc-

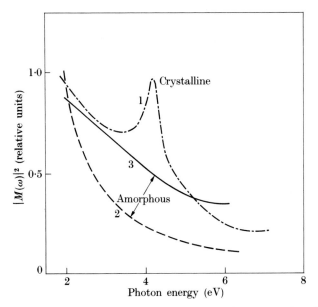

Fig. 6.51. Energy dependence of matrix elements for amorphous germanium: (1) crystal; (2) amorphous (using convolution of crystalline density of states); (3) amorphous (using convolution of amorphous density of states). (From Stuke 1970b, Maschke and Thomas 1974b.)

ture although, as will be seen in later chapters, gross features are frequently preserved.

There are several approaches to an understanding of the ε_2 spectra of amorphous semiconductors. One is to start with the crystalline band structure and to introduce modifications as seem appropriate. Thus we might attempt to generate the ε_2 spectrum by first relaxing the k-conservation rule, allowing all states in the conduction band to be accessible from all states in the valence band with matrix elements independent of energy. For amorphous Ge this procedure results in a smoothed-out absorption curve (Fig. 6.52(a)), which peaks at too low an energy to account for the experimental results, as shown by Herman and Van Dyke (1968). Better agreement is obtained by using matrix elements that depend on energy. Maschke and Thomas (1970b) (see also Stuke (1970b)) have proposed a smoothed version of that appropriate for the crystal (Fig. 6.51).

As will be shown later, the density of states of amorphous Ge has been determined by photoemission and significant differences from the crystalline density of states were found. Using these results the non-direct constant matrix element model is able to account for the ε_2 spectrum of amorphous Ge. The problem is therefore transferred to that of explaining the density of states (see § 6.8.2 and § 7.8).

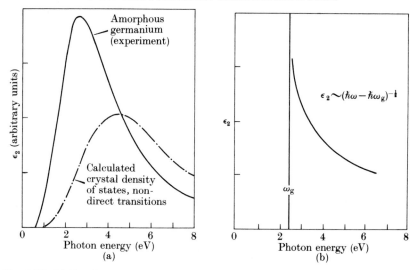

Fig. 6.52. (a) Imaginary part of the dielectric constant ε_2 in amorphous germanium as determined experimentally (solid curve) and as calculated from the crystalline density of states, assuming non-direct transitions and constant matrix elements (chain curve). (b) Expected form of ε_2 on the Penn model (see text).

The *gross* features of the ε_2 spectrum of amorphous Ge, and indeed other materials, can be understood on an isotropic Penn model (Penn 1962, Bardasis and Hone 1967). The Penn or average gap between valence and conduction bands is assumed to be retained and little changed from its value in the crystal. On this model the absorption spectrum has the form shown in Fig. 6.52(b). By introducing Lorentzian broadening, Phillips (1971) demonstrates how the experimentally determined spectrum of amorphous germanium can be approximated.

A very high energy (50–70 eV) absorption experiment on selenium by Cardona *et al.* (1970), in which electrons were excited from a narrow low-lying d level into the conduction band, revealed a spectrum (Fig. 6.53) that was, within experimental error, the same in both crystalline and amorphous forms. Furthermore, fairly good agreement was obtained with a theoretical calculation using the density of states appropriate to the crystal (see Chapter 9). This result suggests that, in selenium at any rate, the total width of the conduction band is approximately unchanged on going to the amorphous state. Nevertheless exciton effects play a role in these spectra (Shevchik, Cardona, and Tejeda 1973).

Similar absorption experiments have been made on silicon (Brown and Rustgi 1972, Brown, Bachrach, and Skibowski 1977), As, and As_2Se_3 (Bordas and West 1976). These are discussed in later chapters where some problems associated with interpretation are emphasized.

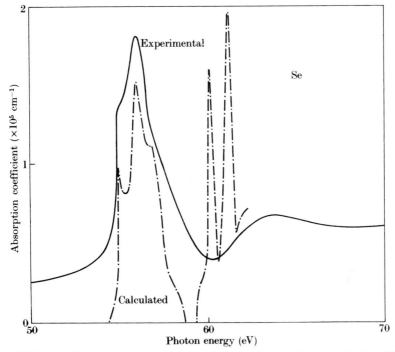

Fig. 6.53. Optical absorption in amorphous selenium using synchrotron radiation. (From Cardona *et al.* 1970.)

6.7.5. *Intraband absorption*

Fig. 6.54 shows the absorption near the fundamental edge in amorphous As_2Se_3 as measured by Edmond (1966). In the solid glassy state (curves (1)–(3)) the absorption coefficient obeys the spectral Urbach rule (§ 6.7.1) and there is a parallel shift to lower photon energies as the temperature is raised. The absorption in the neighbourhood of 10^{-1} cm^{-1} is presumed to be residual and depends on the conditions of sample preparation. In the liquid state (curves (4)–(11)) strong absorption occurs at much lower energies, and it may be considered that the absorption edge is broadening (as well as shifting) as the temperature is raised, in accordance with Urbach's rule. However, the magnitude of α at a fixed photon energy (0·5 eV ~ 4000 cm^{-1}) was found to be proportional to the d.c. electrical conductivity, at least for temperatures above 450°C. This is shown in Fig. 6.55. There is thus a strong possibility that free-carrier absorption is being observed.

It should be noted that free-carrier absorption in crystalline semiconductors is normally observed on the low-energy side of the absorption edge as a component that, at a fixed temperature, increases with decreasing photon energy, i.e. α is proportional to λ^2. This of course simply derives

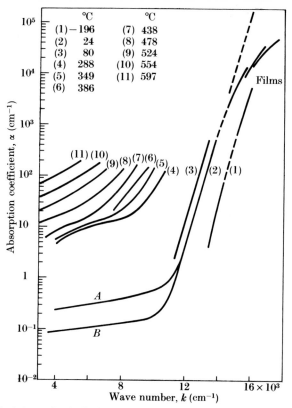

Fig. 6.54. Optical absorption in liquid and glassy As_2Se_3 as a function of temperature. (From Edmond 1966.)

from the Drude formula for a.c. conductivity:

$$\sigma(\omega) = \frac{\sigma(0)}{1 + \omega^2 \tau^2} \qquad \alpha(\omega) = \frac{4\pi}{n_0 c} \sigma(\omega). \tag{6.74}$$

The application of this formula to free-carrier absorption in semiconductors relies essentially on the existence of a relaxation or scattering time as intraband transitions are quantum-mechanically forbidden. As discussed in § 6.4.5 the equation is not applicable for amorphous semiconductors with very small values of τ.

We suggest that the increase in absorption with increasing photon energy shown in Fig. 6.55 arises from an increasing density of available final states. Using the notation of Fig. 6.7(b) we calculate this absorption as follows. Following the arguments of § 2.2 we write

$$\sigma(\omega) = \frac{2\pi^3 e^2 \hbar^3 a}{m^2} \int_{E_A}^{\infty} \frac{N(E - \hbar\omega)N(E)\{f(E - \hbar\omega) - f(E)\} \, dE}{\hbar\omega}. \tag{6.75}$$

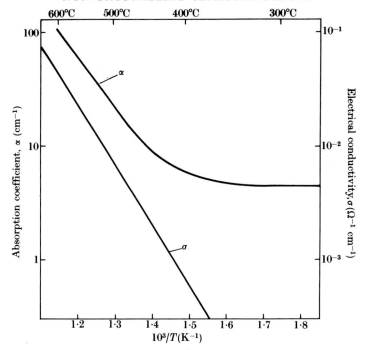

Fig. 6.55. Absorption coefficient at 4000 cm^{-1} from Fig. 6.54 and electrical conductivity as functions of inverse temperature in liquid As$_2$Se$_3$. (From Edmond 1970.)

With $f(E) = \exp\{-(E - E_F)/kT\}$ and $N(E) = N(E_C)(E - E_A)/\Delta E$ the maximum of $f(E)N(E)$ occurs at $E = E_A + kT$. Thus

$$\sigma(\omega) = \frac{2\pi^3 e^2 \hbar^3 a}{m^2} \left\{ \frac{N(E_C)}{\Delta E} \right\}^2 \frac{kT(kT + \hbar\omega)}{\hbar\omega} \int_{E_A}^{\infty} \{f(E - \hbar\omega) - f(E)\} \, dE \quad (6.76)$$

which for $\hbar\omega > kT$ becomes

$$\sigma(\omega) = \sigma_{min} \left(\frac{kT}{\Delta E}\right)^2 \exp\left\{-\frac{(E_A - E_F)}{kT}\right\} \exp(\hbar\omega/kT) \quad (6.77)$$

where (compare eqns (2.14) and (2.17) with $L \sim a$)

$$\sigma_{min} = \frac{2\pi^3 e^2 \hbar^3 a}{m^2} \{N(E_C)\}^2.$$

Eqn (6.77) predicts an absorption that is proportional to the d.c. conductivity and increases exponentially with photon energy. Assumptions that are made here are that the density of states is linear with energy and that the matrix elements for transitions from the localized states at the bottom of the band to extended states higher in the band are the same as those used earlier in this book. A similar expression has been obtained by

Hindley (1970). It should be noted that eqn (6.77) does not predict correctly the slope of the curves shown in Fig. 6.54 and further work on this problem is necessary.

Edmond (1970) has subsequently observed similar behaviour to that reported above in solid glasses of the system $As_2(Se, Te)_3$. In all cases the high-frequency conductivity at $\sim 7 \times 10^{14} \, s^{-1}$ was found to be about 20 times the d.c. value. Bishop *et al.* (1971) have also reported similar observations in $Tl_2SeAs_2Te_3$. The temperature dependencies of the absorption coefficient and the d.c. conductivity for this material are shown in Chapter 9. Identical activation energies of 0·35 eV are found. However, in this glass no dependence of α on wavelength was discovered. In view of the smaller photon energies involved compared with As_2Se_3 and the limited wavelength range studied, a large variation is not to be expected. Again $\sigma(\omega)/\sigma(0)$ was found to be greater than unity; in this case the ratio is about 8.

6.7.6. Photoluminescence

Photoluminescence has been observed in glow-discharge-deposited silicon (Engemann and Fischer 1974*a*,*b*, 1976, and Chapter 7, § 7.7) in various glassy chalcogenides as discussed fully in Chapter 9, and in amorphous arsenic (Bishop, Strom, and Taylor 1976*b*). Reviews have been given by Street (1976) and by Bishop, Strom, and Taylor (1977). In this section we give a description of various mechanisms that have been proposed.

Radiation can be emitted either by an excited defect, or by an electron–hole pair in the conduction and valence bands bound together by their Coulomb attraction (an exciton). In discussions on quantum efficiency in § 6.5.2, we have treated the conditions under which an electron–hole pair will separate, leading to photoconduction, and the conditions under which they will recombine. For a discussion of photoluminescence, we have to ask whether they will recombine with or without the emission of radiation.

Photoluminescence in chalcogenides is thought to occur in defects and these defects are probably the charged dangling bonds of § 6.4.1. The strongest evidence that the centres are charged is that the efficiency of luminescence falls off rapidly with temperature. The electron (or hole) escapes from the *neutral* excited centre and finds a non-radiative recombination channel. The emitted radiation shows a large Stokes shift. Fig. 3.7(b) is therefore applicable. The radiative transition probability, $\sim 10^4 \, s^{-1}$ in this case, must therefore be faster than any of the non-radiative mechanisms discussed in Chapter 3. Excitation of the defect centre can occur directly, or by production of an electron–hole pair in the neighbourhood which then diffuses to the defect (§ 9.6).

Of interest is the fact that in chalcogenides the direct recombination of an electron–hole pair is normally non-radiative, presumably according to the mechanism of Dexter *et al.* (1955) (§ 3.5). We also have to ask whether a free electron–hole pair (exciton) can be self-trapped, and if so whether there is a barrier to self-trapping as in Fig. 3.7. Since we know (a) that the exciton can wander to a defect and (b) that eventually it must disappear by the mechanism of Dexter *et al.*, we conclude that the barrier must exist.

In glow-discharge-deposited silicon, however, we have seen in § 6.5.2 that radiation normally produces an electron–hole pair, that if they separate they normally find a non-radiative recombination path, but that below ~100 K they do not separate and then normally recombine with emission of radiation. There appears to be little Stokes shift; that is (presumably) why the mechanism of Dexter *et al.* does not operate.† The photoluminescence spectrum, particularly at higher temperatures, includes not only the band-to-band peak (at ~1·23 eV) but also peaks corresponding to transitions from conduction and valence band states to the defect states identified from other measurements. This is shown in Fig. 7.61, and in Chapter 7 photoluminescence in this material is discussed in greater detail.

The presence of these transitions shows that, particularly at higher temperatures, the electron–hole pair can wander some distance to a defect and be trapped there. This is remarkable especially because the decay of the radiation is very fast ($\sim 10^8 \, \text{s}^{-1}$). It is also remarkable that these centres give radiative transitions at 150 K and below, in that at higher temperatures they are invoked (§ 6.5.2) to give the channel for non-radiative decay. The present authors consider it possible that, while a pair diffuses to the nearest defect which normally does not contain hydrogen and so radiates, in photoconductivity a quasi-equilibrium is set up between holes and centres at E_Y, some of which contain hydrogen, and those *with* hydrogen give radiationless recombination.

6.7.7. *Vibrational spectra. Density of phonon modes*

The optical properties of crystalline and amorphous semiconductors in the infrared region of the spectrum are dominated by interaction with vibrational modes. In crystalline material conservation of momentum restricts infrared absorption or Raman scattering to processes involving phonons with wavevector close to zero, i.e. the centre of the Brillouin zone, and to those which have associated with them a dipole moment. Such processes can occur in covalent as well as ionic materials if there are three or more atoms per unit cell, if there are defects or impurities present, or in any event by multiphonon processes in which one vibrational mode induces instantaneous charges on the atoms and a second mode simultaneously causes a vibration of the induced charges. In the latter case absorption occurs at sum and difference frequencies of the two coupled phonons.

† Street (1978) gives further evidence that Si and the chalcogenides differ in this way.

Selection rules for first-order (one-phonon) processes in infrared and Raman activity are complementary.

In amorphous materials lattice absorption processes are retained to a degree which depends on the material. Generally speaking fine structure present in the spectra of corresponding crystalline materials is lost. However, a breakdown in selection rules allows coupling to modes that are not active in crystals, so that Raman and infrared activity extend over the entire vibrational spectrum.

Fig. 6.56 compares the infrared absorption spectra of amorphous and crystalline As_2Se_3, as determined by Austin and Garbett (1971). Although much fine structure is lost, gross features near 100 cm^{-1} and 220 cm^{-1} are retained. The reflectivity spectra (Mitchell, Bishop, and Taylor 1972, Felty, Lucovsky, and Myers 1967, Zltakin and Markov 1971, Onomichi, Arai, and Kudo 1971) and Raman spectra (Zallen et al. 1971a, b) confirm the retention of these dominant bands. They are associated (Austin and Garbett 1971, Lucovsky 1974) with local modes of molecular-like units consisting of $AsSe_3$ tetrahedra cross-linked to each other via a single Se atom. (The infrared activity arises from a dynamic rather than a static charge; the bonding is covalent but charge flows out of the bond when under tension and into the bond under compression.) This is an example of a 'molecular' material in which the dominant vibrational modes are retained on disordering; in fact they persist in the liquid state up to 673 K.

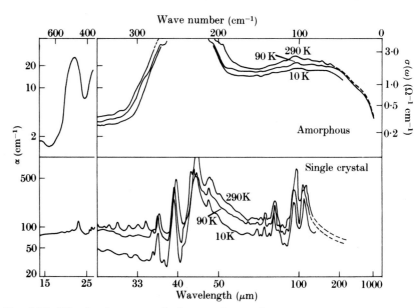

Fig. 6.56. Vibrational spectra of glassy and crystalline As_2Se_3. (From Austin and Garbett 1971.)

Another example of a molecular material is Se, the infrared and Raman spectra of which are shown in Fig. 6.57. The results for amorphous Se seem to bear more of a correspondence to those for the monoclinic than to those for the trigonal crystal, a feature which led to early identification of Se_8 rings as being responsible for the principal activity (Lucovsky *et al.* 1967, Ward 1972). However, studies of Raman spectra in Te (Brodsky *et al.* 1972*a*), for which no crystalline form containing rings exists, has shown that the Raman lines shift on disordering as a consequence of changes in the interchain interaction. In both amorphous Se and Te the molecular species is probably a short chain (see Lucovsky 1972, 1974, Mort 1973*a*, Meek 1976*a*, Martin, Lucovsky, and Helliwell 1976).

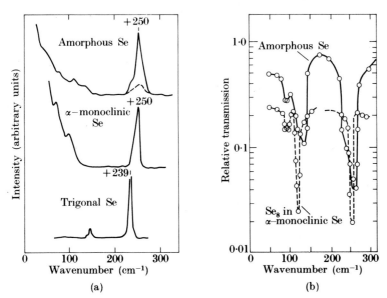

Fig. 6.57. (a) Raman and (b) infrared spectra of amorphous and crystalline forms of selenium. The broken curve in (a) corresponds to a different direction of the polarization of the incident light ($\mathscr{E}$ vector parallel to the direction of observation). ((a) From Ward 1972; (b) from Lucovsky 1969.)

For the glass $Tl_2SeAs_2Te_3$, dominant peaks in the infrared transmission spectrum again indicate the presence of molecular-like units (Taylor *et al.* 1971). However, for this material the peaks disappear abruptly at 480 K, approximately 120 K above the glass transition temperature, suggesting that the molecular units are less tightly bound than in As_2Se_3.

Tetrahedrally connected network structures with no identifiable mole-
.cular units behave in a very different way to the materials described above. Ge, Si, and III–V compounds fall into this category. Raman and infrared spectra for Si are shown in Fig. 6.58(a) (see also Alben *et al.* 1975 and

references therein). The first-order Raman spectrum of crystalline Si exhibits one sharp line at the zone centre TO–LO phonon energy (vertical broken line). In Fig. 6.58(b) the phonon density of states for crystalline Si (broken line, derived from neutron scattering data) has been convoluted with a Gaussian of half-width $30 \, \text{cm}^{-1}$ (solid line) to approximate the situation in amorphous Si in which there is a variation in bond angles. The similarity between the curves in (a) and (b) suggests that for the amorphous

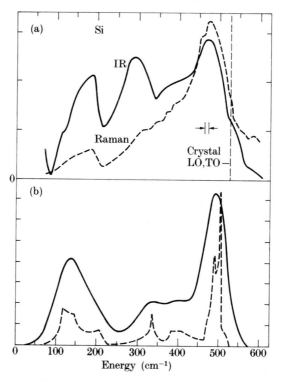

Fig. 6.58. (a) Raman and infrared spectra for amorphous silicon. (b) Phonon density of states for crystalline silicon as determined from neutron scattering. The broken curve has been Gaussian broadened to produce the solid curve. (From Smith *et al.* 1972.)

material all vibrational modes take part in infrared absorption and Raman scattering. However, the relative intensities do not match and the shifts in peak positions indicate differences between the phonon density of states in the crystalline and amorphous forms. One of the problems in attempting to deduce the density of phonon modes in amorphous germanium and silicon from infrared or Raman spectra is lack of knowledge of the energy dependence of the coupling constants. A more direct determination can in principle be made from inelastic neutron scattering data assuming a sufficient bulk of material can be obtained. Such measurements on germanium (Axe *et al.* 1974) indicate a transverse acoustic (TA) peak (the one at the lowest energy in Fig. 6.58), in shape not unlike that for the crystal but shifted slightly to *lower* energies. A displacement of the TA modes is also deduced from specific heat measurements (§ 6.8.1).

Calculations of the vibrational spectra (i.e. the energy dependence of the density of phonon modes) of various CRNs simulating the structure of amorphous germanium (or silicon) have been made by Meek (1976*b*) using the 'recursion method' of Haydock, Heine, and Kelly (1972, 1975). The results do *not* show a shifted TA peak but suggest some dependence of the shape of this peak on the ring statistics of the various models. The transverse optical (TO) peak (the one at the highest energy in Fig. 6.58) shows more sensitivity to bond-angle distortions than to ring statistics. Using a version of an adiabatic bond-charge model (Weber 1974) to represent short-range forces, Meek (1977*a*) suggests that the shift of the TA peak to lower energy indicates a reduction of the bond charge in amorphous germanium relative to that in the crystal.

As a final example of vibrational spectra in amorphous solids we show an interesting comparison of the infrared absorption spectra in amorphous Se, As, and Ge which have two-, three- and four-fold co-ordinations respectively (Fig. 6.59). Se and As show features in two frequency regimes and

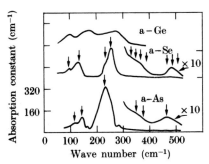

Fig. 6.59. Infrared absorption spectra of amorphous arsenic, selenium, and germanium. The solid arrows indicate first-order modes; the broken arrows indicate overtones and combinations of the first-order modes. (From Lucovsky and Galeener 1976.)

Ge in three. As discussed by Lucovsky and Knights (1974) and Lucovsky and Galeener (1976) the spectrum for Se is best understood in terms of a molecular model, that for Ge in terms of a density-of-states approach, and As, lying somewhere in between, offers useful insight into bridging the gap between the two theoretical approaches (see also Meek 1977*b* and Davis *et al.* 1979).

The far-infrared absorption in amorphous SiO_2 and similar glasses has been discussed by Stolen (1970) and by Wong and Whalley (1970). Bell, Bird, and Dean (1968) and Bell and Dean (1970) have computed the vibrational spectra for various disordered lattices and have shown the importance of localized modes in SiO_2 for example. Raman spectra of GeO_2 and SiO_2 have been obtained by Galeener and Lucovsky (1976); several previously unexplained features in these spectra have been assigned by these authors to longitudinal optical (LO) vibrations which suggests the importance of including long-range Coulomb forces in theoretical studies of vibrational properties of many glasses.

6.8. Other measurements

6.8.1. Thermal conductivity and specific heat

Measurements of the temperature dependence of thermal conductivity and specific heat of several non-crystalline materials (Zeller and Pohl 1971, Pohl 1976) have revealed a low-temperature behaviour that differs considerably from that in crystalline materials. Fig. 6.60 shows that the thermal conductivities and their temperature dependencies below about 2 K are essentially the same for many non-crystalline materials. The T^3 dependence observed in crystalline SiO_2 (quartz) and common to other crystalline materials arises when the phonon mean free path l is constant and equal to the sample size, the thermal conductivity K then having the

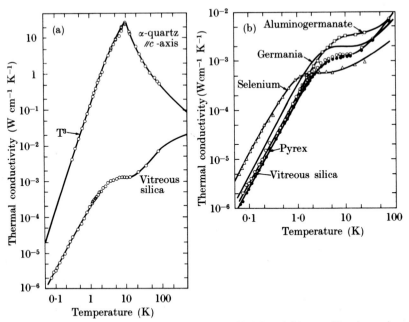

Fig. 6.60. Temperature dependence of thermal conductivity of (a) crystalline (quartz) and vitreous SiO_2 and (b) several other non-crystalline materials. (From Fritzsche 1973, Zeller and Pohl 1971.)

same Debye dependence on T as the specific heat C_v:

$$K = \tfrac{1}{3}C_v l v \qquad C_v = 234 \, NkT^3/\rho\Theta^3 \qquad (6.78)$$

where v is the velocity of sound, N the number density of atoms, ρ the density, and Θ the Debye temperature. As T is raised above 10 K, phonon scattering reduces l and K falls as shown in Fig. 6.60. It should also be mentioned that the absolute values of the thermal conductivity of crystalline materials vary considerably according to the degree of their chemical and structural perfection.

Returning now to the non-crystalline materials, K is several orders of magnitude lower than values generally found in crystals, is similar for all glasses measured to date, and exhibits a weaker temperature dependence ($K \propto T^n$ where $1 \cdot 8 < n < 2$) than found in crystals. Assuming that the heat is carried entirely by Debye phonons, one can use eqns (6.78) to calculate the temperature dependence of the phonon mean free path. This is shown for crystalline and vitreous SiO_2 as well as for other glasses in Fig. 6.61. At the top of this figure are scales showing the wavelength and frequency of

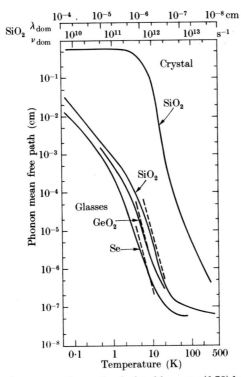

Fig. 6.61. Average phonon mean free path calculated from eqn (6.78) for crystalline SiO_2 and vitreous SiO_2, GeO_2, and Se as a function of temperature: solid curves, experiment; broken curves, 'isotropic' scattering. (From Zeller and Pohl 1971.)

the 'dominant' Debye phonons in SiO_2 at the temperature concerned (roughly $h\nu = 5\,kT$). Below about 1 K, the mean free path in these glasses seems to be about '100 phonon wavelengths'. In view of the anomalously high specific heats to be described below, one might question whether these calculated mean free paths are meaningful. Use of a C_v value larger than that given by Debye theory would yield smaller values of l. However, independent determinations of l, using ultrasonic attenuation and also Brillouin scattering, yield values close to those deduced from the thermal conductivity; at high power levels ultrasonic attenuation experiments actually suggest *higher* values (see below). Further confirmation that the magnitude of l deduced from the Debye theory is correct comes from the observation of a reduced thermal conductivity below 1 K in etched soda silica fibres of diameter 6×10^{-3} cm in which Casimir boundary scattering occurs (Pohl, Love, and Stephens 1974).

The temperature dependence of the specific heat in several glasses is compared with that in quartz in Fig. 6.62. The behaviour in glasses follows

$$C_v = AT + BT^3. \tag{6.78}$$

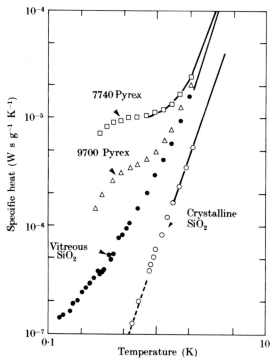

Fig. 6.62. Temperature dependence of specific heat in crystalline and vitreous SiO_2 and in pyrex. (From Zeller and Pohl 1971.)

Not only is there an additional linear term in the specific heat, which is of comparable magnitude in many non-crystalline materials, but the normal T^3 term is usually larger than that calculated from the elastic constants.

Several models have been proposed to explain the linear temperature term in the specific heat.[†] A purely electronic mechanism in which the dependence arises from a finite density of one-electron states at the Fermi level is possible in principle, but the required density $(\sim 10^{20-21} \text{ cm}^{-3})$ would not be compatible with the optical transparency. This latter objection is perhaps overcome in the model of two-electron (bipolaron) states proposed by Anderson (1975) (and also in its adaptation by Mott, Davis, and Street (1975) to specific defects); the optical energy for one-electron excitation can be considerably greater than the thermal excitation energy for the two-electron system. However, according to Phillips (1976), the relaxation (hopping) time for bipolaron excitation is calculated to be too long except perhaps for a small fraction of the states that are considerably closer than the average.

It might be relevant to note here that the linear term in C_v has been observed only in glasses that contain elements having a low co-ordination number in the solid, e.g. a chalcogen or oxygen atom. It is *not* present, or is at least much smaller, in glassy As which has three-fold co-ordination (Phillips and Thomas 1977, Jones, Thomas, and Phillips 1978). This suggests that a likely explanation for the required additional excitations might lie in configurational defects, i.e. places in the glass where the atoms can sit more or less equally well in two or more locations having comparable total energies. The tunnelling-states model of Anderson *et al.* (1972) and of Phillips (1972) is based on the existence of a quasi-continuous distribution of such structural configurations and was introduced in § 6.4.5 in the discussion of a.c. conductivity. We shall not develop this model further but refer the reader to the original papers.[‡] It is sufficient to say here that it not only accounts for the T dependence of C_v, but also provides a possible explanation for the thermal conductivity behaviour. The assumption here is that the excitations do not contribute to the *transport* of heat but act as resonant phonon scatterers; the mean free path is predicted to vary as T^{-1} and hence the thermal conductivity as T^2. Furthermore, at high power levels (in the ultrasonic attenuation experiments for example) the centres are saturated, the scattering is diminished, and the mean free path rises.

[†] Measurements down to 0.02 K in 'Suprasil' have indicated proportionality with $T^{1.2-1.3}$ (Lasjaunias *et al.* 1975) and this behaviour may be more representative of glasses than the linear dependence. Wenger, Amaya, and Kukkonen (1976) also find departures from eqn (6.79) and in addition a negligible change in the low-temperature specific heat of silica after quenching from 1150°C, a treatment that induces considerable internal strain.

[‡] For further developments of this model, as applied to the low-temperature sound velocity and specific heat of vitreous silica, see Piché *et al.* (1974).

The linear term in the specific heat of glasses is common enough to suggest that its origin lies in some intrinsic feature of the non-crystalline state. However, the observations that the magnitude of the term in As_2Se_3 (Stephens 1976) and in SiO_2 (Lasjaunias *et al.* 1975) can be considerably reduced if care is taken to remove water of hydrogenation from the glass, does raise the possibility of the effect being extrinsic. The absence of a linear term in C_v for As may then be related to the purity rather than the two-fold co-ordination of this material.

Turning our attention now to the T^3 term in the specific heat of amorphous materials and its difference in magnitude from that in crystalline states of the same material, we present in Fig. 6.63(a,b) results for Ge and

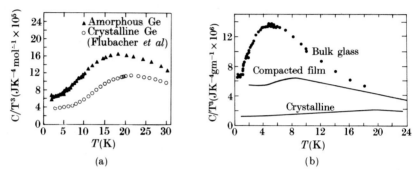

Fig. 6.63. Temperature dependence of specific heat in (a) crystalline and amorphous germanium, (b) crystalline and amorphous arsenic. ((a) From King *et al.* 1974, (b) from Jones *et al.* 1978.)

As. The data are plotted as C/T^3 *versus* T which emphasizes departures from the Debye theory. The intercept at $T = 0$ gives the Debye temperature (see eqn (6.79)); an upward trend in C/T^3 indicates a density of phonon states proportional to higher powers of frequency than predicted by Debye theory ($\propto$(energy)^2), and a peak indicates a maximum in the phonon spectra. Clearly differences between non-crystalline and crystalline materials have their origin in differing distributions of TA modes. In Ge the Debye temperature is 374 K for the crystal and 315 K for the amorphous form. This 16 per cent reduction for the amorphous form continues to higher temperatures, i.e. higher phonon frequencies. A bodily shift of the TA mode to lower frequencies could be accounted for by a reduction in the bond-bending force constants, although it is not obvious why this should occur. A lower Debye temperature suggests weaker bonds. In As the discrepancy is greater ($\Theta_{cryst} = 278$ K, $\Theta_{amorph} = 169$ K) (Wu and Luo 1973, 1974). Furthermore bulk As and compacted films exhibit very different behaviour (Jones, Thomas, and Phillips 1978) and indicate a relative increase in the density of low-frequency vibrational modes in the former (Lannin, Eno, and Luo 1978).

6.8.2. *Photoemission and density of states*

Photoemission is one of the most valuable experimental techniques available in the study of the electronic band structures of semiconductors. The method involves the photoinjection of carriers from valence or core states into the vacuum and an analysis of the energy distribution of the emitted electrons as a function of photon energy. The experiments are relatively difficult to perform properly, and skill is needed to interpret the energy distribution curves (EDCs).

For photons of energy $\hbar\omega$ the energy distribution of emitted electrons is given by

$$N(E, \hbar\omega)\,dE = \left\{\frac{K(\hbar\omega)}{\alpha(\hbar\omega)}\right\} T(E)S(E, \hbar\omega)N_c(E)N_v(E - \hbar\omega)\,dE$$

(6.80)

where K is a scale factor which includes parameters of the experimental apparatus, α is the absorption coefficient, T is an escape function, and S is the fraction of excited electrons of energy E lost because of scattering. Thus, like optical absorption measurements, photoemission gives the combined or joint density of states in the valence and conduction bands. The above formula assumes that the k conservation selection rule is not important.

There are, however, two major advantages that photoemission measurements have over optical absorption. The first is that the energy at which a peak in N_cN_v occurs can be related to some fixed point in the band structure, such as the top of the valence band. In optical absorption experiments only energy differences are determined. In photoemission a determination of absolute energies can, in principle, be made with a knowledge of the electron affinity of the semiconductor and if a retarding-potential method is used to measure the energy distributions of the collector. In practice, however, the measured retarding potential can be related to the energy at the top of the valence band by noting the maximum value of the retarding potential for a given $\hbar\omega$ at which electrons appear in the energy distributions. The second advantage can be seen by considering the schematic density of states shown in Fig. 6.64 in which the valence-band density of states is shown raised by $\hbar\omega$ to obtain the product $N_c(E)N_v(E - \hbar\omega)$. Multiplication by the escape function T yields, if S is constant, a spectrum proportional to the measured energy distribution of emitted electrons. For another value of $\hbar\omega$, a different spectrum results. Peaks in the distribution that do not vary their position with $\hbar\omega$ are to be associated with maxima in the conduction-band density of states; those peaks in the distribution that do vary their position with photon energy originate from maxima in the valence-band density of states. Thus one can, in principle, determine both N_c and N_v separately. However, as conduction-band states

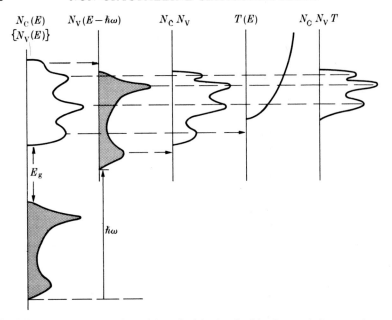

Fig. 6.64. Schematic illustration of the principles involved in photoemission experiments.

below the vacuum level of the semiconductor are inaccessible to the technique, it is normally not easy to obtain much information on $N_c(E)$. In practice a layer of, say, caesium can be evaporated on to the surface in order to lower its work function and hence increase the accessible range.

If the photon energy is raised above, say, 20 eV the structure associated with the conduction band disappears and the EDCs then reproduce the energy dependence of the density of valence states with some modification due to slowly varying photoionization cross-sections. At even higher energies core-level spectra are obtained. Photoemission experiments using high photon energies are referred to as XPS (X-ray photoemission spectroscopy) or ESCA (electron spectroscopy for chemical analysis).

Photoemission spectroscopy in amorphous Ge and Si has been reviewed by Spicer (1974), in III–V compounds by Shevchik (1974), and in some chalcogenides by Davis (1974).

The EDCs of crystalline and amorphous Ge at two different photon exciting energies are shown in Fig. 6.65 (from Spicer 1974). In § 7.17 the changes in the density of valence states (DOVS) which these results suggest will be discussed in more detail. Here we wish to emphasize the differences between results using different exciting energies. Although not completely understood they may be related to differences in photoelectron cross-sections, to matrix elements effects, or to differences in escape depths which are very short for $h\nu \sim 25$ eV.

Some differences associated with conditions of specimen preparation have been revealed by photoemission studies on amorphous Si (Fig. 6.66).

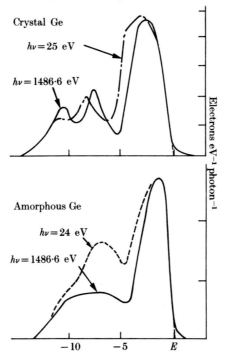

Fig. 6.65. Comparison of EDCs at different exciting energies for crystalline and amorphous germanium. The leading edges have been superimposed and a correction has been made for scattered electrons. (From Spicer 1974.)

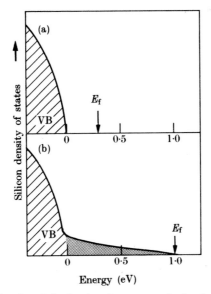

Fig. 6.66. Schematic drawing of the density of states near the band edge found by (a) Pierce and Spicer (1972) and (b) Fischer and Erbudak (1971). (From Spicer 1974.)

As it is difficult to detect state densities below about $10^{19}\,\text{cm}^{-3}$ by this technique, the tailing of the valence band into the gap deduced by Fischer and Erbudak (1971) is very large and may be related to the small source–substrate distance used in the evaporation of their films (see Orlowski and Spicer 1972). The tail tended to disappear with annealing.

Photoemission-derived DOVS for amorphous III–V compounds, As, and the chalcogenides will be described in Chapters 7, 8, and 9 respectively.

A difference in the binding energy of core levels between crystalline and amorphous phases of a material may be expected if the environment of individual atoms is different. In the point-ion approximation the chemical shift of an atomic core level is related to the atom's environment through the local Madelung sum $E_j = \sum_{i \neq j} q_i / R_{ij}$ (Segbahn et al. 1967) where R_{ij} is the separation between atoms i and j and q_i is the charge on atom i. In a typical III–V compound the first term in the series is ~3 eV and, if in the amorphous phase 'wrong bonds' are present, a shift of this magnitude could in principle occur. However, there is the possibility that a redistribution of charge could occur to compensate partly or completely structurally induced core shifts (see Robertson 1975). With this proviso the results for the In 4d level in InP, InAs, and InSb shown in Fig. 6.67 suggest the presence of wrong bonds in InP but not in the other materials (Shevchik, Tejeda, and Cardona 1974a,b). However, this conclusion has been questioned by Raman and DOVS studies (§ 7.7).

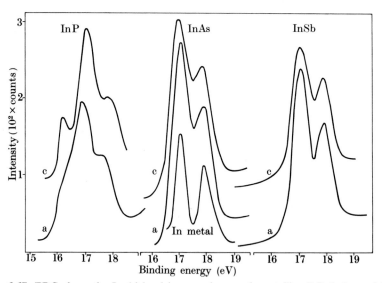

Fig. 6.67. EDCs from the In 4d level in amorphous and crystalline InP, InAs, and InSb obtained with photons of energy 40·8 eV. The binding energies are referred to the top of the valence band. (From Shevchik et al. 1974a.)

6.8.3. Electron spin resonance

Electron spin resonance (e.s.r.) has been extensively used in amorphous semiconductors to investigate defect centres which carry a spin. In addition e.s.r. has been found to enhance hopping conduction in doped and compensated crystalline silicon (§ 4.3.3) and to increase the rate of recombination in amorphous silicon, as discussed below. E.S.R. may be observed in deposited or annealed materials; sometimes, when it is absent or small, it can be enhanced by visible or infrared radiation. The g value, breadth, and shape of an e.s.r. line can often give information about the location of the electron not obtainable in any other way.

Early work of Brodsky and Title (1969) reported e.s.r. signals from amorphous films of Si, Ge, and SiC. From a comparison of the g value, linewidth, and shape of the signal with those detected on cleaved, single-crystal surfaces of silicon by Haneman (1968) it was suggested that the signals arose from dangling bonds. The high density of spins ($\sim 10^{20}$ cm^{-3}) was found to be a true bulk property and it decreased by a factor ~ 10–100 on annealing. The later observation that voids exist in amorphous Si and Ge provided in a natural source of internal surfaces on which such dangling bonds could reside (see Brodsky et al. 1972b). Several studies (e.g. Paesler et al. 1974) produced a fairly good correlation between the free spin density and such properties as porosity, internal stress, and density, thus strengthening this hypothesis (see Fig. 7.15).

A study by Connell and Pawlik (1976) in which hydrogen was incorporated into films of amorphous Ge during the sputtering process indicates that there is approximately only one spin per 100 dangling bond sites in unhydrogenated material. This suggests almost complete pairing of dangling bonds on void surfaces. Those remaining can bond to incorporated hydrogen, leaving a spin signal associated with dangling bonds at small defects such as single dangling bonds. The observation of zero-spin signal density in amorphous Si prepared by the glow-discharge decomposition of SiH$_4$ (Spear 1974b) suggests that even these dangling bonds can be hydrogenated in this material.

More recent studies by photoemission (Wagner and Spicer 1972, Spicer 1974, Eastman and Grobman 1972a,b, Kaplan et al. 1975) of the surface state density and its distribution in energy on crystalline silicon surfaces have now cast doubt on the original correlation between e.s.r. signal strength and density of surface dangling bonds, and consequently the association of spin density with void surfaces in amorphous material has become uncertain and will be considered further below.

DiSalvo, Bagley, and Clark (1974) (see also, DiSalvo, Bagley, and Hutton 1976) first obtained evidence for ordering of spins at low temperature in amorphous Ge, and Hudgens (1976) investigated the absolute value of the magnetic susceptibility of amorphous Ge and Si obtained by evaporation, glow-discharge decomposition, and sputtering. This deviates from a Curie–Weiss law at low temperatures, suggesting antiferromagnetic coupling; the spin density decreases with annealing and hydrogen doping and the Néel temperature drops from ~ 8 K to below $1 \cdot 5$ K.

Thomas and Kaplan (1976) investigated the e.s.r. signal from ultra-high-vacuum-evaporated silicon, in particular the dependence of linewidth on the concentration and temperature dependence of the susceptibility. They came to the conclusion that the spins are randomly distributed and are perhaps single dangling bonds. For glow-discharge-deposited specimens Brodsky and Title (1976) conclude that spins are clustered. If we are right in thinking that dangling bonds are all hydrogenated in this material and that spins must arise from overlap of E_X and E_Y levels, then the amount of overlap could depend critically on disorder, which would favour clustering.

The conclusion that spins have this origin is reinforced by the effect of doping on the photoconductivity lifetime, discussed in § 6.5.2. No e.s.r. signal has at the time of writing been observed for undoped glow-discharge material, but Knights, Biegelsen, and Solomon (1977) observe a weak signal on doping. They also find that a signal ten times stronger can be optically induced *during* the radiation. The concept of a negative Hubbard U is invoked to explain this. We consider that Spear's E_Y band, which is ~ 10 stronger than his E_X, must be mainly due to some kind of deep donor (hole trap) which does not act as an acceptor. A possible model to explain the enhancement of spin by radiation is that, if this defect is denoted by D^0, the reaction

$$2D^+ \rightarrow D^{2+} + D^0 \tag{6.81}$$

is exothermic and is rapid, so that D^+ produced by illumination have only a very short lifetime. Thus p-type doping would produce D^{2+} centres which would not have a spin.

Knights *et al* (1977) see in the case of boron-doped specimens under illumination two lines, a 'central line' with width $\sim 7 \cdot 5$ G and a broad line about 20 G wide shifted to a $0 \cdot 1$ per cent higher g value. Only the former appears in the dark signal. A possible hypothesis is that the narrow line is due to a charged divacancy and the broad one to the unstable D^+.

Solomon, Biegelsen, and Knights (1977) observe spin-dependent photoconductivity (see below) in n- and p-type glow-discharge-deposited silicon, the signal showing a strong maximum near the undoped composition, as shown in Fig. 6.68. We could interpret this by saying that for p-type material the recombination is monomolecular, electrons falling into the spinless D^{2+} states, and for the n-type material it is bimolecular, electrons falling into the states resulting from the capture of holes, which, according to (6.81), must be spinless.

In evaporated and sputtered specimens of Si and Ge, and perhaps in

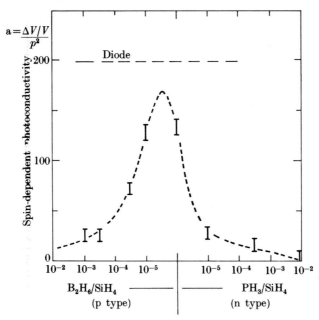

Fig. 6.68. Spin-dependent photoconductivity extrapolated to infinite microwave power as a function of phosphine/silane and diborane/silane volume ratio. The magnitude of the signal for intrinsic silicon, deduced from a p–n junction experiment, is marked 'diode'. (From Solomon *et al.* 1977.)

glow-discharge-deposited specimens also, spins on the surfaces of voids have to be considered. It is known from the work of Kaplan *et al* (1975) and Lemke and Haneman (1975) that a cleaved silicon surface does *not* give an e.s.r. signal, the electrons forming doubly occupied and empty bonds which alternate (Phillips 1977). In crystalline material the number of spins on a void must be even, so no unpaired spins will result. Spins on voids must therefore have one of the following origins.

(a) In the amorphous material odd numbers of spins are possible. Connell and Pawlik (1976) do in fact conclude, from an estimate of the void surface obtained from the optical absorption coefficient, that in sputtered germanium there is not more than one spin per void.

(b) Hydrogen or oxygen contamination leaving some isolated spins.

(c) Donors or acceptors present in the material, as in doped glow-discharge material. These could transfer their electrons to or from surface states on the void, producing states with spin.

At the time of writing, on the available evidence we do not feel able to distinguish between these mechanisms, or to discuss the relationship between the number of spins (N_L) and the density of states ($N_L/\Delta E$) deduced from hopping conductivity. Voget-Grote *et al.* (1976), however, have found an interesting relationship between the two phenomena for differently prepared ion-bombarded samples of amorphous silicon. A close relationship exists between the temperature-dependent part of the line-width and the hopping conductivity (§ 2.7). For a review see Stuke (1977).

As already mentioned, Solomon *et al.* have observed spin-dependent photoconductivity in amorphous silicon (Fig. 6.68). The phenomenon was first observed in crystalline silicon (Lepine 1972) and by Solomon (1976) at a p–n junction and at dislocations in silicon (Lepine *et al.* 1976). A simple theory is as follows. One particle (say a hole) is captured by a spinless centre, which then acquires a spin. If the centre and the electron have the same spin, no recombination is possible. If the field produces spin polarizations p, P, then the recombination time τ is increased according to the equation

$$1/\tau = (1/\tau_0)(1 - pP).$$

For 3000 G and 300 K, pP is 10^{-6}. A saturated e.s.r. signal removes this term. The experiments suggest that the effect is orders of magnitude greater. Solomon gives a tentative explanation in terms of rather complex centres. An alternative (Kaplan, Solomon, and Mott 1978) is to suppose that the electron and hole (in a centre) are bound together by a weak Coulomb force. Then in the ground state of the bound exciton, if spins are parallel, they form a triplet and do not recombine. Thus (except at very low T), they will normally separate. But often they should separate to a distance less than the Onsager radius $e^2/\kappa kT$, so that, while the Hund's rule interaction is negligible, normally the *same* pairs can try again to recombine. The e.s.r. signal, by switching the spin direction, can allow the electron and hole pair to recombine.

We turn now to chalcogenides. Those in the annealed state are diamagnetic, and in general the diamagnetic susceptibility is greater than for the corresponding crystal (for a review see Matyáš 1976). According to White and Anderson (1972) this is because the paramagnetic Van Vleck term is smaller. The absence of Curie paramagnetism at low temperatures and still more the absence of an e.s.r. signal was for a long time one of the puzzles of the subject, since it seemed not to be compatible with a pinned Fermi energy (Fritzsche 1973). Some of the evidence is reviewed in § 9.4. Fritzsche (1976) reviews the situation for a large number of materials in which the spin density would have been observed if it was greater than 10^{15}–3×10^{16} cm^{-3}. In § 6.4.1 we have shown how a model of 'charged dangling bonds', introduced by Street and Mott (1975) making use of Anderson's (1975) model of a negative Hubbard U, was able to reconcile these concepts. According to this model, which is developed further in Chapter 9, 'dangling bonds', such as chain ends in selenium, are normally positively or negatively charged (denoted D^+, D^-) and so do not carry a spin, though they do pin the Fermi energy. The model predicts, however, that if radiation ejects an electron from a D^- into the conduction band or into a D^+ from the valence band, a state with a spin will be produced which will be metastable at low temperatures.

The extensive work of Bishop, Strom, and Taylor (1975, 1976a,b; for a review 1977) on optically induced paramagnetic states can at any rate in part be interpreted in terms of this model. This work is mentioned in connection with photoluminescence of chalcogenides in Chapter 9. In chalcogenides and also in arsenic, irradiation with light of energy corresponding to the Urbach tail of the absorption edge $(\alpha \simeq 100\ \mathrm{cm}^{-1})$ excites photoluminescence, which fatigues (decays) during continuing excitation. The fatigue is accompanied by the appearance of a growing e.s.r. signal not present before illumination, together with optically induced absorption. In contrast to the case of glow-discharge-deposited silicon, the signal is stable below 80 K. Some of the phenomena established were as follows:

(1) The signal saturated at $\sim 10^{17}$ spins cm^{-3} in As_2S_3, As_2Se_3, and As, and 10^{16} spins cm^{-3} in Se.

(2) No spin signal could be induced in crystalline As_2Se_3, suggesting a density of spins less than 10^{16} spins cm^{-3}.

(3) Analysis of the e.s.r. spectra identified a signal due to an electron missing from a non-bonding line pair chalcogen orbital. This we believe to be our D^0 centre, though the observation would be compatible with a self-trapped hole in the valence band if self-trapping turned out to be possible. In glasses containing arsenic another broader line corresponds to a hole localized in an arsenic p orbital.

7

TETRAHEDRALLY BONDED
SEMICONDUCTORS—AMORPHOUS
GERMANIUM AND SILICON

7.1. Methods of preparation
7.2. Structure of amorphous Ge and Si
7.3. Voids, impurities and other defects in amorphous Ge and Si
7.4. Electrical properties of amorphous germanium
7.5. Electrical properties of amorphous silicon
7.6. Optical properties of amorphous germanium
7.7. Optical properties of amorphous silicon
7.8. Density of states in the valence and conduction bands of amorphous germanium and silicon

7.1. Methods of preparation

In the liquid state amorphous germanium and silicon have six-fold co-ordination and are metallic. Bulk glasses cannot be prepared by normal methods of cooling the melt, although for Ge there have been isolated reports of material obtained by splat-quenching (Davies and Hall 1974, Vučič, Etlinger, and Kunstelj 1976). Amorphous films, produced by some form of deposition, have, like the crystalline phases, four-fold co-ordination and are semiconductors. The way in which the film is prepared has a major effect on the electrical and optical properties, as will be seen. The most important methods are as follows.

(a) *Vacuum evaporation* (see for example, Tauc, Grigorovici and Vancu 1966, Clark 1967, Grigorovici, Croitoru, and Dévényi 1967, Walley 1968a,b, Walley and Jonscher 1968, Croitoru and Vescan 1969, Spicer and Donovan 1970, Chopra and Bahl 1970, Thèye 1970, Pierce and Spicer 1971, Spicer, Donovan, and Fischer 1972, Lewis 1972, Bahl, Bhagat, and Glosser 1974). Evaporation in a vacuum can produce films up to 20 μm in thickness, but such thick films tend to break up because of internal strains and most measurements have been made on films of a few microns thickness. Ideal conditions appear to be ultra-high vacuum ($\sim 10^{-10}$ torr), pure starting material, a large source-to-substrate separation (to achieve almost normal incidence), a fairly slow evaporation rate (a few microns per hour), ultra-clean, smooth substrates, and a high substrate temperature (250–300°C). Not all films of course are prepared under such conditions and studies of the properties as a function of one or other of the variables have been quite informative. Films produced by evaporation in a poorer vacuum undoubtedly contain some oxygen and this tends to stabilize the films against crystallization on heating. Films deposited onto substrates held at different temperatures provide materials with a range of properties, as do films deposited at room temperature and subsequently annealed. The density of germanium films prepared by vacuum evaporation has been reported to be up to 30 per cent lower than that of the crystal ($5 \cdot 35 \mathrm{~g~cm}^{-3}$) but, on the average, densities are perhaps 10–15 per cent lower, and Donovan, Ashley, and Spicer (1970) report that evaporation of germanium on to a quartz substrate held at 250–300°C produces amorphous films of density close to that of the crystal.

(b) *Sputtering*. Preparation of films by sputtering in argon has been described by Tauc *et al.* (1970), Moss, Flynn, and Bauer (1973), Hauser (1973), and Paul *et al.* (1973). The starting material is normally a pre-fabricated, polycrystalline, or compressed powder target and with, say, 10^{-3} torr of Ar in the chamber, rates of deposition are a few microns per hour. It is difficult to avoid incorporation of Ar, at a level of perhaps 1 per cent, but the concentration of other impurities can, with care, be kept much lower. As with high-vacuum evaporation, sputtered films have a density that increases with substrate temperature and can be within a few per cent of that of the crystals. The properties of such films are similar to those of evaporated material (at least when considered relative to films produced by glow-discharge—see below). Films of germanium and silicon prepared by sputtering in an argon/hydrogen mixture have been described by Lewis *et al.* (1974), Connell and Pawlik (1976), Lewis (1976), Hauser (1976), Anderson, Moustakas, and Paul (1977), and Moustakas and Paul (1977).

(c) *Electrolytic deposition* (Szekely 1951, Tauc *et al.* 1970). Thick (~30 μm) films of amorphous germanium can be prepared by deposition on to a copper cathode immersed in an electrolytic solution of $GeCl_4$ in $C_3H_6(OH)_2$. Such films, which contain a few per cent of oxygen and copper as impurities, can be obtained unsupported by subsequently dissolving the copper substrate in chromic–sulphuric acid. Only a very limited number of measurements have been reported on films prepared by this method.

(d) *Glow-discharge decomposition*. This method of producing amorphous films employs an electrodeless radiofrequency discharge in germane (GeH_4) or silane (SiH_4) gas. Developed at the STL Laboratories (see Chittick, Alexander, and Sterling 1969, Chittick 1970), and subsequently by Spear and his colleagues (see Le Comber, Madan, and Spear 1972, Spear and Le Comber 1977, and other references cited in this chapter) and by Knights (1976a,b), this method produces films that are several orders of magnitude more resistive than films produced by other methods; the concentration of defects is lower and the band gap is higher. Why this should be so is not yet absolutely clear. According to some authors (see Connell and Pawlik 1976, Hudgens 1976, Hauser 1976, Fritzsche 1977) it is because of the presence of hydrogen, the incorporation of which, even in a concentration less than 1 per cent, undoubtedly serves to neutralize the role of dangling bonds in an otherwise fully co-ordinated network. According to others (see Spear 1974b, Le Comber et al. 1974) the different properties of these films are associated with the particular nature of the deposition process itself. Estimates of the hydrogen content vary and indeed the amount incorporated depends on the deposition temperature and the deposition rate (Brodsky, Frisch, Ziegler, and Lanford 1977). Brodsky, Cardona, and Cuomo (1977) (see also Knights, Lucovsky, and Nemanich 1978), from a detailed study of infrared and Raman spectra on glow-discharge-deposited Si, report concentrations of bonded hydrogen ~35–50 at. per cent and ~15–25 at. per cent in films deposited at room temperatures and 250°C, respectively. Tsai et al. (1977) have studied the effusion of hydrogen on heating; above 600°C most of the hydrogen can be driven out, but the resulting films then have a conductivity and optical gap similar to films prepared by evaporation or sputtering.

In spite of the likely high concentration of hydrogen in amorphous Si and Ge prepared by glow-discharge decomposition, the low density of active defect sites has allowed a comprehensive investigation of the 'intrinsic' properties† of such films. The first reports of substitutional doping of an amorphous semiconductor were made on films deposited by glow-discharge decomposition of SiH_4/PH_3 and SiH_4/B_2O_6 gaseous mixtures (Spear and Le Comber 1975, Spear and Le Comber 1976, Le Comber and Spear 1976, Knights 1976a,b, Spear and Le Comber 1977, Spear 1977).

7.2. Structure of amorphous Ge and Si

The principal experimental method used for determining the structure of these tetrahedrally co-ordinated semiconductors is X-ray or electron diffraction. From an analysis of the angular distribution of the scattered intensity, a radial distribution function (RDF) can be derived which can be compared with those determined from structural models built either by hand or on a computer. For Ge and Si (particularly the former) this procedure has led to a detailed description of the short-range order present

---

† By this we mean the properties of the conduction and valence bands, including the position of the mobility edge, which do not depend to any major extent on states in the gap. For films containing large amounts of hydrogen it might be preferable to refer to them as Si:H alloys.

in amorphous films. Though variations in conditions of preparation lead to changes in electrical and optical properties (see later sections) and also to recognizable changes in the RDF, the latter are relatively insignificant if one is interested in gross features of the structure.†

Experimentally determined RDFs of amorphous silicon (by Moss and Graczyk (1969, 1970) using electron diffraction) and amorphous germanium (by Temkin, Paul, and Connell (1973) using X-ray diffraction) are compared with those of the crystalline phases in Figs. 7.1 and 7.2

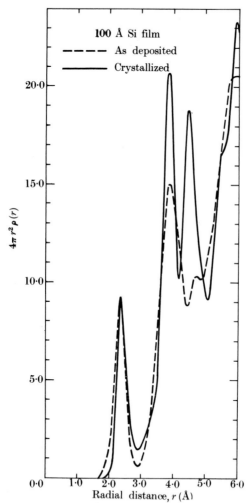

Fig. 7.1. Radial distribution function (RDF) of amorphous (evaporated) and crystalline silicon as determined from analysis of electron diffration data. (From Moss and Graczyk 1970.)

† Voids and other imperfections are discussed in § 7.3.

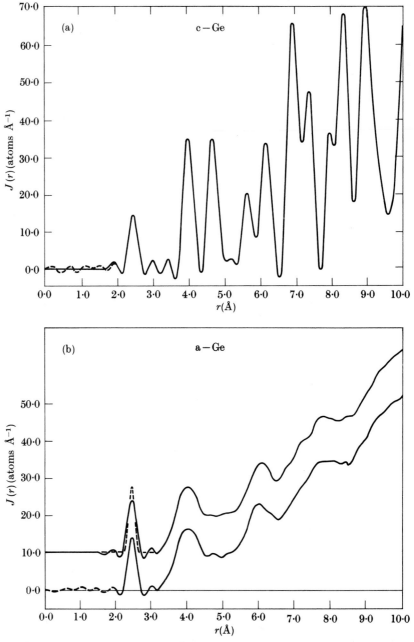

Fig. 7.2. Radial distribution function of (a) crystalline germanium and (b) two films of amorphous germanium prepared by sputtering on to substrates held at 150°C (lower curve) and 350°C (upper curve). Ripples on either side of the first main peak are associated with termination effects. These have been corrected for in the upper curve of (b). (From Temkin *et al.* 1973.)

respectively. Similar curves have been obtained by other workers; see, for example, Richter and Breitling (1958), Coleman and Thomas (1967, 1968), Grigorovici and Manaila (1967, 1969), Shevchik and Paul (1972), Gandais *et al.* (1973), and Graczyk and Chaudhari (1973*a,b*). The RDFs of both amorphous germanium and silicon show that the basic tetrahedral arrangement of the diamond cubic structure is preserved, there being four (the area under the first peak) nearest neighbours at a separation within a few per cent of the crystalline bond length and twelve next-nearest neighbours at an average separation of $\sqrt{(8/3)}$ times that length. The breadth of the second peak, after correction for thermal broadening, indicates bond-angle distortions of about $\pm 10°$ (r.m.s.) off the normal tetrahedral angle of $109° 28'$. However, it is the dramatic loss of the third-neighbour peak present in the crystalline RDFs that marks the first most striking departure from the diamond structure.

The relative orientation of triads of bonds emanating from two nearest neighbours is known as the dihedral angle ϕ (Fig. 7.3(a)). When this is $60°$, the bonds are said to be in the staggered configuration, and when it is zero the bonds are said to be eclipsed. If all bonds are staggered and further units added, one is led to the diamond cubic structure (Fig. 7.3(b)); if they are staggered in three directions but eclipsed in the fourth, one derives the wurtzite lattice; in both of these structures all atoms lie in puckered rings consisting of six atoms. If all bonds are eclipsed and a small distortion ($1° 28'$) of bond angles allowed, pentagonal dodecahedra containing only five-fold planar rings of atoms are created (Fig. 7.3(c)), but these units (called amorphons) cannot fill space completely.

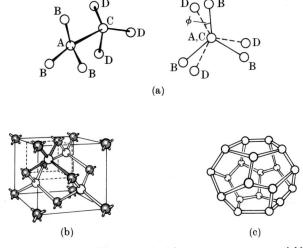

(a)

(b)   (c)

Fig. 7.3. (a) Illustration of the dihedral angle ϕ between next-nearest-neighbour bonds. (b) The diamond cubic structure—all bonds staggered ($\phi = 60°$). (c) An 'amorphon'—all bonds eclipsed ($\phi = 0°$).

The first serious attempt to model a-Ge or a-Si was made by Grigorovici (1968) (see also Grigorovici and Manaila 1969, Grigorovici and Balu 1972, Grigorovici 1973) using a mixture of amorphons and diamond-like units, simulated for the purposes of building by modified Voronoi polyhedra. This model produced an RDF in fair agreement with experiment and achieved the important feature of reducing the third peak which is present in the RDF of crystalline Ge or Si but not in that of the amorphous forms (Figs. 7.1 and 7.2). However, detailed comparison shows that the bond-length fluctuations were overestimated and the bond-angle distortions underestimated in this model.

Polk (1971) (see also Turnbull and Polk 1972) built the first continuous random network that allowed the dihedral angle to take up all values.

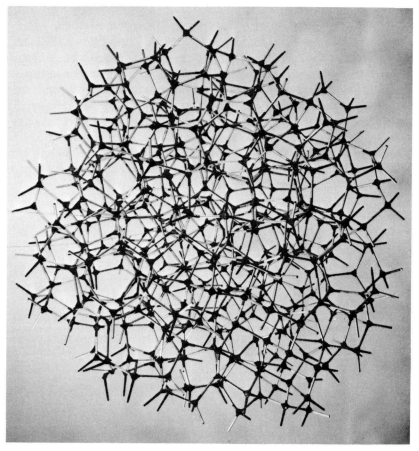

Fig. 7.4. The 440-atom continuous random network (CRN) built by Polk (1971) to simulate the structure of amorphous germanium or silicon.

Starting with a core of five- and six-membered rings, additional atoms were added in such a way that the following conditions were satisfied: (1) there were no unconnected bonds in the interior of the model; (2) the bond length variations were less than 1 per cent; (3) the bond-angle distortions were within about $\pm 20°$ of the tetrahedral bond angle (109°); (4) there was a minimum of strain. The strain minimization was found to be operationally equivalent to requiring that the surface density of dangling bonds remained constant as the model size increased. The resulting model (Fig. 7.4) contained 440 atoms and appeared capable of indefinite extension. Its RDF, determined by direct measurement of atomic separations from some 16 centrally located atoms, shows a good fit to experiment (Fig. 7.5). The

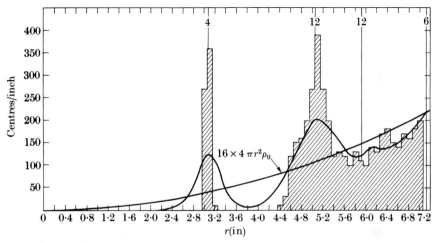

Fig. 7.5. Comparison of the RDF (histogram) of the Polk model with the (scaled) experimental RDF of amorphous silicon (shown in Fig. 7.1). The parabola represents the average density of the model, and the vertical lines and corresponding numbers represent the position and number of neighbours in a diamond cubic crystalline structure with a bond length equal to the average of the model. (From Polk 1971.)

density was determined to be 93 ± 2 per cent of the diamond cubic structure, in reasonable agreement with that of many amorphous samples. The model contains five-, six-, and seven-membered rings of atoms and has a continuous distribution of dihedral angles.

The original Polk model was extended to 519 atoms by Polk and Boudreaux (1973), and the atomic co-ordinates of the whole model adjusted by an iterative technique on a computer until the standard deviation of all interatomic spacings was less than 0.2 per cent. Using the refined co-ordinates, the RDF (Fig. 7.6(a)) as well as the mean tetrahedral bond angle and its standard deviation ($108° \pm 9.1°$), the density (99 per cent of diamond cubic) and the dihedral angle distribution (Fig. 7.6(b)) were

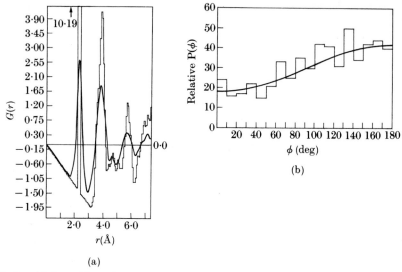

(a)

Fig. 7.6. (a) Comparison of the reduced RDF,

$$G(r) = 4\pi r\{\rho(r) - \rho_0\}$$

for the Polk–Boudreaux 519 atom model (after refinement of the co-ordinates and broadening of the histogram to reduce statistical fluctuations) with the experimental $G(r)$ of amorphous silicon (derived from Fig. 7.1). (b) Dihedral angle distribution of Polk–Boudreaux 519 atom model. (From Polk and Boudreaux 1973.)

determined. The latter shows that the staggered configuration of bonds is roughly twice as likely as the eclipsed.

Steinhardt, Alben, and Weaire (1974) and Duffy, Boudreaux, and Polk (1974) have further refined this 519-atom model by adjusting the atomic positions so as to minimize the Keating expression (Keating 1966, see also Moss *et al.* 1973/74) for the elastic energy in terms of the bond lengths, bond angles, and the bond-stretching and bond-bending force constants. Steinhardt *et al.* (1974) also built a new 201-atom model starting from a seed of 21 atoms and relaxing, by computer, the positions of newly added groups of atoms so as to minimize the energy of the network as it grew. This relaxation procedure, which determines both the connectivity as well as the co-ordinates, was not found to lead to a model with structural characteristics significantly different from the relaxed 519-atom model. Both of the relaxed models had densities within 1 per cent of crystalline Ge, bond-length fluctuations ~1%, and bond-angle distortions with an r.m.s. value ~7°, this being rather smaller than the value (~10°) determined experimentally. Apart from this latter feature, the RDF of both modes as well as that of Duffy *et al.* (1974), who used a somewhat different relaxation procedure, agreed very well with experiment.

A very good fit to the experimental RDF is of course a necessary criterion to be met by any model. However, it may be insufficient for defining the structure uniquely. With this in mind, Connell and Temkin (1974a,b) undertook to build a tetrahedrally co-ordinated continuous random network which contained no odd-membered rings and therefore had a significantly different topology than that of the Polk and similar models. Another reason for attempting this was to model the III–V compounds which are also tetrahedrally co-ordinated and yield RDFs similar to Ge and Si but for which it may be desirable to avoid bonds between like atoms. A 238-atom model was built following the same general procedure as Polk but allowing atoms to lie in rings containing only an even number of atoms. No difficulty in construction was encountered. The RDF of this model, after adjustment of its co-ordinates to minimize bond-length variations, is compared with that of the Steinhardt 201-atom model in Fig. 7.7(a) and with that of the Polk–Boudreaux 519-atom model and experimental data in Fig. 7.7(b). Fig. 7.8 shows the dihedral angle distributions of the Connell–Temkin and the Steinhardt models, and

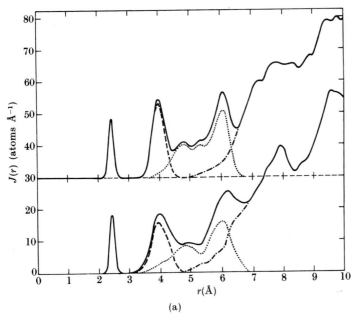

(a)

Fig. 7.7. (a) RDFs of the Connell–Temkin 238-atom even-membered ring model (lower curve) and the Steinhardt 201-atom model (upper curve). The partial RDFs $J_n(r)$ are also shown, $n = 2$(broken curve), $n = 3$(dotted curve), $n \geq 4$(chain curve); r has been scaled to the bond length of germanium. (From Connell and Temkin 1974.) (b) (p. 330) Comparison of the RDFs of the Connell–Temkin model (dotted curve), the Polk–Boudreaux unrelaxed 519-atom model (broken curve), and one of the experimental curves (solid curve) for amorphous germanium shown in Fig. 7.2. (From Paul and Connell 1976.)

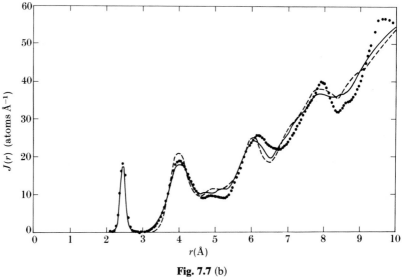

Fig. 7.7 (b)

demonstrates that eclipsed or nearly eclipsed bonds tend to occur less frequently in the even-only over the mixed-ring model. Connell and Temkin note that the experimental RDF of amorphous germanium appears to lie somewhere between that characteristic of the two models. Beeman and Bobb (1975) have used the co-ordinates of the Connell–Temkin model to derive a series of continuous random networks (CRNs) having various proportions of odd- to even-membered rings.

Other CRN models for amorphous tetrahedrally co-ordinated materials have been computer built. Shevchik and Paul (1972) have described the construction of a 1000-atom model that attempted to simulate the actual

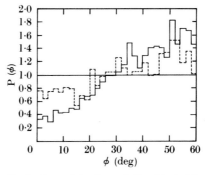

Fig. 7.8. Dihedral angle distributions of the Connell–Temkin (solid curve) and the Steinhardt (broken curve) models. (From Connell and Temkin 1974*b*.)

deposition process by allowing atoms added to a seed of 15 atoms to seek out locations in which, as a first choice, three of its bonds could be satisfied, or failing this just two, and as a final choice, one. In this way surface mobility, which is possible under certain conditions of film deposition, is simulated. The resulting network had about 5 per cent of bonds unsatisfied (the first co-ordination number was 3·8), which is certainly an overestimate for carefully prepared films, but in other respects provided a good fit to the RDF.

Another approach to computer modelling has been followed by Henderson and Herman (1972) and Henderson (1972). Their method involved random displacements of the positions of 64 atoms initially in an f.c.c. crystalline arrangement until, after several hundred moves, the experimental RDF was approximated. The resulting structure, although not satisfactory in several respects, had periodic boundary conditions making it favourable for calculations of vibrational and electronic properties. Other approaches to computer construction of tetrahedrally co-ordinated networks are described by Duffy *et al.* (1974) and Guttman (1976).

Attempts to model the structure of amorphous Ge and Si using microcrystallites have been stimulated by the observation of contrast and fringes in high-resolution electron micrographs. The presence of these features has been taken to infer the existence of coherently diffracting regions of linear dimensions up to 15 Å (Rudee 1971, 1972, Rudee and Howie 1972; see also Howie, Krivanek, and Rudee 1973, Chaudhari, Graczyk, and Herd 1972, 1973, Chaudhari and Graczyk 1974, Herd and Chaudhari 1974, Gaskell and Howie 1974). Whether random network models can account for the observed micrographs has been a subject of some controversy (see Cochran 1973, 1974, Berry and Doyle 1973). Although some random networks have, in certain directions, accidental planar correlations that can produce structure in electron micrographs (Alben, Cargill, and Wenzell 1976), the predicted contrast appears to be too low. Selective filtering in the microscope, particularly in off-set bright-field configurations, could in principle provide an explanation of the experimental data, and this possibility seems likely even in the case of axial geometry used by Krivanek, Gaskell, and Howie (1976) and Freeman *et al.* (1977).

The main problem with construction of microcrystallite models and determination of their scattering properties is that of the connecting tissue which, for small crystallites, can occupy 50 per cent of the volume. Normally this tissue will include elements characteristic of random networks, although a model built by Gaskéll (1975), in which small tetrahedral modules of diamond-cubic symmetry are packed so that {111} faces are in contact and are joined by planes of eclipsed bonds, seems to achieve a good connectivity without resort to such elements. A portion of this

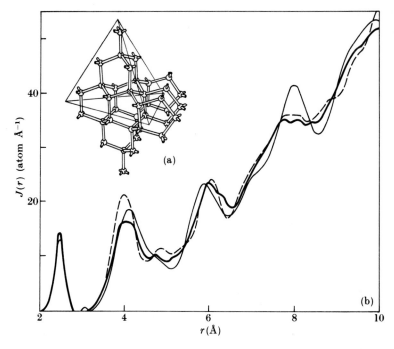

Fig. 7.9. (a) Portion of a polytetrahedral model built by Gaskell. The structure in each tetrahedron is diamond cubic and eclipsed bonds occur across each of the four {111} faces. (b) RDFs: thick solid curve, amorphous Ge (experiment); thin solid curve, polytetrahedral model; broken curve, relaxed Polk–Boudreaux model. (From Gaskell *et al.* 1977.)

model and its associated RDF (Gaskell, Gibson, and Howie 1977) are shown in Fig. 7.9.

Other microcrystalline models have been based on polytypes of the diamond structure, e.g. the wurtzite, ST-12, or BC-8 lattices (see § 7.1.8) or on various clathrate structures. In an attempt to see whether such structures have diffraction properties in accord with those observed for amorphous Ge, Weinstein and Davis (1973/74) calculated the diffraction function $F(s)$ for several types of crystallites of various sizes. Their results (Fig. 7.10) show that none of these structures provides a good fit to the experimental data. Agreement is improved as the crystallite size is reduced—in fact all structures then tend to diffract similarly—but it is precisely under these conditions that the need to include connecting tissue becomes stronger. Weinstein (1974) has reported that a statistical mixture of 60 per cent BC-8 and 40 per cent clathrate I microcrystallites gives an acceptable fit to $F(s)$ but the agreement is still poor compared with that calculated using the Polk–Boudreaux continuous random network (Fig. 7.11) (Chaudhari and Graczyk 1974).

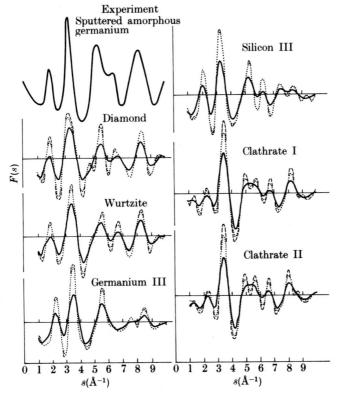

Fig. 7.10. Diffraction functions $F(s)$ calculated for microcrystallites of germanium having various structures. Three theoretical curves are shown for each structure, corresponding to microcrystallites of small (solid curve), medium (dotted curve) and large (broken curve) radii. The small radii were about 4 Å and the large about 6·5 Å. (From Weinstein and Davis 1973/1974.)

7.3. Voids, impurities, and other defects in amorphous Ge and Si

The models described in the preceding section attempt to simulate the structure of 'ideal' amorphous Ge or Si, by which is meant a material free of voids, impurities, and other defects, and, in most cases, containing no unsatisfied bonds. It is normally assumed that there exists a well-defined metastable phase to which all real films will approach on annealing. While it is clear that impurities, gross density deficits, crystalline inclusions, and so on can be regarded as defects which can be avoided, it is not so obvious that the presence of a certain proportion of dangling bonds or even small voids could not release some of the distortion energy of an otherwise fully connected structure, thereby achieving a more stable amorphous state. While we know of no evidence that this is so, the possibility should not be ruled out.

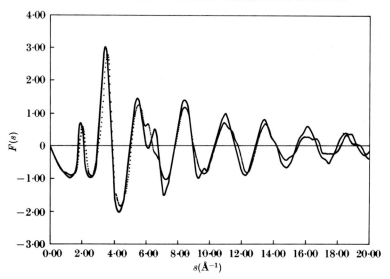

Fig. 7.11. Comparison of diffraction function $F(s)$ derived from the Polk–Boudreaux model (solid curve) and obtained experimentally (points) by Graczyk and Chaudhari (1973a). (From Chaudhari and Graczyk 1974.)

Films of amorphous Ge and Si have densities ranging normally from 3 to 15 per cent below that of the crystals. Continuous random networks and microcrystalline models are normally associated with density deficits of only 1–3 per cent. Taken in conjunction with the bond length and the co-ordination number deduced from the RDF, measurements of the density of real films can therefore provide information on the presence or otherwise of voids in the structure. An equally important method is the observation of small-angle scattering of electrons or X-rays (Guiner *et al.* 1955, Moss and Graczyk 1969, 1970, Shevchik and Paul 1972, 1974, Temkin *et al.* 1973). Temkin *et al.* (1973) estimate that, within the practical limits of this technique, spherical voids ranging in diameter from 3 to 250 Å can be detected. By observing small-angle scattering of electrons from films of amorphous Si of thickness of 100 Å, Moss and Graczyk (1969) concluded that these films contained regions of distinctly deficient density and suggested that these were voids. Shevchik and Paul (1974) measured the intensity of X-rays scattered at angles between 2° and 12° from amorphous Ge films prepared by electrodeposition, by sputtering, and by evaporation. Their results, after applying collimation corrections, are shown in Fig. 7.12.

The distribution functions $N(r)$ for the radii r of the voids and the normalized void distribution $V(r)N(r)$ (assuming that the voids are spherical, $V(r) = 4\pi r^3/3$) as deduced by Shevchik and Paul (1974) from

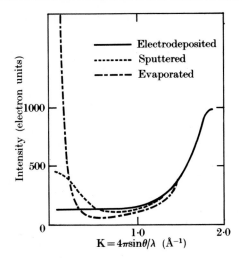

Fig. 7.12. Small-angle X-ray scattering in amorphous germanium prepared by sputtering, evaporation, and electrodeposition. (From Shevchik and Paul 1974.)

their data on evaporated films of both Ge and Si are shown in Fig. 7.13(a,b). The density deficit determined from these data is 4–5 per cent, while weighing and measuring the dimensions of the films yielded a deficit ~10 per cent. This difference might be attributable to uncertainties in the extrapolation to zero wavevector, to the assumption that the voids are spherical, or to the presence of very small voids (monatomic vacancies) or

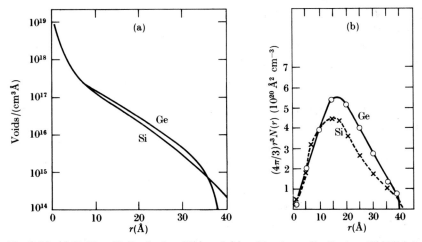

Fig. 7.13. (a) Void radii distribution $N(r)$ and (b) void volume distribution $V(r)N(r)$ for evaporated germanium and silicon as deduced from small-angle X-ray scattering. (From Shevchik and Paul 1974.)

very large voids (say $r > 40$ Å) that were outside the limits of detection in these experiments.

Although Fig. 7.12 implies a lower density of voids in sputtered and electrodeposited films compared with evaporated films, this conclusion should be treated with caution. Not only were the actual density deficits higher than inferred from the scattering data in all cases, but annealing caused the small-angle scattering to *increase*. While it is possible that this arose from a relaxation involving condensation of small voids into larger ones, annealing is normally associated with void removal. Moss and Graczyk (1970) have in fact reported a decrease in the small-angle scattering on annealing evaporated films.

The presence of voids in amorphous Ge deposited by evaporation on to substrates at temperatures below 150°C has been demonstrated convincingly by transmission electron microscopy. Donovan and Heinemann (1971) obtained the micrographs shown in Fig. 7.14. The voids form a network structure penetrating the whole specimen and, as expected, the density of such films is low, up to 15 per cent below the crystalline value. Such films are easily contaminated; although annealing considerably reduces the void content because some of the voids coagulate and move to the surface, it presumably does not remove any gaseous impurities that may have found their way into voids and reacted with the sample. There is little doubt that many measurements reported in the literature have been made on films containing a significant concentration of voids and gross (several atomic per cent) contamination. Films deposited on to substrates held at a temperature above about 150°C appear to have a low void content (Fig. 7.14), although the presence of a fairly significant concentration of small voids cannot be dismissed. Substrate temperature is not the only parameter of importance. Rate and angle of evaporation and the nature of the substrate also play an important role (see for example Chopra, Rastogi, and Pandya 1974, Pandya, Rastogi, and Chopra 1975, Bahl, Bhagat, and Glosser 1974, Fuhs, Hesse, and Langer 1974, Beyer and Stuke 1975a,b, Barna et al. 1976).

Many physical properties of amorphous Ge and Si are changed by annealing, and it would seem that, in the majority of cases, voids and associated contamination (particularly by oxygen) play a dominant role in determining these properties. Paesler et al. (1974) have studied the effects of annealing on the porosity, the free-spin density, the diamagnetic susceptibility, and the internal stress in amorphous Ge films prepared by evaporation onto 300 K substrates (at rates between 10 and 50 Å s^{-1} in a vacuum of 10^{-5} torr). Their results (Fig. 7.15) show that the porosity P, free-spin density N_s, and internal stress σ_s decrease together as the annealing temperature is raised until they all vanish prior to crystallization ($T_c = 500$°C). The susceptibility, however, anneals more slowly, the ratio χ_a/χ_c

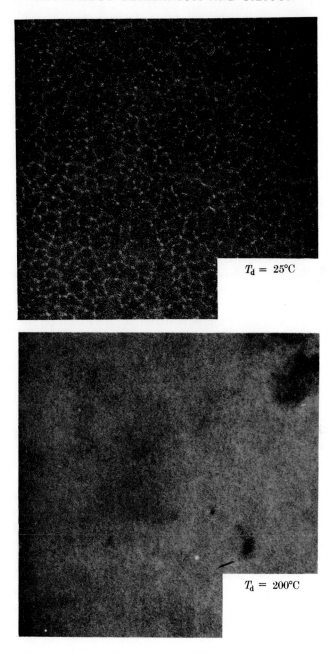

$T_{\rm d} = 25°{\rm C}$

$T_{\rm d} = 200°{\rm C}$

Fig. 7.14. Bright-field electron micrographs of germanium deposited by evaporation in a vacuum better than 2×10^{-8} torr onto cleaved KCl substrates held at 25°C and 200°C. Density-deficient (light) areas are associated with a crack-like network of voids and are present in films deposited below $T_{\rm d} < 150°$C. (From Donovan and Heinemann 1971.)

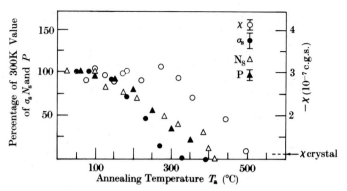

Fig. 7.15. Dependence of porosity P, free spin density N_s, internal stress σ_s and diamagnetic susceptibility χ of amorphous germanium on annealing at various temperatures T_a. (From Paesler *et al.* 1974.)

of the amorphous and crystalline susceptibilities remaining at $2 \cdot 5 \pm 0 \cdot 5$ (Hudgens 1973) up to about 320°C, above which it decreases towards unity at crystallization. Paesler *et al.* (1974) conclude that P, N_s, and σ_s are all associated with voids, whereas the enhanced diamagnetism is related to the details of the bonding in the bulk material. They deduce, for the freshly evaporated films, an internal surface area $\sim 10^6$ cm^2 associated with about 10^{18} voids cm^{-3} having an average radius of 30 Å. It was assumed that the free spins of density $N_s \sim 10^{20}$ cm^{-3} all resided on the internal surfaces of voids. However, this may not be a valid assumption, particularly in view of the insensitivity of N_s to exposure to water vapour or for that matter to oxygen or certain other gases (Agarwal 1973). The likelihood of spin pairing on surfaces (Kaplan *et al.* 1975) makes the earlier suggestion that voids were largely responsible for the free spins (Brodsky *et al.* 1970) slightly suspect (see Title, Brodsky, and Cuomo (1977) and § 6.8.3).

Although Paesler *et al.* (1974) did not measure the electrical conductivity of their films, Fuhs, Niemann, and Stuke (1974) have studied the effect of a network of crack-like voids on electrical transport. Films deposited onto KBr and Formvar substrates showed, by transmission electron microscopy, the presence of such a network in the former but not in the latter. Surprisingly, however, the temperature dependence of conductivity for the two films was virtually identical and furthermore the absolute magnitude of the conductivity was decreased by less than a factor of 2 as a result of the microcracks. Evidently the current manages to circumvent voids, their principal effect being a change in the effective sample geometry. The role of voids too small to be detected by microscopy has not yet been clarified, at least as far as evaporated films are concerned. However, for films prepared by sputtering the situation is clearer, as will be described later.

Evidence that voids affect the ultraviolet reflectivity of amorphous Ge prepared by evaporation has been obtained by Galeener (1971) and by Bauer and Galeener (1972). Their results (Fig. 7.16) have been interpreted in terms of voids with an oblate spheroid shape, the small axis lying in the plane of the film. Such voids of size ~ 10 Å have been shown to produce a type of Maxwell–Wagner effect with a dispersion at frequencies corresponding to the UV portion of the electromagnetic spectrum. The results for films deposited onto a substrate at 160°C are taken to be representative of essentially void-free material; the decrease in reflectivity of films deposited at lower temperatures is then associated with an increasing number of voids.

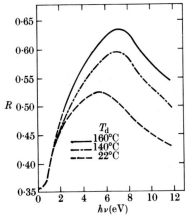

Fig. 7.16. Spectral dependence of the reflectance R of germanium films prepared by evaporation onto substrates held at temperatures T_d varying from 22°C to 160°C. (From Bauer and Galeener 1972.)

A similar sensitivity of the reflectivity on exposure to air of films deposited in ultra-high vacuum has been reported by Helms, Spicer, and Pereskokov (1974) and interpreted as arising from oxygen diffusing to substantial depths into the open morphology of the surface. For obliquely deposited films the void network is expected to be even more pronounced and in such films deposited in an oxygen atmosphere Ma and Anderson (1974) have detected clusters of GeO_2. A detailed study of oxygen uptake and determination of depth profiles for films prepared under different conditions has been made by Knotek (1975a) using Auger electron spectroscopy. He concludes that deposition and maintenance in ultra-high vacuum is essential to avoid oxygen contamination.

The properties of amorphous Ge films prepared by r.f. sputtering have been extensively characterized by Paul et al. (1973), Temkin et al. (1973), and Connell, Temkin, and Paul (1973). Like evaporated films, the properties of sputtered films depend on the conditions of preparation. An increase of the substrate temperature or annealing, for example, increases the density and co-ordination number and reduces the bond-angle distortion; the optical absorption edge shifts to higher energies and the d.c. conductivity is reduced. These effects are, as for evaporated films, associated with void removal and subtle rearrangements of the structure.

A detailed study of hydrogen (and deuterium) incorporation into sputtered films has been made by Lewis *et al.* (1974), Connell and Pawlik (1976), Lewis (1976) and Moustakas and Paul (1977). The effects of hydrogenation on the optical absorption edge, refractive index, conductivity, thermopower, and spin density are in many ways equivalent to those of raising the substrate temperature or of annealing. However, the RDF is not altered as much by hydrogenation as by annealing. It appears that hydrogen is incorporated both on void surfaces and in the bulk, but in neither environment does it cause gross structural rearrangements in the surrounding Ge network. Fig. 7.17 shows the absorption edge shift with hydrogen incorporation. The curve on the extreme right corresponds to approximately 8 at. per cent of hydrogen and exhibits, at lower energies, absorption peaks corresponding to the modes for Ge–H bond bending

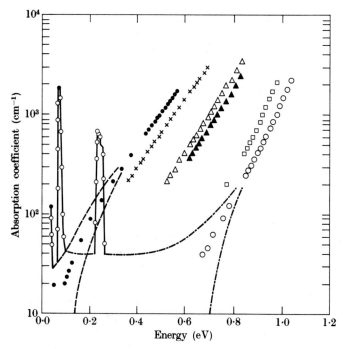

Fig. 7.17. Optical absorption in films of $Ge_{1-x}H_x$ for $x = 0$ (●), $x = 0.01$ (×), $x = 0.028$ (△), $x = 0.03$ (▲), $x = 0.051$ (□), and $x = 0.08$ (○). Relative values of x are more reliable than absolute values. The broken curves indicate error estimates. (From Connell and Pawlik 1976.)

(0·07 eV)† and bond stretching (0·23 eV). These bands had been seen earlier in films prepared by other means (Chittick 1970, Tauc *et al.* 1970) and interpreted differently; positive identification as Ge–H vibrations in the work of Lewis *et al.* (1974) was made by noting that the energies of equivalent peaks in hydrogenated and deuterated films scaled by √2. Under high resolution both these peaks were seen to be doublets (Fig. 7.18),

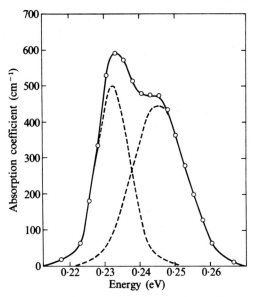

Fig. 7.18. Magnification and deconvolution of absorption peak associated with hydrogen incorporation into amorphous germanium (see Fig. 7.17). (From Connell and Pawlik 1976.)

suggesting two unequivalent sites for the incorporated atoms. As the hydrogen content was increased, the peak at 0·23 eV saturated at a value corresponding to about 3·5 at. per cent hydrogen whereas the peak at 0·245 eV continued to rise. Lewis *et al.* (1974) suggest therefore that the lower-energy peak is associated with Ge–H bonds on the internal surfaces of voids and the higher-energy peak with Ge–H bonds incorporated else-where in the network. These assignments are supported by measurements on crystalline Si by Becker and Gobeli (1963) who found that the Si–H bond stretching vibration occurs at a smaller energy for hydrogen atoms on the surface of crystalline Si than for isolated atoms in the bulk, in the ratio 0·96:1.

A good correlation was found by Connell and Pawlik (1976) between the strength of the 0·23 eV peak and the spin density deduced from e.s.r. The

† Brodsky *et al.* (1977) identify the 0·07 eV band with a 'wagging' rather than a bending mode.

latter fell from an initial value of $1 \cdot 6 \times 10^{19} \, \text{cm}^{-3}$ in unhydrogenated material by a factor of 10 at the highest level of hydrogen content. Assuming the spin density to be proportional to the number of unpaired dangling bonds on the internal surfaces of voids, it was deduced that most of the dangling bonds to which hydrogen eventually became attached were initially paired, only about 1 per cent contributing to e.s.r. in unhydrogenated material. Using an earlier conclusion (Temkin *et al.* 1973) that voids are typically 5 Å in diameter with centres 10 Å apart, there are of the order of 10–20 potential dangling bonds on each void surface but only one spin per 5–10 voids. The spins are thus about 20 Å apart, which is in fairly satisfactory agreement with the estimate of about 40 Å obtained from the spin density directly.

The strength of the absorption peak at $0 \cdot 245$ eV was found to correlate well with the density of the material, allowing an estimate to be made of the volume associated with the hydrogen incorporated into the network. This

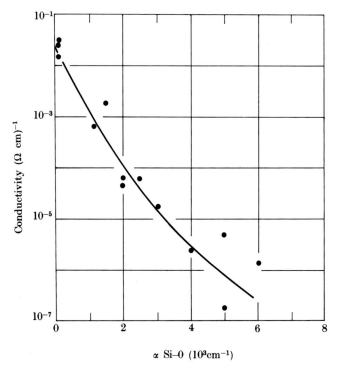

α Si–0 (10^3cm^{-1})

Fig. 7.19. (a) Room-temperature conductivity of evaporated films of amorphous silicon as a function of the infrared absorption coefficient in the Si–O peak near 10 μm. (b) (p. 343) The e.s.r. spin density N_s *versus* deposition rate R for amorphous silicon deposited in partial pressures of oxygen as indicated. The point marked He refers to a film deposited in helium gas at 10^{-4} torr. (From Le Comber *et al.* 1974.)

was found to be $\sim 32\ \text{Å}^3$, which may be compared with the volume associated with each Ge atom in a fully co-ordinated network, namely $24\ \text{Å}^3$.

The behaviour of amorphous Si films on annealing and incorporation of impurities closely parallels that of amorphous Ge. A detailed study has been made by Le Comber et al. (1974), with particular regard to why Si prepared by the glow-discharge decomposition of silane has properties significantly different from those of films prepared in other ways (see § 7.5). The experiments involved the deliberate introduction of oxygen or hydrogen during and after deposition of evaporated, sputtered, and glow-discharge films. The properties monitored were electrical conductivity, e.s.r., spin density, and optical absorption.

Fig. 7.19(a) shows the dependence of the conductivity of films evaporated in various partial pressures of oxygen, the abscissa being the strength of an absorption peak at 9–10 μm associated with the Si—O vibrational mode. The spin density N_s was found to depend both on the partial pressure of oxygen and on the rate of evaporation (Fig. 7.19(b)). For a low partial pressure of oxygen equal to 10^{-6} torr and at high rates of evaporation, N_s reaches 10^{20}–10^{21} cm^{-3}, a value similar to that observed earlier by Brodsky et al. (1970). For higher partial pressures the spin density was reduced until, at 10^{-4} torr of oxygen, no e.s.r. signals could be

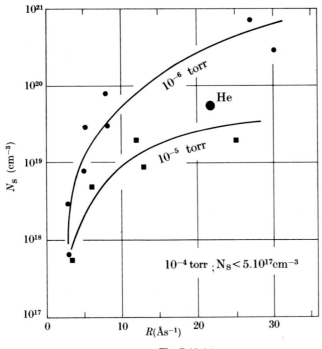

Fig. 7.19 (b)

detected ($N_s \sim 5 \times 10^{17}$ cm^{-3}). It is interesting to note here that no e.s.r. signals were observed in any samples of glow-discharge-deposited amorphous Si. Hudgens (1976) finds a Curie-type paramagnetism, with a small Θ that decreases below 1·3 K as the spin density is annealed out.

The variation of the optical absorption edge with oxygen content is shown in Fig. 7.20. The curves marked E1, E2, and E3 are marked with the room-temperature conductivity which can be correlated with the relative oxygen content from Fig. 7.19(a), while E1′ corresponds to the sample E1 after annealing to 650 K. For comparison three other abosrption edges are shown; a sample glow discharge deposited at 500 K, an amorphous SiO

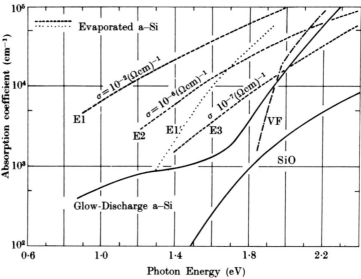

Fig. 7.20. Optical absorption edges of amorphous silicon. Curves E1, E2, and E3 refer to evaporated films with room-temperature conductivities as indicated. Curve E1′ refers to specimen E1 measured after an anneal at 650 K. The glow-discharge film was deposited at 500 K. Curve VF refers to the edge proposed by Brodsky *et al.* (1972) for a void-free sample. The edge for a SiO amorphous film is from Phillip (1972). (From Le Comber *et al.* 1974.)

film (Philipp 1972), and a curve (VF) which Brodsky, Kaplan, and Ziegler (1972) obtained by extrapolation from other absorption edges to one that they consider would correspond to zero free-spin density, i.e. void-free Si. Le Comber *et al.* (1974) claim that these results indicate that oxygen incorporation and annealing show different effects. Oxygen incorporation produces a parallel shift to higher energies towards the edge of SiO because one is introducing strong Si—O bonds. Annealing, however, causes a change of shape of the edge as well as a shift and finally parallels that of the glow-discharge-deposited and the VF edge; the implications here are (i) that annealing is more effective than oxygen in reducing voids although both reduce the spin signal and (ii) the glow-discharge-deposited films are essentially void free and do not contain appreciable amounts of oxygen. This latter argument was supported by the lack of any absorption in the infrared at the frequency of the Si—O bond and also by back-scattering experiments using α particles. An attempt to detect hydrogen in these glow-discharge films using the same two methods failed to reveal any of this contaminant either.

A similar study to that above for evaporated films has been made by Bahl and Bhagat (1975) (see also Bahl, Bhagat, and Glosser 1973, 1974) but including, in addition, measurements of the conductivity as a function of temperature. Although the findings are similar, these authors find some change in the shape of the absorption edge with oxygen incorporation and conclude that the effects of annealing and of oxygen are basically the same and are to be attributed to internal surfaces being rendered inactive.

In view of the results described in this section on the effects of preparation conditions, voids, and contamination on the properties of amorphous Ge and Si, it may be considered rash to describe properties of these materials in any further detail. However, we shall proceed to do this, fully aware of the dangers in the generalizations or comparisons we shall make, and attempt to mitigate the problem somewhat by giving details of the conditions under which the films were processed.

7.4. Electrical properties of amorphous germanium

The electrical properties of amorphous germanium depend, as discussed in the previous section, both on the method and conditions of preparation as well as on subsequent annealing processes. We present here some representative results to emphasize the principal features of the conduction processes, without describing all the available data.

Fig. 7.21 shows the temperature dependence of the d.c. conductivity obtained by Mell (1974a) on a film evaporated at 100 Å s^{-1} on to a quartz substrate held at 300 K in a vacuum of 2×10^{-6} torr. The parameter T_a is the annealing temperature which was maintained for 15 min prior to obtaining the corresponding curves. A temperature-independent activation energy is observed only at high temperatures in the most highly annealed state. On a $T^{-1/4}$ scale the results plot linearly below the transition temperatures indicated by arrows, lying between 200 and 260 K. Annealing decreases the conductivity over the whole of the temperature range measured here. Qualitatively very similar results have been obtained by many workers, although discrepancies exist between reported values of the absolute conductivity, the limiting slope at high temperatures, and the accuracy of the fits to $T^{-1/4}$ behaviour at low temperatures.

Measurements of conductivity and thermopower (Beyer and Stuke 1974) using films prepared under slightly different conditions from those in

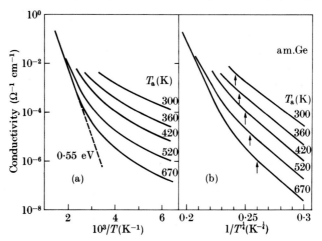

Fig. 7.21. Temperature dependence of electrical conductivity of an amorphous germanium film deposited at 300 K under conditions described in the text and annealed for 15 m at each of the temperatures T_a indicated. (From Mell 1974a.)

Fig. 7.21 are shown in Fig. 7.22. The curves marked 1a–1e correspond to the annealing of a sample deposited at 240 K at a rate of 50 Å s^{-1} in a vacuum of 2×10^{-6} torr and those marked 2a–2c to a sample deposited at 300 K at a rate of 100 Å s^{-1} in a vacuum of 6×10^{-8} torr. The more rapidly deposited film has a slightly higher initial conductivity and the effect of subsequent annealing is less pronounced. At low temperatures both samples, independent of the degree of anneal, exhibit a small negative thermopower. At higher temperatures the more rapidly deposited film has a positive thermopower which increases on annealing, whereas the other sample displays similar but more pronounced behaviour in the negative direction.

The solid lines in Fig. 7.22 are theoretical curves by Beyer and Stuke (1974) based on the assumption that three conduction mechanisms operate in parallel. Using the notation of Chapter 6 and Fig. 6.16, these are hopping of electrons with energies at $E_A(\sigma_n)$, hopping of holes at E_B (σ_p), and variable-range hopping at E_F (σ_h). The fitting procedure was as follows. First the $T^{1/4}$ behaviour in the conductivity was extrapolated to higher temperatures and subtracted from the measured conductivity. The difference was assumed to represent conduction by hopping of carriers activated to E_A or E_B and the relative contribution of these at a particular temperature was obtained from the thermopower; a hopping energy of 0·12 eV was taken for both electrons and holes in band-edge states and for all states of anneal. In addition, for simplicity, the thermopower associated with hopping at E_F was assumed independent of temperature and equal to the small negative value observed at low temperatures ($-60 \mu\text{VK}^{-1}$). The

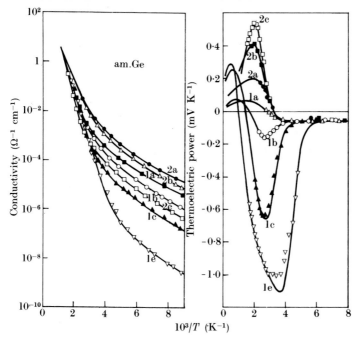

Fig. 7.22. Temperature dependence of conductivity and thermopower of two germanium films deposited and annealed under conditions described in the text. Values of the annealing temperatures can be read off Fig. 7.23(c). (From Beyer and Stuke 1974.)

intercept at infinite temperature of the thermopower was taken as a constant equal to $0.16\,\mathrm{mV\,K^{-1}}$. Fig. 7.23(a,b) show schematically the relative contributions of the three processes to the conductivity σ_h, σ_n, and σ_p and to the thermopower S_h, S_n, and S_p; the total conductivity is $\sigma_h + \sigma_n + \sigma_p$ and the total thermopower is the sum of the individual contributions weighted according to the contributions of the partial conductivities to the total. Fig. 7.12(c) gives the variation of the conductivity activation energies E_n^σ and E_p^σ for the two samples as a function of annealing temperature. This latter figure shows that for the rapidly evaporated sample (2), E_n^σ and E_p^σ both increase almost together with annealing, i.e. the gap $E_A - E_B$ increases and E_F stays close to the centre of the gap; for sample (1), $E_A - E_B$ also increases on annealing but the Fermi level shifts by about $0.15\,\mathrm{eV}$ into the upper half of the gap. The difference in behaviour of the two samples was attributed to the presence of more oxygen in the more slowly evaporated films which, on annealing, tends to eliminate defect states in the lower half of the gap. Finally the intercepts at infinite temperature of σ_n and σ_p rise from $\sim 10^2\,\Omega^{-1}\,\mathrm{cm}^{-1}$ for the unannealed samples to $\sim 10^3$–$10^4\,\Omega^{-1}\,\mathrm{cm}^{-1}$. In spite of the assumptions

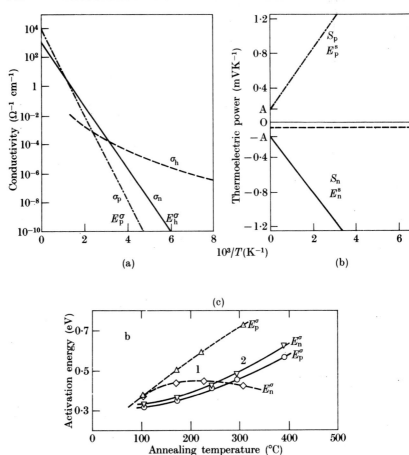

Fig. 7.23. (a), (b) Schematic illustration of the temperature dependence of the components contributing to the conductivity and thermopower curves of Fig. 7.22. The total conductivity σ is the sum of σ_p, σ_n, and σ_h and the total thermopower is given by

$$S = \frac{\sigma_p}{\sigma}\left(\frac{E_p^\sigma}{eT}+A\right) - \frac{\sigma_n}{\sigma}\left(\frac{E_n^\sigma}{eT}+A\right) + \frac{\sigma_h}{\sigma}S_h$$

where E_p^σ and E_n^σ are the activation energies shown, A was taken as $0\cdot16\ \mathrm{mV\,K^{-1}}$, S_h as $-60\ \mu\mathrm{V\,K^{-1}}$, and $E_{p,n}^\sigma - E_{p,n}^S$ as $0\cdot12\ \mathrm{eV}$ for both samples. (c) Annealing behaviour of conductivity activation energies for the two samples shown in Fig. 7.22. (From Beyer and Stuke 1974.)

made, the theoretical fits are good and may be taken as evidence that the conduction processes have been correctly identified.

A similar decomposition of the electrical transport properties of a–Ge evaporated at a pressure of 10^{-8}–10^{-9} torr, but subsequently exposed to air, has been made by Seager, Knotek, and Clark (1974). In this case the

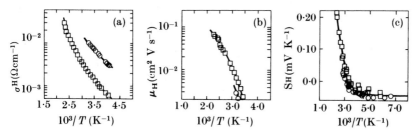

Fig. 7.24. (a) Conductivity, (b) Hall mobility and (c) thermopower *versus* reciprocal temperature for an amorphous germanium film deposited at 300 K (○) and subsequently annealed at 450 K (□). The solid lines are computer fits assuming two conduction paths as described in the text. (From Seager *et al.* 1974.)

Hall mobility μ_H was measured as well as the conductivity and thermopower (Fig. 7.24(a,b,c)) and the results were analysed in terms of two contributions; hole conduction in the valence band at E_B and variable-range hopping at E_F. The dependence on temperature of both μ_H and S (Fig. 7.24(b)) is attributed to the transition between these two modes of conduction. The solid curves are theoretical using the following parameters:

$$\mu_H^P = 1 \cdot 0 \exp(-0 \cdot 09/kT)$$

$$\sigma_h = 1 \cdot 2 \times 10^6 \exp(-78/T^{-1/4}) \qquad S_h = -0 \cdot 04 \text{ mV K}^{-1} \qquad \mu_H^h = 0.$$

Slightly different parameters were used to fit the data on the annealed sample. The most important features to be noted here are the n–p anomaly (S was positive and μ_H was negative throughout the temperature range measured) and the assumption that the Hall mobility associated with hopping at E_F is zero (or negligibly small). These features are consistent with the ideas outlined in § 6.4.7, according to which the activation energy in the Hall mobility (0·09 eV) would represent $(E_B - E_V) - w$, where w is the activation energy associated with hopping conduction at E_B.

It is of interest that it does not appear possible to raise the temperature of amorphous Ge, prepared by evaporation, to the point where transport in extended states at E_C or E_V dominates the conductivity. Such conduction can be identified by an equality of E^σ and E^S or alternatively by a temperature-independent Hall mobility. In most films crystallization or at least partial crystallization sets in first (Knotek 1974, Laude, Willis, and Fitton 1974). However for glow-discharge-deposited Ge, Jones, Spear, and Le Comber (1976) have found that $E^\sigma = E^S$ at high temperatures. As for glow-discharge-deposited Si (see § 7.5), the conductivity of films produced in this way does not exhibit $T^{-1/4}$

behaviour, as may be seen by comparison with an evaporated film in Fig. 7.25(a). The glow-discharge-deposited films presumably contain a lower concentration of active defect sites; furthermore, observation of equality of the activation energies for conductivity and thermopower suggests that these samples have a smaller range of localized states at the conduction-band edge, thereby allowing conduction at E_C at a lower temperature than for evaporated films. At lower temperatures the activation energy for conduction falls, indicating the onset of hopping at E_A. This interpretation is substantiated by the behaviour of the thermopower (Fig. 7.25(b)) which proceeds via a transition region to a line of slope $E_A - E_F$. The parameters used to calculate the (solid) theoretical curves are tabulated in Jones *et al.* (1976); the quantity $\Delta E = E_C - E_A$ is given as between 0·23 and 0·27 eV (for samples deposited on to substrates held at a high temperature) and the

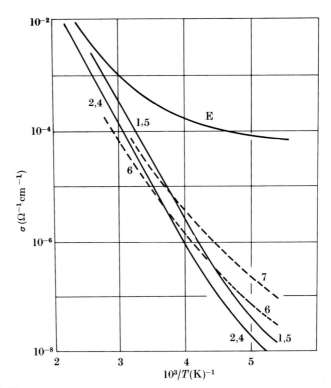

Fig. 7.25. Temperature dependence of (a) conductivity and (b) (p. 351) thermopower of germanium films glow-discharge deposited at various substrate temperatures T_d. Curves 1–5 refer to samples deposited at $T_d = 500$ K, curve 6 to a sample deposited at $T_d = 400$ K, and curve 7 to a sample deposited at $T_d = 300$ K. Curve E in (a) refers to an unannealed evaporated sample and the broken curve in (b) to an evaporated sample annealed at 310°C and shown as curve 1e in Fig. 7.22. The deconvolutions into two conduction paths is described in the text. (From Jones *et al.* 1976.)

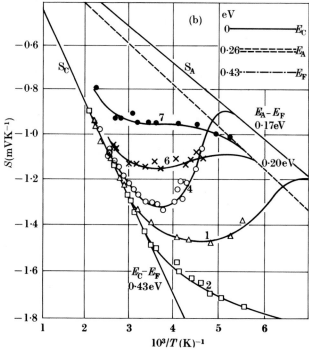

Fig. 7.25 (b)

hopping energy w as ~ 0.1 eV. A similar two-mode conduction process has been successful in explaining similar data for amorphous As (Chapter 8) and amorphous chalcogenides (see Chapter 6).

With regard to conduction at low temperatures by variable-range hopping near the Fermi energy, several workers have reported departures from the theoretical $T^{-1/4}$ behaviour, a feature that is perhaps not surprising if the density of states near E_F is not independent of energy and E_F itself depends on temperature, and if there is a temperature dependence of the pre-exponential term. However, earlier suggestions that the temperature at which $T^{-1/4}$ is observed is sometimes too high (say 260 K) to be consistent with variable-range hopping processes has been refuted by Pollak *et al.* (1973) and by Knotek, Pollak, and Donovan (1974), using the argument that percolation analysis shows a heavy weighting towards an energy range of states near E_F, a range that is much less than might be inferred from the critical impedance of a random impedance network. From such an analysis these authors deduce that, in their evaporated Ge films, E_F lies in a peak of states of bandwidth ~ 0.1–0.3 eV and containing approximately 10^{17} states cm^{-3}. The evidence given was that at the temperature above which $T^{-1/4}$ behaviour breaks down, the direction of the departure is to *lower* values of the conductivity indicating a *decrease* in the number of hopping states available.

In all cases where $T^{-1/4}$ behaviour has been observed, the density of states at E_F deduced from the slope of such plots (§ 6.4.3) has been found, assuming $\alpha^{-1} = 10\,\text{Å}$, to lie in the range $8 \times 10^{17} - 5 \times 10^{18}\,\text{eV}^{-1}\,\text{cm}^{-3}$, the lower value applying to well-annealed samples. The relative invariance of this figure might imply that a certain proportion of defect states, perhaps dangling bonds, are necessary to stabilize the structure of amorphous Ge. Alternatively, one might suspect the analysis. However, an extremely important set of experiments by Knotek *et al.* (1973), Knotek *et al.* (1974), and Knotek (1974), in which variable-range hopping conduction was measured as a function of film thickness, has provided convincing evidence that the analysis is not grossly incorrect.

Fig. 7.26 shows results of the above authors on Ge films evaporated in ultra-high vacuum on to high-temperature substrates under conditions in which voids and contamination are expected to play an insignificant role.

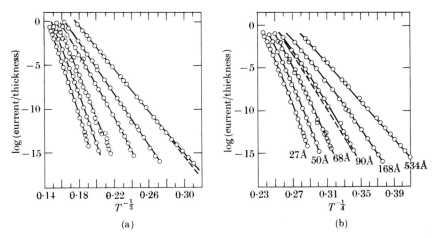

Fig. 7.26. Thickness and temperature dependence of the conductivity of amorphous germanium films deposited by evaporation in ultra-high vacuum showing the transition from three-dimensional ($T^{-1/4}$) to two-dimensional ($T^{-1/3}$) hopping conduction. The broken lines are straight. (From Knotek *et al.* 1974.)

For films of thickness greater than $\sim 500\,\text{Å}$ the conductivity plots linearly on a $T^{-1/4}$ basis and the slope is independent of the thickness. Thinner films show a departure from this behaviour and the temperature dependence of the logarithm of the conductivity is proportional to $T^{-1/3}$ as predicted for two-dimensional variable-range hopping (§ 2.7). Further-more, the slope of these lines is proportional to the inverse cube root of the

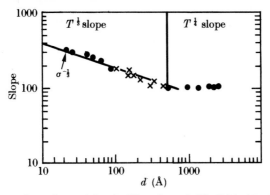

Fig. 7.27. Thickness dependence of slopes of lines shown in Fig. 7.26 with additional data (×). (From Knotek 1974.)

thickness as expected. The slopes of both the $T^{-1/4}$ and $T^{-1/3}$ plots are shown as a function of film thickness in Fig. 7.27. Inspection of eqns (2.63) and (2.64) shows that the complete set of data can be used to determine α^{-1} and $N(E_F)$ independently, which in these films are 10 Å and 10^{18} cm^{-3} eV^{-1} respectively. The variable-range hopping distance in the thick films lies (deduced from the equations of § 2.7) between 120 and 170 Å in the temperature range of these measurements and so the transition to two-dimensional hopping occurs at a film thickness that is three to four times this value. An extrapolation of the $T^{-1/3}$ plots to $T = \infty$ requires use of a theory for hopping for a film of arbitrary thickness, because the hypothetical hopping distance must eventually become small enough so that any film, no matter how thin, exhibits three-dimensional behaviour. Using such a theory (Shante 1973), the extrapolations shown in Fig. 7.28 have been made. The pre-exponential term is independent of film thickness d down to $d = 90$ Å. The divergence in thinner films is probably due to the fact that the diameter of the wavefunction ($2\alpha^{-1} \sim 20$ Å) is then comparable with d.

Anisotropy in the electrical conductivity of evaporated films has been reported by Hauser (1972). A higher conductivity was measured perpendicular to the plane of the films when the thickness was less than about 4000 Å. These data have been interpreted by Pollak and Hauser (1973) without the need to invoke any anisotropy in the structure of the films.

Some difficulties are encountered in attempting to explain the magnitude of the conductivity associated with variable-range hopping (Pollak 1976, Butcher 1976a,b). Most theoretical treatments yield pre-exponential factors that are several orders of magnitude smaller than those observed experimentally.† Whether this is a theoretical or experimental problem is not clear. Any inhomogeneities in the films, particularly voids and cracks,

† Butcher and Hayden (1977b) find a smaller discrepancy.

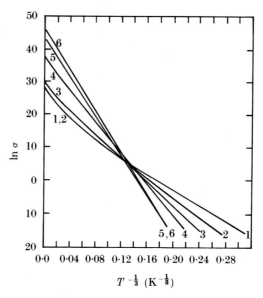

Fig. 7.28. Extrapolation of conductivity data similar to those shown in Fig. 7.26 on $T^{-1/3}$ plots through the $T^{-1/4}$ regime to $T = \infty$. The film thicknesses are as follows: curve 1, 534 Å; curve 2, 166 Å; curve 3, 90 Å; curve 4, 40 Å; curve 5, 27 Å; curve 6, 22 Å. (From Knotek 1974.)

might be expected to lead to a measured conductivity that is lower than that calculated in their absence, although if transport occurred *via* such channels the opposite would be the case. Related to this question is the *change* in the pre-exponential factor observed on annealing (Stuke 1976) or by hydrogenation; the factor is reduced but not sufficiently to bring it in line with theory. Furthermore, it is changed by rather more than might be inferred from the reduction in the density of defect states determined from the decrease in the slope of the $T^{-1/4}$ plots. Thus, experimentally, rather parallel displacements of the curves are observed.

An exception to this anomalous behaviour is provided by the results of Apsley *et al.* (1977) on evaporated films subjected to bombardment by 100 keV Ge^+ ions at a low temperature and subsequently annealed (Fig. 7.29(a)). In these experiments the parameters $N(E_F)$ and α^{-1} were determined independently by taking measurements as a function of electric field (see Fig. 7.29(b) and below). After bombardment to saturation, $N(E_F)$ is calculated to be $\sim 1 \cdot 6 \times 10^{18} \, cm^{-3} \, eV^{-1}$; on annealing, the conductivity of the films is restored approximately to its initial value, the value of $N(E_F)$ falling to $3 \cdot 3 \times 10^{17} \, cm^{-3} \, eV^{-1}$. The pre-exponent is seen to remain sensibly constant at all stages of the anneal, during which α^{-1} was calculated to fall from 15 Å to 13 Å. The above values of $N(E_F)$ are rather lower and those of α^{-1} are somewhat higher than estimated from other types of analysis.

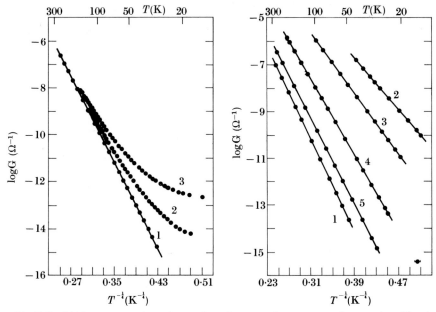

Fig. 7.29. (a) Temperature dependence of conductance of an evaporated germanium film: 1, as deposited; 2, after bombardment with approximately 10^{15} 100 keV Ge^+ ions as ~14 K and annealing at 50 K; 3, 4, and 5, after annealing at 100 K, 200 K, and 300 K respectively. (b) Electric-field dependence of the conductance–temperature plot for the sample after annealing at 330 K. Values of the electric field were as follows: 1, $\leqslant 6 \times 10^3$ V cm^{-1}; $2 \cdot 3 \times 10^4$ V cm^{-1}; $3 \cdot 7 \times 10^4$ V cm^{-1}. A similar field dependence was observed after each annealing stage in (a). (From Apsley *et al.* 1977.)

The excellent linearity of plots of the conductivity *versus* $T^{-1/4}$ found for many films of amorphous Ge, in some cases extending over eight orders of magnitude, and the $T^{-1/3}$ behaviour in thin films is at variance with the argument (Emin 1974) that there should be a strong temperature dependence of the hopping rate from the pre-exponent owing to multiphonon processes. A discussion of this point is given in § 3.5.

Non-ohmic conduction (high-field effects) in a-Ge has been reported by many workers (e.g. Morgan and Walley 1971, Connell, Camphausen, and Paul 1972, Telnic *et al.* 1973). We present in Fig. 7.30(a) results of Elliott *et al.* (1974). In contrast to chalcogenides (Chapter 9) the conductivity of a-Ge is independent of field at sufficiently low fields, say $\leqslant 10^3$ V cm^{-1} (i.e. the behaviour is ohmic), but increases with field to a degree that depends on the temperature (and as might be expected on the mode of preparation, the anneal state, etc.). Although several theoretical expressions can be used to describe the data over particular ranges of field and temperature (§§ 2.7, 3.12), the following relationship is approximately obeyed:

$$\sigma = \sigma_0 \exp\left(\frac{elF}{kT}\right) \qquad (7.1)$$

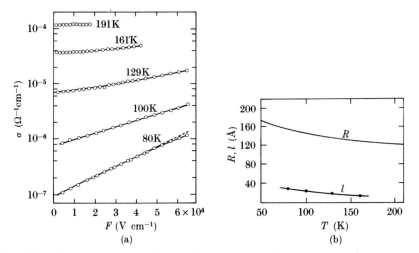

Fig. 7.30. (a) Variation of conductivity σ with electric field F for a germanium film deposited in ultra-high vacuum at 80 K, annealed at 413 K, and measured at the temperatures indicated. (b) Temperature dependence of the characteristic length l deduced from the data in (a). The hopping length R and hopping energy w are also shown. (From Elliott *et al.* 1974.)

where l has the dimensions of length and is a weak function of temperature. Fig. 7.30(b) shows the temperature dependence of l as determined from the data of Fig. 7.30(a). For comparison, the variable-range hopping length R, deduced from the slope of $T^{-1/4}$ plots at low fields using eqn (2.62) is shown on the same diagram. Although the temperature dependence of l is approximately $T^{-1/4}$, the coefficient of proportionality is much larger than that of R. In addition it is clear that l is considerably smaller than R. Pollak and Riess (1976) have considered this problem and conclude that the field changes the percolation paths in such a way that the critical impedance is lowered. They find, making certain approximations,

$$\sigma \propto \exp\left[-0.9\xi_m(0) + \frac{0.17eRF}{kT}\right] \qquad (7.2)$$

for fields $F \ll 2kT\alpha/e$. Here $\xi_m(0)$ is given by

$$\xi_m(0) = \left[\frac{N(E_F)kT}{\alpha^3}\right]^{-1/4} \qquad (7.3)$$

Fitting this expression to the high-field data of Fig. 7.30(a) yields a characteristic length l that is ~1/6 times the low-field hopping length R, approximately as observed.

For fields $F \gg 2kT\alpha/e$, Pollak and Riess find a dependence of the conductivity on field of the form $\ln \sigma \propto (F/F_0)^{1/4}$. The physical process underlying this behaviour is that for values of eFR larger than the average hopping energy, an electron can move by downward hops only, emitting a phonon at each hop (Mott 1970, Shklovskii 1973b). The condition for this process is that

$$F \geqslant \frac{3}{4\pi e N(E_F) R^4}. \qquad (7.4)$$

For $F > F_c$, R should decrease as $F^{-1/4}$ and writing $\sigma \propto \exp(-2\alpha R)$ the above result is obtained. The departure from the simple exponential behaviour seen in Fig. 7.30(a) at 80 K and high fields is likely to represent the onset of this behaviour.

A theory of high-field conductivity has also been advanced by Apsley and Hughes (1974, 1975, 1976). Their expressions, although differing in several respects from those of Pollak and Riess (1976), are simpler and reproduce the main features observed. Detailed comparison with the data of Fig. 7.29(b) and other results has been made (Apsley et $al.$ 1977) enabling independent determinations of $N(E_F)$ and α^{-1}.

Measurements of magnetoconductivity under high-field conditions have been reported by Mell (1974b), who, following Clark et $al.$ (1974), concluded that the non-ohmic behaviour is to be attributed to the hopping process and not to the onset of a different conduction mechanism, such as excitation of carriers out of charged centres by the Poole–Frenkel effect.

Another method of studying variable-range hopping processes near the Fermi level is by measurement of the a.c. conductivity $\sigma(\omega)$ as a function of frequency. Data on evaporated Ge films have been obtained by Chopra and Bahl (1970), Kneppo, Luby, and Červenák (1973), Hauser and Standinger (1973), Mell (1974b), Agarwal, Guha, and Narasimhan (1975), and Arizumi et $al.$ (1974). Results due to the latter authors are shown in Fig. 7.31. The conductivity rises approximately as $\omega^{0.8}$ once the frequency is

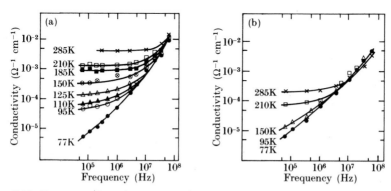

Fig. 7.31. Frequency dependence of conductivity of an evaporated film of germanium measured at the temperatures indicated (a) as deposited and (b) after annealing at 250°C for 30 min. (From Arizumi et $al.$ 1974.)

above a certain value, and the magnitude falls as the temperature is lowered. It is important to note that we assume that these data represent the frequency and temperature dependence of a single (hopping) mode of conduction, unlike the situation in chalcogenide glasses (Chapters 6 and 9) for which a similar frequency-independent behaviour at low frequencies arises from extended-state conduction. Although some theoretical treatments of a.c. conductivity include the d.c. limit, we shall consider here only the $\omega^{0.8}$ region. Interpreted in terms of the formula given in § 6.4.5 and taking a value of $\alpha^{-1} = 8$ Å, the data from all the above mentioned authors yield values of $N(E_F)$ ranging from 3×10^{19} to 2×10^{21} cm^3 eV^{-1}. Although these values could be reduced by a factor of 10 by choosing a (high) value for α^{-1} of 20 Å, there still remains a serious discrepancy between the values of the state density deduced from these measurements and those deduced from the temperature dependence of the d.c. conductivity. Furthermore the spread in values amongst the various samples is at least an order of magnitude larger than can be achieved by annealing or even by hydrogenation if, again, the values deduced from the slopes of $T^{-1/4}$ plots are considered reliable.

This discrepancy has been discussed by Abkowitz *et al.* (1976) who have plotted the density of states deduced from both the d.c. and a.c. data *versus* the d.c. conductivity at 80 K (see Fig. 7.32). $N(E_F)$ was calculated using the equations of § 6.4.5, assuming a value for α^{-1} of 8 Å. The above authors prefer to accept the values of $N(E_F)$ deduced from the a.c. data. They base their case mainly on the following.

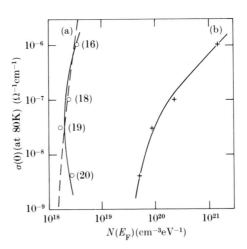

Fig. 7.32. Density of states at E_F deduced from (a) d.c. hopping conductivity and (b) a.c. hopping conductivity *versus* the d.c. conductivity measured at 80 K. The numbers refer to the following references: (16) Chopra and Bahl (1970), (18) Arizumi *et al.* (1974), (19) Hauser and Staudinger (1973), (20) Agarwal *et al.* (1975). (From Abkowitz *et al.* 1976.)

(1) Field-effect experiments on evaporated Ge by Hirose, Tariguchi, and Osaka (1976) and by Malhotra and Neudeck (1975) have been analysed to yield $N(E_F) \sim 10^{19}$–10^{21} cm^{-3} eV^{-1} depending on the sample.

(2) Field-effect experiments on *glow-discharge-deposited* Ge (Jones *et al.* 1976) have been similarly analysed to yield $N(E_F) \sim 2 \times 10^{18}$–$2 \times 10^{19}$ cm^{-3} eV^{-1} depending on the deposition temperature. Furthermore the magnitude of the a.c. conductivity in these samples is considerably lower than for evaporated samples and yield values for $N(E_F)$ similar to those deduced from the field effect (see Fig. 7.47). No variable-range hopping is observed in these samples (see Fig. 7.25(a)).

(3) The values of $N(E_F)$ deduced from the d.c. data show little dependence on preparation conditions and are too low when considered in the light of field-effect and tunnelling studies (Osmun and Fritzsche 1974, Hauser 1974*a,b*).

However, it should be noted that the variation of $\sigma(0)$ with $N(E_F)$, where the latter is deduced from the variable-range hopping formula, is self-consistent as seen by the dotted curve in Fig. 7.32, which is a plot of

$$\sigma(0) = \sigma_0 \exp\left[-\left\{ \frac{16\alpha^3}{kTN(E_F)} \right\}^{1/4} \right] \tag{7.5}$$

with $\sigma_0 = 10^9 \, \Omega^{-1} \, \text{cm}^{-1}$, $\alpha^{-1} = 8 \, \text{Å}$, and $T = 80 \, \text{K}$.

A variation of $N(E_F)$ by a factor of 40, as suggested by the a.c. data, would lead to a variation of $\sigma(0)$ by many more orders of magnitude. Furthermore, a density of states as high as 10^{21} cm^3 eV^{-1} would almost certainly lead to delocalization and a metallic-like conductivity.

Since our conclusion is the opposite of that of Abkowitz *et al.* (1976), namely that the values of $N(E_F)$ deduced from the variable-range hopping formula are more reliable than those deduced from a.c. conductivity (see also Butcher and Hayden 1977*a*), it is necessary to consider why the latter might be in error. Apart from approximations on which the formula for $\sigma(\omega)$ is based (see § 6.4.5), one might consider the possibility that the a.c. process may sample portions of the specimen within which the density of localized states is considerably higher than the average. An estimate of the hopping length R_ω appropriate to a.c. conditions can be made from eqn (2.96). At $\omega = 10^4 \, \text{s}^{-1}$ and taking $\nu_{ph} = 10^{12} \, \text{s}^{-1}$, $\alpha R_\omega \sim 9$ and for $\alpha^{-1} = 8 \, \text{Å}$, $R_\omega \sim 70 \, \text{Å}$. Thus regions of inhomogeneity of considerable size would have to be present, a situation that can hardly be ruled out and indeed might account for the large dependence of $\sigma(\omega)$ on the conditions of sample preparation and history.

The effect of a magnetic field on both d.c. and a.c. conduction in films of amorphous Ge evaporated at a low rate $(5$–$10 \, \text{Å s}^{-1})$ on to a room-temperature substrate has been studied by Mell (1974*b*). Fig. 7.33 shows results at 80 K. The conductivity σ and its change $\Delta\sigma$ in the presence of a magnetic field of 1 kG are seen to have similar frequency dependencies, although $\Delta\sigma$ does not rise quite as steeply as σ and the increase sets in at a slightly higher frequency. Fig. 7.33(b) shows the corresponding phase angle

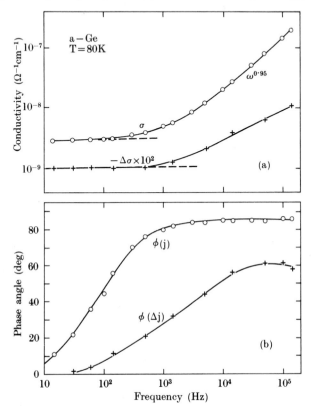

Fig. 7.33. (a) Frequency dependence of conductivity σ and its change in a magnetic field of 1 kG for an evaporated film of germanium; (b) phase angle ϕ between current j and applied voltage and also between Δj in a field of 1 kG and applied voltage. Measurements at 80 K. (From Mell 1974*b*.)

between current and voltage. Although magnetoresistance associated with hopping transport is not yet completely understood (§ 6.4.8), these data do make 'purely atomic' models of a.c. conductivity seem unlikely (§ 6.4.5), as both the d.c. and a.c. processes are similarly affected by the magnetic field.

More detailed measurements of magnetoresistance under d.c. conditions have been reviewed by Mell (1974*a*). The behaviour is complicated (Fig. 7.34); the magnetoresistance can be of either sign depending on the temperature and magnetic field but is *independent* of the angle between the current and the magnetic field. A decomposition of the data taken at 175 K into positive $P(B)$ and negative $N(B)$ components is shown on the right of Fig. 7.34. After an initial increase at low magnetic fields, $P(B)$ saturates whereas $N(B)$ increases as a power law. Both components show only a weak temperature dependence in the $T^{-1/4}$ range of behaviour of the conductivity.

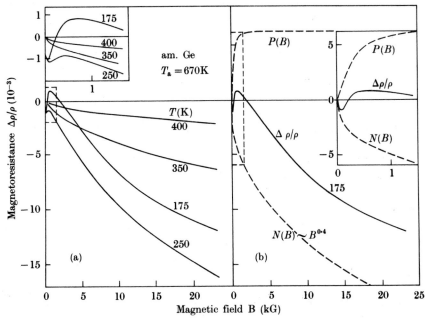

Fig. 7.34. (a) Magnetoresistance $\Delta\rho/\rho$ of amorphous germanium (prepared and treated as for the film in Fig. 7.21) as a function of magnetic field; (b) separation of $\Delta\rho/\rho$ into positive $P(B)$ and negative $N(B)$ components. (From Mell 1974a.)

The shrinkage of localized wavefunctions in a magnetic field, such as applies to impurity conduction (§§ 4.3, 4.6), is unlikely to account for this data, although Maschke *et al.* (1974) using a computer simulation of hopping conduction have suggested that such an effect on the hopping rate still significantly influences hops over *long* distances and could be important. An alternative suggestion, that in a magnetic field spin alignment will cause an electron to hop further than it would in the absence of a field, could be responsible for the negative magnetoresistance (Movaghar and Schweitzer 1977)—see § 6.4.8. Other measurements and interpretations have been given by Clark *et al.* (1973) and by Kubelik and Triska (1973), the former considering spin-flip scattering of extended states by localized spins analogous to the Kondo effect. An unphysically large value ($g \sim 1000$) of the magnetic moment was required.

7.5. Electrical properties of amorphous silicon

As for amorphous germanium, it is convenient to distinguish between the properties of amorphous silicon prepared (1) by conventional evaporation or sputtering techniques and (2) by glow-discharge decomposition of silane. Even within these two categories, variations in properties depending on the deposition conditions (particularly substrate temperature) are found, but the relatively low concentration of active defect sites in glow-discharge-deposited material compared with silicon prepared by most other methods makes the distinction a useful one. Whether or not the presence of hydrogen plays a crucial role, the extensive work of Spear and co-workers, which has resulted in a detailed and self-consistent picture of the electronic states and transport in glow-discharge-deposited films, has been of great value in the development of our subject.

(a) *Films prepared by evaporation.* Amorphous silicon films deposited by evaporation have much higher conductivities than glow-discharge samples deposited at the same temperature. As in evaporated films of germanium, electrical transport normally occurs by hopping at the Fermi level as evidenced by an (approximately) $T^{-1/4}$ dependence of the logarithm of the conductivity and small values of the thermopower. Annealing increases both the magnitude of the thermopower and the activation energy for conduction, leading to band conduction but not, if we regard the inequality of E^{σ} and E^{S} as a reliable indicator, in extended states.

Conductivity and thermopower measurements by Beyer and Stuke (1974), similar to those obtained for amorphous germanium (Fig. 7.22), are shown in Fig. 7.35 for two films (1 and 2) as a function of annealing (a → c and a → d). Since film 1 had been deposited at 16 Å s^{-1} and film 2 at 2 Å s^{-1}, differences in the conductivity were attributed predominantly to a larger concentration of oxygen incorporated into the latter. Additional measurements (Beyer, Stuke, and Wagner 1975), in which samples were bombarded with helium, oxygen, and hydrogen, showed relative changes in the contribution of Fermi-level hopping and band-conduction processes (both of which are evident in Fig. 7.35) according to whether the number of defects was increased (bombardment with He) or decreased (with O or H). Detailed analysis of the results for unbombarded films can be made as for germanium (§ 7.4) but, for the bombarded films, several sets of defect levels, some of which were attributed to surface states at voids, were invoked by Beyer *et al.* (1975).

It appears that amorphous silicon films are much more likely than germanium to take up large quantities of oxygen during evaporation. This apparent disadvantage has been used by Le Comber *et al.* (1974) and by

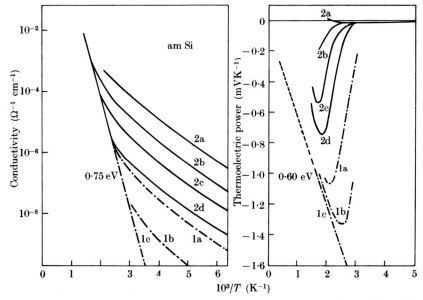

Fig. 7.35. Temperature dependence of conductivity and thermopower of two silicon films deposited at 500 K by evaporation in a vacuum of 2×10^{-6} torr. Film 1 was deposited at 2 Å s^{-1} and annealed at 260°C (1a), 290°C (1b), and 365°C (1c). Film 2 was deposited at 16 Å s^{-1} and annealed at 230°C (2a), 310°C (2b), 365°C (2c), and 450°C (2d). (From Beyer and Stuke 1974.)

Bahl and Bhagat (1975) in studies which in some ways parallel those by Connell and Pawlick (1976) on hydrogenation of amorphous germanium. In the experiments of Bahl and Bhagat (1975), the oxygen content was varied simply by changing the evaporation rate and/or base pressure, the relative concentration being determined from the strength of the Si—O vibrational band absorption. With increasing oxygen content, the conductivity and spin density decreased and the energy gap increased. Similar results were found by Le Comber *et al.* (1974) (some of which have been discussed in § 7.3), although the shape of the optical absorption edge differed markedly in the two investigations.

In situ measurements of the electrical conductivity in the hopping regime for silicon films of varying thickness deposited in ultra-high vacuum have been made by Knotek (1975*b*). As with amorphous germanium (§ 7.4), a transition from $T^{-1/4}$ to $T^{-1/3}$ behaviour was observed as the film thickness was reduced to values comparable with the calculated hopping length. These results are shown in Figs. 7.36 and 7.37. The fit of three- and two-dimensional variable-range hopping theory to these results is even better than for germanium and yields for the radius of the localized wavefunctions $\alpha^{-1} = 3$ Å and for the density of states at the Fermi level $N(E_F) = 3 \times 10^{19} \, \text{eV}^{-1} \, \text{cm}^{-3}$. Compared with values obtained for

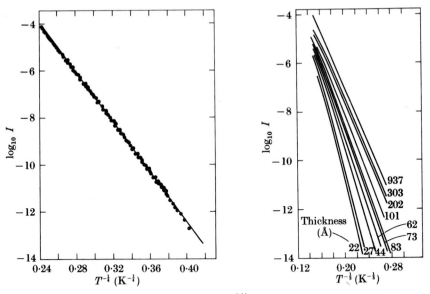

Fig. 7.36. (a) Logarithm of the current I *versus* $T^{-1/4}$ for three films of amorphous silicon of thickness >500 Å prepared by evaporation in ultra-high vacuum at 300 K. Conversion to resistivity is given by $\rho = 7 \cdot 19 \times 10^{-2}\, I^{-1}\, \Omega$ cm. (b) Logarithm of the current I *versus* $T^{-1/3}$ for 10 samples of thickness shown. The 937 Å film shows $T^{-1/4}$ behaviour and is one of the three films shown in (a). Preparation conditions and resistivity conversion as for (a). (From Knotek 1975b.)

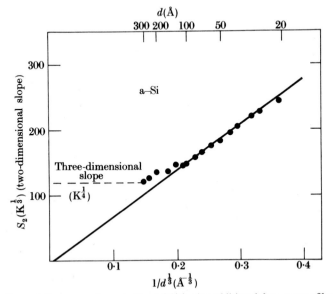

Fig. 7.37. Slope of the curves shown in Fig. 7.36(b), plus additional data, *versus* film thickness d. A dependence as $d^{-1/3}$ is found for $d < 300$ Å as theoretically expected for two-dimensional variable-range hopping. (From Knotek 1975b.)

germanium ($\alpha^{-1} = 10$ Å, $N(E_F) = 2 \times 10^{18}$ eV^{-1} cm^{-3}) the radius is smaller and the density higher. However, just the opposite trend in these two parameters between silicon and germanium was obtained by Abkowitz *et al.* (1976) in a study of a.c. conductivity on glow-discharge-deposited films (see below).

Studies of heterojunctions between crystalline and evaporated silicon have been reported by Fuhs *et al.* (1974) and by Döhler and Brodsky (1974). Magnetoresistance in amorphous silicon (Mell 1974*a*) shows very similar behaviour to that in amorphous germanium (§ 7.4).

(b) *Films prepared by glow-discharge decomposition.* The principal variable in films prepared by this method is the temperature of deposition (T_d). As shown in Fig. 7.38(a) (from Spear 1974*b*), the measured room

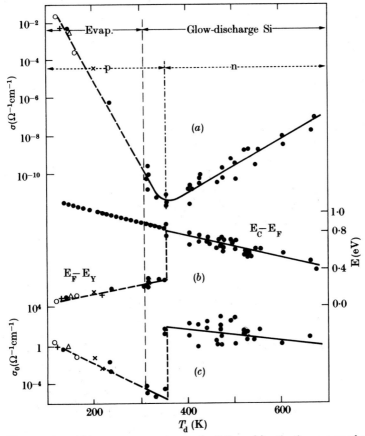

Fig. 7.38. Variation of (a) room-temperature conductivity σ, (b) activation energy of conductivity (see text), (c) pre-exponential factor σ_0 for glow-discharge films of amorphous silicon, with deposition temperature T_d. Points on the left of this figure refer to evaporated films for which T_d has no significance (see text). (From Spear 1974*a*.)

temperature d.c. conductivity changes by five orders of magnitude as T_d is raised from 300 to 640 K. Drift mobility, field effect, and thermopower measurements, to be described later, confirm that the minimum conductivity obtained for a value of $T_d \sim 250$ K is associated with a change in conduction mechanism from, at high T_d, transport by electrons above the mobility edge E_C to, at low T_d, transport by holes hopping in localized states near the valence band. The activation energy of the conductivity (Fig. 7.38(b)) is seen to undergo a discontinuous transition at the same temperature. For n-type samples the activation energy is shown as $(E_C - E_F)$ and for p-type samples as $(E_F - E_Y)$; in the latter case a hopping energy of 0·08 eV has been subtracted from the measured slopes. The pre-exponential factor σ_0, i.e. the intercept on the $1/T = 0$ axis, drops by over seven orders of magnitude at the n–p transition (Fig. 7.38(c)). Similar results to these had been reported earlier by Chittick (1970). The points on the dotted lines on the left of Fig. 7.38 were obtained on films prepared by evaporation and can be fitted rather well onto the extrapolated behaviour of glow-discharge-deposited films; T_d has no direct significance for these points.

Early measurements by Le Comber and Spear (1970) of electron drift mobility in a sample deposited at $T_d = 500$ K (Fig. 7.39(a)) had shown the presence of a 'kink' at about 250 K above which the activation energy was 0·19 eV and below which it was 0·09 eV. The plot of conductivity against inverse temperature (Fig. 7.39(b)) also exhibited a change of slope at the same temperature (240 K). These observations were interpreted in terms of transport in extended states for $T > T_c$, the activation energy in the drift mobility then arising from trapping in the range of shallow localized states at the conduction-band edge, and from hopping in the same states for $T < T_c$. As discussed in § 6.6 and elsewhere, the changes in slope in both μ_D and σ are expected to be the same; this, from the inset to Fig. 7.39(b), is seen to be approximately the case. It should perhaps be noted here that, according to § 6.4.3, the activation energy of 0·09 eV is probably an overestimate since hopping at the band edge is expected to be a variable-range process, the activation energy decreasing with decreasing T.

Subsequent measurements on samples prepared at different values of T_d have confirmed this earlier picture (see Le Comber, Madan, and Spear 1972, 1973). The temperature dependence of the drift mobility is shown in Fig. 7.40. Apart from the absolute magnitude of μ_D, the transition temperature and the slopes above and below it are essentially independent of T_d. One can conclude that the shallow localized states are not greatly affected by T_d and do not therefore arise from defects.† As will be shown, this is not the case for the deeper states, the density of which changes

† If the concentration of hydrogen in the films is dependent on T_d (Brodsky *et al.* 1977), then the presence of this does not affect the shallow localized states either.

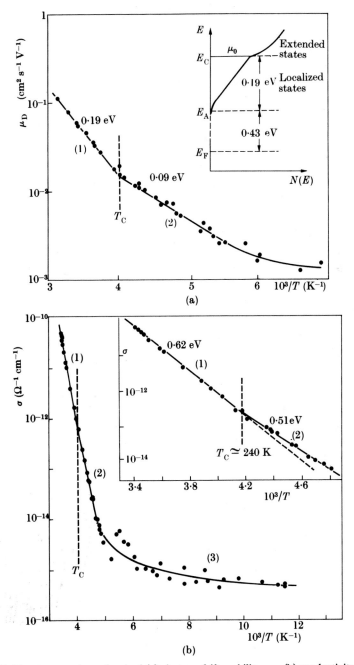

Fig. 7.39. Temperature dependence of (a) electron drift mobility μ_D, (b) conductivity σ in a glow-discharge film of silicon deposited at 500 K. (From Le Comber and Spear 1970.)

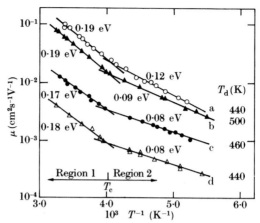

Fig. 7.40. Temperature dependence of the electron drift mobility for four samples of glow-discharge silicon deposited at the temperatures indicated. (From Le Comber *et al.* 1973.)

dramatically. From the above data, a mobility for electrons (in extended states) lying between 10^{-1} and 10^{-2} cm^2 V^{-1} s^{-1} was inferred. Somewhat similar results were obtained by Moore (1977).

The density of localized states throughout the energy gap has been probed by the field-effect technique described in §6.4.9. Results from Spear (1974*b*) are shown in Fig. 7.41, which supersede earlier results of Spear and Le Comber (1972). The dotted lines on this figure are extrapolated densities; in any one sample it is not possible to probe the complete spectrum. The extremities of the mobility gap (~ 1.55 eV) are denoted by E_C and E_V and the density of states near the former has been deduced from the drift mobility data described above. The arrow marks the position of the Fermi level for zero applied field and, as T_d is reduced, it moves from the n-type side of the minimum in the density of states to the p-type side. The information provided by these results is able to account for the data shown in Fig. 7.38, in particular for the change in the activation energy for conduction and the transition from conduction by electrons at E_C to holes near E_V. Results for two evaporated samples are shown; E_1, for which no modulation could be observed, and a sample of lower conductivity ($\sim 10^{-6}$ Ω^{-1} cm^{-1}) before (E_2) and after (E_2') annealing. A subsequent paper (Madan, Le Comber and Spear 1976) describes the effects of annealing on the curves for the glow-discharge-deposited samples. In all the glow-discharge samples, E_F lies near a minimum in the density of states between two bands of defect levels, E_X, directly probed by the field effect, and E_Y, the existence of which is suggested by these results but whose presence is more directly confirmed by photoconductivity and luminescence studies (to be described below). The nature of these states has been discussed by Spear (1974*b*). An analogy is made with a well-

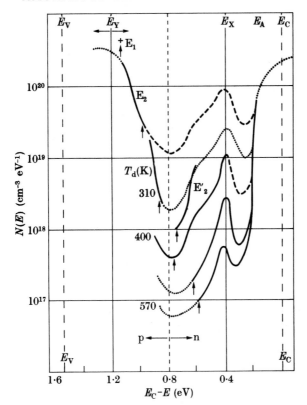

Fig. 7.41. Density of state distributions $N(E)$ for evaporated (E_1, E_2, E'_2) and glow-discharge silicon films, the latter being deposited at the indicated temperatures. Solid curves, $N(E)$ obtained from field-effect measurements; broken curves, extrapolations. The arrows indicate the position of the field-free Fermi level. E'_2 was obtained after annealing E_2 at 540 K. (From Spear 1974b.)

characterized defect in crystalline Si, namely the divacancy (Fig. 7.42). This defect is known (Watkins and Corbett 1965) to give rise to three levels in the forbidden gap—a donor state 0·32 eV above the valence band, a singly charged acceptor state 0·54 eV below the conduction band, and a doubly charged acceptor state 0·15 eV higher. Assuming a preponderance of divacancies in amorphous Si, Spear (1974b) suggests that the E_Y and E_X levels might correspond to the donor and acceptor levels of such defects. However, the inequality of the densities of the two levels, greater in E_Y than in E_X, shows that this interpretation is too simple and other defects should be considered. Although the lowest-order defect in crystalline Si contains four dangling bonds (the monovacancy, which is unstable in the crystal), in an amorphous network defects consisting of any number of orbitals, from a single dangling bond (see frontispiece) upwards, are

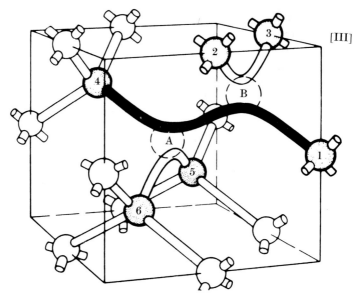

[III]

Fig. 7.42. Configuration of the orbitals in a divacancy in crystalline silicon. A and B show the locations of the missing atoms. (From Watkins and Corbett 1965.)

possible. For the E_Y band, a centre that can act as a donor but not an acceptor is required.

The first successful 'doping' of tetrahedrally co-ordinated amorphous semiconductors by incorporation of trivalent and pentavalent impurities was achieved with glow-discharge-deposited silicon (Spear and Le Comber 1975, 1976, Spear 1977). The generally observed insensitivity of the conductivity of amorphous germanium and silicon (as of other non-crystalline materials) to the presence of traditional doping elements has sometimes been attributed to the likelihood that elements with valency different from the host had all their electrons taken up in bonding, rendering them electrically neutral (§ 2.10), and this is probably true for glasses. It is still not clear whether for films prepared by evaporation this is in fact the case or whether the high density of states at the Fermi energy ensures that any excess electrons condense there, resulting in a shift of E_F small compared with the gap (§ 2.10). Nevertheless, samples deposited by glow discharge in a mixture of SiH_4 and PH_3 (or B_2H_6) are found to have high n- (or p-) type conductivities with the corresponding Fermi levels shifted from the centre of the gap to within $\sim 0 \cdot 2$ eV from the respective mobility edges, depending on the amount of incorporated dopant. Results as a function of phosphine or diborane concentration on samples deposited at 500–600 K are shown in Figs. 7.43 and 7.44. The centres of these diagrams refer to undoped

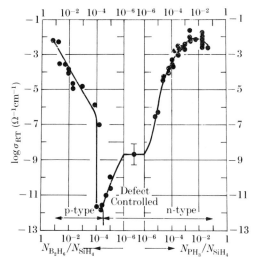

Fig. 7.43. Room temperature conductivity of n- and p-type amorphous silicon plotted as a function of the gaseous composition from which the films were deposited. (From Spear and Le Comber 1976.)

samples which are n type with a Fermi level situated ~ 0.6 eV below E_C. A transition from n- to p-type conduction occurs at a diborane/silane concentration of about 10^{-5}, when E_F is driven through the density-of-states minimum situated at an energy ~ 0.8 eV below E_C (see Fig. 7.41). Although a few parts per million of phosphine are sufficient to raise the conductivity by two orders of magnitude, the rates of rise of conductivity, on either side, slow off considerably as the Fermi level moves into an increasingly higher density of gap states. The variation of conductivity with doping is in good agreement with the density of states distribution shown in

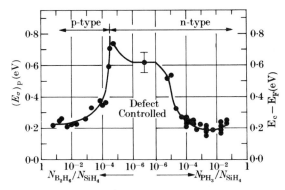

Fig. 7.44. Activation energy of the conductivity for the same films shown in Fig. 7.43. (From Spear and Le Comber 1976.)

Fig. 7.41, lending support to the results obtained from the field-effect data and also indicating that this distribution is not grossly changed by doping. The levels introduced by doping in fact appear to be mixed in with the localized states at the band edges, quite likely having a similar spread of energies and preventing the activation energy for conduction (Fig. 7.44) from falling below ~ 0.2 eV.

Ion-probe analysis of phosphorus-doped samples (Spear and Le Comber 1976) revealed that approximately one-half of the phosphine molecules in the gaseous mixture lead to a phosphorus atom in a given sample; the total number of phosphorus atoms which can be incorporated was found to saturate at about 3×10^{19} cm^{-3}. Calculations of the expected increase of conductivity showed that 30–40 per cent of the phosphorus atoms incorporated act as donors. Boron appears to be even more efficient in providing acceptor levels.

Amorphous silicon can be doped with arsenic by glow-discharge decomposition of a gaseous mixture of silane and arsine (Knights 1976a). The behaviour of the room-temperature conductivity and the position of the optical absorption edge as a function of concentration in the Si–As system are shown in Fig. 7.45. The conductivity of pure silicon (on the left) increases by seven orders of magnitude on incorporation of ~ 0.1 per cent arsenic. Incorporation of more arsenic leads to a

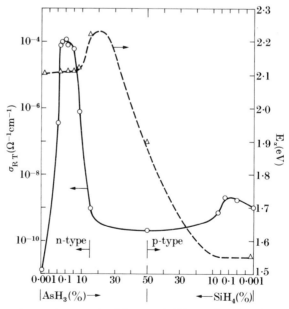

Fig. 7.45. Room-temperature conductivity σ_{RT} (solid curve) and optical gap E_α (broken curve) for glow-discharge-deposited Si–As films as a function of the gaseous composition used in their preparation. (From Knights 1976a.)

reversion to material with a low conductivity $\sim 10^{-9}-10^{-10}\,\Omega^{-1}\,cm^{-1}$, a value which persists all the way to pure amorphous arsenic. This interesting result suggests a transition from impurity doping to alloy behaviour; although silicon can be doped with arsenic, arsenic cannot be doped with silicon. The behaviour is probably associated with the lower co-ordination of the arsenic network (see Chapter 8), the flexibility of which allows silicon to be four-fold co-ordinated. Although such alloying behaviour may seem reasonable, one might then expect the Fermi level to adopt its usual position near the centre of the gap. That this cannot be the case over the whole of the composition range is evident from the observation (Fig. 7.45) that the conductivity remains relatively invariant above about 15 per cent arsenic even though the optical gap, after passing through a maximum at about 20 per cent arsenic, falls by almost $0.7\,eV$. Thermopower measurements were used by Knights (1976a) to determine the sign of the charge carrier; although Fig. 7.45 indicates p-type behaviour for arsenic-rich samples, it should be noted that Mytilineou and Davis (1977) report an n-type thermopower in amorphous arsenic (see § 8.4).

Hall-effect measurements by Le Comber et al. (1977) on doped glow-discharge-deposited silicon reveal, perhaps more dramatically than for any other material, the reversal of sign between the thermopower and the Hall effect. The results of Fig. 7.46 have been interpreted in terms of a decreas-

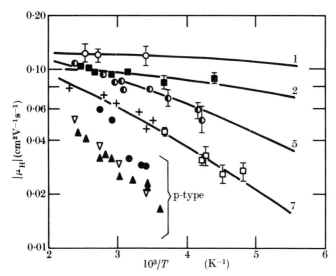

Fig. 7.46. Temperature dependence of Hall mobility for four n-type samples (1, 2, 5, and 7) and three p-type samples of glow-discharge-deposited silicon. The samples were prepared from silane containing the following volume parts per million of B_2H_6 (for the p-type films) and of PH_3 (for the n-type films): ●, 2.3×10^4; ▲▽, 5×10^4; 1, 98; 2, 304; 5, 2×10^3; 7, 3×10^4. The solid curves are theoretical fits assuming two conduction paths, one in extended states at the mobility edge and the other hopping in a donor band. (From Le Comber *et al.* 1977.)

ing contribution to transport by conduction in extended states as the temperature is lowered, accompanied by an increasing contribution from carriers hopping in a donor (or acceptor) band for which the Hall effect is assumed negligible. The sign of the Hall effect is discussed in § 6.4.7.

Field-effect measurements on doped films have considerably extended the range of the experimentally determined density of gap states. The E_Y peak and the rapid rise in $N(E)$ at E_A (see Fig. 7.41) have been verified in measurements by Spear and Le Comber (1976). A new self-consistent analysis did not give markedly different results for the profiles shown in Fig. 7.41, but the overall level was determined to be a factor of 2 or 3 higher. As in earlier work, Spear and Le Comber (1976) neglected the possible influence of surface states, the presence of which would of course lead to $N(E)$ being overestimated. A lower density of gap states (see Knights 1977) would imply rather lower efficiencies of doping than given above.

A study of thermoelectric power in doped silicon has been reported by Jones *et al.* (1977) (see also Friedman 1977) and of photoconductivity by Anderson and Spear (1977). Phosphorus doping leads to an extremely high photosensitivity with the photoconductive gain exceeding unity (for an electric field of $3 \times 10^3 \, \mathrm{V \, cm^{-1}}$) once the Fermi level is moved to within $\sim 0.65 \, \mathrm{eV}$ of E_C.

Measurements of the a.c. conductivity of undoped silicon glow-discharge deposited onto substrates held at various temperatures T_d has revealed a behaviour in good agreement with the predictions of § 6.4.5 (Abkowitz *et*

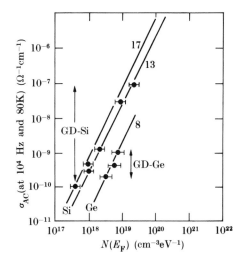

Fig. 7.47. a.c. conductivity at 10^4 Hz and 80 K for various films of glow-discharge-deposited silicon and germanium plotted against the density of states $N(E_F)$ at the Fermi level (as determined from field-effect data). The lines are the theoretical variation according to eqn (6.16) with values of α^{-1} given (in Å) by the number on the curves. (From Abkowitz *et al.* 1976.)

al. 1976). At a sufficiently low temperature $\sigma(\omega)$ was found to be proportional to ω^s. Although $s \sim 0.95$, rather too high (see § 6.4.5) for a reasonable value of ν_{ph}, a clear dependence of $\sigma(\omega)$ on T_d was found as shown in Fig. 7.47 which also includes the data on glow-discharge-deposited germanium discussed in § 7.4. The magnitude of $\sigma(\omega)$ measured at 10^4 Hz and 80 K is plotted against the density of states at the Fermi level determined independently from field-effect experiments. The predicted proportionality between $\sigma(\omega)$ and $N(E_F)$ is seen to be obeyed; α^{-1} was taken as an adjustable parameter and all data lie between the lines defined by $\alpha^{-1} = 13$ Å and $\alpha^{-1} = 17$ Å. Relative to glow-discharge-deposited germanium ($\alpha^{-1} = 8$ Å), the localized states through which hopping occurs are more extended spatially and the value of $N(E_F)$ is an order of magnitude lower. For evaporated films of germanium and silicon, studies of d.c. hopping conduction led to just the opposite trend for both of these parameters as mentioned earlier in this section.

7.6. Optical properties of amorphous germanium

It is difficult to find in the published literature on amorphous semiconductors a greater diversity than that existing for the form of the optical absorption edge in amorphous germanium. Fig. 7.48 collects together early

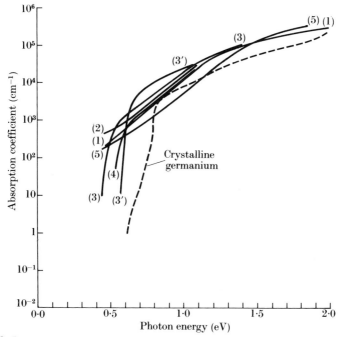

Fig. 7.48. Room-temperature optical absorption edges in evaporated films of germanium as reported by various authors: (1) Chittick (1970), (2) Clark (1967), (3) and (3') Spicer and Donovan (1970a), (4) Chopra and Bahl (1970), (5) Tauc (1970a). The broken curve for crystalline germanium is from Dash and Newman (1955).

data on evaporated films obtained by Tauc (1970a), Clark (1967), Spicer and Donovan (1970), Chopra and Bahl (1970), and Chittick (1970). The edge in crystalline germanium with the diamond structure, shown as a broken curve (Dash and Newman 1955), arises from transitions across an indirect gap of 0·66 eV and a direct gap of 0·8 eV corresponding to the transitions $\Gamma_{25'} \to L_1$ and $\Gamma_{25'} \to \Gamma_{2'}$ respectively (see Fig. 7.49).

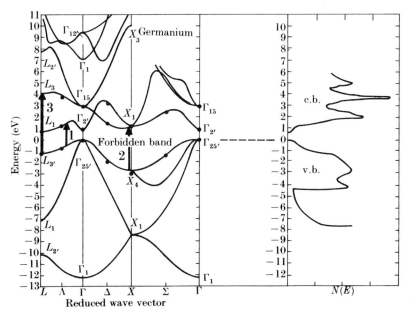

Fig. 7.49. Electronic band structure of crystalline germanium (from Herman *et al.* 1967) showing some principal interband transitions and the density of states in the valence and conduction bands.

There is no doubt that most of the variations evident in Fig. 7.48 arise from differences in methods of preparation and subsequent treatments of the films. The most important point of difference lies in the behaviour of the absorption coefficient at low photon energies. Whilst most workers find a gradual tailing of the edge, Spicer and Donovan (curves 3 and 3′) and Chopra and Bahl (curve 4) report sharp falls in α near 0·5 eV. Thèye (1970, 1971) also reports sharp edges, but near 1·0 eV, in highly annealed films (see Fig. 7.50(a)). It should be emphasized that the thickness limitation imposed on films prepared by vacuum evaporation make it extremely difficult to obtain accurate absorption data for values of α below 10^2 cm^{-1}. Spicer and Donovan (see also Donovan *et al.* 1970) were able to measure down to $\alpha = 10$ cm^{-1} using films up to 2 μm in thickness with faces accurately parallel. Although the position of the sharp edge was found to

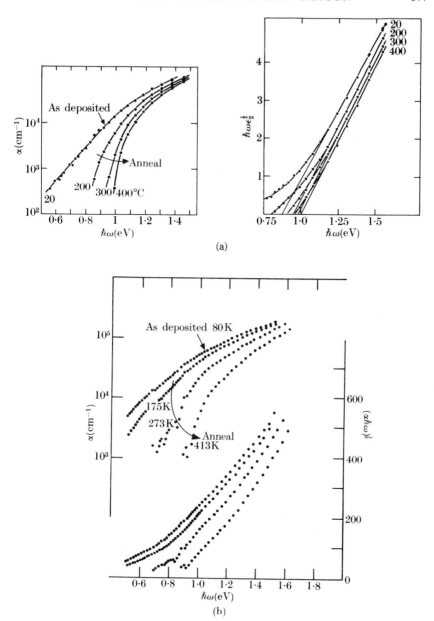

Fig. 7.50. Optical absorption edges in amorphous germanium prepared and treated in various ways. The data in each case are plotted as ln α *versus* $\hbar\omega$ and as $\hbar\omega\varepsilon^{1/2}$ or $(\hbar\omega\alpha)^{1/2}$ *versus* $\hbar\omega$: (a) from Thèye (1974); (b) from Elliott *et al.* (1974); (c) (p. 378) from Connell *et al.* (1973). In (c) curves (1) and (2) refer to films deposited at 25°C and 350°C respectively and curve (3) to a film deposited at 25°C and annealed at 150°C for 100 h; small and large dots in curve (1) refer to transmission and ellipsometry measurements respectively.

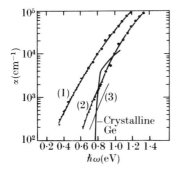 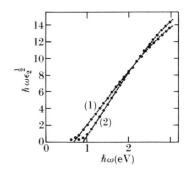

Fig. 7.50 (c)

be sensitive to evaporation conditions and the nature of the substrate, they found $\alpha < 10$ cm^{-1} in the photon energy range from $0\cdot1$ eV up to the sharp edge. Later measurements (Donovan, Ashley, and Spicer 1970) confirmed these observations and correlated the position of the sharp edge with the temperature of deposition and also the film density. For substrate temperatures less than 250°C, Donovan, Ashley, and Spicer (1970) found the position of the edge to lie below $0\cdot6$ eV and the film density to be $\sim 4\cdot7$ g cm^{-3}, while for substrate temperatures in the range 250–300°C the edge shifted abruptly to $\sim 0\cdot7$–$0\cdot8$ eV and the film density increased to within 2 per cent of the value for the crystal ($5\cdot35$ g cm^{-3}).

Connell and Lewis (1973) have criticized these results, as well as those of Chopra and Bahl (1970) and Thèye (1970, 1971), claiming that the sharpness of the edge in each case resulted from errors in measurement or analysis. These criticisms were answered (Thèye 1974, Donovan 1974), but not, in our opinion, sufficiently to justify an unambiguous claim for sharp edges in amorphous germanium.

There is, however, undisputed evidence that the absorption edge shifts to higher energies and sharpens to some extent as either the substrate temperature is raised or the films are annealed. Three sets of results illustrating this are shown in Fig. 7.50(a,b,c). The data of Thèye (1974) were obtained on material deposited by evaporation in a conventional vacuum onto a substrate held at room temperature and the film subsequently annealed step by step up to 400°C. The film of Elliott *et al.* (1974) was prepared by evaporation in ultra-high vacuum at a substrate temperature of 80 K and subsequently annealed to 413 K. Fig. 7.50(c) refers to data of Connell, Temkin, and Paul (1973) on a sputtered film deposited at 25°C and annealed at 150°C, in addition to one deposited at 350°C. In (a) and (c) measurements were made at room temperature; in (b) they were made at 80 K. For comparison the edge in crystalline germanium is shown in (c). Also plotted in each case are the corresponding variations with photon energy of $\hbar\omega\varepsilon_2^{1/2}$ (or $(\hbar\omega\alpha)^{1/2}$), the extrapolations of which can be used as a measure of the gap (see § 6.7).

There are several possible explanations for the shift and sharpening of the absorption edge on annealing.

(i) Diffusion of oxygen into the bulk of the material where it forms Ge—O bonds either in place of Ge—Ge bonds or by reacting with dangling bonds in a defective network. Since the absorption coefficient of GeO_2 occurs at about 6 eV, incorporation of Ge—O bonds may be expected to raise the gap in a manner similar to that observed and calculated for SiO_x compounds (Phillip 1971). If the absorption edge of amorphous germanium is to be ascribed to transitions involving defect states arising from dangling bonds, then their saturation with oxygen would also lead to a shift of the edge to higher energies.

(ii) A *healing* of the network, in the sense of an atomic rearrangement during or after deposition, that reduces the number of dangling bonds and less specific defects. Pertinent to this is the diminution of the network of voids on annealing (§ 7.3) and, presumably, vacancies of atomic size could also be removed. Associated with such topological changes in the structure would be an increase in the average co-ordination number and possibly more subtle changes in the remaining network such as are considered in (iii).

(iii) A *relaxation* of a fully connected network resulting, for example, in changes in the average bond length, the bond-angle distortions, and the dihedral-angle distributions. The effects of these parameters on the band gap can be inferred from calculations for crystalline polytypes of germanium (§ 7.8). A decrease in the bond length or in bond-angle distortion is expected to increase the gap; the effect of changes in the dihedral-angle distribution is less clear.

(iv) A reduction in the magnitude or extent of internal electric fields which are known to broaden absorption edges. This could be an indirect consequence of any of the processes (i)–(iii).

With regard to the first of these possibilities, Connell *et al.* (1973) give convincing arguments, based on the measured oxygen contamination (<0·5 per cent), against this explanation for their data on sputtered films. The results of Elliott *et al.* (1974) on films prepared by evaporation in ultra-high vacuum support this conclusion that oxygen contamination is not an essential requirement for shifting and sharpening the absorption edge. However, this is not to say that incorporation of oxygen into the network cannot shift the edge; indeed Koc, Závětová, and Zamek (1972) and Knotek and Donovan (1973) provide evidence that it does, and Moss, Flynn, and Bauer (1973) have found significant structural changes with ~10 per cent oxygen contamination.

Some of the curves presented in Figs. 7.48 and 7.50 show that the absorption edge obeys a spectral Urbach rule over a limited range of photon energies. The slope observed is approximately one-half to one-third of that found in chalcogenide amorphous semiconductors. According to eqn (6.58), based on the model of Dow and Redfield (1970), which ascribes exponential tailing of the absorption edge to exciton effects in the presence of electric fields, the slope of the Urbach tail decreases as the field increases and the effect is more pronounced for materials of high dielectric constant. Thus a shallow Urbach tail may be expected for amorphous germanium even if the conduction-band and valence-band edges have sharp cut-offs. In addition, annealing may be expected to reduce random fields arising from dangling bonds, voids, or even bond-angle distortions, with a consequent sharpening of the edge.

Detailed structural studies by Temkin et al. (1973), using a differential X-ray scattering technique to determine the changes in small-angle scattering and radial distribution functions for sputtered films, have led these authors to deduce that the densification, which accompanies an increase in the substrate temperature (see Fig. 7.51(a)), arises from a reduction in the number of voids having a diameter less than 7 Å. This is accompanied by increases in the first two co-ordination numbers, a

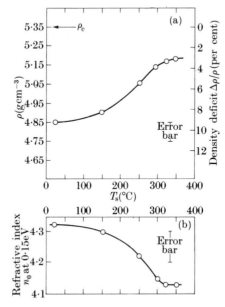

Fig. 7.51. Density ρ and refractive index n_0 (at 0.15 eV) for sputtered films of germanium deposited at various substrate temperatures T_s. In (a) the density deficit relative to the crystalline value ρ_c is shown. Data from Paul et al. (1973) and Connell et al. (1973).

decrease in bond-angle distortions, and a preference for more bonds to move towards the staggered configuration (§ 7.2). No variation in bond length occurred. It is clear that these changes are associated with a removal of dangling bonds as well as some relaxation of the surrounding network. Directly or indirectly they lead to a shift of the optical absorption edge towards higher energies.

Although Donovan et al. (1970) report no change in the value of the low-energy refractive index n_0 of evaporated germanium with annealing or substrate temperature and give a value equal to that of the crystal (4·00), Thèye (1971), Wales, Lovitt, and Hill (1967), and Connell and Paul (1972) find a decrease from about 4·4 to 4·2 on annealing. The variation with substrate temperature T_s for sputtered films is similar and is shown in Fig. 7.51(b). The rather sudden drop occurring principally between $T_s = 150°C$ and 300°C correlates with the increase in density and also with changes in other optical properties to be discussed below.

The frequency dependence of the optical constants (n, k, ε_1, and ε_2) of amorphous germanium through and above the fundamental edge are shown in Fig. 7.52(a,b,c,d). The variation with preparative conditions in this region of the spectrum, although significant, is less obvious than in the controversial edge region. Data presented here have been obtained by Connell et al. (1973) and by Donovan et al. (1970). Thèye (1974) has also made measurements in this region of the spectrum, as also have Jungk (1971) and Bauer and Galeener (1972) (see Fig. 7.16). We comment mainly on the data of Connell et al. (1973).

The lower value of the refractive index at long wavelengths for films deposited at a high substrate temperature is reversed at energies above ~1·2 eV. Likewise the extinction coefficient k, which is initially lower in the high T_s film, has a somewhat higher value in the vicinity of its maximum near 3·8 eV. The variation with photon energy can be fitted to a simple dispersion relation for a damped Lorentzian oscillator of natural frequency ω_0:

$$\varepsilon = \varepsilon_1 + \varepsilon_2 = 1 + \frac{\omega_p^2}{\omega_0^2 - \omega^2 - i\gamma\omega} \tag{7.6}$$

where $\omega_p^2 = 4\pi n_v e^2/m$, for n_v charges per unit volume, and m is the electron mass. The plasma frequency ω_p is also given by

$$\omega_p^2 = \varepsilon_2^{max}\gamma\omega_0 \tag{7.7}$$

where γ is the damping factor and ω_0 is the frequency at which $\varepsilon_1 = 1$. Although the above equation does not accurately describe the data, it can be used to determine ω_0 and ω_p, quantities that have some physical significance. The energy $\hbar\omega_0$ for low T_s films is ~3·3 eV, increasing with higher T_s to the value for diamond crystalline germanium, namely

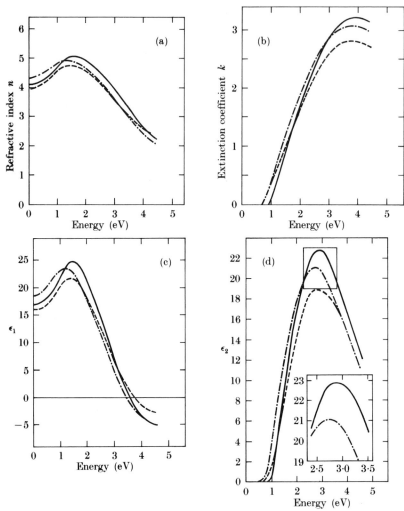

Fig. 7.52. Spectral dependence of (a) refractive index n, (b) extinction coefficient k, (c) real part ε_1 of the dielectric constant, and (d) imaginary part ε_2 of the dielectric constant for sputtered films of germanium. Chain curve $T_s = 25°C$; solid curve, $T_s = 350°C$; broken curve, film prepared by evaporation (Donovan *et al.* 1970). (From Connell *et al.* 1973.)

$\sim 3\cdot5$ eV. The damping factor γ is, from the width of the ε_2 curve, approximately equal to ω_0, giving a value of $\hbar\omega_p \sim 16$ eV which agrees well with the plasma energy deduced from direct energy-loss measurements by Zeppenfeld and Raether (1966) and is slightly lower than the value for crystalline germanium (16·6 eV).

On a simple two-band model of a semiconductor with a spherical Jones zone (Penn 1962, Phillips 1971), the low-energy refractive index n_0 is related to the plasma frequency by (Cardona 1971[†])

$$n_0^2 = 1 + \tfrac{2}{3}(\omega_p^2/\omega_g^2) \tag{7.8}$$

In real materials the Penn gap $\hbar\omega_g$ represents an average separation between valence and conduction bands and is a measure of the covalent bond strength. Since ω_p is slightly smaller and n_0 is somewhat higher in amorphous germanium compared with crystalline germanium, a lower average separation between bonding and antibonding states is inferred. This is indeed seen from photoemission data to be presented below. In principle it could arise from an increased average bond length r_1 (Van Vechten (1969) has shown $\omega_g \propto r_1^{-2.5}$), a reduced average co-ordination number (Phillips 1971), or bond-angle distortions. Paul *et al.* (1973), using structural data of Temkin *et al.* (1973), deduce that the first, but principally the second, of these effects dominates over the third. As the bond length in amorphous germanium does not change with T_s, the increased value of ω_g inferred for the high T_s film presumably arises mainly from the increase in average co-ordination number.

The effective number n_{eff} of free electrons per atom contributing to the optical absorption up to an energy $\hbar\omega'$ is given by the plasma sum rule (Phillip and Ehrenreich 1963)

$$n_{\text{eff}} = \frac{m}{2\pi^2 N e^2} \int_0^{\omega'} \omega \varepsilon_2(\omega)\, d\omega \tag{7.9}$$

where N is the atomic density. Fig. 7.53 shows a plot of $N n_{\text{eff}}$ *versus* $\hbar\omega'$ for sputtered films deposited at $T_s = 25°C$ and $350°C$. Below about 3 eV

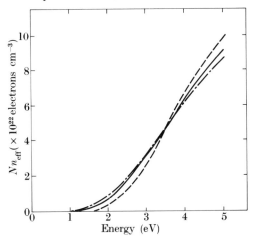

Fig. 7.53. The effective number of electrons involved in transitions up to an energy $\hbar\omega$ for sputtered films of germanium: chain curve, $T_s = 25°C$; solid curve, $T_s = 350°C$; dashed curve, crystalline germanium (Philipp and Ehrenreich 1963). (From Connell *et al.* 1973.)

† M. Cardona, 1971, School of Physics 'Enrico Fermi', Varenna (unpublished).

this is higher than in crystalline germanium owing to a shift of states from both bands towards the gap, but at higher energies the trend is reversed. This then continues all the way to the plasma frequency, as revealed by the data of Bauer and Galeener (1972) and Bauer (1974a,b). In Fig. 7.54 their

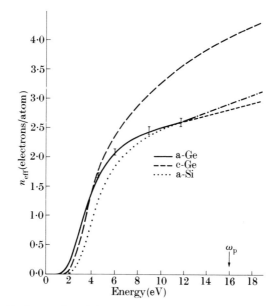

Fig. 7.54. The effective number of electrons per atom involved in transitions up to an energy $\hbar\omega$ for evaporated films of germanium and silicon and for crystalline germanium. The plasma frequency ω_p is shown. (From Bauer 1974a.)

results, which at low energies differ somewhat from those in Fig. 7.53, are plotted in terms of n_{eff}. It is seen that, unlike crystalline Ge for which n_{eff} reaches 4 electrons/atom at the plasma frequency, n_{eff} for amorphous films is only 3 at the same energy. Bauer has suggested that this surprising result arises from a more localized character for the deeper-lying electrons (see below) compared with those in the crystal, leading to reduced matrix elements for transitions from them.

7.7. Optical properties of amorphous silicon

The optical absorption edge of amorphous silicon is, like that of amorphous germanium, dependent in position and shape on the nature and conditions of preparation. A measurement by Beaglehole and Zavetova (1970) on an evaporated film approximately 1 μm thick is compared with the edge of crystalline silicon (Dash and Newman 1955) in Fig. 7.55(a). In the crystal the absorption edge at about 1·1 eV corresponds to indirect transitions from $\Gamma_{25'}$ to the minimum in the conduction band along the Δ

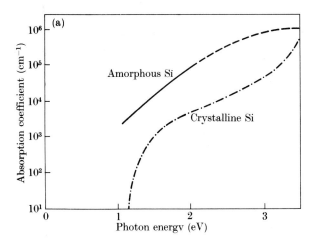

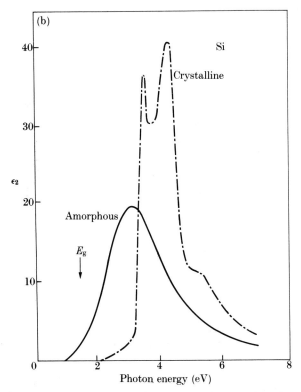

Fig. 7.55. (a) Comparison of room temperature optical absorption edges in evaporated (Beaglehole and Zavetova 1970) and crystalline (Dash and Newman 1955) silicon. (b) ε_2 spectra of amorphous and crystalline silicon. (From Stuke 1970*b*.)

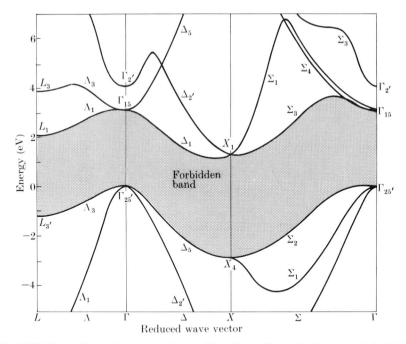

Fig. 7.56. Electronic band structure of crystalline silicon. (From F. Herman and J. P. Van Dyke, unpublished.)

axis of the Brillouin zone (Fig. 7.56). Direct transitions to Γ_{15} do not occur until a photon energy of about 3 eV is reached (Fig. 7.55(b)). The high values of α and ε_2 observed for amorphous silicon in the range from 1–3 eV thus provide good evidence for a relaxation of the k-conservation selection rule. The ε_2 spectrum of amorphous silicon (Fig. 7.55(b)) shows, as for amorphous germanium (§ 7.6), none of the fine structure present in the spectrum for the crystal (Pierce and Spicer 1972).

The edge in amorphous silicon prepared by evaporation has also been measured by Grigorovici and Vancu (1968), Brodsky, Kaplan, and Ziegler (1972), Fischer and Donovan (1972), Lewis (1972), Loveland, Spear, and Al-Sharbaty (1973/74), and Le Comber *et al.* (1974); on sputtered films it has been measured by Brodsky *et al.* (1970) and Loveland *et al.* (1973/74); on glow-discharge-deposited films it has been measured by Chittick (1970) and Loveland *et al.* (1973/74). The variations between different films are even greater than those found for amorphous germanium.

Some of these absorption edges were presented in Fig. 7.20 and discussed in § 7.3. As for amorphous germanium, annealing or the presence of oxygen shifts the edge to higher energies and makes it steeper. The presence of hydrogen in glow-discharge-deposited samples may well have a similar effect; although this has not been studied quantitatively, it seems likely that the relative position of the two edges labelled a and b in Fig. 7.57 (from Loveland *et al.* 1973/74) is associated with a higher concentration of hydrogen incorporated into the sample deposited at the lower temperature. The other edges shown in Fig. 7.57 are for evaporated and sputtered films, as indicated in the caption.

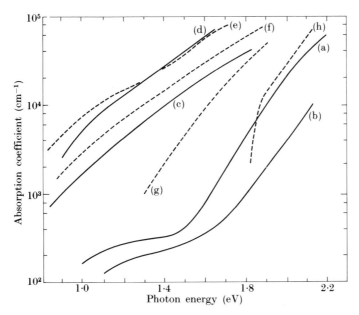

Fig. 7.57. Optical absorption edges in amorphous silicon prepared by different methods: (a) glow-discharge films deposited from 500 to 600 K; (b) glow-discharge films deposited at ~300 K; (c) and (f) sputtered films; (d) and (e) evaporated films; (g) annealed sputtered film; (h) 'extrapolated' edge (Brodsky, Kaplan, and Ziegler 1972, see § 7.3). (From Loveland *et al.* 1973/74.)

A series of absorption edges obtained by annealing, and eventually crystallizing, sputtered films of silicon as measured by Brodsky *et al.* (1970) are shown in Fig. 7.58(a). For unannealed samples, Brodsky *et al.* (1970) find that, in the spectral range $\hbar\omega = 1\cdot4-2\cdot4$ eV, a relation of the form $\alpha \sim (\hbar\omega - E_0)^2$ with $E_0 = 1\cdot26$ eV fits the absorption data. For annealed films the extrapolated gap is several tenths of an electronvolt higher (see also Lewis 1972) and E_0 approaches that deduced for glow-discharge-deposited films, namely $1\cdot5-1\cdot6$ eV.

The decrease of refractive index with annealing found by Brodsky *et al.* (1970) and shown in Fig. 7.58(b) for sputtered films was not observed by Fischer and Donovan (1972) for evaporated films. However, Schwidefsky (1973) finds that the refractive index depends sensitively on the deposition

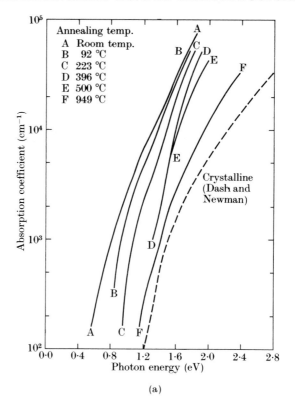

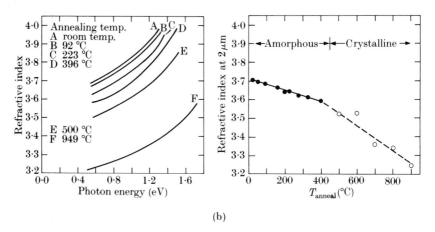

Fig. 7.58. Dependence of (a) absorption edge, (b) and (c) refractive index of a sputtered film of silicon on annealing for 2 h at the temperatures indicated. The film crystallized during the 500°C anneal. (From Brodsky *et al.* 1970.)

temperature and attempts to account for the variation in terms of an increased polarizability in samples containing a large concentration of dangling bonds.

It is of interest to consider the spectral dependence of photoconductivity in relation to that of the absorption coefficient. Whilst Fischer and Donovan (1972) find, for evaporated films, an almost exact coincidence between the absorptance and photoconductivity, indicating an energy-independent quantum efficiency, Loveland et al. (1973/74) find, for glow-discharge-deposited samples, this not to be the case. Figs. 7.59 and 7.60 reproduce

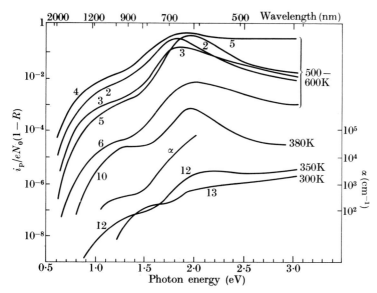

Fig. 7.59. Spectral dependence of photoconductivity in glow-discharge films of silicon deposited at the temperatures indicated. The ordinate represents the number of charge carriers flowing around the circuit per photon absorbed by the specimen. α is the absorption coefficient of a film deposited at $T_d = 500$ K. (From Spear 1974b.)

the data of the latter authors. The structure in α (right-hand scale of Fig. 7.59) is evident in all the photoconductivity curves. The temperatures refer to those of the substrate during deposition. The ordinate is derived from the equation for the photocurrent.

$$i_p = eN_0(1 - R)\{1 - \exp(-\alpha d)\}\frac{\eta \tau}{t_t} \qquad (7.10)$$

where N_0 is the number of incident photons per second, R the reflectivity, d the film thickness, η the quantum efficiency, τ the recombination lifetime, and t_t the carrier transit time. For weakly absorbed light ($\alpha d < 0.4$) therefore, the ordinate represents $\alpha d \eta \tau / t_t$ and using the absorption

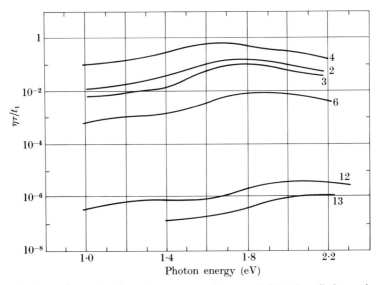

Fig. 7.60. Dependence of $\eta\tau/t_t$ on photon energy for several of the glow-discharge-deposited samples of Fig. 7.59. (From Loveland *et al.* 1973/74.)

data the curves of Fig. 7.60 can be derived. The fall of $\eta\tau/t_t$ with decreasing photon energy, although not as marked as that of η for amorphous selenium for example (see § 6.5.3), is considered to arise from the same process, namely an increasing probability of 'geminate' recombination. A further discussion of recombination processes in photoconductivity in silicon is given in § 6.5.2.

The onset of photoconduction in glow-discharge-deposited silicon depends somewhat on the deposition temperature but from Fig. 7.59 is seen to lie between ~0·6 and 0·8 eV. According to Loveland *et al.* (1973/74) and Spear (1974*b*) this corresponds to excitation from states at the Fermi level (which shifts with T_d, see Fig. 7.41) to the mobility edge E_C in the conduction band or perhaps to tail states at E_A, although the probability for transitions to the latter might well be low because of small overlap between initial and final states, both being localized. This interpretation differs from that of Fischer and Donovan (1972) who propose that the onset they observed at ~0·7 eV corresponds to band-to-band transitions; however, for their evaporated films the absorption edge lay at much lower energies than for glow-discharge-deposited films, closely resembling, in fact, that of curve A in Fig. 7.58(a) for unannealed sputtered silicon. The shoulders between 1·1 and 1·3 eV in Fig. 7.59 are interpreted by Loveland *et al.* (1973/74) as arising from the maximum in the density of states (E_Y) deduced from field-effect measurements on these samples (Fig. 7.41) and the subsequent rise from band-to-band transitions. The temperature dependence of the photocurrent in these films is discussed in §6.5.2.

Although photoluminescence has been observed in silicon prepared by evaporation in ultra-high vacuum (Engemann, Fischer, and Mell 1977), most studies have been made on glow-discharge-deposited material. Results from Engemann and Fischer (1974b) taken from films deposited at $T_d \sim 450$ K are reproduced in Fig. 7.61. Excitation was at 647·1 nm

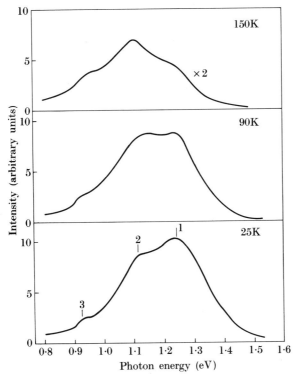

Fig. 7.61. Luminescence spectra of glow-discharge silicon deposited at 450 K and measured at 25, 90, and 150 K. (From Engemann and Fischer 1974b.)

($\sim$1·9 eV) with a 0·5 W krypton laser. The luminescence band is wide and displays structure, the relative heights of the three peaks varying with temperature. The total luminescence intensity is, however, independent of temperature up to $\sim$100 K above which it falls off sharply (see Fig. 7.62(a) below). In contrast to chalcogenides (see Chapters 6 and 9) the luminescence is not believed to have a *large* Stokes shift, and assignment of the three peaks has been made in terms of the three transitions (see Fig. 7.41), $E_A \rightarrow E_B$ (1·25 eV), $E_A \rightarrow E_Y$ (1·10 eV), and $E_X \rightarrow E_B$ (0·92 eV).

Subsequent measurements by Engemann and Fischer (1976) showed that, although samples deposited at a lower temperature exhibited a decreased luminescence intensity and a somewhat different spectral shape (Engemann and Fischer 1974a), the temperature dependence of the intensity remained unchanged, and furthermore the photoconductivity and luminescence were complementary. Fig. 7.62(a) shows both these results. In addition they found that the photoconductivity and luminescence exhibited very different decay times (the luminescence decayed with a time constant $\sim$20 ns, approximately 10^5 times faster than the photoconductivity); the luminescence depended linearly on light intensity over the whole temperature range; finally the luminescence could be quenched by application of an electric field (Fig. 7.62(b)). The decay time (20 ns) is about what one would expect for an optical transition if the electron and hole orbitals are strongly localized; the contrasting case for an exciton trapped at a charged centre is discussed in § 9.6 in connection with chalcogenides.

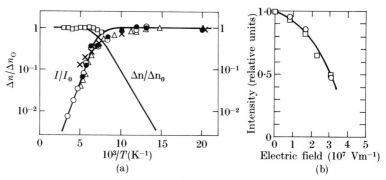

Fig. 7.62. (a) Temperature dependence of normalized photoluminescence intensity for glow-discharge silicon deposited at 520 K ($\bigcirc$), 440 K ($\triangle$), 400 K ($\bullet$), and 320 K ($\times$). The photoconductivity ($\square$) is independent of temperature in the temperature range where the luminescence intensity varies strongly. (b) Electric field quenching of luminescence. (From Engemann and Fischer 1976.)

Although monomolecular recombination might be expected for those carriers recombining through defect levels E_X and E_Y, band-to-band recombination (the transition at 1·25 eV) is monomolecular only for the special case of geminate recombination. Engemann and Fischer (1976) therefore propose that photogenerated electron–hole pairs either separate from their mutual Coulomb potential, in which case they lead to photoconductivity, or else they recombine emitting radiation. The solid lines in Fig. 7.62(a) are theoretical curves based on this model, the activation energy of 0·12 eV in the luminescence being a measure of the binding energy of the electron–hole pair following thermalization. The fact that the luminescence intensity is lower for samples deposited at a low temperature is probably explicable in terms of an increased density of centres through which the carriers decay non-radiatively. The photocurrent is also lower in such samples. The electric-field quenching of luminescence shown in Fig. 7.62(b) is attributed to a lowering of the potential barrier for escape as described in § 6.5.3, where the same model was used to explain an electric-field *enhancement* of quantum efficiency for photogeneration, and in § 6.7.6.

Electric-field quenching of luminescence and the complementarity of luminescence and photoconductivity in glow-discharge-deposited silicon are evident in data obtained on doped samples (Rehm *et al.* 1976). With increasing concentrations of either boron or phosphorus the spectral dependence of the luminescence band shows a shift to lower photon energies, the peak occurring at ~0·8 eV for high doping levels (Fig. 7.63).

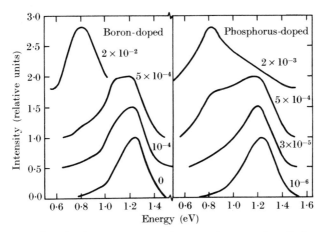

Fig. 7.63. Normalized luminescence spectra for p- and n-type glow-discharge-deposited silicon. Doping concentrations as shown. (From Rehm *et al.* 1976.)

For either dopant the total luminescence intensity falls with increasing concentration as shown in Fig. 7.64. Rehm *et al.* (1976) attribute this to an increased probability for carrier pairs to separate on account of the internal electric fields produced by charged donors or acceptors.

Photoluminescence in pure and doped glow-discharge-deposited silicon has also been studied by Nashashibi, Austin, and Searle (1977*a*). Their results, which include data on the excitation spectra, are shown in Fig. 7.65. Although the shift of the luminescence peak with doping appears to be considerably less than that measured by Rehm *et al.* (1976), Nashashibi *et al.* (1977*b*) report the growth of the 0·8 eV band with increasing temperature.

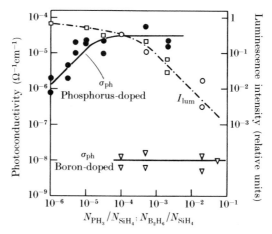

Fig. 7.64. Luminescence intensity and photoconductivity in glow-discharge-deposited silicon as a function of doping concentration (gaseous ratio). (From Rehm *et al.* 1976.)

The excitation spectra look, at first sight, similar to those found for chalcogenide glasses (Chapter 9), but a noteworthy difference is that the peaks lie at photon energies corresponding to much higher values of the absorption coefficient. As for the chalcogenides the fall-off of the excitation spectrum on the low-energy side is associated simply with the reduced

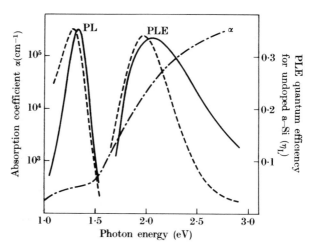

Fig. 7.65. Photoluminescence (PL) and excitation spectra (PLE) at 10 K for undoped (solid curve) and n-type (broken curve) glow-discharge-deposited silicon. The quantum efficiency (left-hand scale) is absolute for the PLE spectrum of the undoped sample; the PLE spectrum of the doped sample is normalized to this curve. Intensities for the PL curves are plotted on an arbitrary (but linear) scale. Also shown in the absorption coefficient α for undoped glow-discharge-deposited silicon at 300 K. (From Nashashibi *et al.* 1977a.)

level of absorption; the photoluminescence efficiency is, in fact, independent of photon energy up to the maxima. The fall-off on the high-energy side probably occurs for a different reason than in chalcogenides, namely an increased probability of carrier-pair separation (with consequent non-radiative recombination) when the carriers thermalize to a relatively larger separation in their mutual Coulomb field, as indeed they are expected to do if they start with a higher excess kinetic energy (see § 6.5.3). The enhanced fall-off on the high-energy side of the excitation spectrum for doped samples is probably associated with the above mentioned internal fields arising from charged impurities.

Nashashibi *et al.* (1977*a,b*) also report on the temperature dependence of the luminescence intensity (Fig. 7.66). For doped samples the temperature dependence is reduced but for doped and compensated samples the

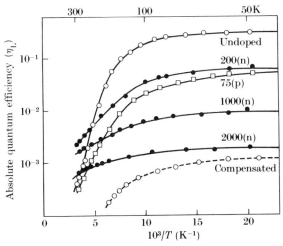

Fig. 7.66. Temperature dependence of luminescence in undoped and n- and p-type doped glow-discharge silicon. The doping levels (in gaseous parts per million) are indicated. The broken curve is for a compensated sample prepared with 1700 ppm of phosphine and 1840 ppm of diborane in the silane. Excitation was at 2·3 eV. (From Nashashibi *et al.* 1977*a*.)

steeper fall observed in undoped samples at high temperatures is partially restored. The results shown here refer to the total luminescence; Nashashibi *et al.* (1977*b*), however, find very different temperature dependencies for the 0·8 eV and 1·3 eV bands, the former varying least.

Electroluminescence from forward-biased p–i–n diodes fabricated from glow-discharge-deposited silicon has been reported by Pankove and Carlson (1976). The luminescence peaks at ~1·27 eV which is very close to the value found in photoluminescence. Pankove and Carlson (1977) also report that photoluminescence diminishes in intensity and shifts to lower energies as hydrogen is evolved from glow-discharge-deposited silicon by heating.

7.8. Density of states in the valence and conduction bands of amorphous germanium and silicon

The density of valence-band and conduction-band states in amorphous germanium and silicon has been determined by ultraviolet and X-ray photoemission spectroscopy (UPS and XPS respectively), as well as by other techniques.

Early results on germanium (Donovan and Spicer 1968, Donovan 1970, Ribbing, Pierce, and Spicer 1971) and on silicon (Pierce and Spicer 1972) using UPS are compared with the calculated $N(E)$ for the diamond crystalline structures in Fig. 7.67. For both materials, the uppermost (p-like)

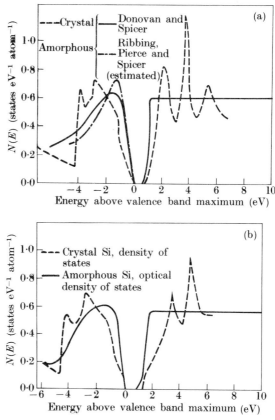

Fig. 7.67. Density of electron states in evaporated films of (a) germanium and (b) silicon as determined by UPS. The broken curves are the calculated densities of states for the diamond cubic crystalline phases. (From Spicer 1974.)

valence bands (from 0 to $\sim -4\,\text{eV}$) are similar in width to those in the crystals but are relatively featureless, have steeper leading edges, and peaks at energies about 2 eV closer to the band edge. For the conduction bands, all the structure calculated for the crystalline phases is apparently lost; however, according to Spicer (1974) the possibility of a weak maximum in $N(E)$ near the conduction-band edge, or a monotonic increase of decrease in $N(E)$ for $E > 2\,\text{eV}$, could not be ruled out from these experiments.

XPS allows one to probe deeper into the valence bands but with somewhat diminished resolution. Results on cleaved crystalline and on evaporated germanium and silicon were first obtained by Ley *et al.* (1972). In Fig. 7.68 we reproduce essentially similar results on germanium due to Eastman, Freeouf, and Erbudak (1974) which also include the density of

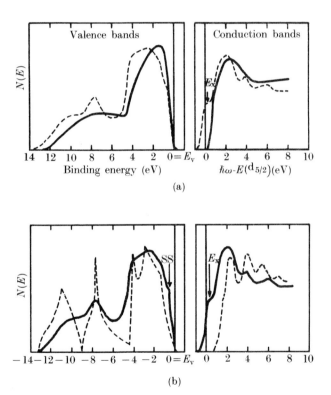

Fig. 7.68. (a) Density of states in amorphous (solid curve) and crystalline (broken curve) germanium as determined by XPS (excitation energy 25 eV). For the valence band a secondary electron background has been subtracted. For the conduction band, $N(E)$ was deduced from secondary emission and the edge has been sharpened by 0·5 eV to allow for experimental 'broadening effects'. (b) The crystalline spectrum of (a) compared with a calculated (EPM) spectrum by Shaw. (From Eastman *et al.* 1974.)

conduction-band states obtained from a 'partial photoelectric yield spectra'[†] following excitation from the $3d_{5/2}$ core level.

We first discuss the results obtained for the crystal. The density of valence states (broken line on the left-hand side of Fig. 7.68(a)) consists of a fairly flat-topped band about 5 eV wide with some evidence of the fine structure present in the calculated spectrum (Fig. 7.68(b)). The small shoulder marked SS at 0·7 eV below the top of the band is ascribed to surface states (Eastman and Grobman 1972a,b). Separated from this p-like band by a well-defined minimum at 6 eV lie the s–p bands of total width ∼ 7 eV. Compared with theory (Fig. 7.68(b)), these bands are not so well resolved into their two components, but the expected dip in the vicinity of 9 eV is clearly observed. The experimental density of conduction states, although again in rather poor agreement with theory, does display the three predicted peaks near 2 eV, 4 eV, and 6 eV measured with respect to the top of the valence band. In addition, however, there is a pronounced shoulder marked E_X which is believed to be of excitonic origin owing to the localized nature of the 3d core hole left behind on excitation. Such exciton effects have not been included in the theoretical curve.

The results for amorphous germanium (solid curve, Fig. 7.68(a)) show several important features when compared with those for the crystal. As in Fig. 7.67 the uppermost p-like valence band is somewhat narrower, has a steeper leading edge, and has its maximum shifted towards the gap. The lower-lying s–p bands are merged into one and shift bodily ∼1 eV towards higher energies (lower binding energy). For the conduction band one notices again the loss of the three peaks present in the crystalline spectrum but, in contrast to the results of Donovan and Spicer (1968) (Fig. 7.67), a broad maximum centred at ∼2·3 eV is left. Perhaps surprisingly the core excitonic feature, observed for the crystal, is lost.

Eastman et al. (1974) have calculated the ε_2 spectrum of amorphous germanium using the above densities of states, assuming matrix elements independent of energy and relaxing the k-conservation rule. This result is compared in Fig. 7.69 with the experimental curve of Donovan, Spicer, Bennett, and Ashley (1970) derived from reflectivity data and shown earlier in Fig. 6.50. The agreement is good, in view of the assumptions made, and should be compared with that calculated using the crystalline density of states (and the same assumptions) shown in Fig. 6.52(a).

There have been several different theoretical approaches to the calculation of the density of states of amorphous materials, and reproduction of the experimentally determined density of valence states (DOVS) of amorphous germanium and silicon have become somewhat of a testing

---

[†] These spectra represented the spectral distribution of secondary Auger electrons, escaping from a depth ∼ 15–25 Å, the number of which should be proportional to the number of photoexcited holes and hence to the distribution of final states.

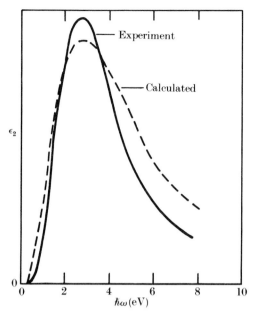

Fig. 7.69. ε_2 spectrum for amorphous germanium as deduced from $N(E)$ of Fig. 7.68(a) compared with that derived from reflectivity data. (From Eastman *et al.* 1974.)

ground for these. Here we summarize the present state of achievement for these materials.

Early methods based on modifications to the band structure of diamond crystalline germanium are now mainly of historical interest. Thus the 'dilated crystal' approach of Herman and Van Dyke (1968), the 'Lorentzian-broadened density of states' model of Brust (1969*a,b*), the Phillips 'Penn-gap model' (Phillips 1971), and the 'complex band structure' (CBS) method of Kramer (1970*a,b*, 1971) and of Maschke and Thomas (1970*a*) all obtained, with somewhat *ad hoc* adjustable parameters, fair reproduction of the ε_2 spectrum of amorphous germanium; however, they all failed to predict the density of states (see Thorpe and Weaire 1974). Subsequently Kramer and Treusch (1974) improved the CBS method by including short-range order of forms different from that occurring in the diamond structure. This refinement accounted in part for the coalescing of the s–p band peaks shown in Fig. 7.68(a) but not the change in shape and position of the p band.

It is now clear that approaches that do not take into account the different topology of amorphous as compared with crystalline germanium (as exemplified for example by ring orders different than six) are not expected to be successful. It is the short-range order rather than the absence of long-range order that is of prime importance in determining

DOVS. However, loss of translational symmetry removes sharp singularities arising from Brillouin zone effects and produces selective broadening; for optical transitions the energy dependence of matrix elements is smoothed (Fig. 6.51) and the restriction of k conservation is lifted.

Calculations based on crystalline polytypes of germanium, which have various different types of short-range order, have been very helpful in understanding the role that local bonding arrangements play in determining the density of states. Fig. 7.70 shows empirical pseudopotential calculations of the density of states for the diamond cubic (FC-8), wurtzite

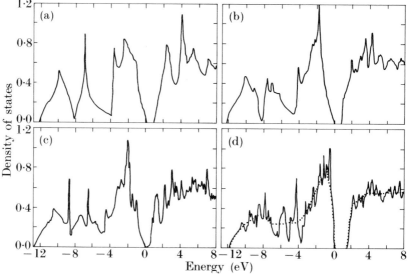

Fig. 7.70. Density of states calculated by Joannopoulos and Cohen (1973a) for germanium in (a) diamond cubic, (b) wurtzite, (c) Si III, and (d) Ge III structure. A dotted line is drawn through the spectrum in (d) to eliminate the fine structures arising from Brillouin zone effects. (From Thorpe and Weaire 1974.)

(2H), Si III(BC-8) (also known as GeIV), and GeIII (ST-12) structures by Joannopoulos and Cohen (1973a,b). Henderson and Herman (1972), Ortenburger, Rudge, and Herman (1972), Henderson, Herman, and Ortenburger (1974), Henderson and Ortenburger (1973), and Ortenburger and Henderson (1972, 1973, 1974) have also calculated $N(E)$ for these structures as well as for undistorted and distorted 4H and 6H polytypes and various clathrate structures. All these structures have tetrahedral bonding (distorted in the case of BC-8, ST-12, and the clathrates) but rather different ring statistics. As the size and complexity of the unit cell increases, the density of states shows more fine structure but less gross structure. For ST-12 (Fig. 7.70(d)), when smoothed (dotted line) to simulate the loss of Brillouin zone effects, most of the principal effects observed for amorphous germanium are obtained, i.e. the uppermost p-like valence band is skewed to higher energies, the dip in the s–p band is smoothed out, and the conduction band is relatively featureless.

Tight-binding models of amorphous germanium fall into several classes. The first is based on a simple Hamiltonian including interaction integrals V_1 between tetrahedral sp^3 orbitals belonging to the same atom and V_2 between orbitals on the same band joining nearest neighbours (Weaire 1971, Weaire and Thorpe, 1971, 1973, Thorpe and Weaire 1971, 1974). Perfect tetrahedral co-ordination (but implied topological disorder) is assumed, but quantitative disorder corresponding to a spread in bond angles (and hence in V_1) or next-nearest-neighbour interactions (V_2) is neglected. It has been proved (Weaire 1971, Heine 1971) that for *all* structures with tetrahedral co-ordination this Hamiltonian yields a band gap equal to or greater than $2|V_2|-4|V_1|$. Fig. 7.71 shows the density of

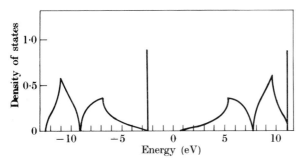

Fig. 7.71. Density of states using a tight-binding Hamiltonian for crystalline silicon/germanium. The delta functions in each band have the same weight as the rest of the band ($V_1 = 2·5$ eV, $V_2 = -6·75$ eV). (From Thorpe and Weaire 1974.)

states calculated for the diamond crystalline structure. The conduction band ($E > 0$) is inadequately described because of the omission of the d states, but the gross features of the valence band are obtained, namely the two peaks in the lower part and a separate p band shown as a delta function but having a weight equal to the rest of the band. The known breadth of this p band arises from next-nearest-neighbour interactions (see Ziman 1971) which have been neglected in this model. The shape of the lower bands depends on V_1/V_2 and on the topology; it is shown for four structures in Fig. 7.72(a). The Bethe lattice is an infinitely branched tree-like structure with no closed loops, whereas the Husimi cactus lattices are made up of only closed loops, two to each atom (see Fig. 7.72(b)). These pseudolattices are not periodic and results obtained using them serve to illustrate the importance of ring statistics on the valence-band density of states. Thus the absence of even-membered rings or the presence of odd-membered rings acts so as to erode the central dip. A calculation using

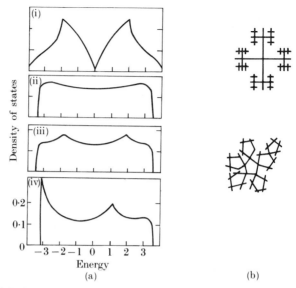

Fig. 7.72. (a) Valence-band density of states using a one-band Hamiltonian for various silicon/germanium lattices: (i) diamond cubic, (ii) Bethe lattice, (iii) six-ring Husimi cactus, (iv) five-ring Husimi cactus. (From Thorpe and Weaire 1974.) (b) Illustration of a Bethe and a five-ring Husimi cactus lattice. (From Thorpe and Weaire 1974.)

the topology of the Polk 201-atom continuous random network (CRN) model with randomly joined surface atoms is shown in Fig. 7.73. Joannopoulos and Cohen (1973a,b) have applied the model to the polytype structures mentioned above. Uda and Yamada (1975) using the same method have considered the effects of dangling bonds in a Bethe lattice with the interesting result that states appear near the centre of the band gap.

Tong (1974), Tong, Swenson, and Choo (1974), and Choo and Tong (1976) have calculated the density of states appropriate to a variety of tetrahedrally co-ordinated atomic clusters using extended Hückel theory (EHT) and obtained all the principal differences between the crystalline

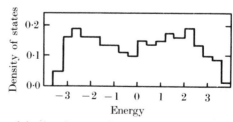

Fig. 7.73. Valence-band density of states using a one-band Hamiltonian for the connectivity of the 201-atom Polk model. (From Alben *et al.* 1973.)

and amorphous densities of states. However, computer limitations restrict the number of atoms that can be handled using this method. Larger models are more conveniently treated by the chemical pseudopotential method introduced by Anderson (1968, 1969)—see also Weeks, Anderson and Davidson (1973), Bullett (1974, 1975a,b,c, 1976), Haydock et al. (1972, 1975). Using this method, which defines a 'local density of states' in terms of the near-neighbour configuration of bonding orbitals, Kelly (1974), and Kelly and Bullett (1976a) have calculated the valence-band density of states for two silicon crystalline lattices, diamond and ST-12, as well as for two CRNs, the Boudreaux–Polk and Connell–Temkin models (see § 7.2). The results of these calculations, which incorporate the site-to-site variation of the interaction integrals V_1 and V_2, are shown in Fig. 7.74. The valley in

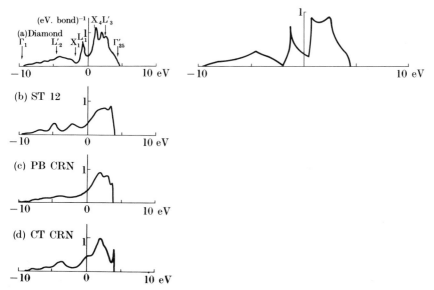

Fig. 7.74. Valence-band density of states for silicon atoms arranged on (a) the diamond lattice, (b) the ST 12 lattice, (c) the Polk–Boudreaux CRN, and (d) the Connell–Temkin CRN as calculated by the chemical pseudopotential and recursion method. The result for the diamond lattice obtained by a Brillouin zone sum of the bands derived by using the same interactions, is shown on a larger scale on the right. (From Kelly and Bullett 1976a.)

$N(E)$ for the diamond structure referred to several times above and shown here at about -2 eV is virtually eroded for the Boudreaux–Polk model but reappears for the Connell–Temkin model. As the latter contains no odd-membered rings but does have bond-angle distortions, the importance of ring statistics is once again seen.

DOVS spectra obtained by Meek (1977c) using essentially the same recursion method as Kelly and Bullett (1976) for a series of CRNs having

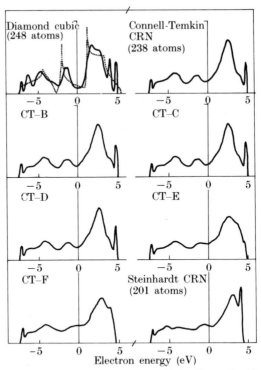

Fig. 7.75. Valence-band density of states for silicon in the diamond cubic structure and for various CRNs as calculated using the recursion method. The ring statistics of the various CRNs are shown in Table 7.1. (From Meek 1977c.)

an increasing proportion of odd-membered rings are shown in Fig. 7.75. Five of the networks were derived from the Connell–Temkin model by a restructuring procedure (Beeman and Bobbs 1975). The progressive filling of the s–p dip and the steepening of the leading edge of the valence band as the proportion of odd-membered rings increases is clearly evident from these calculations. (See also Alben *et al.* (1975).)

TABLE 7.1

CRN	Number of atoms	Rings per atom		
		5	6	7
Connell–Temkin	238	0	2·432	0
C-T B	238	0·059	2·270	0·156
C-T C	238	0·158	2·090	0·313
C-T D	238	0·189	1·955	0·469
C-T E	238	0·336	1·506	0·808
C-T F	238	0·398	1·287	1·008
Steinhardt	201	0·430	0·889	0·989

A related method of calculating the DOVS is the cluster–Bethe–lattice method (CBLM), (Joannopoulos and Yndurain 1974, Yndurain and Joannopoulos 1975, 1976), which yields results in full agreement with the above conclusions (see Joannopoulos and Cohen 1976).

The status of the theory of the multiple scattering within clusters of tetrahedral atoms (McGill and Klima 1972, Keller 1971) has been reviewed by Greenwood (1973) and by Thorpe and Weaire (1974). The importance of the local atomic arrangement in terms of staggered and eclipsed bond configurations is seen in the density of states near the pseudogap (Fig. 7.76). A real gap is not obtained by this method; there are also problems associated with boundary conditions and with proper representation of the correlations between clusters. The method may be

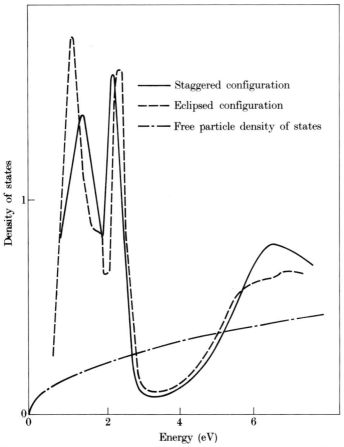

Fig. 7.76. Density of states (valence and conduction bands) for clusters of eight silicon sites in the staggered and eclipsed configurations and the free-particle density of states, as calculated by multiple-scattering theory. (From McGill and Klima 1972.)

useful if developed (see Jones 1974), but at present it has not been successfully applied to the lower valence bands, i.e. below the 'muffin-tin zero' of the atomic potential.

The densities of states in the conduction bands of amorphous germanium and silicon as deduced from photoemission data were shown in Figs. 7.67 and 7.68(a). In Fig. 7.77 soft X-ray absorption spectra of crystalline and amorphous silicon obtained by Brown and Rustgi (1972) are compared. In both cases the spectra have been decomposed into two components separated by 0·6 eV, corresponding to a spin–orbit split 2p-core level which is the initial state for the transition. For crystalline silicon (Fig. 7.77(a)) three features are observed which may correspond to those in the

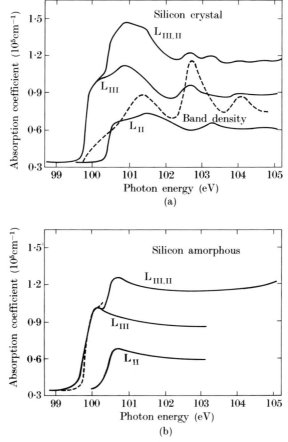

Fig. 7.77. High-resolution X-ray absorption spectra of (a) crystalline and (b) amorphous (evaporated) silicon. The spectra are resolved into L_{III} and L_{II} components. In (a) the broken curve is the theoretical conduction-band state density according to Kane. In (b) the broken curve reproduces part of the crystalline spectrum of (a). (From Brown and Rustgi 1972.)

theoretical density of states of the conduction band (shown dotted and positioned to line up with the L_{III} threshold). If this is what is being observed then there appears to be considerable enhancement near the threshold owing to exciton effects (Altarelli and Dexter 1972). The results have been discussed further by Brown, Bachrach and Skibowski (1977). There is the possibility that some of the structure may arise from EXAFS effects (§ 6.3). In any event the spectrum for the amorphous film (Fig. 7.77(b)) shows none of this structure and indeed looks similar to that presented in Fig. 7.68(a). An attempt to calculate the conduction-band density of states in amorphous germanium using plane waves orthogonalized to valence states has been made by Bullett (1974).

8

ARSENIC AND OTHER THREE-FOLD
CO-ORDINATED MATERIALS

8.1. Introduction
8.2. Forms and preparation of arsenic
8.3. Structure of amorphous arsenic
8.4. Electrical properties of amorphous arsenic
8.5. Optical properties of amorphous arsenic and the density of states in the bands
8.6. States in the gap of amorphous arsenic
8.7. Amorphous antimony, phosphorus, and related materials

8.1. Introduction

There are several elements of group V having three-fold co-ordination in the amorphous state, arsenic and antimony being the two on which most information exists. Amorphous phosphorus is also three-fold co-ordinated, whereas amorphous bismuth is not and probably has a close-packed structure. Although crystalline GeTe and GeS(GeSe) have structures like rhombohedral and orthorhombic arsenic respectively, the atoms in the amorphous forms of these materials are not three-fold co-ordinated but rather each constituent adopts an environment appropriate to its valency, i.e. Ge is four-fold co-ordinated and the chalcogen two-fold co-ordinated; these alloys are therefore included in Chapter 9.

In many ways the three-fold co-ordinated elements have properties that are intermediate between the four-fold co-ordinated materials (Ge and Si, Chapter 7) and those that are two-fold co-ordinated (Se and Te, Chapter 10). They are also traditional 'glass formers' and are common constituents in binary, ternary, and multicomponent chalcogenides (Chapter 9). Most of this chapter is devoted to arsenic; antimony and phosphorus are discussed in a final section.

8.2 Forms and preparation of arsenic

Table 8.1 gives some details of the various allotropes of arsenic that have been identified. The symbols are those chosen by Stöhr (1939) and Krebs, Holtz, and Worms (1957). Rhombohedral As, the familiar semimetal, can be grown as single crystals (see Taylor, Bennett, and Heyding 1965); orthorhombic As can be prepared from glassy arsenic by heating with

TABLE 8.1

Forms of arsenic †

Allotrope	Structure	Density $(g \, cm^{-3})$	Electrical nature	Comments
α-arsenic	Rhombohedral (A7)	5·72	Metallic	Familiar semimetal
ε-arsenic	Orthorhombic	5·54	Semiconducting gap ~0·3 eV	Mineral form— arsenolamprite
Yellow arsenic	Uncertain	2·07?	Semiconducting gap ⩾2·5 eV	Unstable
β				
γ arsenic	Amorphous	4·73	Semiconducting gap ~1 eV	Different forms
δ		−5·18		described but not properly classified

† Thin-film variants of amorphous form not included.

mercury (see Smith, Leadbetter, and Apling 1975). The preparation of yellow As is described, for example, by Mellor (1929). Bulk glassy As cannot be prepared in the normal way by quenching from the melt because it sublimes at atmospheric pressure. Stöhr (1939) produced the various forms of glassy As, indicated in the table, by heating rhombohedral As between 650 and 700 K in a sealed evacuated ampoule placed in a temperature gradient. Bulk glass has also been prepared by the pyrolisis of arsine AsH_3 (Vallance 1938) and is available commercially.†

In the following sections measurements made on rhombohedral and orthorhombic As will be compared with those made on both bulk glass and thin films of amorphous As. Although thin films can be prepared by electrodeposition (see Breitling and Richter 1969), evaporation (Moss 1952), reactive d.c. sputtering (Holland 1963), or glow-discharge decomposition of arsine (Knights and Mahan 1977), the most comprehensive set of data exists for films prepared by r.f. sputtering (Greaves, Knights, and Davis 1974, Greaves 1975, Greaves, Davis, and Bordas 1976a).

Pin-hole-free films of amorphous As with good surfaces and stability, having thicknesses in the range 200 Å–50 μm can be prepared by r.f. sputtering from polycrystalline or glassy targets in various gaseous ambients. Although hydrogen allows fast sputtering rates, approximately 100 times the physical rate for this gas (Moore 1960) suggesting reactive sputtering, films so produced show strong infrared absorption arising from As—H bond vibrations. Non-reactive sputtering in argon allows rates ~1–2 Å s^{-1} and leads to what are probably the purest films. Unlike deposition by evaporation, no cooling of the substrate is necessary.

† From MCP Electronics Ltd, Alperton, Wembley, Middlesex HA0 4PE. Throughout this Chapter we refer to this material as a glass even though it is not prepared by supercooling the liquid but by vapour transport in hydrogen.

8.3. Structure of amorphous arsenic

The structures of the double layers present in the two stable crystalline allotropes of As are illustrated in Fig. 8.1(a,b). In both structures the atoms are incorporated into two-dimensional networks of puckered six-fold rings, the bonding between the three-fold co-ordinated atoms being mainly covalent. Adjacent layers are situated at distances larger than the covalent intralayer bond length but shorter than the van der Waals distance. In rhombohedral As the bond length r is 2·51 Å, the bond angle θ is 97·2°, and all pairs of atoms have their bonds in the staggered configuration. In

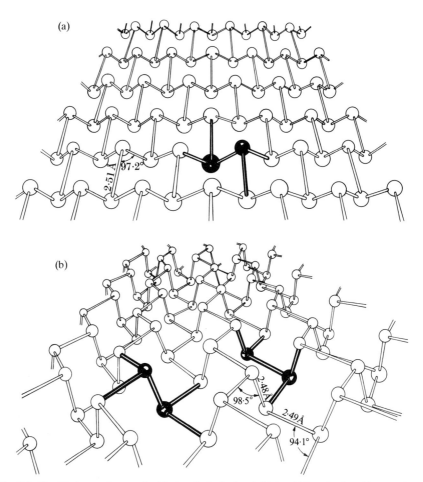

Fig. 8.1. Double layer of atoms in (a) rhombohedral and (b) (p. 411) orthorhombic arsenic. Bond lengths and angles taken from Wyckoff (1963) and Smith *et al.* (1975). (From Greaves 1975.)

orthorhombic As (which is isomorphous with black P) $r = 2\cdot48(9)$ Å, $\theta =$ 94·1° or 98·5°, some pairs of atoms having their bonds in the staggered configuration and some being semi-staggered. The layers in orthorhombic As are more puckered and the interlayer separation is greater. The unit cells of the two allotropes are shown in Fig. 8.2.

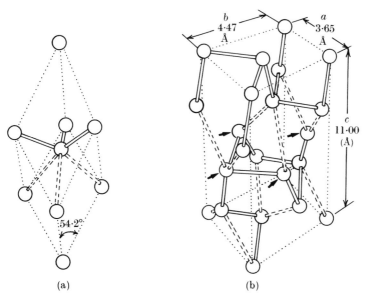

(a)   (b)

Fig. 8.2. Unit cells of (a) rhombohedral and (b) orthorhombic arsenic. The interlayer bonding is shown by broken lines. (From Greaves 1975.)

Attempts to model the structure of amorphous As have emphasized the covalent bonding at the expense of the interlayer bonding. That this is necessary can be seen from the RDFs and structure factors $S(k) = I(k)/f^2(k)$ for the amorphous and rhombohedral forms shown in Figs. 8.3 and 8.4. In the RDF for rhombohedral As, peaks occur at 2·5 Å (three one-bond intralayer neighbours), 3·1 Å (three one-bond interlayer neighbours), 3·7 Å (six two-bond intralayer neighbours), 4·0 Å (two-bond interlayer neighbours), 4·5 Å (three-bond intralayer neighbours), etc. In amorphous As the first intralayer peak is retained but the first interlayer peak is missing or at least shifted. The second peak in the RDF of amorphous As embraces peaks observed for rhombohedral As due to two-bond intralayer and interlayer neighbours, but the next two intralayer peaks are missing. Unwanted interlayer correlations have been minimized in various microcrystalline models of amorphous As by moving the layers apart and

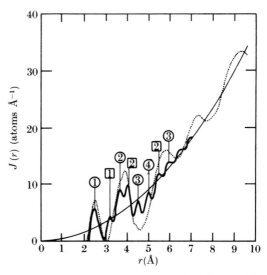

Fig. 8.3. RDF of arsenic. The dotted curve was obtained for precipitated (amorphous) material and the solid curve for the same material after annealing at 250°C which causes partial crystallization to the rhombohedral phase. The peaks marked ○ and □ are associated with intralayer and interlayer coordinations respectively, and the numbers in the symbols indicate whether they are 1, 2, 3, etc. bond neighbours. The average density parabola $4\pi r^2 \rho_0$ is shown by a continuous curve. (From Breitling 1972.)

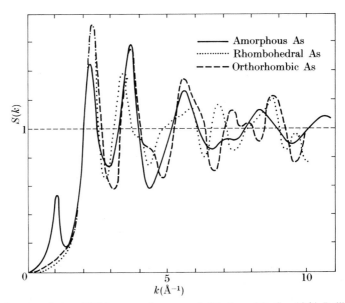

Fig. 8.4. Structure factors $S(k)$ for amorphous arsenic (Krebs and Steffen 1964, Bellisent and Tourand 1976) compared with those expected from 12 Å diameter microcrystallites of rhombohedral (dotted) and orthorhombic (broken) arsenic. (From Smith *et al.* 1975).

rotating them relative to each other (Richter and Gommel 1957, Krebs and Schultze-Gebhardt 1955), by mixing layers of rhombohedral and ortho-rhombic As (Krebs and Schultze-Gebhardt 1955, Krebs and Steffen 1964) or by displacing atoms within a given layer (Smith *et al.* 1975). In general, however, microcrystalline models have not produced as good a fit to the experimental $S(k)$ and RDF as the continuous random network model to be described below.

Another reason for minimizing interlayer correlations in models of amorphous As is the dramatic change in electrical properties. The semi-metallic nature of rhombohedral As is attributed (Krebs 1968, 1969) to the overlap between the back lobes of orbitals in adjacent layers and the semiconducting nature of amorphous As to the loss of this overlap. The importance of this interlayer bonding in determining electrical properties is evident in orthorhombic As in which the interlayer spacing is increased over that in rhombohedral As, producing a small-band-gap semiconductor.

In Fig. 8.5 two RDFs obtained by Krebs and Steffen (1964) for what these authors called β- and γ-arsenic (both amorphous) are given. From the shape and position of the first peak in the RDFs it is deduced that each As atom is surrounded by three nearest neighbours at $\sim 2 \cdot 5$ Å and that most, and probably all, of the bond-length disorder is thermal in origin. The position of the second peak at $3 \cdot 75$ Å gives an average bond angle close to $97°$. The area under this peak, however, amounts to nine to ten atoms, implying contributions from neighbours other than the six two-bond intralayer neighbours in the crystal. Therefore, unlike the situation in

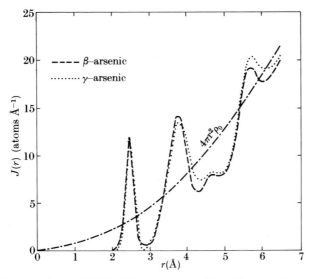

Fig. 8.5. Experimental RDFs of β and γ arsenic. (From Krebs and Steffen 1964.)

amorphous Ge, for which only next-nearest neighbours contribute to the second peak in the RDF, it is difficult to estimate the extent of bond-angle distortion. However, the minimum between the first and second peaks indicates that atoms separated by more than two bonds do not come closer than ~2·9 Å. Furthermore, as there are no features at 3·2 and 4·5 Å, which would indicate preference for staggered or eclipsed bonding configurations, interpretation of the RDFs in terms of microcrystalline materials seems unlikely. A simulation of the structure of amorphous arsenic by a continuous random network (CRN) with a broad distribution of dihedral angles seems to be the natural choice.

A three-fold co-ordinated CRN containing 533 atoms was built by Greaves (see Greaves 1975, Greaves and Davis 1974*a,b*). The construction procedure was similar to that employed for CRNs of tetrahedrally co-ordinated structures (§ 7.2) with the added constraint of keeping apart atoms not directly bonded. This was necessary because of the 'blind' side of the asymmetrical three-bonded units and was achieved with the aid of a spacer of (scaled) length 2·9 Å corresponding to the minimum in the experimental RDFs (Fig. 8.5). The resulting model is rather open and pervaded by caverns, a density deficit (relative to that of the crystalline structures) occurring in spite of complete connectivity. Another feature worth emphasizing is that, although only those bonds contained within layers in the rhombohedral or orthorhombic structures were used, the model is *not* layer like in appearance: the layers bifurcate and reconnect in a manner such that they can be recognized only over small regions.

The Greaves–Davis model contains rings of various orders as shown in Fig. 8.6. The bond and dihedral angle distributions† of the model, after computer refinement of the co-ordinates to bring all bond lengths to within 0·5% of each other (Elliott and Davis 1976) are shown by the dotted lines in Fig. 8.7(a,b). The average bond angle of the model turned out to be 102° with a standard deviation of 9°, but it was found easy to adjust this by a computer relaxation procedure to almost any value selected. The solid lines in Figs. 8.7(a,b) show the bond- and dihedral-angle distributions after adjustments of the co-ordinates to minimize the local strain energy by a method similar to that used by Steinhardt *et al.* (1974) for tetrahedrally co-ordinated structures (Elliott and Davis 1976, Davis *et al.* 1977). The bond-angle distribution for the relaxed model peaks at $98° \pm 6·9°$ and the dihedral-angle distribution shows a preference for staggered over eclipsed bonding configurations by a factor of about 2. The lower histogram in Fig. 8.7(b) is a reconstruction of the dihedral-angle distribution of the model derived solely from the ring statistics and considerations of what dihedral

---

† The dihedral angle ϕ is the average rotation necessary to bring next-nearest-neighbour bonds into the eclipsed configuration.

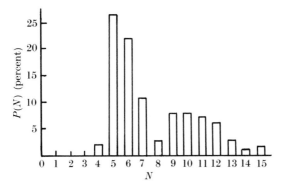

Fig. 8.6. Ring statistics of the Greaves–Davis model. $P(N)$ is the fraction of N-fold rings. (From Greaves and Davis 1974a.)

angles are allowed (or most likely) in rings of various order, subject to bond-angle constraints (Davis *et al.* 1977).

Fig. 8.8(a) shows, for the unrelaxed model, the measured RDF in histogram form and after Gaussian smoothing of $G(r)(=[J(r)/r]-4\pi\rho_0 r)$ to bring the height of the first peak to the value obtained experimentally. Fig.

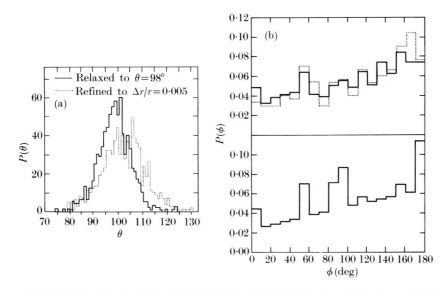

Fig. 8.7. (a) Bond angle $P(\theta)$ and (b) dihedral angle $P(\phi)$ distributions in the Greaves–Davis model. The dotted lines refer to the original co-ordinates after refinement (i.e. equalization of bond lengths) and the solid lines to those after relaxation (i.e. local energy minimization). The lower histogram in (b) is a theoretical reconstruction as described in the text. ((a) From Elliott and Davis 1976; (b) from Davis *et al.* 1977.)

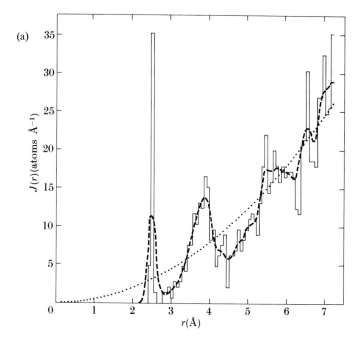

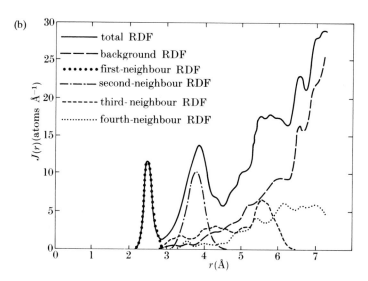

Fig. 8.8. (a) RDF of the unrelaxed Greaves–Davis model scaled to the bond length of arsenic. The broken curve is obtained after Gaussian smoothing the histogram and the dotted curve is the average density parabola. (b) Partial RDFs of the model i.e. $J_n(r)$ with n as indicated. (From Greaves and Davis 1974a.)

8.8(b) shows this smoothed RDF decomposed into n-neighbour contributions. Unlike four-fold co-ordinated CRNs, the second peak contains considerable contributions from neighbours that are connected by more than two bonds.

Greaves (1976) has shown how it is possible to obtain a fair synthesis of the total RDF shown in Fig. 8.8 by calculating $J_n(r)(n = 1, 2,$ and $3)$ purely from a knowledge of the ring statistics and the bond- and dihedral-angle distributions. This step towards interpreting RDFs in terms of the above parameters (also investigated for four-fold structures by Temkin, Paul, and Connell (1973) and Temkin (1974, 1978)) is a significant one towards the desirable reverse procedure of deducing parameters defining the local topology of non-crystalline structures directly from their RDFs.

The RDF of the relaxed Greaves–Davis model is compared with the two experimental RDFs (shown in Fig. 8.5) in Fig. 8.9. The overall fit is good, the average density of the model (which determines the parabola about which the RDF lies) being $4 \cdot 8 \text{ g cm}^{-3}$ when scaled to the bond length in arsenic, compared with $4 \cdot 74$ and $4 \cdot 97 \text{ g cm}^{-3}$ measured for the two amorphous forms. By further altering the co-ordinates of the model to obtain average bond angles of 102° and 96°, appropriate to amorphous P and Sb respectively, the RDFs of these materials have also been simulated (Fig.

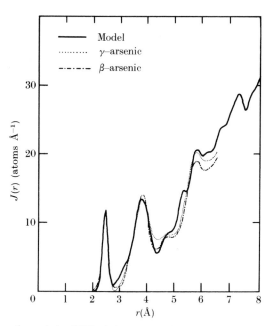

Fig. 8.9. Comparison of the RDF of the relaxed Greaves–Davis model with experimental curves obtained by Krebs and Steffen (1964). (From Davis *et al.* 1977.)

8.10(a,b)). The fits in these cases, however, are not so good as for arsenic, although approximately the correct densities were obtained: $2 \cdot 2 \, \mathrm{g \, cm^{-3}}$ for P (compared with experimental values of $2 \cdot 34$ and $2 \cdot 7 \, \mathrm{g \, cm^{-3}}$ for the red and black amorphous forms respectively) and $5 \cdot 07 \, \mathrm{g \, cm^{-3}}$ for Sb (compared with the experimental value of $5 \cdot 01 \, \mathrm{g \, cm^{-3}}$). A more recent experimental determination of the RDF of amorphous P is described in Krebs and Gruber (1967).

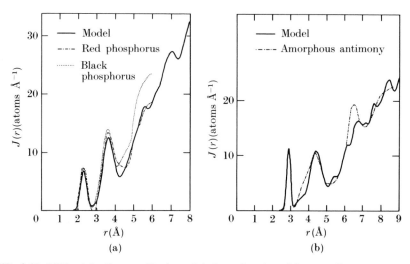

Fig. 8.10. RDFs of the Greaves–Davis model after relaxation of the co-ordinates to average bond angles of 102° and 96° compared with experimental RDFs of amorphous phosphorus (Hultgren *et al.* 1935) and antimony (Krebs and Steffen 1964). (From Davis *et al.* 1977.)

Comparison of $S(k)$ associated with the Greaves–Davis model and experiment (see Bellisent and Tourand 1976) has been made by Greaves and Davis (1974*a,b*) and Davis *et al.* (1977). The peak near $1 \, \text{Å}^{-1}$ shown in Fig. 8.4, present in the amorphous but not the crystalline forms of arsenic, is reproduced by the model. However, termination errors in the Fourier transform can lead to spurious ripples in this range of k and the model is hardly large enough to establish with certainty whether or not the agreement is fortuitous. A similar peak is also observed in amorphous As_2Se_3 and As_2S_3. According to Smith *et al.* (1975) the peak corresponds to a real-space separation of about 6 Å and its area indicates a correlation length of about 20 Å. These authors favour a quasi-crystalline model for arsenic, containing layers similar to those occurring in the orthorhombic structure but disordered, separated by the required 6 Å, and extending over linear dimensions of ~20 Å. For the corresponding peaks in As_2Se_3 and As_2S_3, Bishop and Shevchik (1974*a*) opt for a similar model, while de

Neufville, Moss, and Ovshinsky (1974) and Moss (1974) suggest that it could arise from the presence of As_4Se_6 or As_4S_6 molecules in the respective materials.

8.4. Electrical properties of amorphous arsenic

The temperature dependence of the d.c. electrical conductivity of a sputtered film, a bulk glass, and two crystalline forms of arsenic are shown in Fig. 8.11. The horizontal broken line is the minimum metallic conductivity (Chapter 2) for three-fold co-ordination. From these curves, rhombohedral arsenic is seen to be metallic, orthorhombic arsenic a small-band-gap (~ 0.3 eV) semiconductor exhibiting impurity band conduction at low

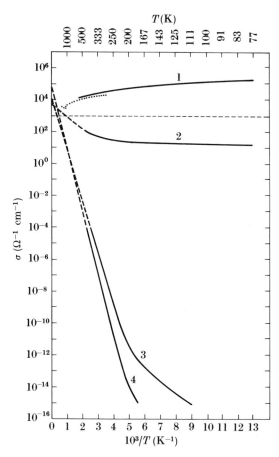

Fig. 8.11. Temperature variation of conductivity σ for (1) rhombohedral, (2) orthorhombic, (3) sputtered, and (4) glassy arsenic. The horizontal broken line refers to the minimum metallic conductivity ($\sim 980\ \Omega^{-1}\ cm^{-1}$) for three-fold co-ordination. (From Greaves and Davis 1974b.)

temperatures, and amorphous arsenic, whether in the form of a sputtered
film or a bulk glass, a semiconductor with a gap ~1·2 eV. The slightly
higher value of the activation energy for bulk glass (~0·74 eV) compared
with the sputtered films is in accord with the slightly higher optical gap of
the former (§ 8.5).

Fig. 8.12 gives some details of the effect that different rates of deposition
have on the conductivity of r.f. sputtered films deposited at 290 K and

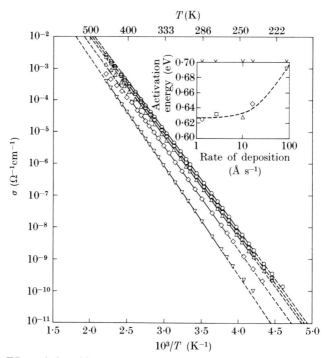

Fig. 8.12. Effect of deposition rate on the temperature dependence of conductivity of
sputtered arsenic deposited at 290 K and annealed for 1 h at 450 K. (From Greaves
et al. 1974.)

annealed for 1 h at 450 K. An increase in the activation energy for
conduction from ~0·63 eV to ~0·7 eV as the sputtering rate is increased is
observed as shown in the inset. At the same time the density was found to
decrease by as much as 10 per cent for a seven-fold increase in deposition
rate. The effect of annealing a film sputtered at a rate ~17 Å s^{-1} is shown
in Fig. 8.13. The electrodes here were in sandwich geometry and the results
show more clearly the departure from linearity below ~220 K evident also
in Fig. 8.11. Annealing at 332 and 353 K decreases the conductivity in this
region, for which log σ plots linearly against $T^{1/4}$, thereby suggesting
variable-range hopping at the Fermi level. (Using eqn (2.63) with a value

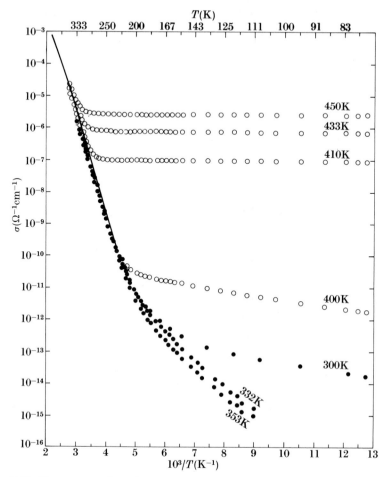

Fig. 8.13. Temperature dependence of conductivity of sputtered arsenic measured in sandwich geometry. On annealing above 353 K the conductivity increases, probably because of partial crystallization leading to short-circuit paths between the electrodes. This effect is not observed with gap geometry. (From Greaves *et al.* 1974.)

of $\alpha^{-1} = 10$ Å, the slopes yield a value for $N(E_F)$ of $\sim 4 \times 10^{17}$ cm^{-3} eV^{-1}.) Further annealing of this film results in the opposite behaviour, i.e. an increase in conductivity. This is attributed to partial crystallization of the films leading to short-circuit paths between the electrodes. The solid curve in Fig. 8.13 is for a film with electrodes in gap geometry; partial crystallization, although probably present, does not manifest itself in this configuration.

The intercept of the conductivity at $1/T = 0$ of all the sputtered films measured in this investigation is $\sigma_0 \sim 2 \times 10^4$ Ω^{-1} cm^{-1} compared with the value for the bulk glass of $9 \times 10^4 \, \Omega^{-1} \, \text{cm}^{-1}$. These values are indicative of transport in extended states. Writing $\sigma_0 = \sigma_{\min} \exp(\beta/2k)$, where β is the temperature coefficient of the optical gap ($\sim 6 \times 10^4 \, \text{eV deg}^{-1}$—see below), $\sigma_{\min}$ is calculated to be $\sim 7 \times 10^2 \, \Omega^{-1} \, \text{cm}^{-1}$ for the sputtered films, and $\sim 3 \times 10^3 \, \Omega^{-1} \, \text{cm}^{-1}$ for the glass; these values bracket the theoretical value of $9 \cdot 8 \times 10^2 \, \Omega^{-1} \, \text{cm}^{-1}$ shown in Fig. 8.11. The room-temperature value of $E_C - E_F$ (the thermopower is negative—see below) is $E(0) - 300\beta/2$, i.e. $0 \cdot 65 \, \text{eV}$ for bulk glass and $0 \cdot 54 - 0 \cdot 61 \, \text{eV}$ (depending on the deposition rate) for sputtered films. The Fermi energy therefore lies close to the centre of the optical gap, which (see § 8.5) is $\sim 1 \cdot 3 \, \text{eV}$ for the bulk glass and $\sim 1 \cdot 1 \, \text{eV}$ for sputtered films.

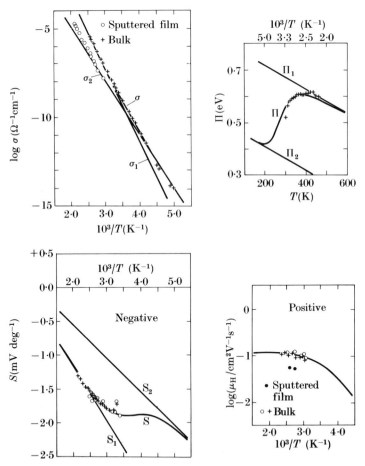

Fig. 8.14. Conductivity σ, thermopower S, Peltier coefficient Π, and Hall mobility μ_H as a function of temperature in amorphous arsenic. The solid lines are theoretical fits assuming two conduction paths, one at E_C and the other at E_A. (From Mytilineou and Davis 1977.)

Hauser, Di Salvo, and Hutton (1977) report that the d.c. conductivity of amorphous arsenic prepared by sputtering onto a substrate held at 77 K displays $T^{-1/4}$ behaviour. After annealing the films at room temperature, T^{-1} behaviour with an activation energy of 0·5 eV was observed.

Measurements of the temperature dependence of the d.c. conductivity σ, the thermopower S, and the Hall mobility μ_{H} in amorphous arsenic have been made by Mytilineou and Davis (1977) (see also Davis and Greaves 1976). The results shown in Fig. 8.14 have been interpreted in terms of an increasing contribution to conduction in band-edge-localized states as the temperature is lowered and analysed according to the procedure outlined in Chapter 6. In the transition region the Peltier coefficient $\Pi = ST$ exhibits a temperature dependence of the opposite sign to that occurring when conduction is predominantly in extended *or* localized states. The Hall effect is positive, i.e. opposite in sign to the thermopower, a sign anomaly of the opposite sense to that customarily observed in, for instance, the chalcogenides. The sign of the Hall effect in amorphous semiconductors has been discussed in §§ 2.14 and 6.4.7 and by Emin (1977*a,b*).

The frequency dependence of conductivity in amorphous arsenic has been measured by Elliott and Davis (1977). Fig. 8.15(a,b) show the

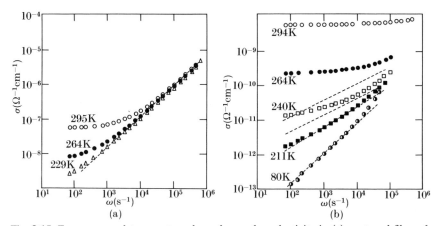

Fig. 8.15. Frequency and temperature dependence of conductivity in (a) sputtered films of arsenic and (b) bulk glassy arsenic. The broken lines are obtained after subtraction of the d.c. conductivity. (From Elliott and Davis 1977.)

frequency and temperature dependence in sputtered films and bulk glass. Although the variation of $\sigma(\omega)$ in the films is as ω^{s}, with $s \sim 0·9$ at a sufficiently low temperature, and interpretation in terms of hopping at E_{F} is possible, the variation in the glass is very different. Not only is the magnitude of $\sigma(\omega)$ several orders of magnitude lower, but the exponent s varies with temperature from a value of $\sim 0·5$ at 264 K to 0·9 at 80 K. A plausible explanation of these results is described by Elliott and Davis (1977).

Studies of the d.c. conductivity of bulk glassy arsenic by Elliott, Davis, and Pitt (1977) and by Elliott and Davis (1977) as a function of applied pressure have revealed dramatic effects. The room-temperature conductivity (Fig. 8.16) increases by about seven orders of magnitude as the

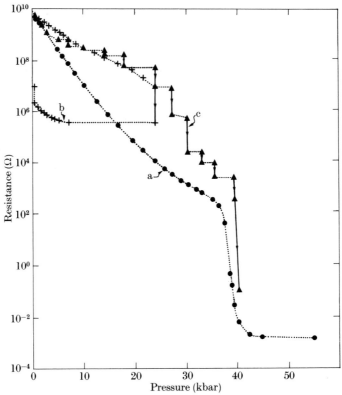

Fig. 8.16. Pressure dependence of the room temperature conductivity of glassy arsenic. Curve a is for a loading rate of $1 \cdot 35$ ton min^{-1}; curve c refers to a loading rate of $0 \cdot 45$ ton min^{-1} which is interrupted at various pressures, and curve b refers to a sample returned to atmospheric pressure from 24 k bar. (From Elliott *et al.* 1977.)

pressure is increased to ~40 kbar, at which pressure the sample irreversibly crystallizes to the rhombohedral phase, a transition accompanied by a further five orders of magnitude increase in conductivity. This sequence happens whatever the loading rate and whether or not the loading proceeds in steps (see Fig. 8.16). A sample released from pressure before the sharp transition (e.g. curve b in Fig. 8.16) recovers somewhat, but its resistance is finally lower than that of the starting material. The temperature variation of the resistance of such a sample is shown in Fig. 8.17. A linear plot on a $T^{1/4}$ scale is indicative of variable-range hopping conduction and the slope of Fig. 8.17(b) yields, under the assumption that $\alpha^{-1} = 10$ Å, a density of pressure-induced defect levels at the Fermi level of $\sim 6 \times 10^{17} \, cm^{-3} \, eV^{-1}$. A model for the states in the gap of amorphous arsenic will be described in § 8.6.

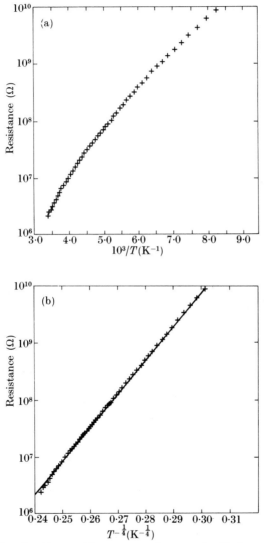

Fig. 8.17. Variation of resistance with temperature for a sample of bulk arsenic released from a pressure of 24 k bar *versus* (a) T^{-1} and (b) $T^{-1/4}$. (From Elliott *et al.* 1977.)

Photoconductivity in two sputtered films after annealing has been reported by Greaves *et al.* (1974). Their results as a function of temperature, shown in Fig. 8.18, are similar in shape to those observed in many

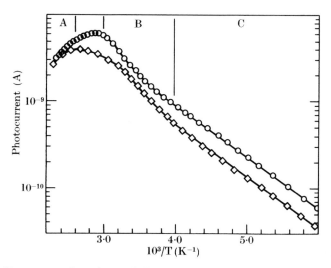

Fig. 8.18. Temperature dependence of photocurrent in two annealed films of arsenic. The circles and diamonds refer respectively to films 1 and 4 of Fig. 8.12. (From Greaves *et al.* 1974.)

chalcogenide glasses and in glow-discharge-deposited Si (see Chapters 6, 7, and 9). In region A the photocurrent is less than the dark current and was found to vary linearly with light intensity. Beyond the maxima, the photocurrent is larger than that in the dark; it was found to vary with light intensity sublinearly (with powers ranging from 0·7 to 0·9) in region B and approximately linearly in region C. The activation energy in region C is ~0·12 eV, which, by analogy with results on glow-discharge-deposited Si, may represent the range of localized states at the conduction-band edge. Its magnitude is rather less than that used in fitting the results of Fig. 8.14. The spectral dependence of photoconductivity in these films shows a fall at the absorption edge but considerable photoconductivity occurs down to ~0·65 eV (Greaves *et al.* 1974). Measurements on bulk glass will be described in the next section.

8.5. Optical properties of amorphous arsenic and the density of states in the bands

The optical constants at room temperature of sputtered films of As within the vicinity of the fundamental optical absorption edge are given in Fig. 8.19. The edge (Fig. 8.19(a)) is exponential with a slope of 12 eV^{-1}. In

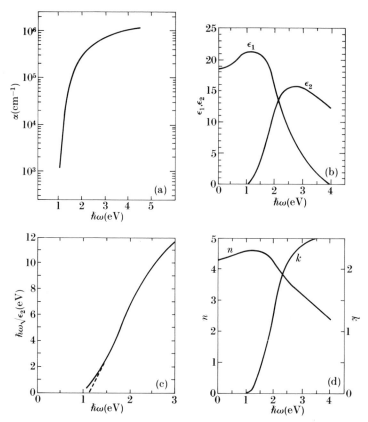

Fig. 8.19. Optical constants *versus* photon energy derived from transmission and reflectivity data on sputtered films of arsenic. (From Greaves, private communication.)

Chapter 6 the slope of this Urbach tail was compared with that in amorphous Ge, Se, and B and a correlation with the valency of these elements noted. Results on bulk glassy As will be presented below when it will be seen that the exponential behaviour continues down to an absorption coefficient $\sim 10^2$ cm^{-1}. Above the exponential tail the results can be plotted as $\hbar\omega\sqrt{\varepsilon_2}$ *versus* $\hbar\omega$ to determine an optical gap. Extrapolation of this line (Fig. 8.19(c)) yields a gap of 1·1 eV which is approximately twice the activation energy for d.c. conductivity, as mentioned above.

The absorption edges in amorphous arsenic prepared in various ways are compared in Fig. 8.20(a,b). The insensitivity of the shape and position of the edge to the nature and conditions of preparation is in striking contrast to the variation found for amorphous germanium and silicon (Chapter 7). The edges for the bulk glass, the film prepared by glow-discharge dissociation of arsine, and the film prepared by sputtering in an argon/hydrogen atmosphere all line up perfectly. All these films could, however, contain some hydrogen and the somewhat lower absorption threshold for films sputtered in pure argon (curve A of Fig. 8.20(a)) might be considered to support this possibility. The displacement of the edge ($\sim$0·2 eV) is almost exactly as inferred from the conductivity data on pure films and bulk glass described in the last section.

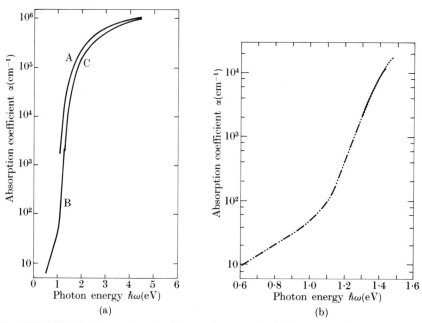

Fig. 8.20. (a) Optical absorption edges in amorphous arsenic: A, films prepared by sputtering in argon; B, bulk glass (Knights 1975); C, film sputtered in 30/70 mixture of H_2 and Ar. (From Greaves *et al.* 1976a.) (b) Optical absorption edges in amorphous arsenic: chain curve, bulk glass (as B above); solid curve, prepared by glow-discharge decomposition of arsine onto substrate at 227°C; dotted curve, film prepared by glow-discharge decomposition of arsine onto substrate at 27°C. (From Knights and Mahan 1977.)

Below about $\alpha = 10^2$ cm^{-1}, the exponential absorption edge undergoes a change of slope, not unlike that observed in some chalcogenides (Chapters 6, 9) and in glow-discharge-deposited Si (Chapter 7). This region of the edge can be observed only for bulk glass, and some variation from sample to sample is evident. It seems quite likely that this tail is associated with defect absorption, a suggestion supported by the photoconductivity data of Knights and Mahan (1977) shown in Fig. 8.21; the spectra associated with bulk and glow-discharge-deposited arsenic, although differently normalized, are quite distinct. The inflection in both curves, however, occurs at a photon energy $\sim$1·1 eV, i.e. close to the onset of the tail absorption seen in Fig. 8.20.

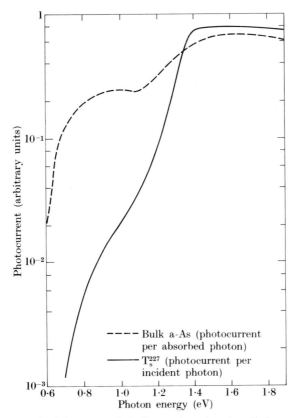

Fig. 8.21. Photoconductivity spectra in bulk arsenic and a glow-discharge-deposited film deposited at 227 K. (From Knights and Mahan 1977.)

A further comparison of the optical properties of amorphous arsenic prepared in various ways is shown in Fig. 8.22 where the reflectivity of sputtered, evaporated, and bulk glassy arsenic from 0.03 to $30\,eV$ is plotted. These data were obtained using a variety of spectrometers and sources, including a synchrotron radiation facility (Greaves *et al.* 1976*a*). The most noticeable difference between the reflectivity spectra of sputtered or evaporated films and bulk glass occurs at low photon energies; below $\sim 3\,eV$, the reflectivity of the films falls and flattens out at ~ 0.4 whereas that of bulk glass falls steeply to a minimum value of 0.29. The reflectivity spectrum of evaporated arsenic (obtained by Raisin *et al.* 1974*a*) is similar to those of the other two forms, except that the peak near $14\,eV$ appears as an extended shoulder. Measurements by Hudgens *et al.* (1976) of the reflectivity of bulk glassy arsenic are similar to those presented here up to about $3\,eV$, beyond which these authors find a sharp drop to

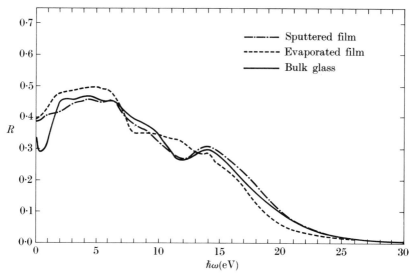

Fig. 8.22. Reflectivity spectra of three forms of amorphous arsenic as indicated. (From Greaves *et al.* 1976*a*.)

$R \sim 0.3$ at 4·5 eV. The reason for this difference is not clear. Over a similar energy range to that of Fig. 8.22, the reflectivities of rhombohedral arsenic (Greaves *et al.* 1976*a*, Raisin, Leveque, and Robin 1974*b*, Cardona and Greenaway 1964) and orthorhombic arsenic (Greaves *et al.* 1976*a*) show considerably more structure; in particular free-carrier effects produce a dramatic rise in R at low energies.

Fig. 8.23 shows a portion of the ε_2 spectrum (derived by a Kramers–Kronig inversion of the reflectivity data of Fig. 8.22) of sputtered arsenic compared with that obtained by Raisin *et al.* (1974*b*) for rhombohedral arsenic. For the metallic crystal there is the expected rise of ε_2 at low energies and considerable fine structure; this is lost in the spectrum for the amorphous form and furthermore there is a small displacement between the centroid of the fine structure and the single peak observed in the film. The peak occurs at a slightly higher energy in amorphous arsenic, suggesting a somewhat larger average gap between bonding and antibonding states.

Further analysis of the optical data, using a procedure similar to that outlined for amorphous germanium in § 7.6, has been made by Greaves *et al.* (1976*a*). The important parameters evaluated are displayed in Table 8.2. We simply note here the similarity of the plasma frequency ω_p deduced for the various forms of arsenic; an essentially similar value, namely 17·8 eV, has also been obtained by direct energy loss measurements on amorphous arsenic (Abreu, private communication). Writing

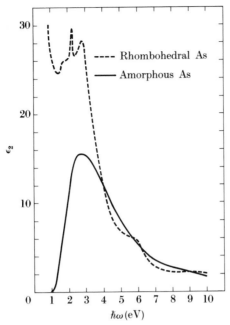

Fig. 8.23. Comparison of ε_2 spectrum of amorphous (sputtered) arsenic with that of rhombo-hedral arsenic. (From Greaves *et al.* 1976*a*.)

$\omega_p^2 = 4\pi n_v e^2/m$ where n_v is the number of charges per unit volume, $\hbar\omega_p = 18$ eV gives a value of approximately 5 for the number of free electrons per atom contributing to the plasma; thus the two s and the three p electrons contribute in all the forms of arsenic considered. There is good agreement between the values of $\hbar\omega_0$, where ω_0 is the natural frequency of a Lorentzian oscillator taken to describe the ε_1 data, and the value of the Penn gap

TABLE 8.2

Parameters extracted from optical data on crystalline and amorphous arsenic†

Method of evaluation	Quantity	Crystalline (rhombohedral)	Amorphous (sputtered film)	Amorphous (bulk glass)
Lorentzian fit to ε_2	$\hbar\omega_p$	—	16·0 eV	18·0 eV
Maximum of Im($-1/\varepsilon$)	$\hbar\omega_p$	17·8 eV	18·0 eV	17·8 eV
Lorentzian fit to ε_1	$\hbar\omega_0$	2·7 eV	3·6 eV	3·9 eV
Use of Penn model (see § 7.6)	$\hbar\omega_g$	2·8 eV	3·5 eV	4·0 eV
	$\varepsilon_1(0)$	26·2	18·4	14·2

† From Greaves *et al.* 1976*a*.

(obtained by using eqn (7.8) in conjunction with the measured low-frequency refractive index). The higher value of $\hbar\omega_g$ in the bulk glass compared with the sputtered film is attributed to more perfect connectivity in the former. The same suggestion was made by Connell *et al.* (1974) to account for a higher value of $\hbar\omega_g$ in sputtered films of germanium when deposited on to a higher-temperature substrate. However, in contrast to the situation in germanium, the value of $\hbar\omega_g$ for crystalline arsenic is considerably *lower* than for either amorphous form. We associate this with *stronger* covalent bonding in amorphous, relative to crystalline, arsenic; as the interlayer bonding is diminished, the intralayer bonding becomes stronger. The bond length in fact decreases slightly from 2·51 Å in rhombohedral arsenic (Wyckoff 1963) to 2·49 Å in amorphous arsenic (Krebs and Steffen 1964), corroborating this suggestion.

The effective number of electrons n_{eff} contributing to optical absorption up to a given energy $\hbar\omega$ of sputtered and glassy arsenic is shown in Fig. 8.24(a). The density of valence-band states $N_v(E)$ for evaporated arsenic, as determined using X-ray photoelectron spectroscopy by Ley *et al.* (1973), is displayed immediately underneath. The valence electrons are seen to occupy two bands. Since at the minimum value of $N_v(E)$ near 8 eV n_{eff} is approximately 3, and at a plasma energy of 18 eV (see above) n_{eff} is approaching 5, it is possible to deduce that the valence electrons are

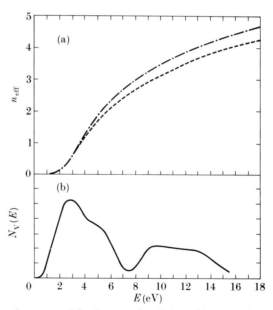

Fig. 8.24. (a) n_{eff} for sputtered (broken curve) and glassy (chain curve) arsenic as determined from optical data (Greaves *et al.* 1976*a*). (b) Density of valence band states, i.e. $N_V(E)$, for evaporated arsenic as deduced from XPS (Ley *et al.* 1973). (From Greaves *et al.* 1976*b*.)

distributed as for the free atom of arsenic, namely three in p-like states and two in deeper-lying s-like states (the s–p splitting in atomic arsenic is 9·8 eV). Hybridization of the bonds is therefore not, on the average, strong.

Kelly and Bullett (1976*b*), using the chemical pseudopotential and recursion method, have calculated the density of states expected for amorphous arsenic, basing their calculations on the co-ordinates of the Greaves–Davis model (§ 8.3) and using separate s and p orbitals. Their results are shown in Fig. 8.25(c) along with others for the full rhombo-hedral crystal (Fig. 8.25(a,b)) and for a single layer of the latter (Fig.

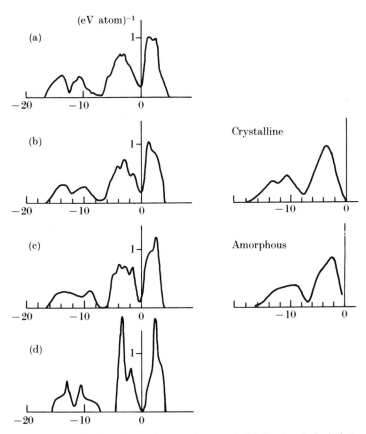

Fig. 8.25. Calculated density of states for arsenic atoms in (a) the rhombohedral structure, as determined by sampling the Brillouin zone, (b) the rhombohedral structure, as determined by the recursion method, (c) the Greaves–Davis CRN, as determined by the recursion method (average over 10 central sites), and (d) a single layer having the same geometry as the rhombohedral structure, as determined by sampling the two-dimensional Brillouin zone. The Fermi level is at $E = 0$. (From Kelly and Bullett 1976*b*.) On the right are the experimental densities of valence band states as determined by Ley *et al.* (1973).

8.25(d)). Experimentally determined distributions from Ley *et al.* (1973) are shown on the right of this figure. Comparing Fig. 8.25(b) and (c) one notices a partial filling in of the s-band dip attributable to the presence of odd-membered rings in the CRN (analogous to the situation in germanium, § 7.8) and also a deepening of the pseudogap at $E = 0$ (a true gap was not obtained in these calculations). Other features of these calculations are discussed by Kelly and Bullett (1976*b*), Greaves *et al.* (1976*b*), and Davis and Greaves (1976). Related calculations and discussions have been given by Robertson (1975), Joannopoulos and Pollard (1976), and Joannopoulos (1976). More recent photoemission data by Schevchik (1977) on sputtered films deposited onto substrates held at 20 and 220°C exhibit rather less change in the form of the p-band density of states in going from amorphous to crystalline arsenic than do the data of Ley *et al.* (1973) presented above.

In Chapter 9 synchrotron radiation absorption measurements by Bordas and West (1976) on amorphous arsenic are presented (Fig. 9.38). Discussion as to what extent these represent the conduction-band density of states is given there, as well as in Bordas and West (1976), Greaves *et al.* (1976*b*), and Liang and Beal (1976).

An 'isomorphism' between the electronic and vibrational density of states in three-fold co-ordinated systems has been emphasized by Joannopoulos and Pollard (1976) and Joannopoulos (1976). Using this feature these authors calculate the vibrational spectra appropriate to the Greaves–Davis CRN and compare it with the infrared spectrum shown in Fig. 6.59. The vibrational spectra in arsenic (as well as antimony) has been deduced from Raman spectra by Lannin (1976, 1977). It has also been measured by inelastic neutron scattering (Salgado, Gompf, and Reichardt 1974, Leadbetter, Smith, and Seyfert 1976). Related theoretical papers are by Chen, Vetelino, and Mitra (1976), Beeman and Alben (1977), Meek (1977*b*) and Davis, Wright, Doran, and Nex (1979).

The absorption in amorphous arsenic in the spectral range 10 to $200 \, \text{cm}^{-1}$ has been measured as a function of temperature by Al-Berkdar, Taylor, Holah, Crowder, and Pidgeon (1977). These authors compare the results with those obtained in the infrared for chalcogenides and deduce that the coupling of photons to vibrational modes is smaller.

8.6. States in the gap of amorphous arsenic

In this book we have proposed two basic models for states in the gap of amorphous semiconductors. Common to both is that they are associated with defects in a network which is otherwise fully connected and in which local valencies are satisfied. For germanium and silicon on the one hand, we have followed Spear and his co-workers in assuming that paired dangling bonds at divacancy-like defects are the principal contributor to the distribution of localized levels in the gap, although the possible importance of single isolated dangling bonds, particularly in the absence of hydrogen, and of larger defects associated with microvoids or incorporated impurities, cannot be dismissed. For Se, Te, and chalcogenides we have thought in terms of under– or over-co-ordinated atoms, such sites carrying a charge. The principal difference between the defects in these two classes of material appears to be associated with (i) a greater degree of flexibility in structure containing two-fold co-ordinated atoms and (ii) the availability of excess (lone-pair) electrons. Taken together, these features allow atomic rearrangements or local configurational changes to take place in the vicinity of defects, thereby favouring bonding situations in which electrons are paired. The effect on states in the gap is to reverse the normal ordering of energy levels associated with different charge states of defects. We have to ask whether one of these models, or a different one, is appropriate for amorphous arsenic.

In many ways the properties of arsenic show a closer resemblance to those of the chalcogenides than to those of germanium and silicon. The Fermi level is located (and probably pinned) near the centre of the gap ($\sim 1 \cdot 2$ eV) but variable-range ($T^{1/4}$) hopping conduction at E_F is not normally observed, except perhaps at low temperatures (see Fig. 8.13). Luminescence occurs with a peak intensity at an energy of about $0 \cdot 55$ eV (i.e. roughly half the gap) during excitation with near band-gap light, and, following photoexcitation with the same wavelength, an induced absorption band extending from about $0 \cdot 45$ eV up to the band edge is observed if the sample is maintained below 77 K. Accompanying the photoinduced absorption band is an e.s.r. signal ($\sim 10^{17}$ spins cm^{-3}) and both can be thermally or optically bleached as for the chalcogenides. The optical results (Bishop et al. 1976b, 1977) are shown in Fig. 8.26. The temperature dependence of the luminescence (Kirby and Davis 1979), although showing some differences from that found, for example, in As_2Se_3 and As_2S_3, is very similar to that observed in $As_2Se_{1.5}Te_{1.5}$—a chalcogenide with virtually the same band gap as arsenic.

The similarities between the behaviour of arsenic and the chalcogenides led Davis and Greaves (1976) to propose that a similar model for the gap states seemed appropriate. This has been questioned by Knights and Mahan (1977) on the basis of field-effect measurements on glow-discharge-deposited arsenic, which suggest a density of states at E_F of about 10^{17} cm^{-3} eV^{-1}, and of the similarity between the photocurrent spectra (Fig. 8.21) of this material and of glow-discharge-deposited silicon.

It is certainly true that there are some differences between the properties of amorphous arsenic and the chalcogenides, which suggest that, at least, some modification of the D^+D^- model is necessary. The observation of $T^{1/4}$ behaviour in the conductivity of samples that have been subjected to and released from high pressure (§ 8.4), in films sputtered onto low-

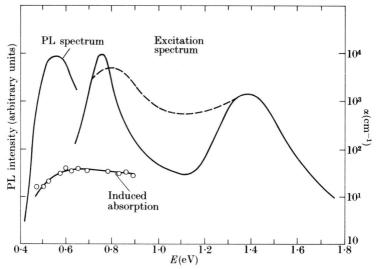

Fig. 8.26. Photoluminescence (PL) spectrum, excitation spectrum (under two different conditions of luminescence fatigue), and photo-induced absorption in glassy arsenic. (From Bishop *et al.* 1976*b*.)

temperature substrates (Hauser, Di Salvo, and Hutton 1977), and also in films that have been bombarded with high-energy ions (Troup and Apsley, private communication), argues against a large polaron energy being involved in hopping between at least some of the gap states. In the bulk glass an e.s.r. signal indicating $\sim 2 \times 10^{15}$ spins cm^{-3} is observed (Bishop *et al.* 1977). The photoexcitation spectrum for luminescence (Fig. 8.26) displays two peaks, in contrast to one for the chalcogenides. The absence of a linear term in the low-temperature specific heat of glassy arsenic (Phillips and Thomas 1977, Jones, Thomas, and Phillips 1978), although not necessarily relevant in a discussion of electronic states, also serves to distinguish arsenic from the chalcogenides (§ 6.8.1).

Possible defect states in arsenic, analogous to the D^+ and D^- states in chalcogenides, are four-fold and two-fold co-ordinated arsenic atoms. The electronic configuration at such sites has been described by Kastner and Fritzsche (1978) and Greaves, Elliott, and Davis (1979). In contrast to the chalcogenides, however, there are no lone-pair electrons; to form the D^+ centre an electron would have to be promoted from an s to a p orbital, so the considerable energy of promotion as well as the Hubbard U would have to be compensated by a configurational energy gain, in order for charged centres to form. However four-fold co-ordinated arsenic is well-known in doped crystalline semiconductors and seems likely to exist in amorphous III–V compounds containing arsenic. Furthermore there is evidence for its presence in doped amorphous silicon (Hayes, Knights, and Mikkelsen 1977) and perhaps in amorphous arsenic itself (Nemanich, Lucovsky, Pollard and Joannopoulos 1978).

The nature of the centres responsible for the dark and photo-induced e.s.r. signals is clearly of importance in any consideration of defect states in amorphous arsenic. Fig. 8.27 (curve (a)) shows the line spectra obtained on

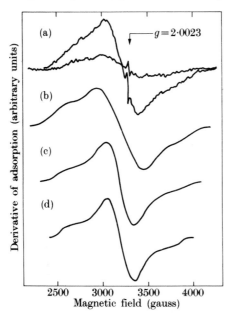

Fig. 8.27. (a) ESR signals in glassy arsenic before and after irradiation with band gap light; (b), (c), (d) theoretical spectra as described in the text (from Bishop *et al.* 1977.)

bulk glass at $4 \cdot 2$ K by Bishop *et al.* (1977) before and after illumination with $1 \cdot 15$ μm light. Although the spectrum after illumination is only approximately five times as intense as that obtained before, the limited penetration of the light (~ 100 μm) means that only about one-tenth of the sample volume was affected, and Bishop *et al.* (1977) estimate the density of optically induced spins ($\sim 10^{17}$ cm^{-3}) to be at least 50 times greater than the equilibrium density in the dark. Subtraction of the two spectra in (a) leads to a signal very similar in linewidth and shape to that observed after irradiation, suggesting that optical excitation simply increases the concentration of some localized paramagnetic centres present before irradiation. However, minor differences between the shapes of the spectra indicate a slight difference in local bonding configuration between sites that are optically sensitive and those that are not. The large breadth of the signals suggests that the lineshape is determined primarily by a large hyperfine interaction rather than by spin–orbit interaction. Furthermore, the sum of the spectra for arsenic and that observed for amorphous selenium reproduce that obtained for amorphous As_2Se_3 (Bishop *et al.* 1977). The average rate at which the optically induced centres in arsenic are created is the same as that of the arsenic component of the As_2Se_3 signal; it grows with a time constant ~11 secs, about 10 times faster than for the Se component. The optically induced absorption (Fig. 8.26) grows at the same rate as the e.s.r. signal in all cases.

Curves (b), (c), and (d) in Fig. 8.27 are theoretical lineshapes calculated for an electron localized on one, two, and three arsenic atoms respectively; for further delocalization the similarity with curve (a) was found to deteriorate rapidly. In calculating these lineshapes the unpaired spin was assumed to exist predominately in a *single* p orbital with only a small (~5 per cent) s admixture. The spin Hamiltonian parameters used are given by Bishop *et al.* (1977).

During the growth of the photo-induced e.s.r. and absorption in amorphous arsenic, the luminescence band (which is excited by the same wavelength light) decreases in intensity; subsequent irradiation with mid-gap light removes the metastable paramagnetic centres and restores the luminescence. This parallels the behaviour in the chalcogenides (Chapters 6 and 9).

We conclude that a model for the gap states in amorphous arsenic similar to that for chalcogenides is likely. Charged defects, created by an exothermic reaction from neutral centres, exist, but a small number of neutral centres remain. Whether these are at sites associated with a positive, rather than a negative, correlation energy, or whether they arise because D^+ states associated with defects in some regions of the material lie at a lower energy than D^- states associated with defects in another (i.e. overlap between the D^+ and D^- bands), is open to question. Certainly the energy separations between levels are scaled down from those in most chalcogenides. Not only is the gap of arsenic relatively small but also the D^- levels probably lie well above the top of the valence band and perhaps close to mid-gap (Kastner and Fritzsche 1978, Greaves *et al.* 1979).

The excitation spectrum of Fig. 8.26 deserves comment. This is the integrated photoluminescence intensity as a function of exciting wavelength. Its shape contrasts sharply with the single-peaked spectra observed for chalcogenides. Bishop *et al.* (1976*b*) propose that the twin peaks arise from strong fatiguing effects: the broken and solid portions of the spectrum in Fig. 8.26 represent conditions of progressively increasing luminescence fatigue. Kirby and Davis (1979) suggest, however, that this explanation is probably not correct. While fatiguing occurs for the higher energy peak there appears to be none associated with the lower energy peak; in fact excitation near 0·7 eV *enhances* the luminescence. In their model, the low-energy peak at ~0·8 eV is associated with direct excitation

from a D^- level lying just below mid-gap. The Stokes shift of the luminescence is then only ~0·2 eV. Illumination at a slightly lower energy
(~0·7 eV) excites the dark-equilibrium concentration of D^0 centres, producing charged centres and hence enhancing the luminescence.

8.7. Amorphous antimony, phosphorus, and related materials

The RDFs of amorphous antimony and phosphorus were shown in Fig.
8.10 and discussed briefly in § 8.3 with regard to the extent to which their
structures might be simulated by three-fold co-ordinated random
networks. Modification of the average bond angle in the Greaves–Davis
model to 96° and 102° (inferred from the positions of the second peaks in
the RDFs of amorphous antimony and phosphorus respectively) was
shown to provide fair, if less than satisfactory, reproductions of the RDFs.
One thing is clear, however—namely that interlayer correlations present

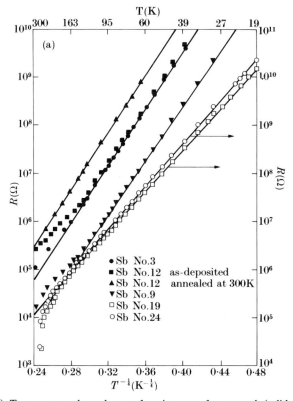

Fig. 8.28. (a) Temperature dependence of resistance of sputtered (solid symbols) and
evaporated (open symbols) antimony. Sample No. 9 was evaporated from pure Sb while No. 24
was evaporated from Pt–Sb. (From Hauser 1974a.) (b) Temperature dependence of resistance
of Sb–Ni alloys. (From Hauser 1975.)

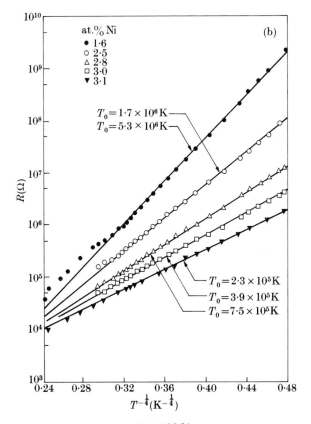

Fig. 8.28(b).

in the crystalline forms of these materials is minimized in the amorphous structures. This feature, as for amorphous arsenic, probably accounts for their semiconducting properties (Moss 1952) and low densities.

The properties of amorphous antimony have been investigated by Sommer (1966) and by Hauser (1974a,b). Hauser (1975) has studied Sb–As, Sb–Bi, Sb–Sn, Sb–Ge, Sb–Ni, and Sb–Ga alloys. Mg–Sb alloys have been investigated by Ferrier and Herrell (1970). Stoichiometric III–V compounds appear to have tetrahedral bonding (Shevchik and Paul 1973).

Fig. 8.28(a) shows the resistance of various sputtered and evaporated films of antimony plotted *versus* $T^{-1/4}$ (Hauser 1974a). The linearity of these plots suggests variable-range hopping at the Fermi energy; assuming $\alpha^{-1} = 8$ Å, the slopes yield $N(E_F) \sim 2 \times 10^{19}$ cm^{-3} eV^{-1} for the sputtered films and $\sim 10^{20}$ cm^{-3} eV^{-1} for the evaporated films. All samples were deposited at 77 K and, although the sputtered films remained amorphous up to room temperature, they exhibited irreversible annealing effects above 160 K. The evaporated films showed evidence for partial crystallization at temperatures as low as 180 K and they completely crystallized at $\sim$270–300 K.

Variable-range hopping conduction has also been observed in films of Sb alloyed with As, Bi, Sn, Ge, Ni, and Ga (Hauser 1975). In the first four of these alloys the slope of the $T^{1/4}$ plot alters very little with changing concentration, whereas addition of Ni leads to a rapidly decreasing slope (Fig. 8.28(b)). Sb_xGa_{1-x} films become metallic for $x \lesssim 0\cdot5$ and are superconductors for x in the range 0–0·42. Sputtered films of pure Ga can have a T_c as high as 8 K (cf. $T_c = 1\cdot07$ K for crystalline Ga). Hauser (1975) has produced superconducting Bi by getter sputtering at 77 K in an argon/hydrogen mixture. The films had a T_c of 6·65 K; crystalline Bi is not a superconductor.

The Raman spectra of amorphous antimony has been studied by Wihl, Stiles, and Tauc (1972) and by Lannin (1977).

9

CHALCOGENIDE AND OTHER GLASSES

9.1. Introduction
9.2. Structure
9.3. Electrical properties of chalcogenide glasses
      9.3.1. Introduction
      9.3.2. d.c. conductivity
      9.3.3. Thermopower
      9.3.4. Hall effect
9.4. States in the gap
      9.4.1. Introduction
      9.4.2. Screening length
      9.4.3. Field effect
9.5. Drift mobility
9.6. Luminescence
9.7. Photoconductivity
9.8. Effect of alloying on the dark current
9.9. Numerical values in the model of charged dangling bonds
9.10. Charge transport in strong fields
9.11. Density of electron states in conduction and valence bands
9.12. Optical properties
9.13. Switching in alloy glasses
9.14. Oxide glasses

9.1. Introduction

In this chapter the properties of amorphous semiconductors containing one or more of the chalcogenide elements, S, Se, or Te, are reviewed. Within certain ranges of composition it is possible to form glasses by combination with one or more of the elements As, Ge, Si, Tl, Pb, P, Sb, and Bi, among others. Of the binary glasses, As_2S_3, As_2Se_3, and As_2Te_3 have been most extensively studied and are often regarded as prototypes of the chalcogenide glasses. Mixed systems such as the As_2Se_3–As_2Te_3 binaries and the As_2Se_3–As_2Te_3–Tl_2Se systems have also been the subjects of detailed investigations. Because of the large variety of such ternary and quaternary systems, classification of these materials becomes difficult, particularly in view of the freedom that is allowed in amorphous systems to depart from stoichiometric proportions of the constituents. Multicomponent glasses of (seemingly) arbitrary composition, for instance $As_{30}Te_{48}Si_{12}Ge_{10}$, have been studied in connection with the phenomenon of electrical switching

(§ 9.13). However, the properties of amorphous semiconductors formed from a wide variety of elements of differing valency may not necessarily be more complex than those of the binaries. Such compositions may favour a fully connected structure with most bonds satisfied and hence, with respect to certain properties, approach an ideal random network of atoms. However, in some systems there may be a greater tendency for phase separation than in others. In many cases there does not seem to be any significant qualitative difference between the properties of amorphous chalcogenide semiconductors of stoichiometric proportions and those of others.

However, use of stoichiometric compositions allows useful comparison with the material in its crystalline phase. For many of the stoichiometric materials discussed in this chapter the crystalline phase has a layer structure.

For any given group of elements it is not normally possible to form glasses for all compositions. The extent of the glass-forming region in several ternary systems was displayed graphically in Fig. 6.3. Table 9.1 (from Owen 1970) indicates the extent of the glass-forming region in some other ternaries (see also Savage and Nielsen 1965a). Well inside a glass-forming boundary, samples can be prepared by cooling from the melt; samples with compositions outside the boundary require deposition by evaporation or a similar technique in order to attain the amorphous phase. Near the boundary, fast quenching of the melt (like splat cooling) is sometimes used to obtain a glass. Even with compositions that readily form glasses it is sometimes useful to prepare specimens by deposition for certain experiments requiring thin films.

Annealing of chalcogenide glasses does not appear to lead to such marked changes of properties such as were described in Chapter 7 for amorphous Ge and Si, results from different laboratories on the same material being often in good agreement. However, marked differences in electric and magnetic properties of sputtered films do follow annealing in certain cases (cf. § 9.4). Our analysis of these properties is based on the assumption that point defects exist in these materials, their concentration depending on the method of preparation; they appear to be responsible for photoluminescence, and to determine the drift mobility and the position of the Fermi energy. Voids, however, do not normally form in deposited films as far as is known; thus Shevchik and Bishop (1974a), using small-angle X-ray scattering, found no evidence for their existence in several chalcogenides.

On the whole, these materials obey the so-called '8-N bonding rule' proposed by Mott (1969a), according to which all electrons are taken up in bonds so that large changes of conductivity with small changes of composition do not occur. The small effect of composition on conductivity was

TABLE 9.1

Extent of glass-forming regions in some ternary chalcogenide systems A—B—C

A	B	Group Ia			Group IIa			Group IIIa				Group IVa		Group Va				Group VIa		Group VIIa		
		Cu	Ag	Au	Zn	Cd	Hg	B	Ga	In	Tl	Sn	Pb	P	As	Sb	Bi	Se	Te	Cl	Br	I
As	S	×	×	×	×	×	×		×	×	○	×	×	×		⊗	×	●	⊗	×	⊗	●
	Se	○	○	×	○	○	○	○	×	×	⊗	○	×	⊗		⊗	×		⊗			⊗
	Te										⊗												⊗
Ge	S				×				×	○	○	×	×	●	●		×		×		×	
	Se					×				×		×		⊗	●	⊗	×		×			
	Te													○	⊗							
Si	S													⊗	⊗	⊗						
	Se														⊗	⊗						
	Te																					

From Owen 1970.

× very small; ○ small; ⊗ moderate; ● large.

established by the pioneering work of Kolomiets and co-workers (summarized by Kolomiets 1964) and is illustrated in, for instance, Fig. 9.9. This seems to be generally the case for glasses such as $As_{2-x}Se_{3+x}$, when x varies about the value zero, and for alloy glasses on the addition of elements such as Ge or Si with four (or more) outer electrons. However the addition of elements with less than four outer s and p electrons often has a major effect on the conductivity. Thus the addition of Cu, Ag, In, and Tl to As_2Se_3 and As_2S_3 causes a marked increase in the conductivity, 5 per cent of copper in As_2Se_3 increasing the room-temperature value by $\sim10^4$. According to Liang, Bienenstock, and Bates (1974), and Hunter, Bienenstock, and Hayes (1977*a,b*) copper promotes local four-fold co-ordination. We discuss in § 9.8 why this affects the conductivity. However, Donikov and Borisova (1963) found that the conductivities of glassy As_2Se_3 is essentially unchanged by concentrations up to 3·8 at. per cent Be or up to 1·9 at. per cent Mg or Ca.

Chalcogenides containing several per cent of elements such as Ni, W, Fe, and Mo can be highly conducting if prepared by co-sputtering (Ovshinsky 1977, Flasck *et al.* 1977). Presumably the temperature of deposition is sufficiently low relative to the glass transition temperature that valence satisfaction or the creation of charged compensating defects is inhibited (see Fritzsche 1977).

The elements Se and Te, although these have some properties similar to those of other chalcogenides, apparently contain structural units with molecular properties, and in view of this are discussed separately in Chapter 10. Chapter 10 also contains experimental results on the effect of small amounts of various elements in Se, together with a theoretical discussion.

In this chapter we shall not attempt to review all the published literature on chalcogenide glasses but shall concentrate on some recent work concerned mainly with electric and optical properties.

9.2. Structure

Compared with the elemental materials discussed in Chapters 7 and 8, the structures of most amorphous chalcogenides are not so well characterized. For any binary system A_xB_{1-x}, analysis of the RDF is complicated by the difficulty of separating contributions from A–A, B–B and A–B bonds; in multicomponent systems identification is even more ambiguous. A review of diffraction studies has been given by Wright and Leadbetter (1976). In principle, EXAFS (§ 6.3) is capable of making the distinction between bond types but, at the time of writing, problems associated with data reduction and analysis have not allowed conclusive results to be obtained except in a few cases. For several systems, infrared and Raman spectroscopy have been found to be the most useful techniques in the study of local structure.

The simplest structural model of a binary system is that of a continuous network in which the 8-N valence co-ordination rule is satisfied for both components and at all compositions; for instance in As_xSe_{1-x} every As atom would be three-fold co-ordinated and every Se atom two-fold co-ordinated at any value of x. Even for such networks, however, the question arises as to the degree of chemical ordering; in other words, to what extent are heteropolar bonds favoured over homopolar bonds? A network in which the distribution of bonds is purely statistical will be called a random bond network (RBN).[†] In an RBN there is no 'chemical' preference for one kind of bond over another, even at compositions corresponding to compounds in the crystalline state. However, if the relative strengths of the various pairs of bonds differ significantly, the heteropolar bond being the strongest, then a chemically ordered bond network (OBN) model may be more appropriate. In an OBN, bonds between unlike atoms occur whenever and wherever these are allowed within the constraints of composition and network connectivity.

The bond distributions predicted for RBN and OBN models (but without consideration of possible constraints imposed by topology) for $Ge_{1-x}X_x$, where X is a two-fold co-ordinated chalcogen atom, are shown in Fig. 9.1 (Lucovsky et al. 1977, White 1974a). Similar curves can be constructed for binaries having co-ordination numbers different from those in the II–IV system given here as an example.

An additional constraint in an OBN may arise from the possible existence of molecular-like species as an integral part of the network; characteristic vibrations associated with such atomic clusters can often be observed and identified by infrared and Raman spectroscopy. The formation of molecular units disconnected from the main network is also possible, for example short chains or closed rings of chalcogen atoms in binaries rich in such elements. Even more extreme are the cases of completely molecular glasses or phase-separated materials.

Lucovsky et al. (1977) have suggested a scheme for characterizing the structure of binary (or indeed multicomponent) glasses that takes into consideration the above possibilities in a logical way. The structure is specified in three stages: (1) the atomic co-ordination of each constituent; (2) the bond distribution; (3) the molecular structure of network-forming groups of atoms.

A few systems are now considered in detail.

Ge_xX_{1-x}. The points shown in Fig. 9.1 are the relative strengths of Ge–S vibrational features observed in infrared spectra of $Ge_{1-x}S_x$ glasses (Lucovsky et al. 1974). The agreement with the predictions of the OBN

[†] Compare CRN, a continuous random network, introduced for elemental amorphous materials. An RBN is in fact a CRN.

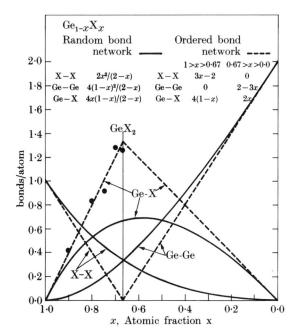

Fig. 9.1. Bond distributions in $Ge_{1-x}X_x$, where X is a chalcogen, for a random bond network (RBN) (solid curves) and an ordered bond network (OBN) (broken lines). The solid points show the composition dependence of the strength of an infrared peak associated with the Ge—S vibration in $Ge_{1-x}S_x$ alloys. (From Lucovsky *et al.* 1977.)

model provides strong evidence for the existence of 8-N co-ordination and chemical ordering on the S-rich side of the Ge–S system. Similar confirmation has been obtained for Ge–Se alloys (Tronc, Bensoussan, and Brenac 1973). It is of interest to note that the known crystalline phases of these systems include not only GeX_2 but also GeX; in the latter the 8-N rule is not satisfied, all atoms being three-fold co-ordinated. The possibility of three-fold co-ordinated structures for amorphous GeS, GeSe, and GeTe has in fact been raised by Bienenstock (1973) and by Arai *et al.* (1976), but later evidence (see Bienenstock 1974) seemed to rule against it.

The molecular species present in Ge chalcogenides depends upon the composition. For chalcogen-rich material one finds GeX_4 tetrahedra, the number of which is simply proportional to the concentration of Ge atoms. The excess chalcogen X is then incorporated as chains joining such tetrahedra or, at higher concentrations of X, an increasing number of ring molecules in solution with the network. For example in $Ge_{0.2}Se_{0.8}$, Se_8 rings have been identified by Raman spectroscopy (Lucovsky *et al.* 1974). For Ge-rich material, GeX_4 molecules again form, but the presence of a larger molecule, namely $X_3Ge–GeX_3$, can be identified. Fig. 9.2 shows the

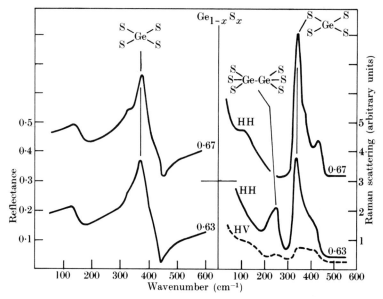

Fig. 9.2. Infrared reflectance and Raman scattering for two alloys in the $Ge_{1-x}S_x$ system ($x = 0.67$ and $x = 0.63$). Schematic representations of molecular configurations assigned to various peaks are shown. The feature associated with the molecule $S_3Ge—GeS_3$ is seen only in Raman scattering from Ge-rich material and is strongly polarized. (From Lucovsky *et al.* 1977.)

strongly polarized Raman mode at $250\ cm^{-1}$ associated with this molecule in $Ge_{1-x}S_x$ with $x = 0.67$ (i.e. GeS_2) and $x = 0.63$. The absence of a corresponding feature in infrared reflectivity is expected for a vibration involving an out-of-phase motion along the Ge bond direction, and also rules out an explanation in terms of three-fold co-ordinated Ge and S as proposed by Arai *et al.* (1976).

Similar behaviour is found in $Ge_{1-x}Se_x$ glasses. Raman spectra for Se-rich alloys (Fig. 9.3) show features corresponding to bond-stretching modes of Se–Se and $GeSe_4$ units. The former gives rise to the peak at $250\ cm^{-1}$ and the latter to the peak at $202\ cm^{-1}$. Tronc *et al.* (1973) have observed similar spectra but proposed an interpretation in terms of vibrations of Ge–Se–Ge triatomic units. The peak at $219\ cm^{-1}$ has also been interpreted in different ways; according to Nemanich *et al.* (1977) it could be associated with a 12-atom ring containing six Ge and six Se atoms. The small peak at $180\ cm^{-1}$ appearing in $GeSe_2$ probably arises from the analogue of the molecule found in the Ge–S system, namely $Se_3Ge—GeSe_3$.

As_xX_{1-x}. Infrared and Raman spectra of glasses in the As–S and As–Se systems are shown in Fig. 9.4. As for the Ge chalcogenides there is good

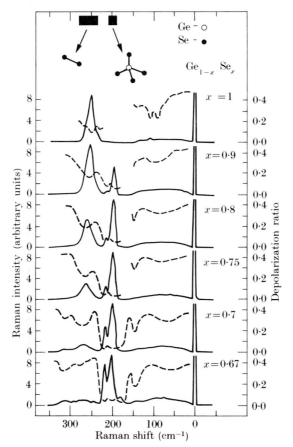

Fig. 9.3. Raman spectra and depolarization ratio from Se-rich $Ge_{1-x}Se_x$ alloys. The solid boxes denote the spectral regions in which features associated with vibrations of Se—Se and $GeSe_4$ occur. Possible assignments of other peaks are given in the text. (From Nemanich *et al.* 1977.)

evidence for satisfaction of the 8-N co-ordination rule and chemical ordering at all compositions. The molecular species present in the $As_{1-x}S_x$ networks are the analogues of those found in $Ge_{1-x}X_x$, namely AsS_3 pyramids and $S_2As–AsS_2$ molecules. For $As_{1-x}Se_x$, however, there seems to be less evidence for molecular ordering. This is probably associated with the small difference in electronegativity between As and Se.

RDFs of a series of As_xSe_{1-x} glasses with x ranging from 0 to 0·5 have been obtained by Renninger and Averbach (1973). Models for the atomic arrangements in these glasses have been computer generated using a Monte Carlo procedure in which the positions of an initially random arrangement of 150 atoms were repositioned until the RDF agreed, within

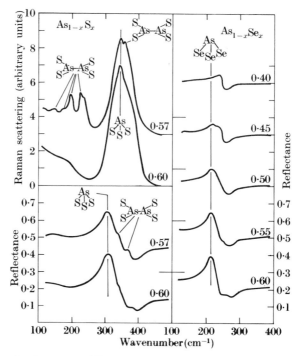

Fig. 9.4. Infrared reflectance and Raman scattering for various $As_{1-x}X_x$ alloys. Molecular configurations assigned to the features are shown. Whereas the 8-N valency rule and the OBN model appear to be appropriate for all these alloys, there seems to be no evidence for molecular ordering in $As_{1-x}Se_x$. (From Lucovsky *et al.* 1977.)

certain limits, with experimental curves (Renninger, Rechtin, and Averbach 1974). Although the 8-N rule was assumed to be favoured, some under-co-ordinated atoms were tolerated. 'Partial' chemical ordering was introduced by not permitting As–As bonds for $x < 0.4$, but As–Se bonds were not considered to be favoured over Se–Se bonds and no molecular ordering was assumed. Cross-sections of a few of these models are shown in Fig. 9.5 and a typical fit to the RDF for As_2Se_3 in Fig. 9.6. An interesting result of this modelling procedure is that, as the As concentration is increased from zero, chains of Se become cross-linked, the interconnection of the network increasing up to As_2Se_3. However, for higher As concentrations, there is a tendency for the atoms to form rings which are still connected to the network (see Fig. 9.5(d)).

For all amorphous materials structural differences are expected between bulk glass and evaporated thin films. For As_2S_3 these have been investigated in some detail (Apling, Leadbetter, and Wright 1977). Raman spectra of an as-deposited film, a film annealed at 180°C, and a bulk glass are

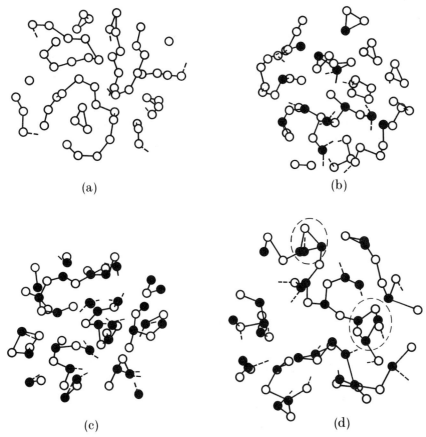

(a)   (b)

(c)   (d)

Fig. 9.5. Cross-sections of computer-generated models of $As_{1-x}Se_x$ glasses. Open circles are Se atoms and solid circles are As atoms. The broken lines correspond to bonds to atoms lying out of the slices shown (a) $x = 0$; (b) $x = 0.79$; (c) $x = 0.60$; (d) $x = 0.50$. On the As-rich side of As_2Se_3 a tendency for the network to break up into rings is shown. (From Renninger *et al.* 1974.)

compared in Fig. 9.7 (Solin and Papatheodorou 1977). The sharp additional bands in the spectrum for the as-deposited film suggest the presence of molecular units which polymerize into the network on annealing. De Neufville *et al.* (1974) suggested, from X-ray diffraction data, that the structure of freshly evaporated films of As_2S_3 could be described as a dense random packing of As_4S_6 molecules; however, the Raman spectra imply the existence of a considerable amount of polymerized network as well as other types of molecular units in the films. For example the sharp peaks at 180 cm^{-1} and 230 cm^{-1} which remain on annealing may be associated with As_4S_4 units. These, as well as As_4, S_2, and other molecules occur in the

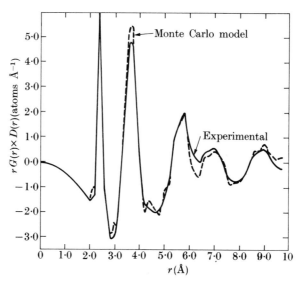

Fig. 9.6. Comparison of the experimentally determined reduced radial distribution function with that generated from the computer model of As_2Se_3 (Fig. 9.5(d)). (From Renninger *et al.* 1974.)

vapour phase, although this does not of course necessarily imply their preservation on deposition (Leadbetter, Apling, and Daniel 1976).

The structure of As–Te glasses and films has been investigated and discussed by Cornet and Rossiter (1973) and Cornet (1977). These authors propose a breakdown in the $8-N$ valency rule in Te-rich alloys, some Te atoms becoming three-fold co-ordinated and some As atoms six-fold co-ordinated. EXAFS studies (Pettifer, McMillan and Gurman 1977), however, suggest that some degree of chemical ordering occurs at compositions near to that of stoichiometry.

9.3. Electrical properties of chalcogenide glasses

9.3.1. Introduction

The d.c. conductivity σ of most of the chalcogenide glasses near room temperature obeys the relation $\sigma = C \exp(-E/kT)$. Fig. 9.8 shows some typical logarithmic plots of σ against $1/T$ for chalcogenides with E varying from about 0·3 eV to more than 1 eV. Although the values of $2E$ lie close to the photon energy corresponding to the onset of strong optical absorption, intrinsic conduction must not be assumed, as we shall see. Therefore we shall not double the observed values of E in order to obtain the band gap denoted by B. In selenium, for which traces of impurity can lead to a change from p- to n-type behaviour, we give in the next chapter an

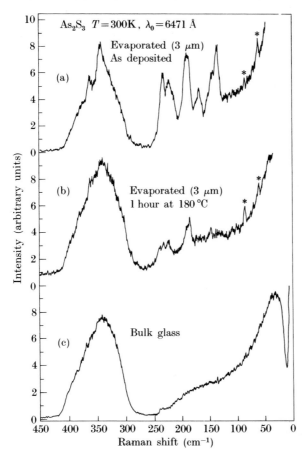

Fig. 9.7. Raman spectra of As_2S_3: (a) as-deposited film; (b) annealed film; (c) bulk glass. The lines marked with a star are instrumental ghosts. (From Solin and Papatheodorou 1977.)

estimate of the band gap (or mobility gap), determined from electrical properties.

As explained in Chapter 6, values of C in the range 10^3–$10^4 \, \Omega^{-1} \, cm^{-1}$ are often thought to indicate conduction at a mobility edge, C being given by $\sigma_{min} \exp(\gamma/k)$. For chalcogenides we believe this to be so, but the conclusion has been questioned, not only because the concept of a minimum metallic conductivity has not been universally accepted (Chapter 2), but perhaps more seriously because of the contention of Emin, Seager, and Quinn (1972) and Emin (1977b) that the current carrier (a hole) forms a polaron, and its motion is activated (cf. Chapter 3). We do not think that this conclusion is justified (see below) for the chalcogenides.

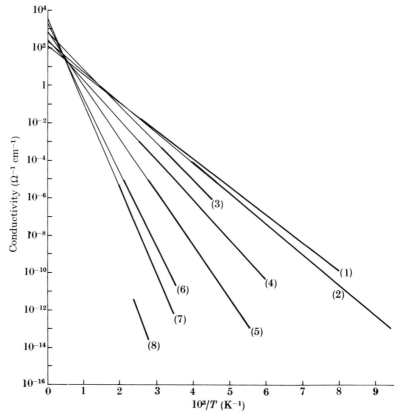

Fig. 9.8. Temperature dependence of electrical conductivity in some amorphous chalcogenide semiconductors, illustrating the relationship $\sigma = C \exp(-E/kT)$. Heavy lines are the experimental results and fine lines are the extrapolation to $1/T = 0$ (note that the actual variation of As$_2$Te$_2$.Tl$_2$Se (Andriesh and Kolomiets 1965); (3) As$_2$Te$_3$ (Weiser and Brodsky 1970); (4) 4As$_2$Te$_2$.As$_2$Se$_3$ (Uphoff and Healy 1961); (5) As$_2$Se$_2$.Tl$_2$Se (Andriesh and Kolomiets 1965); (6) 3As$_2$Se$_3$.2Sb$_2$Se$_3$ (Platakis *et al.* 1969); (7) As$_2$Se$_3$ (Edmond 1968); (8) As$_2$S$_3$. (From Edmond 1968.)

We show in Fig. 9.9 the variation of C and E with composition for some binary alloys. An approximate invariance of C in the Se–Te and As$_2$Se$_3$–As$_2$Te$_3$ systems and of E in the Se–As and As$_2$Se$_3$–As$_2$Se$_3$ systems are apparent. Because of the difficulty of measuring the electrical conductivity of selenium over a wide temperature range, it is difficult to determine E and C in this and other high-resistivity materials with any confidence (see also Chapter 10). The small value of C for As$_2$S$_3$ may perhaps indicate a wide range of localized states, and conduction by hopping.

The electrical properties of chalcogenide glasses may be compared with those of glow-discharge-deposited silicon and germanium (Chapter 7) as

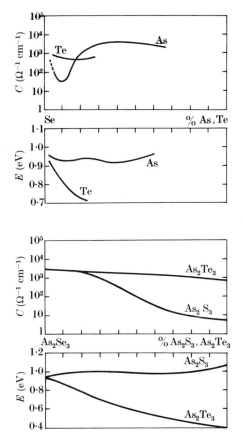

Fig. 9.9. Variation of the pre-exponential factor C and the activation energy for conduction E in the Se–Te, Se–As, As_2Se_3–As_2Te_3, and $AsSe_3$–As_2S_3 systems.

follows. Measurements of thermopower show them to be p type; measurements of drift mobility and of mobility in photoconduction have been interpreted particularly by Owen and co-workers (§ 9.4) in terms of a trap-limited mobility, a quasi-equilibrium being set up between holes in the valence band and those in traps due to point defects, as described in § 6.5.1. In this book we adopt this model. Drift-mobility measurements have not as yet been used to establish a mobility edge in the valence band as they have been for the conduction band in glow-discharge-deposited silicon. The best evidence for a mobility edge comes from the investigations by Nagels *et al.* (1974) on conductivity, thermopower, and Hall coefficient described in § 6.6 and further in § 9.4. These, however, have also been interpreted in terms of the assumption that holes form polarons

(Emin *et al.* 1972). We think that part of the evidence against this is that many of the trap energies determined by Marshall and Owen (1975) can be identified with those established from other experiments, for instance photoluminescence, and can be rationalized in terms of the model of charged dangling bonds introduced in Chapter 6 and described in greater detail in § 9.4. Other arguments are that in the threshold switch in its on state (§ 9.13) electrons and holes appear to have comparable mobilities, and these mobilities are much higher than would be expected for a polaron. In § 3.3 we showed that the formation of a molecular polaron depends critically on the parameters of the system (cf. Fig. 3.5). In SiO_2 (§ 9.14) it appears that holes form polarons, while electrons have a very high mobility. For further discussion of this controversy, see Mott and Street (1977).

In the next section, then, we discuss electrical phenomena which do not depend on gap states (except in so far as these may pin the Fermi energy),

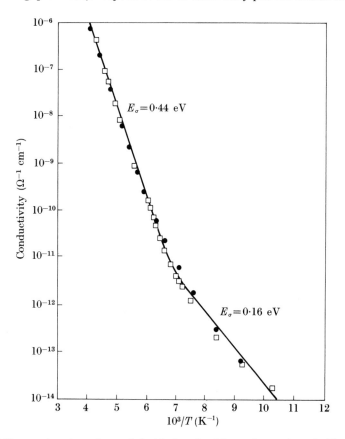

Fig. 9.10. Temperature dependence of electrical conductivity σ of amorphous As_2Te_3. (From Marshall and Owen 1975.)

namely d.c. conductivity, thermopower, and Hall effect, in terms of these models. In § 9.4 we discuss gap states, and their role in pinning the Fermi energy, screening and field effect, and, in determining drift mobility, recombination and conductivity at very low temperatures.

9.3.2. d.c. conductivity

In general, as Fig. 9.8 shows, $\log \sigma$ is a fairly linear function of $1/T$, and variable-range hopping conductivity behaving even approximately as $A \exp(-B/T^{1/4})$ is not generally observed. However, deviations from linearity can occur at low temperatures, and, for sputtered As_2Te_3 and other chalcogenides, $T^{1/4}$ behaviour was found by Hauser and Hutton (1976) and Hauser, Di Salvo, and Hutton (1977). Fig. 9.10 shows results of Marshall and Owen (1975). Croitoru et al. (1970) found similar behaviour at high temperatures with an activation energy of 0·53 eV. Fig. 9.11 shows

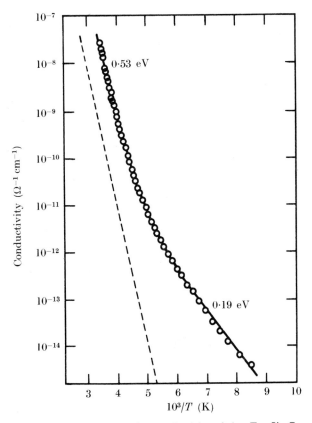

Fig. 9.11. Temperature dependence of d.c. conductivity of $As_{30}Te_{48}Si_{12}Ge_{10}$; circle, r.f. sputtered film; broken line, bulk glass ($\sigma \times 10^{-2}$). (From Marshall and Owen 1975.)

results of Marshall and Owen (1975) for an alloy glass; deviations occurring for the sputtered film are absent for a bulk glass, so their origin is likely to be defects. This is discussed in § 9.4. Fig. 6.12 shows that for many glasses there is little change of slope in going from solid to liquid, though at higher temperatures a gradual transition to metallic behaviour ($\sigma \sim 10^3 \, \Omega^{-1} \, \text{cm}^{-1}$) is normal, as described in Chapters 5 and 6. Fig. 9.12 shows the continuity of slope for As_2Se_3.

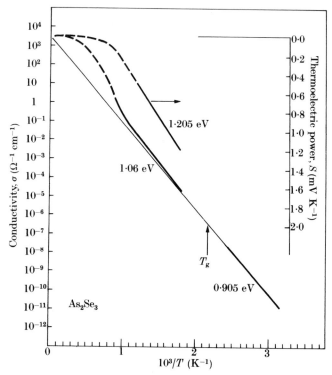

Fig. 9.12. Temperature dependence of d.c. conductivity and thermopower of amorphous and liquid As_2Se_3. (Data from Edmond 1966.)

9.3.3. Thermopower

This, as already stated, is normally positive. Early work was frequently done at temperatures in or near the liquid phase. Thus Fig. 9.12 shows data for As_2Se_3 due to Edmond (1966); the activation energy in the liquid appears greater than in the solid, suggesting that the gap decreases with increasing T, as discussed in Chapter 6. For the solid a main question is whether E_σ, the activation energy for conduction, is the same as that (E_S) for the thermopower. For As_2Se_3 both were measured down to room temperature by Hurst and Davis (1974) and found to be identical. Seager and Quinn (1975) claim a considerable divergence from well above down to room temperature; they give $E_\sigma = 0\cdot91$ eV, $E_S = 0\cdot60 \pm 0\cdot07$. These data, along with others, have been collated and discussed by Mytilineou and Roilos (1978).

As_2Te_3 and $As_2Te_{3-x}Se_x$ have been investigated by many authors, who all conclude that $E_\sigma > E_S$ over the temperature range investigated. These include Seager, Emin, and Quinn (1973), Seager and Quinn (1975), Grant et al. (1974), and Nagels et al. (1974). The difference $E_\sigma - E_S$ is of order $0\cdot15$ eV.

There are, as we have seen in Chapter 6, several possible explanations for such a discrepancy.

(a) That conduction is at E_B (the top of the valence band) and there is a large hopping energy due to disorder. We think this unlikely because the pre-exponential factor in σ is too large ($\sim 10^3 \ \Omega^{-1} \ cm^{-1}$), and because a Hall coefficient is observed, which we do not expect for hopping.

(b) That the carriers are polarons, and the difference $E_\sigma - E_S$ is the polaron hopping energy W_H. This is the proposal of Emin et al. (1972).

(c) That there are fairly long-range inhomogeneities in the potential at the top of the valence band, necessitating the use of classical percolation theory (§ 6.4.2).

(d) That all measurements have been made in the transition region between hopping transport at the band edge and transport at the mobility edge (E_V). Fig. 6.32 shows the behaviour to be expected, which was proposed by Nagels et al. (1974) to explain their own measurements, developed in detail by Mott, Davis, and Street (1975) and described in § 6.6. For a variety of reasons presented there, this is the explanation we favour. Further discussion of our reasons awaits our treatment of the Hall effect in the next section.

9.3.4. Hall effect

The measurements of Male (1967) on the Hall mobility μ_H of various chalcogenide glasses in the solid and liquid states were reproduced in Fig. 6.6; the values of μ_H, and the independence of temperature, interpreted in terms of Friedman's formulae (§ 2.14), provided the earliest evidence that conduction in these materials is at a mobility edge. More recent results are due to Seager et al. (1973), Nagels et al. (1974), Roilos and Mytilineou (1974), and Mytilineou and Roilos (1978) for a variety of chalcogenides. Some of these results are shown in Fig. 6.21(b). They can be summarized as follows.

(a) The Hall coefficient is normally negative for these p-type materials, as Friedman's original theory predicted, or alternatively as we would expect for hopping polarons. Emin (1977a), however, has discussed the conditions under which it could be positive (§ 2.14).

(b) At high temperatures μ_H seems to tend to a constant value.

(c) At low temperatures μ_H drops off exponentially with $1/T$.

For these two explanations have been put forward.

(i) That of Emin and co-workers, who suppose that the polaron formula (§ 3.9), namely $\mu_H \propto T^{-3/2} \exp(-\frac{1}{3}W_H/kT)$ can be used; the flattening off at high temperatures is due to the term $T^{-3/2}$, and the observed activation energy at low temperatures corresponds well with $\frac{1}{3}(E_\sigma - E_S)$.

(ii) That proposed by Nagels et al. (1974), that at high temperatures Friedman's temperature-independent μ_H for charge transport at a mobility edge is applicable, while in the hopping regime the Hall effect is negligible. The model was described in § 6.6. The values of $\Delta E = E_B - E_V + w$ and of the hopping energy w, deduced by these authors from conductivity, thermopower, and Hall effect for As_2Te_3 are $E_B - E_V = 0.11$ eV, $w = 0.03$ eV, and for the pre-exponential factors $\sigma_{min} \exp(\gamma/k) = 3200 \ \Omega^{-1} \ cm^{-1}$ and $\sigma_{hop} = 90 \ \Omega^{-1} \ cm^{-1}$.

For the reasons given in § 9.3.1 we favour the second explanation. In the remainder of this chapter and in our discussions of selenium in Chapter 10, then, it will be supposed that holes in the valence bands of chalcogenides do not form polarons. As we have seen in Chapter 3, a polaron of V_K type, due to bonding between a partly occupied lone-pair orbital on one atom and a full one on another, will be formed in a valence band if it is narrow enough, but if it is not there is *no* polaron formation. The valence band of chalcogenides, as of oxide glasses, is thought to be formed from lone-pair orbitals on Se, Te, or O (Kastner 1972). This means that of the three pairs of p orbitals in the outer shell, two form bonds with neighbouring atoms while one does not. The upper part of the valence band is formed from the latter, which are designated 'lone pair' because the pair of electrons are not associated with a bond. It is reasonable to suppose that this band is narrow, and indeed calculations for selenium (Chapter 10) and observations for chalcogenides (§ 9.11) and for SiO_2 (§ 9.14) show that this is so. Thus the conditions for polaron formation may be approached, and only experiment can show whether they exist or not in a given material.

9.4. States in the gap

9.4.1. Introduction

It has already been stated that chalcogenide glasses appear to obey the 8-N co-ordination rule, whatever the composition. Arsenic will be co-ordinated with three atoms, Se with two, Ge with four, and so on. If constituents with fewer (N) than four outer electrons are present, however, they may form four bonds and may perhaps be negatively charged. An example is provided by the work of Kumeda et al. (1976a) on the incorporation of Mn in an alloy glass; they find that for small concentrations (<0.2 per cent) the e.s.r. signal from the $3d^7$ shell is characteristic of four-fold co-ordinated Mn; a possible hypothesis is then that the charge is $-2e$. This is discussed further in § 9.8. The same principle is well known in oxide glasses; thus aluminium and boron in glasses containing Na^+ ions can have co-ordination number 4 (§ 9.15). Some model for the compensating charge (Na^+ in a soda oxide glass) is thus required.

In this book we have taken the point of view that an ideal continuous random network with a certain entropy due to disorder is the correct first approximation to the properties of a glass, but that in addition, just as for a crystal, there are point defects, and that in the melt they contribute an entropy calculable just as for a crystal† (cf. § 2.10 and Bell and Dean 1968). In a glass they will be quenched in. In evaporated or sputtered films, or in SiO_2 grown thermally on silicon, higher concentrations are to be expected. These defects are responsible for 'states in the gap'.

In chalcogenides the role of these states in the gap, associated with defects with a concentration depending on the method of preparation, is relevant to the following.

(a) The screening length, for instance of the space-charge at an interface.

(b) Pinning or locating the Fermi energy.

(c) Deviations from a straight line plot of log σ versus $1/T$ (Figs. 9.10, 9.11).

(d) Determination of the drift mobility of holes, which, unlike electrons in glow-discharge-deposited silicon, seems to be limited by defects.

(e) Photoluminescence and a small tail to the optical absorption edge.

(f) Providing recombination centres for electrons and holes, for instance in photoconduction.

(g) Possibly responsible for a.c. conductivity (§§ 2.15, 6.4.5).

(h) ESR and Curie paramagnetism are absent in annealed material, but observed sometimes in unannealed specimens (Smith 1972, Hauser and Hutton 1976 and after illumination (Bishop et al. 1975, 1976a,b).

(i) Providing compensating charges for charged impurities (§ 9.6).

We look first at the pinning mechanism which locates the Fermi energy. Fig. 9.13 shows a logarithmic plot of resistivity against optical band gap. This shows for a variety of chalcogenide glasses listed in Table 9.2 that the thermal activation energy E_σ for conduction is proportional to the gap (though E_F must be nearer to E_V than to E_C, the materials being p type). Only As_2S_3 with 1 per cent of Ag appears to be an exception (cf. § 9.6).

The evidence, however, that these materials are normally not intrinsic, but that the Fermi energy is pinned, is extensive. It is well (and emphatic-

---

† In chalcogenides, as in other amorphous materials, we think that the simplest defect is a 'dangling bond', which when neutral carries a spin. For chalcogenides both the chalcogen or arsenic can play this role. In oxide glasses the 'non-bridging oxygen' is a familiar example.

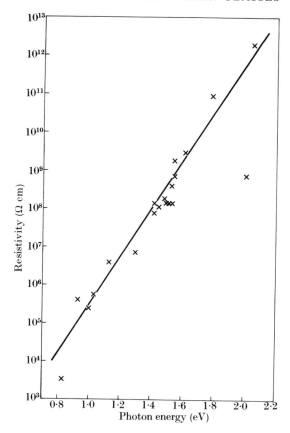

Fig. 9.13. Resistivity of various chalcogenide glasses plotted against photon energy corresponding to an absorption coefficient of about 8 cm^{-1}. (Data from Table 9.2.)

ally) summarized by Fritzsche (1973) who writes 'can we suppose $2E_\sigma$ = bandgap? The answer is *no*'. One argument is that the straight lines of Fig. 9.2 and similar plots are unlikely for an intrinsic material, because, in spite of the 8-*N* rule, it seems likely that there would be *some* donors or acceptors and that this would lead to the kind of curvature in the ln σ *versus* $1/T$ plot observed for sputtered specimens, unless their concentration was less than 10^3 cm^{-3}. It is of course possible that, unlike glow-discharge Si containing phosphorus or boron, the 8-*N* rule is satisfied for all atoms. However, the estimates of the screening length at a metal–semiconductor interface given by Fritzsche (1973), of order 300 Å, are not compatible with intrinsic behaviour. It is not easy, however, to demonstrate whether the Fermi energy is pinned, with a finite value of the density of states there, or whether it lies midway between deep donors and acceptors separated by an energy 2ε, the donors being lower. Such a model is used by Marshall

TABLE 9.2

Positions of room-temperature optical absorption edges and resistivities in several chalcogenide systems

Material	Wave number for 15% transmission for specimen 0.178 cm thick (cm^{-1})	Corresponding energy (eV)	Resistivity	
			at 50°C (Ω cm)	130°C (Ω cm)
As$_{34.25}$Se$_{65.75}$	12 330	1·53		4·10×10^8
As$_{37.6}$Se$_{62.4}$	12 248	1·52		1·57×10^8
As$_{38.7}$Se$_{61.3}$				1·49×10^8
As$_{40}$Se$_{60}$	12 131	1·50		1·54×10^8
As$_{42}$Se$_{58}$	12 049	1·49		1·82×10^8
As$_{50}$Se$_{50}$	12 510	1·55		18·4 ×10^8
As$_{35}$(Se$_2$Te)$_{65/3}$	7892	0·98	1·7 ×10^7	
As$_{40}$(Se$_2$Te)$_{60/3}$	8045	1·00	3·3 ×10^7	2·5 ×10^5
As$_{45}$(Se$_2$Te)$_{55/3}$	8260	1·02	6·9 ×10^8	
As$_2$S$_3$	16 555	2·05		2·0 ×10^{12}
As$_2$S$_2$Se	14 435	1·79		9·4 ×10^{10}
As$_2$SSe$_2$	13 069	1·62		3·1 ×10^9
As$_2$Se$_3$	12 131	1·51	(2 ×10^{11})	1·54×10^8
As$_2$Se$_{2.5}$Te$_{0.5}$	9095	1·13	8·75×10^8	4·05×10^6
As$_2$Se$_2$Te	8045	1·00	3·3 ×10^7	2·5 ×10^5
As$_2$SeTe$_2$	6715	0·83	1·8 ×10^5	3·5 ×10^3
(As$_4$Sb$_2$)Se$_9$	10 461	1·30	2·2 ×10^9	7·4 ×10^6
As$_{40}$S$_{60}$	16 555	2·05		2·0 ×10^{12}
As$_{40}$S$_{60}$Ag$_1$	16 205	2·01		7·9 ×10^8
As$_{40}$Se$_{60}$	12 131	1·51		1·54×10^8
As$_{40}$Se$_{60}$Ag$_1$	11 445	1·42		8·1 ×10^7
As$_{40}$Se$_{40}$Te$_{20}$	8045	1·00	3·3 ×10^7	2·5 ×10^5
As$_{40}$Se$_{40}$Te$_{20}$Ag$_1$	7527	0·93	5·5 ×10^7	4·2 ×10^5
As$_{34.25}$Se$_{65.75}$Ag$_1$	11 702	1·45		1·17×10^3
As$_{40}$Se$_{60}$Ag$_1$	11 445	1·42		0·81×10^8
As$_{50}$Se$_{50}$Ag$_1$	(11 480)	(1·42)		1·48×10^8
As$_{34}$S$_{66}$Ag$_1$				2·1 ×10^9
As$_{40}$Se$_{60}$	12 131	1·51		1·54×10^8
As$_{40}$Se$_{60}$Ge$_5$	12 525	1·55		7·5 ×10^8
As$_{40}$Se$_{40}$Te$_{20}$	8045	1·00	3·3 ×10^7	2·5 ×10^5
As$_{40}$Se$_{40}$Te$_{20}$Ge$_5$	8260	1·03		5·9 ×10^5

From Edmond 1968.

and Owen (1976) in their description of their field-effect studies (§ 9.4.3). It is illustrated in Fig. 9.14.

An outstanding property of these materials is the absence of e.s.r. (cf. § 6.8.3) and Curie paramagnetism. E.S.R. investigations have been made by

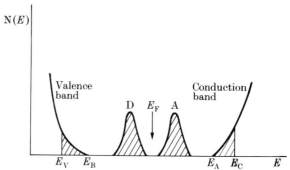

Fig. 9.14. Model of density of states in material containing deep donors (D) below acceptors (A).

Agarwal (1973) and Fritzsche (1973), which showed e.s.r. signals only after heat treatment, as also found by Smith (1972). Hauser *et al.* (1977) found an e.s.r. signal in sputtered As_2Te_3 and other chalcogenides, which disappeared on annealing. It is found after illumination (Bishop *et al.* 1975, 1976*a*,*b*). In annealed specimens, however, a Curie term C/T in the paramagnetism occurs only in the presence of iron impurities (Tauc *et al.* 1973). Fritzsche (private communication) puts an upper limit of 10^{15} spins cm^{-3}.

Variable-range hopping is not normally observed in chalcogenide glasses, though the low-T behaviour of Figs. 9.10 and 9.11 may indicate the onset of this form of conduction. Adler *et al.* (1974) have observed $T^{1/4}$ behaviour in the off state of a threshold switch, formed from an alloy glass, and the work of Hauser and Hutton (1976) and Hauser *et al.* (1977) shows that an e.s.r. signal is to be associated with $T^{1/4}$ hopping at low T, the signal disappearing on annealing. However, a.c. conductivity behaving so that $\sigma(\omega) \propto \omega^s$ ($0\cdot8 < s < 1$) is widely observed in specimens with no e.s.r. signal.

The absence of e.s.r. and Curie paramagnetism in annealed specimens, together with evidence that the Fermi energy is pinned and the density of states finite, was for several years a major puzzle in the understanding of chalcogenide glasses. One way out would be to assume, as already suggested, that the density is in fact zero but E_F is determined by deep donors lying below deep acceptors near mid-gap as in Fig. 9.14. In this case the observed a.c. conduction would have to be ascribed to soft phonons (§ 6.4.5). The other possible explanation is that the true density of states at E_F is finite, but that electrons near the Fermi energy form pairs with antiparallel spins so that the one-electron density $N(E_F)$ is zero. This was first suggested by Anderson (1975). A rather different version of the model was put forward by Street and Mott (1975); this was outlined in Chapter 6 and will be adopted here. This model is preferred to one in which the density vanishes, and in which deep donors and acceptors locate the Fermi energy, simply because it is successful in explaining a wide variety of

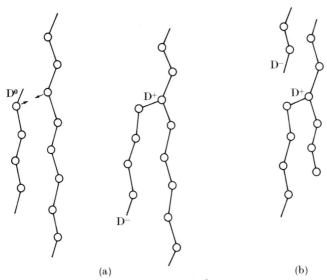

(a)   (b)

Fig. 9.15. (a) D^+ and D^- centres formed from two D^0 centres at the ends of a Se chain. (b) Intimate valence alternation pair.

phenomena. We cannot at the time of writing point to an experiment which uniquely proves that the density of states at E_F is finite, though the cumulative evidence is strong.

We first recapitulate the model.† The defect we have in mind can have three charge states D^+, D^-, and D^0. Since they are each associated with a different local atomic configuration they can be considered as different defects. The negatively charged defect D^- is a dangling bond associated with an under-co-ordinated atom, for example a chalcogen (say Se) bonded to *one* other atom or a pnictide (say As) bonded to two other atoms. When an electron is removed from the dangling bond (forming D^0) it is assumed that there is an attraction of the atom in question towards a fully co-ordinated neighbouring chalcogen atom, one of the lone-pair electrons on the latter being used to form a bonding orbital and the other an antibonding orbital. We do not believe, however, that this bond is as strong as when a second electron is removed, for then both lone-pair electrons from the neighbouring chains are used in bonding and the former singly co-ordinated chalcogen becomes essentially three-fold co-ordinated—the D^+ centre (see Fig. 9.15(a)). It is assumed that the reaction

$$2D^0 \rightarrow D^+ + D^-$$ (9.1)

† Street and Mott 1975, Mott, Davis, and Street 1975, Street 1976, Mott and Street 1977, Adler and Yoffa 1976, Adler 1976, Kastner, Adler, and Fritzsche 1976, Kastner 1977, Kastner and Fritzsche 1978.

is exothermic, that is the total energy (electrons plus lattice) associated with the pair of charged defects D^+ and D^- (both without spin) is lower than that of two neutral defects D^0 (both with spin). The Coulomb repulsive energy between the two electrons at D^- is more than compensated by the lattice energy gained; this is what is meant by a negative effective correlation (Hubbard) energy U for the defect.

Kastner *et al.* (1976) and Kastner and Fritzsche (1978) have provided insight into the above processes using chemical-bond arguments. In their notation D^+ and D^- are denoted by C_3^+ and C_1^-, the C standing for chalcogen and the subscript indicating the co-ordination. In contrast to our description of D^0, however, they consider the neutral centre to be three-fold co-ordinated with the antibonding electron residing symmetrically at the defect, which they therefore designate C_3^0. The creation of a 'valence-alternation pair', C_3^+ and C_1^- is considered to occur in two stages. First a neutral dangling bond, C_1^0 interacts with the lone pair of a neighbouring chalcogen forming a three-fold and a two-fold chalcogen according to the reaction $C_1^0 + C_2^0 \rightarrow C_2^0 + C_3^0$. During this reaction, (i) one of three electrons in lone pair orbitals at C_1^0 is transferred to a lower lying bonding orbital, and (ii) one of the two lone pair electrons at C_2^0 is transferred to a bonding orbital and the other to an antibonding orbital. Secondly two C_3^0 defects convert to C_3^+ and C_3^- by transferring two electrons in antibonding orbitals and two in bonding orbitals into lone pair orbitals—an exothermic reaction. The above-mentioned authors make the important observation that the valence-alternation pair is associated with the same number of bonds as the continuous random network, so that the energy to form it may be quite low.

The energy level of Fig. 9.16 has been proposed by Street and Mott (1975). The D^- and D^+ defects act as shallow acceptors and donors for

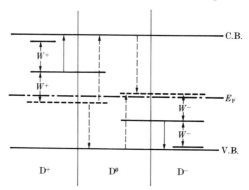

Fig. 9.16. Energy levels associated with D^+, D^0, and D^- according to Street and Mott (1975). Transitions associated with thermal excitation of an electron trapped at D^+ and a hole trapped at D^- to the conduction band (C.B.) and valence band (V.B.), respectively, are denoted by solid arrows. Optical transitions are denoted by dotted arrows: luminescence (downward facing arrows), absorption (upward facing arrows).

trapping processes but, having trapped their respective carriers, they distort to become D^0s with activation energies for release that are enhanced by W^- and W^+ respectively. (Note that $W = 2W_H$.) The Fermi level is pinned midway between these two deeper levels (see § 9.8, Mott, Davis and Street 1975, and Adler and Yoffa 1976), *even if the concentrations of* D^+ *and* D^- *differ quite widely* as can happen if there are other charged centres present (see § 9.8). The dotted levels in Fig. 9.16 are useful for representing optical transitions (which occur in times short compared to that of atomic relaxations), as indicated in the caption.

The energy levels of the D^+, D^- states will doubtless be broadened by disorder. Marshall (1977), from an analysis of dispersion in hole drift mobility in As_2Te_3, estimates a spread of ~ 0.025 eV in the depth of his trapping centres, which we identify with D^-. A spectrum of values of the negative Hubbard U may be responsible for the spread.

The absence of an e.s.r. signal in annealed specimens is strong evidence that all centres then have negative U, that is that the reaction (9.1) is exothermic. However, in sputtered As_2Te_3, e.s.r. signals have been observed before annealing by Hauser and Hutton (1976), showing (in our view) that some centres with positive U then exist. It is significant that these authors also observe variable-range hopping with σ varying as $\exp(-B/T^{1/4})$ at low temperatures. This means that the factor (§ 3.5) $\exp(-W_H/\frac{1}{4}\hbar\omega)$ is not too small to prevent jumps in which D^+, D^0 change places. According to the consideration of Chapter 3, and in the interpretation of photoluminescence given later in this chapter, W_H should be $\sim\frac{1}{4}$ of the Stokes shift observed in photoluminescence and thus $\sim\frac{1}{8}B$, B being the band gap. For $B = 1$ eV as for As_2Te_3 this is ~ 0.125 eV, and if $\hbar\omega = 0.05$ eV our factor is $e^{-8} \simeq 10^{-3}$, which is compatible with the observation of variable-range hopping.

In annealed specimens for which an e.s.r. signal is absent, $T^{1/4}$ behaviour is not observed. We suppose that the polaron energy involved for the exchange of D^+, D^-, which should be $4W_H$ (Phillips 1976), is too large; the factor discussed above will now be 10^{-12}. None the less, a.c. conductivity varying as ω^s has been extensively observed (§ 6.5.4) and ascribed to states in the gap. Mott and Street (1977) suggest that the explanation may be that a high concentration of pairs of D^+, D^- close together may normally be present because of their attraction for each other. If the charge from D^+ overlaps into the antibonding state of D^-, the polarization energy may be greatly reduced and approach that of D^0. Another model, also involving pairs and due to Elliott (1977) is discussed in § 6.4.5.

The deviations from the straight plot of log σ *versus* $1/T$ shown in Figs. 9.10 and 9.11 may possibly be the beginning of variable-range hopping, consequent on overlap between D^+ and D^- bands due to centres with positive U. Another explanation follows, however, from the analysis of Marshall and Owen (1975), who combine it with drift-mobility curves (§ 9.5) to identify a (spinless) hole trap 0.13 eV below E_F, the defect responsible being present in sufficient concentration to allow hopping from one to another. These authors (Marshall and Owen 1976) also identify the level in field-effect studies. Mott and Street (1977) suggest that it may be a D^+ and a D^- centre in close proximity, forming a dipole. Such pairs must indeed be present, and their concentration should depend on the rate of cooling. The concept of 'intimate pairs' of D^+, D^- centres (C_3^+, C_1^-) was introduced by Kastner *et al.* (1976) (see also Kastner and Fritzsche 1978, Tsai *et al.* 1977). We believe, following Street (1977), that such a pair, illustrated in Fig. 9.15(b), can be formed by the absorption of a photon *not* at a defect. The configuration diagram envisaged is shown in Fig. 9.17.

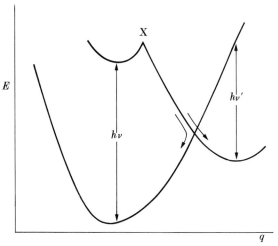

Fig. 9.17. Street's model for photodarkening; $h\nu$ and $h\nu'$ are the absorption energies before and after darkening.

Street proposes that this is the origin of photodarkening.[†] If indeed the configuration diagram does appear as shown, the metastable state should be formed with a quantum efficiency at present hard to calculate, but perhaps several per cent. Moreover, this form of self-trapped exciton is a singlet state; there is clearly no triplet state below it.

9.4.2. Screening length

A discussion of the theory of the screening length when $N(E_F)$ is finite was given in § 6.4.9. In our model of charged dangling bonds the one-electron value of $N(E_F)$ vanishes but the true density of states, due to paired electrons, is high. We denote it $N/\Delta E$, where N is the number of such

† For an alternative explanation, see Tanaka, Hamanaka, and Iizima 1977.

centres and ΔE their (small) spread in energies. The treatment will depend on whether the electron's potential energy E, which has to be screened, is greater or less than this width ΔE of the D^- level. In the former case we may write

$$\frac{d^2V}{dx^2} = \frac{4\pi Ne^2}{\kappa}. \tag{9.2}$$

Thus the potential varies as $2\pi Ne^2x^2/\kappa$ and the screening length is

$$(\kappa E/2\pi Ne^2)^{1/2}. \tag{9.3}$$

In the latter case, instead of (9.2) we have

$$\frac{d^2V}{dx^2} = \frac{4\pi Ne^2 V}{\Delta E}$$

so that V falls off as $\exp(-x/x_0)$ with

$$x_0 = \left(\frac{\kappa\,\Delta E}{4\pi Ne^2}\right)^{1/2}. \tag{9.4}$$

The experimental evidence has been reviewed by Fritzsche (1973). He states that for amorphous semiconductors no deviation from a linear relationship between resistance and thickness is observed down to 1000 Å, which gives an upper limit to x_0. He also describes experiments on photovoltage, which is expected to occur if radiation is absorbed in a space-charge region at a metal–semiconductor interface. For an alloy glass ($Ge_{16}As_{35}Te_{28}S_{21}$), he finds $x_0 \sim 300$–500 Å. From the frequency dependence of the capacity, he finds $x_0 \sim 160$ Å.

Whichever form (9.3) or (9.4) we take, with E or ΔE a few tenths of an electronvolt, these results imply a value for the density of defects N in the range 10^{17}–10^{18} cm^{-3}.

9.4.3. Field effect

Field-effect measurements were first made successfully for a chalcogenide alloy glass by Egerton (1971). More recent investigations are due to Marshall and Owen (1976), who used sputtered films of As_2Te_3 and an alloy glass ($As_{30}Te_{48}Si_{12}Ge_{10}$), and by Mahan and Bube (1977) on As_2Te_3. The former authors were able to apply a gate voltage up to 10^6 V cm^{-1} with negligible leakage current. They obtained the following results, which differ strikingly from those for glow-deposited silicon:

(a) As shown in Fig. 9.18, the field-effect current tends to a value proportional to the gate voltage.

(b) In As_2Te_3 and for an alloy glass the activation energy of the field-effect current is $\sim 0 \cdot 13$ eV smaller than E (the activation energy in the

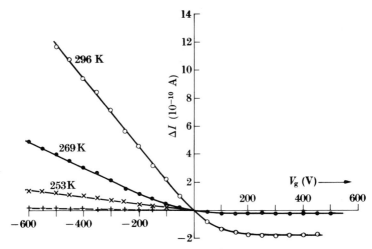

Fig. 9.18. Plot of ΔI against gate voltage v_s for an alloy glass, $As_{30}Te_{48}\ Si_{12}Ge_{10}$ at various temperatures. (From Marshall and Owen 1976.)

bulk). Marshall and Owen thus postulate centres this amount below E_F, which receive the surface charge (Fig. 9.15). These states are already identified from studies of d.c. current (§ 9.4.1).

Mahan and Bube (1977), however, while agreeing that the field-effect current is proportional to the gate voltage, find that the dependence on T is as shown in Fig. 9.19. Below room temperature the charge resides at the Fermi level and above room temperature in the valence (or conduction) band. Mott and Street (1977) suppose, as indicated in § 9.4.1, that pairs of D^+, D^- may give the states 0.13 eV below E_F, and, if present in a considerably higher concentration than unpaired defects, will give the behaviour observed by Marshall and Owen. Otherwise they expect the behaviour seen by Mahan and Bube. Probably the ratio of paired to unpaired states may be sensitive to heat treatment.

9.5. Drift mobility

As for glow-discharge-deposited silicon, measurements of drift mobility have yielded important results, but unlike that material the mobility seems to be dominated by discrete levels, some of which have been interpreted as charged dangling bonds. Drift mobility studies on amorphous As_2S_3 and As_2Se_3 have been made by Owen and Robertson (1970), Kolomiets and Lebedev (1967), Pai and Scharfe (1972). In contrast with the well-defined transits observed in amorphous selenium, for example, hole transport is characterized by a statistical spread in arrival times similar to that found in Se at low temperatures (see Fig. 6.27). If an 'effective' mobility is determined from the minimum of the spectrum of transit times in As_2Se_3, it

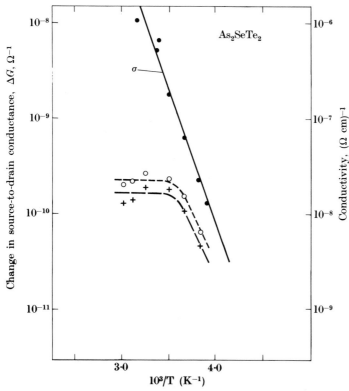

Fig. 9.19. Field effect conductance modulation in As_2SeTe_2 as a function of temperature. + for thickness 13 μm and voltage 182 V, $\bigcirc$ for 8 μm and 470 V. σ shows the conductivity. (From Mahan and Bube 1977.)

is found to be electric-field dependent (Fig. 9.20). The 'zero-field' hole mobility at room temperature is $\sim 5 \times 10^{-7}$ cm^2 V^{-1} s^{-1} and, as a function of temperature, is observed to have an activation energy of about 0·5 eV. Mott, Davis, and Street (1975) propose that this is the energy required to release a hole from a D$^-$ state (see § 9.7).

More recent work by Owen, Marshall, and co-workers includes the effects of high fields (§ 9.9). Fisher, Marshall, and Owen (1976) study the glasses As–Se over a considerable range of composition. By comparing the activation energies for different transit times and also from the investigation of thermally stimulated currents (Street and Yoffe 1972), they conclude that there is a series of traps with fairly discrete energy levels and that different experiments reveal different values because the time to achieve a quasi-equilibrium depends on the experimental conditions.

For the more conducting materials such as As$_2$Te$_3$ and alloy glasses, it has not as yet proved possible to make transport measurements by the

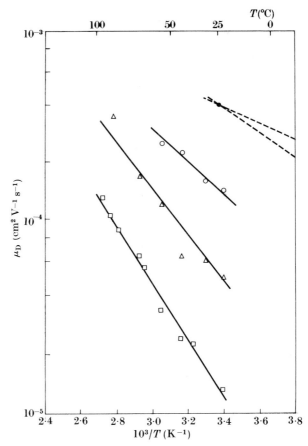

Fig. 9.20. Temperature dependence of hole drift mobility in amorphous As_2Se_3 for different electric fields: $\square$, $9\cdot4\times10^4$ V cm^{-1}; $\triangle$, $18\cdot8\times10^4$ V cm^{-1}; $\bigcirc$, $28\cdot2\times10^4$ V cm^{-1}; $\bullet$, 55×10^4 V cm^{-1}. (From Owen and Robertson 1970.)

time-of-flight method because of the short relaxation times in these materials. Marshall and Owen (1975), however, have obtained values of μ_D in photoconduction in the following way. Using a coplanar configuration for the specimen, the current is measured after the onset of illumination at times when a quasi-equilibrium has been established with 'traps', but for times smaller than the recombination time τ. This current should then be $ne\mu_D F$, where n is the total number of quanta absorbed per unit volume, which is known. Fig. 9.21 shows their results for As_2Te_3. The activation energy $0\cdot22$ eV is interpreted by them as the depth of a trap; again we suppose this to be the D^- state. The behaviour at low temperatures is interpreted by Marshall and Owen as being due to a discrete trap, nearer the Fermi energy and present in low concentration; Mott, Davis,

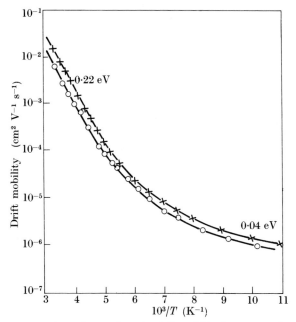

Fig. 9.21. Temperature dependence of carrier drift mobility in two samples of amorphous As$_2$Te$_3$ for low fields. (From Marshall and Owen 1975.)

and Street (1975), however, suggest that it may be due to hopping between D$^-$ and D (i.e. a hole trapped by a D$^-$), a process that would involve an activation energy of polaron type. The interpretation of Marshall and Owen may well be correct, and if so perhaps these traps are paired D$^+$ and D$^-$, as already suggested.

From Fig. 9.20 Marshall and Owen deduce the concentration N of the former defect (D$^-$ in our view); one can write

$$\mu = \frac{\sigma_{min}}{eN} \exp\left(-\frac{\varepsilon}{kT}\right)$$

where σ_{min} could be taken from the extrapolation of the dark conductivity to $1/T = 0$, corrected for the term $e^{\gamma/k}$ (eqn (6.11)). If we take a reasonable theoretical value for conduction at a mobility edge ($\sigma_{min} \sim 10^2 \, \Omega^{-1} \, cm^{-1}$), the results of Fig. 9.13 give $N \simeq 10^{19} \, cm^{-3}$. However, in the temperature range of the experiments the results of Nagels *et al.* (1974) discussed in §9.2.4 indicate that conduction is by hopping at the extremity of the band, and instead of σ_{min} an appropriate factor will be $\sigma_0 \sim 2 \, \Omega^{-1} \, cm^{-1}$. It thus appears that $N \sim 2 \times 10^{17} \, cm^{-3}$ may be a better estimate.

Drift mobility measurements on As$_2$Se$_3$ doped with iodine have been reported by Banerji and Hirsch (1974) and by Pfister, Melnyk, and Scharfe (1977) (see §10.3.1).

9.6. Luminescence

Radiative recombination during photoexcitation of carriers (pho-toluminescence) in amorphous chalcogenide semiconductors has been observed and studied by several groups, some of the earliest work being due to Kolomiets and co-workers. Some of the principal features are illustrated by data on As_2Se_3 in Fig. 9.22 (from Bishop and Mitchell 1973).

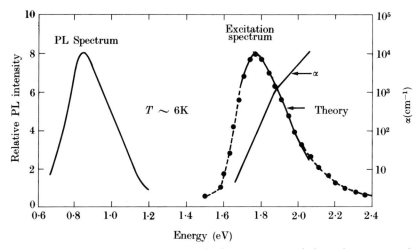

Fig. 9.22. Photoluminescence spectrum, excitation spectrum, and absorption spectrum for glassy As_2Se_3. The solid circles and connecting broken line represent the measured excitation spectrum for a sample $0 \cdot 15$ cm thick. The 'theory' refers to that of de Vore. (From Bishop and Mitchell 1973.)

(1) The luminescence is emitted in a broad peak of width a few tenths of an electronvolt centred at an energy considerably less than the optical gap and, in most cases, quite near to the activation energy for d.c. conduction and thus approximately half the band gap (Street 1976). In at least one case (Kolomiets, Mamontova, and Babaev 1970) a smaller emission peak was observed close to the band-gap energy in crystalline As_2Se_3 and Kolomiets *et al.* (1972) in an alloy glass.

(2) The luminescence is most efficient when excitation occurs in the tail of the optical absorption edge (absorption coefficients $\sim 10-10^2$ cm^{-1}). This feature was largely responsible for the failure of several earlier attempts to observe luminescence and therefore to reproduce the original results of Kolomiets, Mamontova, and Babaev (1970). The excitation spectra shown in Fig. 9.22 is a measure of the integrated luminescence intensity *versus* excitation energy. The *shape* of the luminescence band is found to be independent of excitation energy.

(3) The intensity of the luminescence is a strong function of tempera-ture, increasing by several orders of magnitude on cooling from room

temperature to that of liquid helium and reaching an efficiency ~10–20 per cent.

(4) The decay of luminescence following cessation of excitation is fairly rapid: according to Street *et al.* 1973, Street, Searle, and Austin 1974*a*), 95 per cent of the signal decays within about 1 ms, the remainder decaying with a time constant of 5–10 ms. There is also a decay *during* excitation (fatigue). The decay is, however, slow compared with that in a–Si (cf. §§ 6.7.6, 7.1).

(5) The luminescence can be enhanced by simultaneous excitation with light in a restricted wavelength band (Bishop, Strom, and Guenzer 1974).

With regard to the electronic transitions involved in the luminescence, three models have been proposed. The first (Weiser 1972) involves thermalization of the photogenerated electron–hole pair during which time the carriers emit phonons and fall down through localized states in band tails until they reach 'recombination edges', whereupon they recombine radiatively. A description of this process has been given by Fischer and Vornholz (1975). The second model, first proposed by Kolomiets, Mamontova, and Babaev (1970) (see also Kolomiets *et al.* 1974) involves localized recombination levels situated somewhere near the centre of the energy gap, with either the downward electron or upward hole transition giving rise to the luminescence. The third, put forward by Street and Mott (1975) and Mott, Davis, and Street (1975), supposes that the excitation takes place at the negatively charged dangling bonds already described (D^-) or in some cases D^+ (Mott and Street 1977). These give rise to absorption near the band tail but the shift of the emission frequency to near half the band gap is due to a Stokes shift.

The evidence for the luminescence involving defect levels is fairly conclusive. From experiments concerned with the softening temperature and dissolution properties of As_2Se_3 doped with In and Ge, Kolomiets, Mamontova, and Babaev (1972) conclude that the recombination centres involved could not be associated with impurities and might be identified with broken bonds. More detailed doping studies on Se by Street, Searle and Austin (1974*b,c*) lead to a similar conclusion; the intensity of the luminescence increased markedly with addition of As and Te both of which are expected to increase the number of unsatisfied bonds, the former by branching and the latter by shortening Se chains. Street, Searle, and Austin (1975) give evidence for weak tails to the absorption spectrum, both in crystalline and amorphous As_2Se_3, which can be associated with these defects. Some further details of the luminescence spectrum in Se alloys are presented in Chapter 10.

The observation that the excitation spectrum peaks at a photon energy corresponding to low values of the absorption coefficient has been considered in detail by Bishop (1973), Bishop and Mitchell (1973) and by Street *et al.* (1973), Street, Austin, Searle, and Smith (1974) and by Street (1976). Both groups agree that the drop in excitation efficiency on the low-photon-energy side of the peak arises simply because of decreased absorption of the radiation by the sample; as expected therefore this portion of the excitation spectrum depends on sample thickness as shown in Fig. 9.23 on

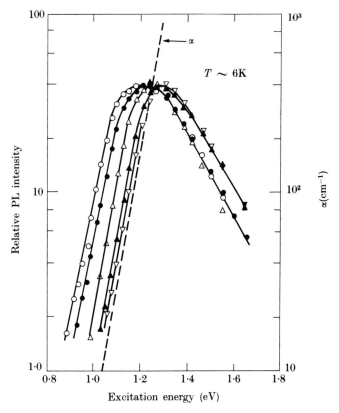

Fig. 9.23. Photoluminescence excitation spectrum for samples of $As_2Se_{1.5}Te_{1.5}$ of varying thickness. All spectra are normalized to the number of incidence photons and to equal peak intensity. The optical absorption spectrum α is also shown. Sample thicknesses are (in cm) ●, 0.1; △, 3.4×10^{-2}; ▲, 1.5×10^{-3}; ▽, 1.0×10^{-3}. (From Bishop and Mitchell 1973.)

$As_2Se_{1.5}Te_{1.5}$. What is more interesting, however, is that for very thin samples ($<15 \mu$m) the peak and the high-energy side of the excitation spectrum also shift. From this and similar observations on As_2Se_3, Bishop and Mitchell (1973) concluded that surface recombination is responsible for the high energy fall-off and, fitting their results to the theory due to De Vore (1956), deduce a diffusion length for the photo-excited carriers of 1 or 2 μm. This diffusion length L appears large for an amorphous material; when combined with a characteristic response time for the luminescence of about 10^{-4} s (deduced from a decrease in luminescence intensity for chopping frequencies above about 250 Hz), a mobility for the diffusing carriers $(\mu = L^2 e/\tau kT)$ of 0.55 cm^2 V^{-1} s^{-1} was inferred, which is orders of magnitude greater than measured drift mobilities in As$_2$Se$_3$ (see § 9.5).

An alternative explanation of the high-energy fall-off was proposed by Street et al. (1973) and Street (1976), who do not find agreement with the De Vore theory, at least in the case of As$_2$S$_3$, and furthermore emphasize the temperature independence of the shape of the excitation spectrum. Instead of surface recombination they propose a quantum efficiency dependent on the excitation energy. These authors assume (1) that emission takes place at defects, later to be identified with charged dangling bonds; (2) that photoluminescence occurs either when one of these defects is excited directly, or when an electron–hole pair is excited in the vicinity and can drift towards it, as described in § 6.7.6; (3) photoluminescence occurs only if radiation can occur before the particles separate. In this model, the greater the absorption frequency, the more likely are the electron–hole pairs to separate and then recombine by a non-radiative path. This model is in accord with the observations that the temperature dependence and the shape of the luminescence spectrum as well as the low-temperature decay time of the luminescence are all independent of the excitation energy.

The temperature dependence of the luminescence efficiency provides further information on the process by which carriers can separate. For almost all chalcogenides so far investigated (see however Kirby and Davis 1979) the intensity follows the relation (Street et al. 1974a, Street 1976)

$$y_L = \text{const. } \exp(-T/T_0) \qquad (9.5)$$

where T_0 is characteristic of the material. This behaviour is illustrated for a few glasses in Fig. 9.24. Values of T_0 vary between 20 and 40 K. Street et al. (1974a) account for this in the following way.

(a) The centre is charged, so that to separate a carrier from it after excitation leaving a D^0, no Coulomb attraction has to be overcome. In terms of the model of § 9.3, these centres are D$^-$ and the carrier is an electron, or the centre is D$^+$ and the carrier a hole.

(b) In the conduction band there is a long-range spatial variation of the potential, probably caused by charged centres (D$^+$ and D$^-$).

After relaxation of the centre (D^0 with an electron weakly bound to it), the electron can either recombine with the emission of radiation or diffuse away, which will lead to a radiationless recombination process to be discussed later. For the second process, Street et al. (1974a) develop a simple model in which the potential fluctuations are represented by parabolic

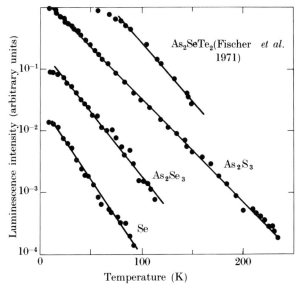

Fig. 9.24. Temperature dependence of the luminescence intensity of various chalcogenide glasses. (From Street 1976.)

wells separated by a distance R_0. These may perhaps be caused by the microfields which lead to an Urbach edge in the model of Dow and Redfield (1972; cf. § 6.7.1). Assuming a tunnelling rate proportional to $\exp(-2\alpha r - w/kT)$ and a tunnel length r depending on the activation energy w according to the equation $r = R_0 - 2\gamma w^{1/2}$, where γ defines the parabolas, one finds that the energy at which the rate is a maximum is $w = (2\alpha\gamma kT)^2$. Thus the tunnelling rate is proportional to $\exp(T/T_0)$, where $T_0^{-1} = 4\alpha^2\gamma^2 k$. The luminescence efficiency η can be written in terms of the probabilities for radiative (p_r) and non-radiative (p_{nr}) paths for low values of η by

$$\eta = \frac{p_r}{p_r + p_{nr}} \sim \frac{p_r}{p_{nr}}.$$

Under these conditions the lifetime τ is given by

$$\tau^{-1} = p_{nr}$$

and since the temperature dependence of the lifetime and the efficiency are observed to be the same (see below), Street $et\ al.$ deduce that p_r is independent of temperature and therefore that

$$\eta \propto \exp(-T/T_0).$$

From the observed values of T_0, one can deduce a separation between neighbouring wells given by $R_0 \sim 10/\alpha$, and thus of 50 Å for a tunnelling constant α^{-1} of 5 Å.

Most of the above experimental results were obtained under pulsed conditions on account of the decay of the luminescence during excitation. This 'fatigue effect' has been investigated particularly by Mollot, Cernogora and Benoit à la Guillaume (1974) and by Cernogora, Mollot, and Benoit à la Guillaume (1973, 1974), and an explanation in terms of the progressive conversion of D^- to D^0 centres has been given by Street (1976). Of interest also is the decay of luminescence *following* excitation. Using a xenon flash lamp with a pulse duration of 10 μs, Street *et al.* (1974a) have investigated this decay in As_2S_3 at various points on the emission spectrum and as a function of temperature. Fig. 9.25 shows a typical decay curve; there is no unique time constant. The decay depends slightly on the emission energy (Street 1976), but the decay lifetime at 10 K is found to be independent of excitation energy.

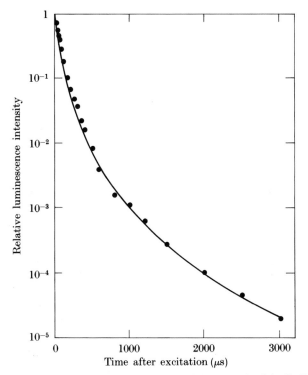

Fig. 9.25. Decay of luminescence at 7 K of a melted glass sample of As_2S_3. Excitation energy 2·34 eV, emission energy 1·24 eV. (From Street 1976.)

In Figs. 9.26(a,b) we show some results of Bishop *et al.* (1974) on enhancement spectra in As_2Se_3 and As_2S_3 glasses. It was found that irradiation of the samples with continuous monochromatic light enhanced the luminescence and that the enhancement spectrum extended nearly all

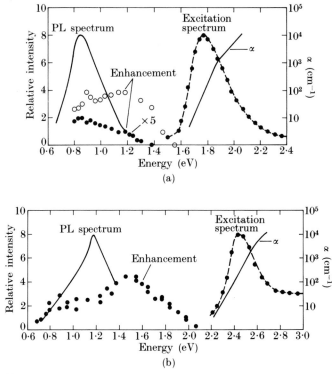

Fig. 9.26. (a) Photoluminescence excitation, enhancement, and absorption spectra for glassy As$_2$Se$_3$ at 6 K. The excitation and enhancement spectra are normalized to the number of incident photons. (b) The same for As$_2$S$_3$. (From Bishop *et al.* 1974.)

the way from the low-energy limit of the luminescence to the low-energy limit of the excitation spectrum. For As$_2$Se$_3$ two enhancement spectra corresponding to two different spectral distributions of interband exciting light are shown. The enhancement reached 40 per cent and had a response time of several seconds. It would appear that interband photoexcited carriers which become trapped and whose ultimate fate would be non-radiative recombination can be re-excited from these centres back into the radiative channel.

Next we compare the luminescence observed in amorphous materials with that in the crystals. As first reported by Kolomiets the luminescence spectra in amorphous and crystalline As$_2$Se$_3$ and As$_2$S$_3$ are very similar, lending support to the idea that defect centres are involved and further-more that these are of the same kind in the two phases. In contrast the excitation spectra and the temperature dependence of the luminescence are quite dissimilar, and there are no observable fatiguing effects. Fig. 9.27 compares the two phases of As$_2$Se$_3$. The most obvious difference is in the

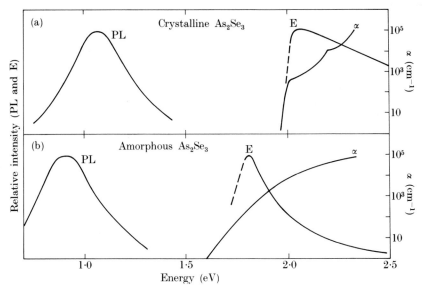

Fig. 9.27. Excitation and photoluminescence spectra and absorption coefficient of crystalline and amorphous As₂Se₃. (From Street 1976.)

excitation spectrum, which for the crystal does not exhibit the high-energy fall-off characteristic for glasses. The higher mobility expected in the crystal appears to ensure that the radiative path is followed even if excitation is not in the vicinity of the recombination centres. This is supported further by the observation that the luminescence efficiency in the crystal is high, approaching unity at low temperatures.

The temperature dependence of the luminescence intensity also differs between the two phases as shown in Fig. 9.28. Instead of the dependence as $\exp(-T/T_0)$ displayed by the glass, the behaviour in the crystal is more accurately described by two well-defined activation energies.

The linewidths of the photoluminescence spectra are very similar in crystalline and amorphous chalcogenides, suggesting that the breadth of the luminescence is independent of disorder. Street (1976) (see also Street, Austin, and Searle 1975) propose that the breadth is due to coupling with phonons by the mechanism illustrated in Fig. 3.7.

To summarize, then, we suppose that photoluminescence is due to excitation in D^- centres (or possibly D^+) which are present in a concentration of 10^{17}–10^{18} cm^{-3}; a charged bound exciton is then formed. The centre relaxes; the Stokes shift leads to re-emission of a broad line with frequency such that $h\nu$ is about half the band gap and broadened according to eqn (3.36). This is confirmed by the similarity between the behaviour of crystals and glasses. Unless the temperature is low, the charged exciton

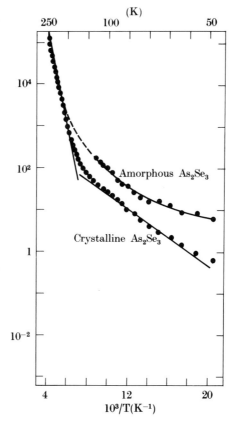

Fig. 9.28. Temperature dependence of photoluminescence intensity of crystalline and amorphous As_2Se_3. The quantity plotted is $y^{-1} - 1$ where y is the intensity normalized to unity at the low-temperature limit. (From Street 1976.)

dissociates into a D^0 centre and a free electron, which recombines without radiation by one of the mechanisms described in the next section.

This model is supported by more recent results of Bishop *et al.* (1975, 1976*a,b*, and 1977 for a review). These authors find an optically induced e.s.r. signal and optical absorption due to trapped holes in several chalcogenide glasses illuminated below 15 K. The assumption is that neutral D^0 centres are created, and that at low temperatures these are stable. Some discussion of the e.s.r. behaviour is given in § 6.8.3. The concentration of defects responsible is estimated to lie in the range 10^{17}–10^{18} cm^{-3}.

Doping with In or Ge changes the luminescence spectrum of As_2Se_3 (Kolomiets, Mamontora, Smorgonskaya, and Babaev 1972), as do traces of O, S, and Te in amorphous Se (see Street 1976). This is discussed in § 9.8; doping is expected to change drastically the relative concentration of the

D^+ and D^- centres, so that the sign of the centre responsible for the photoluminescence may change (Mott and Street 1977).

The model supposes that when an electron and a hole escape from each other's field, recombination is always non-radiative. It is known that a free electron–hole pair recombines non-radiatively (§ 6.7.6), so it is supposed that the mechanism of Dexter *et al.* (1955, cf. § 3.5.2) must be operative, because a larger distortion occurs when the electron is in a *small* orbit (Mott 1977*a*). Another mechanism will be described in the next section.

9.7. Photoconductivity

There have been many observations of photoconductivity in As_2Se_3, As_2Te_3, and alloy glasses.[†] In general, these show, when photocurrent is plotted against $1/T$, the three regimes of Fig. 6.31; these are the high-temperature regime I in which the photocurrent i_p is less than the dark current and proportional to G (the number of quanta absorbed per $cm^{-3} s^{-1}$). The rate of recombination is then determined by the *dark* concentration of carriers. Next comes regime II, where the number of carriers is greater than the dark concentration, so that $1/\tau$ is proportional to the number of photoexcited carriers and $i_p \propto G^{1/2}$. Thirdly in regime III there is little dependence on temperature and i_p is proportional to G. Direct measurements of the rate of recombination have also been made, and these will be discussed below.

Experiments on photoconductivity can measure the total charge generated; if carriers are formed in a thin layer adjacent to one surface of a film, a strong enough pulsed field will extract them. A possible dependence of quantum efficiency on frequency of the radiation and on temperature is discussed in § 6.5.3. In the experiments described here it is believed that the quantum efficiency normally approaches unity.

We discuss first direct measurements of the decay time τ. This has been measured for As_2Te_3 as a function of temperature by Moustakas and Weiser (1975), and Fig. 6.33 shows their results for three intensities. While in regime I the lifetime τ, due to recombination of holes with *thermally* generated electrons, does not depend on the intensity of the radiation and decreases with increasing concentration of the electrons, in regime II it seems to be independent of T. Fig. 6.34 shows similar results for As_2Se_3 due to Main; there is now a small activation energy (0·08 eV) in τ.

A value of τ that depends little on temperature is to be expected for a multiphonon process at fairly low temperatures, though $1/\tau$ should eventually rise as T increases. The process we envisage for chalcogenides is the following: first the electrons are captured by D^+ centres, the holes are in

---

[†] Fagen and Fritzsche (1970), Weiser *et al.* (1970), Arnoldussen *et al.* (1974), Kolomiets and Lyubin (1973), Grant *et al.* (1974), Main (1974), Taylor and Simmons (1974), Moustakas (1974), and Moustakas and Weiser (1975).

quasi-equilibrium with the D^- centres, and trapping and release determines their mobility. For recombination, the rate-determining step is

$$2D^0 \to D^+ + D^-$$

which can be treated by the method of Chapter 3 as a multiphonon process.

If τ is independent of T, or if its temperature dependence is known, the mobility of electrons can be deduced from the slope of the plot of $\ln i_p$ against $1/T$ in regime II. There has been controversy about how this is to be interpreted. Moustakas and Weiser (1975) and earlier writers note that for As_2Te_3 the slope (~ 0.15 eV) corresponds well with the difference between E_σ and E_S, which supports a polaron (or other hopping) model for the mobility. If the considerations of § 9.3.3 are valid, however, the cor-responsible for the photocurrent in regime II as for the dark current. The process for the generation of holes may be split up into

To summarize, then, it seems likely that recombination occurs through trapping by D^+, D^- states, with tunnelling between them by the process $D^0 \to D^+ + D^-$ as the rate-determining step. This is, at most, weakly dependent on T and the slope of the curve of $\ln i_p$ versus $1/T$ is mainly determined by the trap-limited mobility.

In regime III it seems probable that carriers which fall into the traps (D^+, D^-) do not get out again before recombination, as was proposed by Simmons and Taylor (1974). Some measurements at helium temperatures (Jenkins, Levy, and Hodby 1976) are discussed by Mott, Davis, and Street (1975).

A consequence of this model is that the same sign of a carrier must be responsible for the photocurrent in regime II as for the dark current. The process for the generation of holes may be split up into

$$D^+ + D^- \to 2D^0$$

followed by

$$D^0 \to hole + D^-.$$

For the generation of electrons the second process becomes

$$D^0 \to electron + D^+.$$

For a p-type material, the second process must have the larger energy. However, these last two equations are those that determine the activation energies in the drift mobility, and thus that in i_p in regime II.

9.8. The effect of alloying on the dark current

The conductivity of chalcogenides is in general insensitive to the addition of small amounts of elements such as Si and Ge. This property, already referred to earlier, was first established by Kolomiets and co-workers (see

Kolomiets 1964) and is usually explained by the 8-N rule, according to which each element is surrounded by 8-N neighbours (N being the number of electrons outside a closed shell) so that all electrons are taken up in bonds. Even at *alloying* concentrations such that the optical gap changes, the Fermi level remains pinned close to the centre of the gap (see Fig. 9.13).

In some systems, one can pass right through stoichiometric proportions without any discontinuity in the conductivity. Fig. 9.29 shows the room-

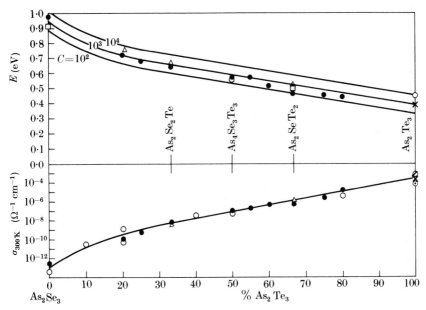

Fig. 9.29. Variation of the activation energy for electrical conduction and the room-temperature conductivity in the As$_2$Se$_3$–As$_2$Te$_3$ system. Data from ○ Uphoff and Healy (1961), ● Vengel and Kolomiets (1957), × Weiser and Brodsky (1970), ⊘ Rockstad (1970), □ Edmond (1966), △ Male (1970), and ⊙ Croitoru *et al.* (1970).

temperature conductivity σ and activation energy E in the system As$_2$Se$_3$–As$_2$Te$_3$. If we write $\sigma = C \exp(-E/kT)$ and calculate the variation of E from the measured variation in conductivity assuming $C = 10^2$, 10^3, and $10^4 \, \Omega^{-1} \, \text{cm}^{-1}$, it is seen (top three curves) that over the whole composition range C lies near $10^3 \, \Omega^{-1} \, \text{cm}^{-1}$, a value appropriate for transport at a mobility edge. The optical gap in this system also varies smoothly. In other systems, extrema in conductivity and optical gap occur at stoichiometric proportions (see Rockstad and de Neufville 1972, Hurst and Davis 1974), but the pinning of the Fermi level near mid-gap seems to be quite general. We have suggested that pinning is associated with the presence of point defects D$^+$ and D$^-$.

There are however exceptions to the rule. The effect of oxygen in increasing the conductivity of amorphous selenium is well known, and Cu, Mn, and many other elements are known to increase the conductivity of chalcogenides by decreasing the activation energy for conduction, and often without an associated decrease in the optical gap. Our theme in this section is that they act by unpinning the Fermi energy.

In a normal n-type semiconductor with uncompensated donors ε_1 below the conduction band, one places the unpinned Fermi energy, in the limit as $T \to 0$, at $\frac{1}{2}\varepsilon_1$ below the conduction band; the introduction of compensation displaces it to ε_1 below. If an upper Hubbard band (Chapter 4) is taken into account at an energy $\varepsilon_1' = \varepsilon_1 - \varepsilon_2$ below the conduction band, it would be more accurate in the uncompensated case to place the Fermi level midway between ε_1 and ε_1'. Normally $\varepsilon_1 - \varepsilon_1'$ is the Hubbard U. If as in chalcogenides distortion takes place, so that $\varepsilon_1 - \varepsilon_1'$ decreases, it is clear that the Fermi energy remains midway between them, and when the condition is reached that the reaction $2D^0 \to D^+ + D^-$ is exothermic or involves no energy, the Fermi energy becomes pinned, as defined in § 2.10. In this section we propose that if the specimen contains negatively charged impurities, the Fermi energy becomes unpinned again, and that this is the cause of the high conductivity observed in certain cases.

With the model of charged dangling bonds, the energy W to produce a pair of holes is that of the reaction

$$D^+ \to D^- + 2p$$

where p denotes a hole. Normally the activation energy for conduction, E_σ, will be $\frac{1}{2}W$. This may be seen by the principle of detailed balancing as follows. The rate of generation of holes will be proportional to $\exp(-W/kT)$. As long as D^- and D^+ are present in comparable quantities, the recombination rate which involves a collision between two holes and a D^- will be proportional to p^2. Thus

$$p \propto \exp(-\tfrac{1}{2}W/kT).$$

It has been suggested by one of us (Mott 1976d), however, that in glasses containing charged impurities (e.g. Mn^{2-}) the compensating charge will be D^+ centres, so that in the melt the law of mass action will ensure a very small concentration of D^-. If N^+, N^- are the concentrations of the D^- centres,

$$N^+N^- = C \exp(-E/kT) \tag{9.6}$$

where E is the energy to form the pair. If this is so, the concentration of D^- may be negligible at room temperature and equal to $\frac{1}{2}p$ at finite temperatures. The rate of recombination is then proportional to p^3, so that

$$p \propto \exp(-\tfrac{1}{3}W/kT). \tag{9.7}$$

We thus expect a drop in the activation energy by $\frac{1}{3}$. The Fermi energy is now no longer pinned. This behaviour is analogous to that of a doped and compensated semiconductor, where as we have seen E_F is pinned at ε_1 below a conduction band, and with the removal of compensation rises to $\frac{1}{2}\varepsilon_1$ below it.

The same result can be obtained from Boltzmann statistics as follows. We suppose that, at $T = 0$, there are N positive and M negative centres per unit volume. At finite temperature there are $N - \frac{1}{2}p$ positive and $M + \frac{1}{2}p$ negative centres and p free holes. The free energy is therefore

$$\tfrac{1}{2}pW - kT \ln\left\{\frac{(M+N)!N_v!}{(M+\frac{1}{2}p)!(N-\frac{1}{2}p)!\,p!(N_v-p)!}\right\}$$

where $N_v = N(E_v)kT$ for carriers at a mobility edge. This is a minimum when

$$-\frac{1}{2}\frac{W}{kT} = \ln\left\{\frac{p}{N_v-p}\left(\frac{M+\frac{1}{2}p}{N-\frac{1}{2}p}\right)^{1/2}\right\}$$

and thus, if $N_v \gg p$ and $N \gg \frac{1}{2}p$, when

$$p(M+\tfrac{1}{2}p)^{1/2} = N_v N^{1/2} \exp\left(-\frac{1}{2}\frac{W}{kT}\right). \qquad (9.8)$$

In the normal case $(M > p)$

$$p = N_v\left(\frac{N}{M}\right)^{1/2} \exp\left(-\frac{1}{2}\frac{W}{kT}\right). \qquad (9.9)$$

For $M \ll p$

$$p = 2^{1/2}N_v^{2/3}N^{1/3} \exp\left(-\frac{1}{3}\frac{W}{kT}\right). \qquad (9.10)$$

As T increases, there should be a transition from one regime to the other, the slope of the plot of $\log \sigma$ versus $1/T$ decreasing, as shown in Fig. 9.30.

This model has been applied to the large conductivity increases shown by the addition of Cu, Ag, Ga, In, and Tl in As_2Te_3 and Cu and Ag in As_2S_3 (Mott 1976d, Mott and Street 1977). References are Danilov and Myuller (1962), Danilov and Mosli (1964), Edmond (1968), Kolomiets, Rukhyl-adev, and Shilo (1971), Owen (1967). Detailed measurements of E_σ are lacking, but they can be estimated from the values of σ. For As_2Se_3 the activation energy for (p-type) conduction is 0.9 eV, so if the D^+ centres are removed the activation energy should be reduced by 0.3 eV and the conductivity enhanced by $\exp(0.3\,\text{eV}/kT)$, which is $\sim 10^5$ at room temperature. This is similar to the enhancement (10^4) observed on the addition of 5 per cent of Cu according to results reported by Liang et al. (1974). The much smaller shift in the absorption edge might be due to the

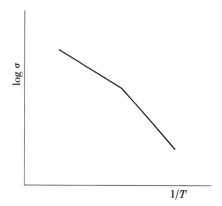

Fig. 9.30. Plot of log σ *versus* $1/T$ for the model discussed in the text (schematic).

four-co-ordinated copper with its associated D^+ centres, which could act as a deep donor near the valence-band edge.

Watanabe, Inagaki, and Shimizu (1976) find that in amorphous $Ge_{0.42}S_{0.58}$ the elements Ca and Ag (but not Zn, Cd, Al, In, or I) bring about a marked drop in the activation energy for conduction, from 1·13 to ~0·75 eV for about 0·2 per cent Ag, which is almost exactly $\frac{1}{3}$. They suggest that Ag acts as a chain terminator; if so it may be attached to D^- centres, leaving D^+ intact. Kolomiets, Mamontova, Smorgonskaya, and Babaev (1972) showed that about 0·2 per cent In shifts the photoluminescence line in As_2Se_3 from 0·9 to 0·77 eV and intensifies it (see also Street 1976). The In should remove the D^-, normally responsible for photoluminescence, and put in a high concentration of D^+. We think these must be responsible for the displaced line (Mott and Street 1977). Perhaps the quantum efficiency is much less (owing to easier escape of a hole), so it does not normally appear in the undoped material.

Some of the concepts used here have been criticized by Fritzsche (1977) and Kastner and Fritzsche (1978). Four-fold co-ordination certainly does not necessarily involve a negative charge. Moreover, it is difficult to understand that the concentration of D^- quenched from the glass transition temperature could be small enough to give the behaviour of Fig. 9.30.

9.9. Numerical values in the model of charged dangling bonds

This section gives estimates of some of the quantities involved in the model used in this chapter. The first is the concentration (N cm^{-3}) of the defects to be expected. The following evidence is available.

(a) The absorption tail identified by Street *et al.* (1974*a,b*) leading to photoluminescence. This suggests 2×10^{17} cm^{-3} for As_2S_3 and 10^{18} cm^{-3} for As_2Se_3.

(b) Screening lengths (Fritzsche 1973) indicate for As_2Te_3 about 10^{17} cm^{-3}.

(c) The behaviour of the threshold switch. If the theory of the on state given in § 9.13 is correct, in the alloy glass used N this must be less than about 10^{18} cm^{-3}.

(d) The absolute magnitude of the drift mobility. According to § 9.4, this indicates $\sim 2 \times 10^{17}$ cm^{-3} in As_2Te_3.

The second quantity is the energy W_1 required to take a hole from a D^0 centre into the valence band. This is the activation energy of the photocurrent in regime II and in the drift mobility. Street (1976) gives (in eV)

$$W_1 = 0 \cdot 29 \ (As_2Te_3)$$

$$W_1 = 0 \cdot 55 \ (As_2Se_3).$$

The energy of the reaction $D^+ + D^- \to 2D^0$ will be denoted by $2E$. Then $E + W_1$ is the activation energy for conduction (E_σ). Thus we find for E (in eV), using observed values of E_σ,

	E_σ(obs)	E
As_2Te_3	0·46	0·17
As_2Se_3	0·90	0·35

We note that ΔE, the spread of values of E, must be less than E if the bands are not to overlap, giving an e.s.r. signal.

Next, we look at W_2, the energy required to form an electron in the conduction band. This must be $B - E_\sigma$, where B is the band gap. Estimating B from optical data, we have (in eV)

	B(estimated)	W_2
As_2Te_3	0·8	0·34
As_2Se_3	1·8	0·9

We see that E_F is near to mid-gap. In the next chapter we reach a similar conclusion for selenium, for which more detailed data are available.

We next ask what hopping energies are expected for the following processes.

(1) The movement of an electron allowing a D^0 centre to exchange with D^+ and D^- centres.

(2) Charge transport due to exchange of D^+ and D^-.

For the former, one expects a hopping activation energy of about one-quarter the Stokes shift (Chapter 3), and thus one-eighth the band gap, $\sim 0 \cdot 1$ eV in As_2Te_3 and $0 \cdot 2$ eV in As_2Se_3. Such small energies allow unactivated motion at low temperatures of polaron type (eqn 3.35), and one expects therefore a small term in the current of this kind at low temperatures, with activation energies equal to E, the energy to form a D^0 centre, namely 0.35 eV and 0.17 eV for As_2Se_3 and As_2Te_3 respectively, if the concentration of defects is great enough to allow tunnelling from one to another, perhaps along percolation channels. This may possibly be the origin of the 'tail' to the $\log \sigma$ *versus* $1/T$ plot shown in Fig. 9.10.

In contrast, the hopping energy for the exchange of D^+ and D^- (motion of a bipolaron) should be four times as large and thus 0.68 eV even in As_2Te_3. The bipolaron should therefore be practically immobile at and below room temperature, as pointed out by Phillips (1976).

9.10. Charge transport in strong fields

An account of various theoretical models is given in § 3.12. For the chalcogenides, both the conductivity and the drift mobility have been investigated by various authors. Thus in glassy As_2Se_3 de Wit and Crevecouer (1972) find that the conductivity obeys the form $\sigma = \sigma_0 \exp(F/F_0)$ up to fields F of the order of 4×10^5 V cm^{-1} for various thicknesses of specimen between 10^{-3} and 10^{-1} cm, though above this field there is a rapid increase. A remarkable feature of these results is the apparent absence of any ohmic region, or region in which the conductivity varies as $\sinh(F/F_0)$. This behaviour is confirmed by the work of Marshall and Miller (1973) and Marshall, Fisher, and Owen (1974). These workers express the conductivity and the drift mobility in the form

$$\sigma(F) = \sigma(0) \exp(eaF/kT)$$

$$\mu_D(F) = \mu_D(0) \exp(eaF/kT)$$

and find that a, which decreases with increasing temperature, is the same for both. No full explanation of these results has been given. If the model of charged dangling bonds is accepted, then the drift mobility of holes is limited by trapping by D^- states and Poole–Frenkel behaviour is expected. As regards the conductivity, if in the dark conductivity we treat the generation of a carrier by two consecutive reactions

$$D^+ + D^- \rightarrow 2D^0 \quad \text{and} \quad D^0 \rightarrow \text{hole} + D^-$$

and suppose that the first process is unaffected by the field, the second is just the same as that which determines the drift mobility. Thus the equilibrium number of carriers will depend on the same parameters as in the drift mobility. One might expect the change in the number of free carriers to be given by a Poole–Frenkel mechanism. This is certainly not in agreement with the low-field results, with $\log \sigma$ increasing as F. Mott and Street (1977), however, show that by applying the Onsager (1938) mechanism as developed by Pai and Enck (1975), some measure of agreement may be obtained.

The electric-field dependence of drift mobility in As_2Se_3, for which the transient response is very dispersive, has been described and analysed by Pfister (1977a,b).

9.11. Density of electron states in conduction and valence bands

As described in Chapter 6, information on the energy spectrum of electron states in the valence and conduction bands (as opposed to edge and gap states) can be obtained by a variety of optical techniques. Absorption or reflectivity spectra, for example, can be analysed in terms of a joint or convoluted density of states. Assumptions concerning the energy dependence of interband matrix elements (which depend on the nature, and hence energy, of the initial state as well as on the energy of the exciting radiation) need to be made; frequently the assumption of 'constant matrix elements' is made, leading to a joint density of states (JDS) proportional to $\omega^2 \varepsilon_2(\omega)$, although, as pointed out by Liang and Beal (1976), it might be more reasonable to assume a 'constant oscillator strength' in which case the JDS is proportional to $\omega \varepsilon_2(\omega)$ (see Chapter 6). Use of high-energy radiation, for instance from a synchrotron, can be used to excite from narrow core levels and should, in principle, give the conduction-band density of states, although (as will be seen below) there are problems with interpretation associated with matrix elements and exciton effects. Photo-emission data can be analysed to yield the valence and conduction bands separately and, in particular, X-ray photo-emission spectroscopy (XPS) should give, after suitable correction of the raw data, the density of valence states (DOVS) alone. It is the latter that is most directly related to the structure, i.e. to local atomic configurations, but interpretation in these terms is less advanced for the chalcogenides than for Ge, Si, or As (Chapters 7 and 8).

Reflectivity spectra of crystalline and amorphous As_2S_3 and As_2Se_3 by Zallen *et al.* (1971) are shown in Fig. 9.31. Fine structure present in the (polarization-dependent) edges for the crystals is lost for the glasses, but the division into two bands separated by a minimum at about 7–8 eV is retained. This division is associated with a split valence band as will be evident from XPS spectra presented below. By a Kramers–Kronig analysis of such reflectivity data, the ε_2 spectra can be obtained (see Drews *et al.* 1972).

For XPS spectra we shall assume that the electron energy distribution curves (EDCs) have been properly corrected for inelastically scattered electrons and instrumental effects, so that the curves represent, at least approximately, the DOVs. In most cases the overall spectra have been determined using Al$K\alpha$ X-rays (1486·6 eV), and a typical resolution is ~0·5 eV. Increased resolution (~0·1–0·3 eV) is obtained using ultra-violet excitation (e.g. HeII line 40·8 eV, HeI line 21·2 eV, Ne line 16·9 eV), but a

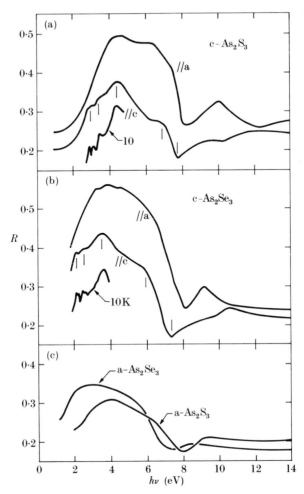

Fig. 9.31. Reflectivity spectra of crystalline and amorphous As_2S_3 and As_2Se_3. (From Zallen *et al.* 1971.)

problem associated with ultra-violet photo-emission spectroscopy (UPS) is a very low photo-excitation cross-section for s states. Composite spectra (XPS for s states and UPS for p states) are sometimes presented in the literature.

Fig. 9.32(a,b,c) shows XPS spectra for As_2S_3, As_2Se_3, and As_2Te_3 (Bishop and Shevchik 1975). The remarkable similarity between the spectra for the crystalline and amorphous forms of these materials (data for crystalline As_2Se_3 is not included) indicates little change in the energy distribution of valence states. For As_2S_3 and As_2Se_3 this is perhaps not unexpected, as the co-ordination numbers of both As and Se atoms (3 and

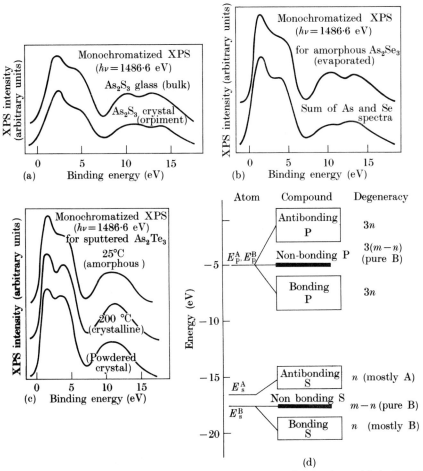

Fig. 9.32. XPS intensities for valence bands of crystalline and amorphous: (a) As_2S_3, (b) As_2Se_3, and (c) As_2Te_3. (From Bishop and Shevchik 1975.) (d) Bonding schemes in arsenic chalcogenides. (From Shevchik and Bishop 1975.)

2 respectively) remain unchanged. The lower curve in Fig. 9.32(b) is the sum of spectra obtained for the individual elements, As and Se, weighted according to their proportions in As_2Se_3; it thus appears that the orbitals of the constituent atoms interact in a similar way even if 'wrong' bonds are present. For As_2Te_3 the similarity is at first sight surprising, because, although the co-ordination in the glass is 3–2, in the crystal all the Te atoms are three-fold co-ordinated, while the As atoms occupy both tetrahedrally and octahedrally co-ordinated sites.

The three principal structural features in all the above spectra are associated with I non-bonding p orbitals, II bonding p orbitals, and III s orbitals. The way that these arise has been described by Shevchik and Bishop (1975). Fig. 9.32(d) illustrates the general situation in a compound $A_n B_m$ for which p and s atomic orbitals are important and well separated in energy. The antibonding p band forms the conduction band; the remainder are valence states. Bishop and Shevchik (1975) have confirmed this general picture by a simple tight-binding calculation based on a 'model' As_2Se_3 layer. The preservation of non-bonding orbitals in crystal-line As_2Te_3 (in which Te is three-fold co-ordinated), is associated with the fact that there are insufficient As p orbitals to interact with all of the Te p orbitals. If hybridization with s orbitals can be neglected and the crystal-field splitting of the p states is small compared with the bonding–antibond-ing interaction, a separate non-bonding band is obtained (Shevchik and Bishop 1975).

The s bands deserve some comment. It was shown in earlier chapters that for elemental Ge and to some extent for As, the dip present in the s band for the crystal is filled in for amorphous films, a feature that was explained by the presence of odd-membered rings in the structures. For the binary chalcogenides discussed here, the dip is to be ascribed more to the splitting of the s band into bonding states associated with the chalcogen and antibonding states associated with the metal (see Fig. 9.32(d)). Topological disorder appears to be insufficient to fill in the gap and perhaps a consider-able degree of chemical order is preserved. The absence of the dip in both crystalline and amorphous forms of As_2Te_3 arises because the As and Te s orbitals are nearly degenerate in energy and the 'chemical gap' disappears.

Bullett (1976) has calculated the electronic density of states of As_2S_3, As_4S_4, As_2Se_3, As_4Se_4, and As_2Te_3 crystals using a localized-orbital approach. The method is applicable to amorphous networks. Electron levels expected for certain defect sites, e.g. a three-fold co-ordinated Se atom, a two-fold co-ordinated As atom, and a Se dangling bond, were calculated, but severe self-consistency problems prevented treatment of thermal relaxation effects around such defects.

Fig. 9.33 shows the XPS spectra of amorphous and crystalline GeTe reported by Shevchik *et al.* (1973*a,b*). As for As_2Te_3 there is a difference in the short-range order between the crystalline and the amorphous form. Crystalline GeTe has a distorted rocksalt structure with a nearest-neigh-bour separation of 3 Å, while that of the amorphous form is believed to be a 4–2 coordinated CRN (Bienenstock 1974) with an interatomic spacing of $\sim$2·7 Å. The smearing of the Te 5s and Ge 4s levels is to be expected from a breakdown in chemical order (i.e. the presence of Ge—Ge and Te—Te bonds). The uppermost (lowest binding energy) p-like band exhibits more noticeable differences: the shape in this region can, however, be simulated, as for As_2Se_3 (Fig. 9.32(b)), by a simple addition of the DOVS found in elemental Ge and Te as shown in the lower half of the figure. The peak

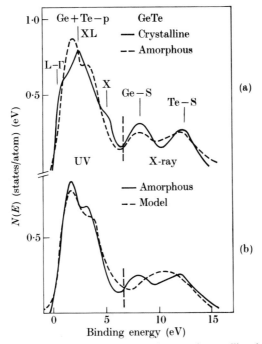

Fig. 9.33. (a) Density of valence states of amorphous and crystalline form of GeTe. (b) Comparison of the density of valence band states with the sum of those of amorphous Ge and crystalline Te. (From Shevchik *et al.* 1974*a*.)

near 2 eV is associated with lone-pair orbitals on the Te atoms. This is clear from Fig. 9.34 showing spectra for Ge_xTe_{1-x} alloys, obtained by Fisher *et al.* (1974), in which this peak is seen to disappear as the Te concentration is reduced. The energies of the lone-pair electrons are insensitive to whether the bonding electrons on Te interact with neighbouring Te or Ge atoms.

An example of core-level spectra obtained for GeTe (Shevchik *et al.* 1973*b*), are shown in Fig. 9.35. Both the Ge and Te levels shift by about 0·6 eV to higher binding energies in the amorphous form relative to the crystal, but this arises from a simple shift in the Fermi level (see § 6.4.3). There are no relative shifts between the Ge and Te levels in the two forms, indicating that no large change occurs in the charge distribution around the constituent atoms. Furthermore, the breadths of the core levels are similar in the amorphous and crystalline forms, which is slightly surprising if wrong bonds are present.

Synchrotron radiation absorption experiments on amorphous As_2Se_3 by Bordas and West (1976) are presented in Fig. 9.36. The lower curve shows the spectrum produced by *addition* of the spectra obtained following excitation from the d-core levels of As and Se atoms *in the compound*. The

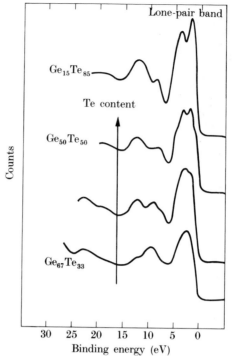

Fig. 9.34. XPS spectra of $Ge_x Te_{1-x}$. (From Fisher *et al.* 1974.)

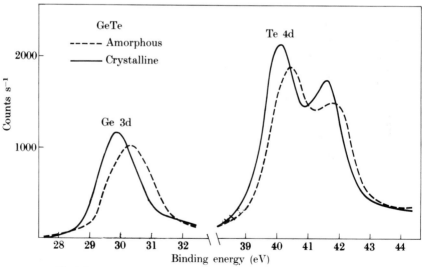

Fig. 9.35. Core-level spectra for GeTe. (From Shevchik *et al.* 1973*b*.)

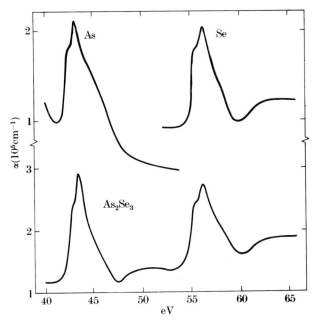

Fig. 9.36. Absorption of synchrotron radiation in amorphous As and Se and (lower curve) in amorphous As_2Se_3. (From Bordas and West 1976.)

upper curve shows the individual spectra for the elements in the amorphous form. The similarity between these curves suggests that in the compound the conduction band of As_2Se_3 is *not* being probed, but rather the sum of final states associated with the excited atoms. Furthermore this behaviour casts doubts on whether the conduction bands of As and Se are being probed in the experiments on the elements. The role of core–exciton effects as well as matrix elements are not sufficiently clear to analyse these data further. However, in contrast to Si (Chapter 7), core absorption does *not* give the conduction band density of states for As_2Se_3.

9.12. Optical properties

The optical absorption edges of the chalcogenides are characterized by an absorption coefficient α that rises exponentially with increasing photon energy up to a value of α in the range 10^3–10^4 cm^{-1}. This spectral Urbach's rule has been discussed in Chapter 6, and some chalcogenide glasses were included in Fig. 6.44. At higher values of the absorption coefficient, the most frequently reported behaviour is $\alpha \hbar\omega = B(\hbar\omega - E_0)^2$, where B lies in the range 10^5–10^6 cm^{-1} eV^{-1} and E_0 can be taken as an optical gap. Examples were shown in Fig. 6.48. Values of E_0 determined in this way generally correspond to actual values of α (on the Urbach edge) lying

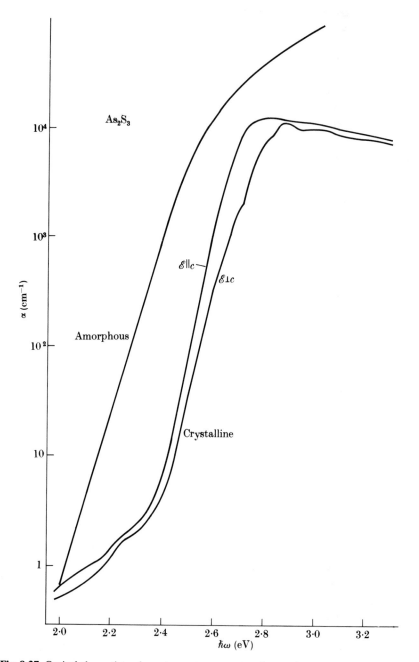

Fig. 9.37. Optical absorption edges at room temperature of amorphous and crystalline As_2S_3. (From Kosek and Tauc 1970.)

between 10^2 and 10^3 cm^{-1}. At low values of α the Urbach behaviour is sometimes observed to break away to another approximately exponential form but of lower slope. In glasses specially prepared for use in fibre optics this tail absorption occurs at very low values of α ($\sim 10^{-5}$ cm^{-1}) and in some cases may not be true absorption but a drop in transmission caused by Rayleigh scattering from macroscopic density fluctuations. In chalcogenides it frequently occurs at α values in the range 10^{-1}–10^2 cm^{-1} and, although in the more conducting glasses it can arise from free carrier absorption, it is more commonly caused by defect or impurity absorption. In many materials the defect centres appear to be the dangling bonds which are important for photoconductivity and luminescence (see §§ 9.4, 9.6, 9.7).

Figs. 9.37 and 9.38 show the optical absorption edges for amorphous As_2S_3 and As_2Se_3 compared with those for the crystalline forms. Fine

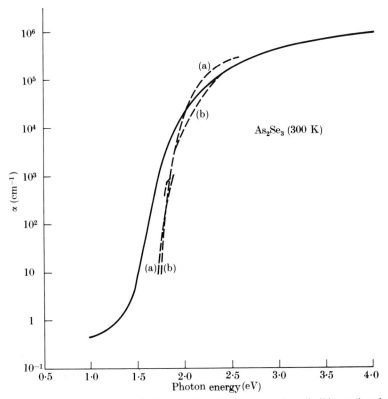

Fig. 9.38. Room-temperature optical absorption edges in amorphous (solid curve) and crystalline As_2Se_3 (broken curves); electric vector (a) parallel and (b) perpendicular to the a axis. Data for the amorphous material from Felty and Myers (private communication) and Edmond (1966), and for the crystalline material from Shaw *et al.* (1970) and Zallen *et al.* (1971a.)

structure present in the (polarization-dependent) edges for the crystals are lost and accurately exponential edges obeying $\alpha = \alpha_0' \exp(\Gamma \hbar \omega)$ are observed. Values of Γ at room temperature are $18\cdot6\,\text{eV}^{-1}$ for As_2S_3 and $20\,\text{eV}^{-1}$ for As_2Se_3; these slopes are not particularly sensitive to the conditions of preparation, a feature which is generally true for chalcogenides.

Fig. 9.39 compares the edge in amorphous As_2S_3 at room and liquid-nitrogen temperature. Curve (a) shows that the slope of the exponential

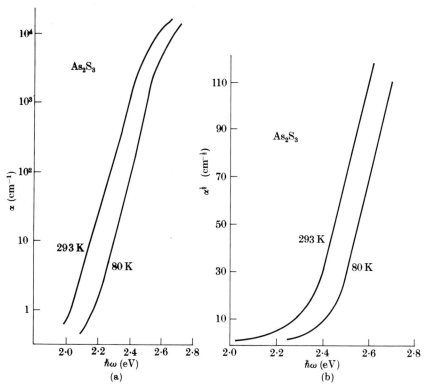

Fig. 9.39. Optical absorption edge in amorphous As_2S_3 at 293 K and 80 K plotted as (a) $\ln \alpha$ against $\hbar \omega$ and (b) $\alpha^{1/2}$ against $\hbar \omega$. (From Kosek and Tauc 1970.)

portion of the edge is increased by only ~13 per cent for this fall in temperature. Curve (b) shows more clearly the behaviour of α at photon energies above the exponential region. Good straight lines are obtained on a plot of $(\alpha \hbar \omega)^{1/2}$ *versus* $\hbar \omega$ and an optical gap E_0 of magnitude $2\cdot36\,\text{eV}$ at room temperature can be obtained by extrapolation. Fig. 9.39(b) yields a temperature coefficient of E_0 of approximately $-10^{-4}\,\text{eV K}^{-1}$ in this temperature range. For amorphous As_2Se_3, similar plots yield $E_0 =$

1·76 eV at room temperature and the edge has approximately the same temperature dependence. In the liquid state both materials exhibit edges of reduced slope having a value for the temperature coefficient about three times higher.

The temperature dependence of the absorption edge in amorphous As_2Te_3, as determined by Weiser and Brodsky (1970), is shown in Fig. 9.40. At room temperature the absorption coefficient follows the relation

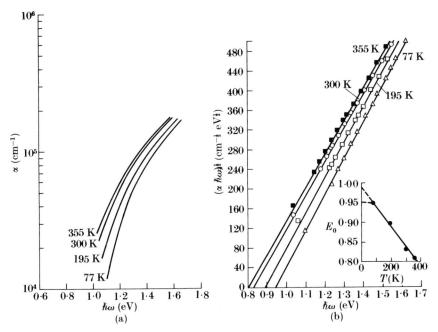

Fig. 9.40. Temperature dependence of the optical absorption edge in amorphous As_2Te_3 plotted as (a) $\ln \alpha$ against $\hbar\omega$ and (b) $(\alpha\hbar\omega)^{1/2}$ against $\hbar\omega$. The inset to (b) shows the temperature dependence of the intercept on the $\hbar\omega$ axis. (From Weiser and Brodsky 1970.)

$\alpha\hbar\omega = 4\cdot7\times10^5 \ (\hbar\omega - 0\cdot83)^2$, and the temperature coefficient of E_0 is -5×10^{-4} eV K^{-1} although, as for other materials, this increase as T is lowered is not expected to continue to $T = 0$ (see dotted lines in inset to Fig. 9.40(b)). At lower values of α Urbach behaviour with $\Gamma \sim 19$ eV^{-1} has been observed by Rockstad (1970) (see Fig. 6.44).

The optical absorption edge of amorphous GeTe is compared with that of the crystal in Fig. 9.41 (from Bahl and Chopra 1969). Although there does not appear to be a large difference in the position of the two edges, crystalline GeTe has a different short-range structure than amorphous films and is, in fact, a small-band-gap ($\sim0\cdot3$ eV) semiconductor; the edge is displaced to higher energies on account of a large Burstein shift. The

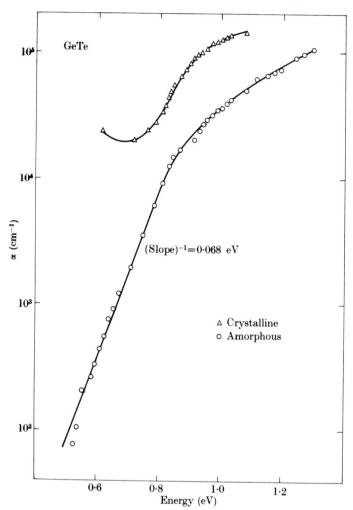

Fig. 9.41. Optical absorption edges in amorphous and crystalline GeTe. (From Bahl and Chopra 1969.)

absorption edge is characterized by an exponential tail, but the slope $(\Gamma \sim 15 \text{ eV}^{-1})$ is lower than that found in the arsenic chalcogenides. At high values of α a quadratic dependence for $\alpha\hbar\omega$ is found, yielding $E_0 \sim 0.8$ eV (Tsu, Howard, and Esaki (1970) report 0.7 eV) with a temperature coefficient $\sim 3 \times 10^{-4}$ eV K^{-1}.

We now turn to the weak tail absorption referred to at the start of this section. For As$_2$S$_3$, Tauc *et al.* (1970) have found the results shown in Fig. 9.42, which they interpret in terms of band tailing but which we prefer to consider as being caused by absorption into fairly discrete dangling-bond

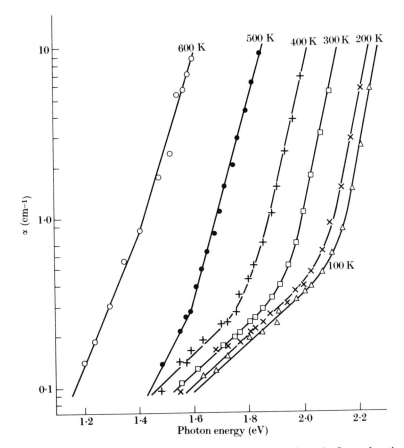

Fig. 9.42. Low-energy tail of the optical absorption edge in amorphous As$_2$S$_3$ as a function of temperature. (From Tauc *et al.* 1970.)

defect states. The evidence for this comes from the work of Street, Searle, and Austin (1975) who have used photoluminescence excitation spectra to probe the tail in this material. The excitation spectra for two thicknesses of As$_2$S$_3$ are shown by the two left-hand curves of Fig. 9.43. They both exhibit tails below about 2 eV. Under the assumption that the quantum efficiency for luminescence is unity in the spectral region of the Urbach tail and below, as shown by the chain curve,† the absorption coefficient α can be calculated from

$$I = \beta\{1 - \exp(-\alpha d)\}$$

where I is the integrated intensity of the luminescence, β is a correction for reflectivity, and d is the thickness. The computed spectral dependence of α

† The fall at higher energies is discussed in § 9.6.

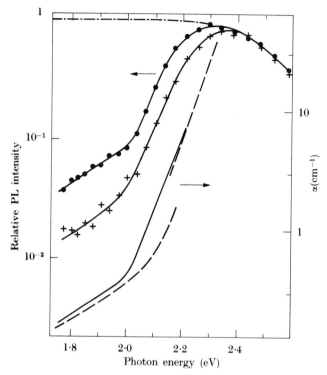

Fig. 9.43. Relative photoluminescence intensity in As_2S_3 as a function of exciting energy, together with absorption coefficient α. The specimen thicknesses are (in cm) $0\cdot21$ (●) and $0\cdot063$ (+). The broken curves give the data of Street *et al.* (1974) (upper curve) and that of Tauc, Menth, and Wood (1970) (lower curve): the full line is computed. (From Street 1976.)

is shown by the solid curve. The broken curves are direct absorption measurements made for the high-energy region by Street, Searle, Austin, and Sussmann (1974) and for the lower energies (also shown in Fig. 9.42) by Tauc *et al.* (1970). The reason for the displacement between these two segments is not clear; however, it does appear that there is a good correlation between the tails in the excitation spectrum and that in α, providing good evidence that the same defect centres are responsible for both.

Fig. 9.44 (after Wood and Tauc 1972) compares tailing in two ternary compounds with that in As_2S_3. Kumeda *et al.* (1976a,b) have observed tailing in $As_{20}Se_{80}$ glasses synthesized with Se of various degrees of purity and some containing trace amounts of oxygen. Their results are shown in Fig. 9.45. The purest sample does not exhibit a tail above $\alpha = 1\ cm^{-1}$ whereas others, particularly those containing oxygen, do. Interestingly an e.s.r. signal is observed in samples (b) and (d) but not in the remainder; a correlation between the presence of the tail and an e.s.r. signal is thus not

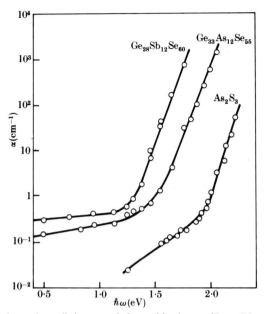

Fig. 9.44. Weak absorption tails in some chalcogenide glasses. (From Wood and Tauc 1972.)

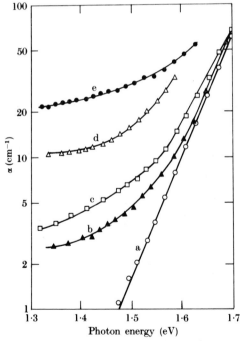

Fig. 9.45. The optical absorption coefficients α of various $Se_{80}As_{20}$ samples: (a) is synthesized using 99·999 per cent pure Se and does not show an e.s.r. signal; (b) is synthesized using 99·99 per cent pure Se and does not show an e.s.r. signal; (d) is synthesized using the same Se as (b) and As with an oxide layer and shows an e.s.r. signal; (e) is synthesized using 99·999 per cent pure Se and As with an oxide layer and does not show an e.s.r. signal. (From Kumeda *et al.* 1976*b*.)

clear. However, it seems likely that defect centres are responsible for the tail in all cases.

In the more conducting glasses tail absorption can result from free-carrier absorption. For $Tl_2SeAs_2Te_3$, Bishop *et al.* (1971) (see also Mitchell *et al.* 1972) have reported, below an Urbach tail, a wavelength-independent absorption which increases with temperature with an activation energy of 0·35 eV. This dependence continues through the glass-transition temperature without change as shown in Fig. 9.46, where $n\alpha$ (n being the

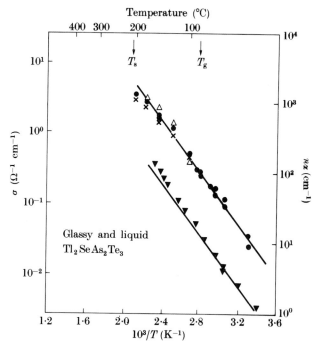

Fig. 9.46. Conductivity at d.c. and optical (infrared) frequencies *versus* reciprocal temperature in glassy and liquid $Tl_2SeAs_2Te_3$. The open triangles, circles, and crosses represent data taken at 5, 10, and 15 μm (2000, 1000, and 66 cm^{-1}) respectively. The solid triangles represent d.c. measurements. (From Mitchell *et al.* 1972.)

refractive index) at three different wavelengths is plotted against $1/T$. The solid triangles in this figure represent the variation of d.c. electrical conductivity, which exhibits the same activation energy. Although there is an (unexplained) factor of about 8 in magnitude between the optical and electrical conductivity, the parallel temperature dependencies confirm free-carrier absorption. Similar data for liquid As_2Se_3 were shown in Chapter 6 (Figs. 6.54 and 6.55) although for this material the absorption is not independent of wavelength.

9.13. Switching in alloy glasses†

The realization that films of chalcogenide alloys show fast and reversible switching from a high to a low resistance state (Ovshinsky 1968) was one reason for the rapid growth of interest in these materials from 1968 onwards. There are of course many forms of switching, which can occur in a wide variety of materials and even in liquid alloys of S, Se, and Te (Busch *et al.* 1970); it is unlikely that the same mechanism is responsible in all cases. The purpose of this section, however, is to examine specifically the behaviour of thin (1 μm) films of alloy glasses in the Te–As–Si–Ge system first developed as switches by Ovshinsky and co-workers, and to ask whether their behaviour depends essentially on their non-crystalline properties, therefore coming within the scope of this book, and whether switching can give information about these materials not easily obtainable in other ways.

A typical glass switching device consists of a layer of the material 1–5 μm thick sandwiched between two electrodes. When low voltages are applied, conduction is ohmic with a resistance at room temperature of order 10^5 Ω. For fields above $\sim 10^4$ V cm^{-1} non-ohmic effects set in, as described in § 9.10; just before switching a very rapid reversible rise in the current sometimes occurs (Buckley and Holmberg 1975, see below). Switching occurs at a critical field, some multiples of 10^5 V cm^{-1}, and is extremely rapid, taking place in less than 10^{-10} s; however, there is a delay time before switching, typically 10 μs but decreasing exponentially with voltage above the minimum switching field. The current–voltage characteristic is shown in Fig. 9.47. The current in the on state depends little on temperature or voltage; the latter is about 1 V, and the current is maintained unless the 'holding current' drops below some critical value.

In the memory switch, constructed from a less stable alloy (e.g. $Ge_{17}Te_{19}Sb_2S_2$), partial crystallization of a conducting channel occurs some milliseconds *after* threshold switching. Whether this is due to heating or to a high density of carriers within the channel is a problem we shall discuss below.

A forming process occurs during the initial switching event; some authors (e.g. Thomas, Fray, and Bosnell 1975) have considered this an essential part of the switching mechanism but according to Adler *et al.* (1974) switches can be constructed in which no forming occurs, and we do not consider it further here.

The main controversy about the mechanism of switching in these devices has been whether it is thermal, a hot conducting channel being formed leading to a negative resistance, or whether some electronic process like double injection is involved. Discussions of thermal instabilities go back to

† For a review see Adler, Henisch, and Mott (1978).

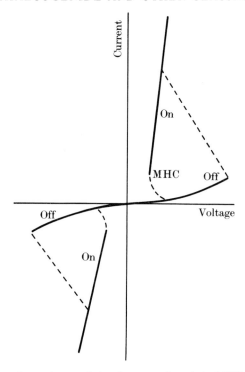

Fig. 9.47. Current–voltage characteristic of an ovonic switch. MHC denotes minimum holding current.

Lueder and Spenke (1935) and Ridley (1963); recent treatments applied to the threshold switch have been given by Fritzsche and Ovshinsky (1970), Warren (1970), Stocker, Barlow, and Weirach (1970), Kroll and Cohen (1972), and Kroll (1974). In addition there are many attempts to interpret the observations in terms of a thermal theory (e.g. Allison, Dawe, and Robson 1972, Robertson and Owen 1972). Calculations such as those of Kroll and Cohen, using the observed dependence of current on voltage, envisage a conducting channel in the on state at a temperature of 500–600°C.

A system in which switching is probably thermal is the vanadate glass switch investigated by Higgins, Temple, and Lewis (1977) and by earlier workers. Here the switching process is related to the metal–insulator transition occurring at 68°C in *crystalline* VO_2, which may be present in sufficient concentration in the glass. However, the evidence, reviewed by Adler *et al.* (1978), suggests strongly that this is not the correct model for the chalcogenide glasses.

The alternative explanation is that in the on state double injection of electrons and holes is taking place, the potential across the filament being

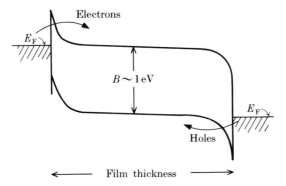

Fig. 9.48. Potential across filament in the on state according to double-injection model.

as in Fig. 9.48. This was first proposed by Mott (1969*b*) and by Henisch (1969). It is supposed that in the conducting channel, which is wider than the film thickness as illustrated in Fig. 9.49, sufficient carriers are injected to ensure the presence of a degenerate electron gas and a degenerate hole gas, of density perhaps 10^{18} cm^{-3} and with the Fermi energy of both on the extended side of the respective mobility edge. There is thus only a small potential drop in the film, and an obvious conclusion of the model is that the holding voltage should be slightly greater than the mobility gap.

If double injection is the correct model, one has to ask how the Schottky-type barriers are maintained. An early paper (Mott 1971) ascribed them to carriers trapped in gap states. Such a model had to be abandoned for two reasons. The first is that, if the estimate of § 9.4.2 is correct, the appropriate screening length of order 5×10^{-6} cm is too great to allow tunnelling;

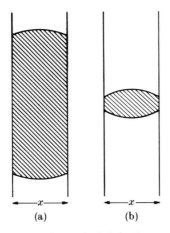

Fig. 9.49. The conducting channel, shown shaded, in the on state: (a) large current, no recombination, constant voltage; (b) small current, recombination, increasing voltage.

the other is that the rapid reversal of the on-state current occurring when the polarity is reversed (Henisch and Pryor 1971) seemed to rule out release and retrapping in deep states. Following a proposal by Lee (1972), Mott (1975a) has given an account of a dynamic mechanism by which the barriers could be maintained, because carriers which have just tunnelled through them are not thermalized and move quickly away, leading to a space charge due to the slowly moving carriers of opposite sign. According to this model (a) the current injected per unit area is a characteristic of the glass, so that, as the current varies, the area of the channel changes proportionately, and (b) the density of carriers in the channel is that at which the electron–hole gas has its minimum energy (cf. § 4.2). Some of this charge might however be in the D^+, D^- states; the trapping probability would be smaller for hot electrons.

Experimental evidence in favour of (a) was first provided by Henisch and Pryor (1971), who found that the rate of decay of the on state was independent of current. This implied that current density is constant. More direct evidence comes from the work of Petersen and Adler (1976), who, by preparing switches in which one electrode is doped silicon, were able to deduce from current saturation effects the cross-section of the channel. Their results are shown in Fig. 9.50; it will be seen that the current is approximately proportional to the square of the dimension of the channel. Moreover the resistance of the channel can be obtained by the 'transient on-state characteristic' (TONC), introduced by Pryor and Henisch (1971), in which a transient pulse is superimposed on the steady on state. The change in the differential resistance is interpreted on the assumption that the area of the channel does not have time to change during the pulse. From these techniques Petersen and Adler deduce the potential drop across the film outside the barriers ($\sim 0 \cdot 1$ eV) and the conductivity in the channel ($\sim 12 \ \Omega^{-1} \ cm^{-1}$). This figure is reasonable for metallic conductivity with E_F near but above the mobility edge and a density of electrons of order $10^{18} \ cm^{-3}$. From the width of the channel they also deduce that the rise in temperature is unlikely to be as great as 100 K, and may well be much less.

Further evidence against a theory that bases the behaviour of the on state on a high temperature comes from the work of Kolomiets, Lebedev, Rogachev, and Shpunt (1972) and of Vezzoli et al. (1974) on the observations of radiation from the on state. The latter authors find that the radiation has a maximum frequency below $1 \cdot 5$ eV and the intensity has a maximum round 1 eV. This fits the assumption that the radiation is due to transitions from band to band. At 600°C the band gap would have nearly disappeared.

An essential point of the double-injection model is that the carriers can cross the film with only small recombination. If the field in the bulk of the

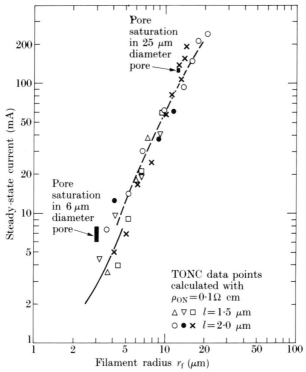

Fig. 9.50. Dependence of filament radius on steady–state current. l refers to thickness of film. Two points indicate the results of pore saturation, the solid line is obtained from saturation effects in heterojunctions and the other points from TONC results. (From Petersen and Adler 1976.)

film is 10^3 V cm^{-1} and the mobility ~ 10 cm^2 V^{-1} s^{-1}, the transit time for a film 1 μm thick is $\sim 10^{-8}$ s. In § 6.5.2 we showed that nonradiative processes gave a decay in photoconduction represented by the equation

$$\mathrm{d}n/\mathrm{d}t = -bn^2$$

with $b \simeq 10^{-10}$ cm^3 s^{-1} and independent of temperature. In the on state all the recombination centres (D^+ and D^- of § 9.4.1) will be occupied, so if N is their concentration we expect the rate of recombination to be bN. If this is to be less than 10^{-8} s^{-1}, N should be less than 10^{18} cm^{-3} (compare estimates elsewhere in this chapter).

Turning now to the memory switch, it is uncertain whether the partial crystallization is due to heating or to the high density of the electron–hole gas, but it is known (cf. Ovshinsky and Klose 1972) that strong illumination produces a similar effect and it seems likely that the breaking of bonds is at least as effective as temperature in promoting crystallization.

As regards the switching process itself, there have been many sugges-tions of its cause, including tunnelling from the electrodes, impact ion-ization, and so on. If, however, the model given above of the on state is correct, what is needed is a mechanism that allows the resistivity in the interior of the glass to drop rapidly so that the potential drop is transferred suddenly to the electrodes and injection of electrons and holes can start. There is evidence that it cannot be the carriers in the conduction and valence band that cause switching in this or any other way, since Henisch, Smith, and Wihl (1974) show that strong illumination, which increases the number of free carriers by an order of magnitude, does not affect the switching voltage nor the delay time (Adler *et al.* 1977). We are not then at the time of writing able to propose a satisfactory model for the switching process itself.

If double injection is the correct model for the on state, the behaviour of chalcogenide switches does strongly suggest that carriers with energies in the extended-state side of the mobility edges do not form polarons.

9.14. Oxide glasses

No attempt will be made in this section to review the extensive experimen-tal work on the structure and optical properties of oxide glasses.† Our aim is to pick out a few of their properties, particularly those of SiO_2, which can be compared with those of the conducting glasses described in the rest of this chapter.

Various calculations exist for the density of valence-band states in SiO_2, in the glassy state (Bennett and Roth 1971, Di Stephano and Eastman 1971a,b, Yip and Fowler 1975, Pantelides and Harrison 1976, Schlüter and Chelikowsky 1977). As for the chalcogenides, the upper part of the valence band is thought to be formed from lone-pair oxygen 2p orbitals and is therefore non-bonding and comparatively narrow, of width 2–3 eV in some earlier calculations, though Schlüter and Chelikowsky give ~5 eV.

The structure of vitreous silica was investigated by Mozzi and Warren (1969) using X-rays with the method of fluorescence excitation. Interpret-ing their results in terms of a random network model, they find that the Si—O—Si bond angle varies all the way from 120° to 180°, with a maxi-mum at 144°; Da Silva *et al.* (1975) amend this to 153°. Bell and Dean (1972) and Bell, Bird, and Dean (1974) constructed a model which gives the mean bond angle at 153°; their histogram shows most bond angles between 140° and 170°, but with tails extending to 120° and 180°.

Several investigations exist of the drift mobility μ_D of electrons photo-excited into the conduction band of glassy SiO_2. According to Hughes (1973) it is comparatively high, $20 \pm 3 \, \text{cm}^2 \, \text{V}^{-1} \, \text{s}^{-1}$, and can be accounted for quantitatively by scattering by optical phonons. In older work Williams

---

† For a comprehensive review, see Wong and Angell (1977).

(1965), from measurements of the drift distance before trapping in thermally grown SiO_2, obtained values of μ_D in the range 17–34 $cm^2\,V^{-1}\,s^{-1}$. Goodman (1967) found the Hall mobility to be little different from μ_D. We are led to conclude, therefore, that the mobility edge in SiO_2 is very near the band edge ($E_C - E_A < kT$), so that the behaviour of electrons is similar to that in a crystalline semiconductor. At the time of writing the reason for this is not clear, particularly in view of the large variation in the Si—O—Si angle. A possible explanation is that the conduction band may be formed mainly from oxygen 3s and silicon 4s orbitals; there is little variation in the Si—O and O—O nearest-neighbour distances, and so, on the analogy of the conduction-band behaviour of liquid rare gases (cf. §§ 2.10, 5.12), one would expect only a very small range of localized states, of width less than kT.

For holes, however, Hughes (1975, 1977) has found that the drift mobility is of the form

$$\mu_D = A\,\exp(-W/kT)$$

and that after about 10^{-5} s the activation energy W increases from 0·13 eV (the 'prompt' mobility) to 0·37 eV. It is suggested (Mott 1977b,d) that 0·13 eV is the energy difference between the band edge and the mobility edge, and that polarons of V_K type form after a delay as described in § 3.5.2. We thus assume, as in As_2Se_3, that self-trapping can occur owing to the attraction between an oxygen with a hole (O^-) towards one without O^{2-}, but that unlike the situation in the chalcogenides a stable polaron can be formed. In Fig. 3.5(i), then, we suppose that W is positive. If the probability of formation of a polaron is $\omega\,\exp(-w/\frac{1}{4}\hbar\omega)$ and $\omega \sim 10^{12}\,s^{-1}$, $w/\hbar\omega = \frac{1}{4}\ln 10^7$, giving $w \sim 0·2$ eV.

Phillip (1966, 1972; see also Tauc 1970a, Revesz 1973) found that the fundamental optical absorption in vitreous SiO_2 is very similar to that in the hexagonal crystal. This is shown in Fig. 9.51. The first peak is, we

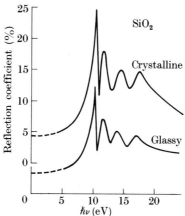

Fig. 9.51. Reflectivity of crystalline and glassy SiO_2. (From Phillip 1972.)

suppose, due to the formation of an exciton, though there has been some controversy about this (Ruffa 1968, Pantelides and Harrison 1976, Mott 1977b,d,e). In the crystal the lowest band-to-band transition is optically forbidden, but according to Schlüter and Chelikowsky (1977) a transition of 0·6 eV greater energy is allowed. The binding energy of a Wannier-type exciton ($m_{\text{eff}} e^4/2\hbar^2\kappa^2$) is about 2 eV, and so such an exciton should be separated from the band edge by ~1·4 eV. Here we have taken $\kappa = 2\cdot5$ and $m_{\text{eff}}/m_e = 0\cdot5$, following Snow (1967) and Haack (1977). Mott (1977b) proposes that the lowest exciton state ($k = 0$) is excited at an energy of 9 eV, but that in the glass, and also in the crystal at room temperature, the k-selection rule breaks down so that the peak at 10·2 eV is due to a maximum in the *exciton* density of states. Using this and other evidence, he estimates the band gap to be 10·6 eV. On the other hand, Di Stefano and Eastman (1971b) show that photoconduction occurs with high quantum efficiency for frequencies above 9 eV, so it seems more likely that the indirect gap lies at 9 eV and overlaps the direct exciton line, which can then dissociate spontaneously (cf. Mott 1978).

Further evidence for an exciton comes from the plots (Fig. 9.52) of the density of valence-band states (due to Ellis, Gaskell, and Johnson 1977) and the observed absorption coefficient ε_2; the lone-pair band 2–3 eV wide is

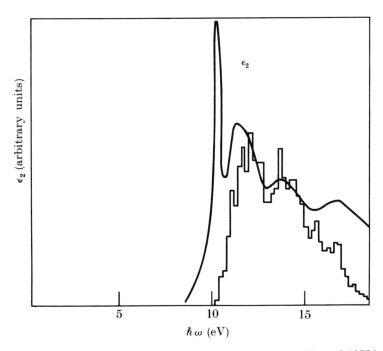

Fig. 9.52. ε_2 for SiO_2 compared with calculated band form. (From Ellis *et al.* 1977.)

well shown, but cannot be identified with the first peak in the absorption, which we ascribe to an exciton.

Since a hole in the valence band of SiO_2 can be self-trapped, the same must be true of an exciton; also, since the hole is localized in the neighbourhood of the electron by Coulomb forces, we might well expect the trapping energy to be greater for the exciton than for the hole. According to the arguments illustrated in Fig. 3.5(iii), an exciton, like a free carrier, may wander a considerable way before self-trapping. The exciton absorption line will *not* have the width marked ΔE in Fig. 3.7(i) owing to interaction with phonons, and its width, as we have suggested, is due to the inherent breadth of the exciton band. After self-trapping, since pure SiO_2 does not show photoluminescence when illuminated near the band edge, we must suppose that radiationless recombination must occur by the mechanism of Dexter *et al.* (1955) described in § 3.5.2. Thus in Fig. 3.7 the point X must lie below D. This mechanism is also invoked by Parke and Webb (1975) to explain the presence or absence of fluorescence in certain glass compositions containing bismuth.

We turn now to the concept of a 'non-bridging oxygen'. This is an oxygen atom bound to a single silicon, so that when neutral it will contain a single unpaired electron in a non-bonding 2p orbital. We believe this point defect is exactly analogous to the D^0 described in § 9.4 (a Se atom bonded to only one As). The non-bridging oxygen is most familiar in soda glasses; the sodium ion Na^+ is present interstitially, compensated by a

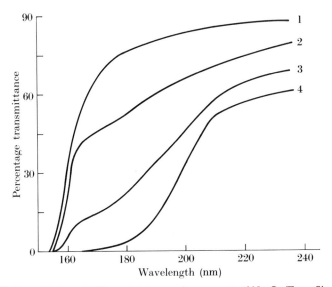

Fig. 9.53. Transmission of SiO_2 containing varying amounts of Na_2O. (From Sigel 1973/74.) 1, undoped; 2, 0·005 mol% Na; 3, 0·2%; 4, 0·5%.

negatively charged non-bridging oxygen in which the lone-pair orbital is occupied by two electrons. The non-bridging oxygen in this state is thought to produce a deep donor level, responsible for the familiar shift of the absorption edge of soda glass to lower frequencies than for SiO_2. Stroud, Schreuffer, and Tucker (1965) put the donor level at 2·4 eV above the valence band, and Sigel (1973/74), through an investigation of SiO_2 containing less than 1 per cent of Na_2O, showed that a very broad 'bound exciton line' displaced ~2·5 eV from the absorption line is formed (Fig. 9.53). Sigel also points out that aluminium in a soda glass is four-fold co-ordinated and therefore negative, the charge being compensated by Na^+; in this case the absorption band due to non-bridging oxygen is absent.

The *bound* exciton produced by excitation of a non-bridging oxygen and an Na^+ ion cannot of course move before distortion of the lattice. Thus in Fig. 3.7 the quantity ΔE should appear as the breadth of the absorption line. This we believe to be the reason why the line is so broad, as is shown in Fig. 9.53.

10

SELENIUM, TELLURIUM, AND THEIR ALLOYS

10.1. Structure of amorphous selenium and tellurium
10.2. Optical properties of amorphous selenium and tellurium
10.3. Electrical properties of amorphous selenium
    10.3.1. Electrical conductivity
    10.3.2. Drift mobilities
    10.3.3. Carrier lifetimes and ranges
    10.3.4. Photogeneration in amorphous Se; xerography
10.4. Some properties of liquid selenium and Se–Te alloys.

10.1. Structure of amorphous selenium and tellurium

The commonly held view of amorphous selenium is that it consists predominantly of a mixture of two structural species, long helical chains and
eight-membered rings, with strong covalent bonds existing between atoms
within the molecular units and weaker forces, perhaps of the van der Waals
type, binding together neighbouring units. However, most of the earlier
evidence for the existence of rings has been questioned and details of the
structure are not as yet resolved.

Attempts to fit the RDF of amorphous Se (which differs in detail according to the method of preparation, for instance vacuum deposition or melt
quenching) have been based on the two structural species found in the
known crystalline polymorphs of Se. Trigonal Se consists of parallel helical
chains of atoms (Fig. 10.1(a)) with a nearest-neighbour spacing of 2·32 Å

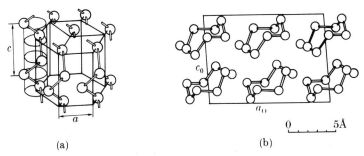

(a)   (b)

Fig. 10.1. Structure of crystalline selenium: (a) trigonal; (b) β-monoclinic (α-monoclinic
differs from (b) in the relative orientation of the rings).

and a bond angle of 105°. The dihedral angle of 102° has the same sign or sense along the whole length of each chain yielding a 'spiral pitch' of three atoms. The chains are packed together as shown with the closest separation between Se atoms on adjacent chains being 3·46 Å. Monoclinic Se exists in either an α or β form (Fig. 10.1(b)). Both structures consist of puckered eightfold rings with covalent bond lengths and angles essentially the same as in trigonal Se but with a dihedral angle that alternates in sign around each ring; the separation between molecules is larger than in trigonal Se.

In the trigonal form Te has a smaller ratio, compared with Se, between the interchain and intrachain separations. The smaller dihedral angle means that alternation of its sign does not lead to closed rings and Te does not exist in the monoclinic structure. This particular point was used in the discussion of a comparison of Raman spectra for amorphous and trigonal Te and Se in § 6.7.7 which concluded that an earlier interpretation of Raman and infrared absorption data in terms of Se_8 rings had been incorrect.

Various models for the structure of amorphous Se have been proposed (Richter and Herre 1958, Kaplow, Rowe, and Averbach 1968, Richter 1972). They have been based on modifications to the two structural units referred to above and variously have involved planar chains (dihedral angle equal to zero), expanded eightfold rings, and rings of lower and higher order. Reproduction of the experimentally obtained RDFs (shown for films deposited at 77 K and at room temperature in Fig. 10.2) in their entirety, is

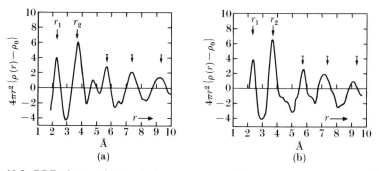

Fig. 10.2. RDF of amorphous selenium evaporated (a) at room temperature and (b) at liquid-air temperature. (From Richter 1972.)

not of course possible without considering how such units are packed together. Some comments can be made, however. In both RDFs the first peak occurs at 2·32 Å and contains two atoms. Thus the bond length and co-ordination of either the trigonal or monoclinic crystalline forms is retained. The second peak (at 3·68 Å in Fig. 10.2(a) and 3·86 Å in Fig. 10.2(b)) is close to the second-neighbour distance within a chain or ring

(3·69 Å), but is too large to be accounted for solely in these terms and one must invoke an intermolecular spacing of about the same value to bolster up the peak. Certainly the interchain spacing of trigonal Se (3·46 Å) is absent. In the region between about 4 Å and the next common peak at 5·75 Å, the two forms of amorphous Se display very different behaviour and one might conclude that this arises from a different stacking of the molecular units (see Richter 1972). Although the three remaining dominant peaks lie approximately at spacings expected for a flat zigzag chain (shown by arrows), this is unlikely to be a unique interpretation.

A 539-atom twofold co-ordinated model consisting of densely packed but convoluted chains has been hand built by Long et al. (1976) with the object of simulating the structure of amorphous Se and Te. Tetrahedral units were used, but while two bonds to each atom were relatively stiff and represented covalent bonds, two were flexible and were used to couple chains, hence simulating the much weaker interchain interaction. Not all of the latter links were utilized in the model, which was constructed so as to avoid groups of parallel chains and large distortions in the covalent bonds and angles. The chains were not allowed to terminate within the model or to close in on themselves. (This latter restriction was made deliberately to avoid ring formation, although the authors claim that rings could have been included without difficulty.) The atomic co-ordinates of the model were computer relaxed by minimizing the local potential as derived from three forces: covalent bond stretching, covalent bond bending, and a van der Waals (Lennard-Jones) interchain interaction, the latter with a cut off at 4·5 Å for Se and 5·1 Å for Te. The relaxed model had (for Se) an r.m.s. covalent-bond-length variation of 0·89 per cent, an r.m.s. bond-angle variation of 3·96° around the crystalline value of 105° and a broad dihedral-angle distribution. The scaled density inside a sphere of radius about two-thirds of that of the model was within 3 per cent of the measured value for amorphous Se (which differs only slightly from that in the monoclinic and trigonal varieties), although the average density of smaller spheres was somewhat lower.

The reduced Gaussian-broadened RDF of this model is compared in Fig. 10.3(d) with the experimental RDF (broken curve) for amorphous selenium deposited at 77 K (from Kaplow et al. 1968). This experimental RDF is also reproduced in parts (a), (b), and (c) of Fig. 10.3, where it is compared with the broadened RDF of α monoclinic, β monoclinic, and trigonal Se respectively. The model RDF shows better agreement with experiment than do any of the crystalline RDFs, but there are discrepancies, notably the failure to reproduce the plateau between 4 and 5·5 Å and a shift of the calculated peaks for $r > 5$ Å towards lower r relative to experiment. With regard to the first of these features one should note that the RDF of trigonal Se (Fig. 10.3(c)) displays, at about 4·5 Å, a

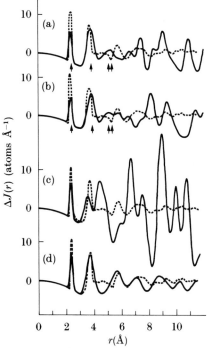

Fig. 10.3. Gaussian broadened ΔJ (solid curve) for (a) α-monoclinic Se, (b) β-monoclinic Se, (c) trigonal Se, and (d) the 363-atom interior sphere from the 539-atom model. In each case, the broken curve is the experimental ΔJ for vapour-deposited amorphous Se at 77 K (Richter 1972). The arrows in (a) and (b) mark the average distances corresponding to different types of pair separations in a single ring. (From Long *et al.* 1976.)

maximum rather than the minimum obtained from the model. This peak therefore arises from the parallel arrangement of chains, suggesting that deliberate avoidance of this in the construction of the model was over-emphasized. It should also be noted that contributions to the RDF from atoms within eight-membered rings (marked by arrows in (a) and (b)) are in the wrong places to account for the plateau.

A problem associated with a more detailed analysis of the model RDF is the similarity of the intrachain and interchain RDFs. Beyond the first-neighbour intrachain peak at 2·32 Å there is an accidental coincidence of distances between atoms on the same chain and atoms on neighbouring chains.

In the light of the evidence from interpretation of XPS/UPS spectra described in § 10.2, it is interesting to enquire as to the effect of the dihedral-angle distribution on the RDF of the above model. By a computer procedure Long *et al.* (1976) altered the co-ordinates of their model so that a peak in the otherwise fairly flat dihedral-angle distribution occurred close to the trigonal value of 102°. The agreement of the resulting RDF with the experimental curve was worsened over that found for the 'freely rotating' chain model. The conclusion that there appears to be no preference in amorphous Se for the bonds to take up the dihedral angle of trigonal Se is in contradiction to the conclusions of Bullett (1975a,c) and Robertson (1976) (see § 10.2).

Application of the above model to the simulation of the structure of amorphous Te (Long *et al.* 1976) indicates that the simple van der Waals interchain potential used for Se is less satisfactory for Te, perhaps on account of an enhanced interaction between chains in the latter.

10.2. Optical properties of amorphous selenium and tellurium

The room-temperature optical absorption edges of amorphous selenium and tellurium are compared with those of the crystalline modifications in Fig. 10.4. The exponential portion of the edge in amorphous selenium is, according to Hartke and Regensburger (1965), described by

$$\alpha = 7 \cdot 35 \times 10^{-12} \exp(\hbar\omega/0 \cdot 058 \text{ eV}) \text{ cm}^{-1}$$

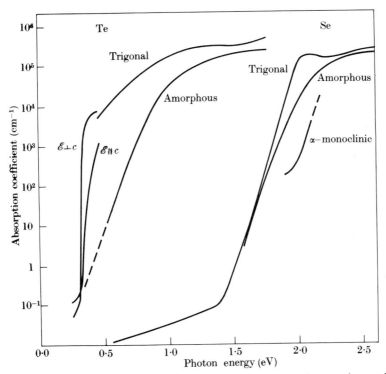

Fig. 10.4. Room-temperature optical absorption edges in amorphous and crystalline tellurium and selenium. (From Stuke 1970a.)

although other workers find slightly different parameters. Its position lies between that of the two crystalline modifications of selenium (Prosser 1961, Roberts *et al.* 1968). The edge in amorphous tellurium has a similar slope (Keller and Stuke 1965) but its position is considerably displaced towards higher energies relative to the steeper edge in trigonal tellurium (Tutihasi *et al.* 1969, Grosse 1969). No monoclinic form of Te is known. The larger displacement of the edge in tellurium may be related to the greater interaction between chains in this material.

The temperature dependence of the edge in amorphous and liquid selenium is shown in Fig. 10.5. In the liquid state (above 400 K) the slope

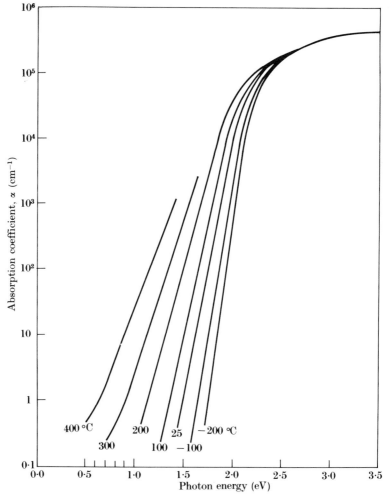

Fig. 10.5. Temperature dependence of the optical absorption edge in amorphous and liquid selenium. (From Siemsen and Fenton 1967.)

of the edge is in accord with Urbach's rule (Chapter 6). Below room temperature the shift is almost parallel (Knights, private communication) with a temperature coefficient of about -7×10^{-4} eV K^{-1}. In trigonal selenium (Roberts *et al.* 1968) Urbach's rule is observed for $\mathscr{E} \perp c$ down to 77 K; for $\mathscr{E} \| c$ the indirect edge is observed and this has a similar temperature coefficient. Rather smaller values have been inferred by Weiser and Stuke (1969) from electroreflectance measurements on both amorphous and trigonal Se.

Above the exponential edge, the form of the absorption coefficient in amorphous selenium obeys the relation

$$\alpha \hbar \omega \sim \varepsilon_2 (\hbar \omega)^2 \propto (\hbar \omega - E_0)$$

with $E_0 = 2 \cdot 05$ eV at room temperature (Fig. 10.6). This relationship, as opposed to the more common variation with $(\hbar \omega - E_0)^2$, is believed to arise

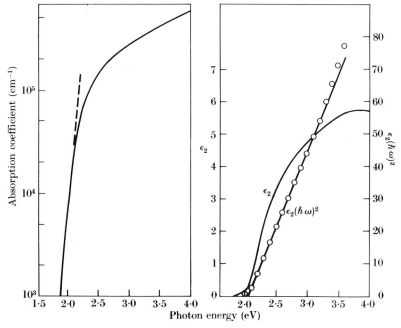

Fig. 10.6. Room-temperature absorption edge in amorphous selenium plotted as α, ε_2, and $\varepsilon_2 (\hbar \omega)^2$ against $\hbar \omega$. (From Davis 1970.)

from a sharp rise in the density of states at the band edges (Davis 1970, Davis and Mott 1970), a possibility suggested by the one-dimensional nature of the chain-like structure.

Considerable insight into the electronic levels and structure of amorphous selenium has been obtained by consideration of the XPS/UPS spectra. Data from Shevchik, Cardona, and Tejeda (1973) are shown in

Fig. 10.7(a) (see also Nielsen 1972). The two lower curves are taken to represent the DOVS in trigonal and amorphous Se and have been obtained by combining XPS and UPS spectra. X-ray excitation is needed to reveal the s bands extending from about 8 to 18 eV; u.v. excitation is needed to resolve the fine structure in the lone-pair (0–3 eV) and bonding (3–8 eV) p bands. Basically the spectra resemble those of the binary chalcogenides described in Chapter 9. The upper curve and boxes in Fig. 10.7(a) are

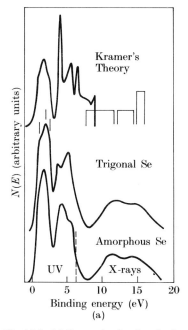

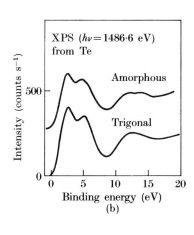

Fig. 10.7. (a) Composite density of valence states of amorphous and crystalline Se. The upper curve shows the density of states calculated by Kramer (1970b); (b) Raw X-ray photoemission data from amorphous and trigonal Te. (From Shevchik, Cardona, and Tejeda 1973.)

obtained from density-of-states calculations for trigonal Se by Kramer (1970a,b, 1971) and Sandrock (1968); the position and widths of the various bands agree well with experiment. XPS data for trigonal and amorphous Te are shown in Fig. 10.7(b). The spectra are virtually identical to each other but the resolution (0·5 eV at best) of the measurements is poor. For trigonal Te fine structure in the bonding p band has been observed by UPS (see Fig. 10.8 and Shevchik, Cardona, and Tejeda 1973), but since similar data for amorphous Te are not available, it is not possible to say in what way this is changed.

    The primary difference between the DOVS of amorphous and trigonal Se lies in the p-bonding band (3–8 eV), the twin peaks of this being reversed in intensity. It is on this feature that several theoretical papers have focused attention. Schlüter, Joannopoulos, and Cohen (1974), using the empirical pseudopotential method (EPM), have calculated the densities of states for trigonal Se and Te. Fig. 10.8(a) shows their results (note that the energy scale is reversed from Fig. 10.7) compared with experimental curves. In order to understand the origin of the two peaks in the p-bonding band, these authors have calculated the electronic charge distributions associated with the states contributing to each peak. Contours of the localized bonding charge (selected from the total by considering only the Fourier components that have wavelengths less than the nearest-neighbour distance) within and in between the chains in trigonal Se are shown in Fig. 10.8(b). It is seen that the

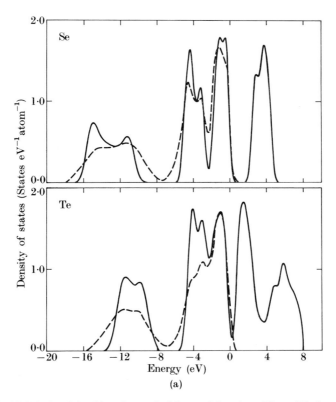

(a)

Fig. 10.8. (a) Calculated densities of states (solid curves) for trigonal Se and Te, broadened by 1·2 eV for the s-like states and by 0·7 eV for the remaining states. Superimposed are the experimental photoemission spectra (broken curves). The scales for the XPS and UPS curves are arbitrary. (From Schlüter *et al.* 1974.) (b) (p. 526) Bonding charge of trigonal Se for (i) lower and (ii) upper p-like bonding states, calculated as described in the text. Only positive contours are shown, with values in units of e/Ω. The contour separation is 0·84 units in (i) and 0·36 units in (ii). On the right are similar results for a model structure in which the interchain distance has been increased by 20 per cent. (From Joannopoulos *et al.* 1975.)

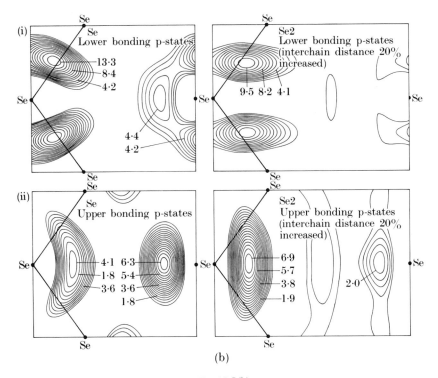

Fig. 10.8(b).

lower p-like bonding state (i.e. the one furthest away from the lone-pair peak) contains states that are mainly involved in *intra*chain bonding, since the charge is well localized between atoms belonging to the same chain, whereas the upper peak contains states principally involved in *inter*chain bonding. The ratio of intrachain to interchain bonding is 1·15; similar results for Te yield 1·25. Joannopoulos, Schlüter, and Cohen (1975) propose that the reversal in intensity of these two peaks for amorphous Se could arise from an increase of about 20 per cent in the interchain separation. Such an increase is, however, not supported by the rather similar densities of trigonal and amorphous Se.

Alternative explanations for the reversal in the intensity of the p-bonding peaks have been proposed by Shevchik (1974*b*), Bullett (1975*a,c*), and Robertson (1976). Shevchik (1974*a,b*), using a Slater–Koster type tight-binding calculation for an isolated Se chain, has shown that the relative heights of the peaks are sensitive to the dihedral angle (i.e. the relative orientation of second-neighbour bonds when projected on to a plane perpendicular to the intermediate bond) and proposes that this is reduced from 102° (the value in trigonal Se) to about 80° in amorphous Se.

This would represent a tightening of the 'pitch' of the helical chain or alternatively might be expected if Se$_6$ puckered rings were present. However, Bullett (1975*a*,*c*) has shown, using chemical pseudopotential theory, that (1) removal of regular interchain interactions destroys the two-peak structure, and (2) a dihedral angle that is unchanged from the value in trigonal Se but the sign of which either alternates or varies randomly as one proceeds from atom to atom along a chain, restores the features but with the asymmetry reversed. An alternating sign of the dihedral angle corresponds to a ring structure, although in the presence of bond-angle distortions the rings need not close; a random sign of the dihedral angle leads to distorted chains. It is not possible to decide between the two alternatives from these calculations. Robertson (1976), using a different method of calculation, reaches the same conclusions and furthermore rules out large changes in the magnitude of the dihedral angle on the grounds that the potential barrier opposing such a change is large (on account of

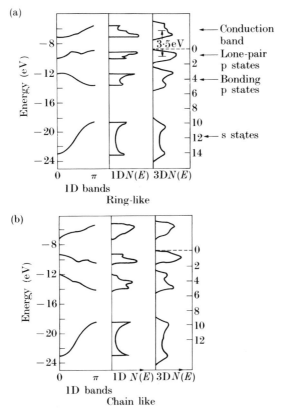

Fig. 10.9. The one-dimensional bands, 1D DOS, and 3D DOS (including inter-unit inter-actions) for (a) chain — and (b) ring — short-range order. (From Robertson 1976.)

repulsions between lone-pair electrons on adjacent atoms) being ~0·4 eV per atom, i.e. about 10 times the crystallization energy. The results of Robertson's density-of-states calculation (in one and three dimensions), are shown in Fig. 10.9, from which the importance of the sense of the dihedral angle can be seen; however, it should again be stressed that the similarity of the ring-like density of states with that of amorphous Se should not be taken to infer the presence of Se_8 ring molecules.

The form of the optical absorption (ε_2) of Se and Te is shown in Fig. 10.10. The main features of the spectra for the crystals (trigonal in both

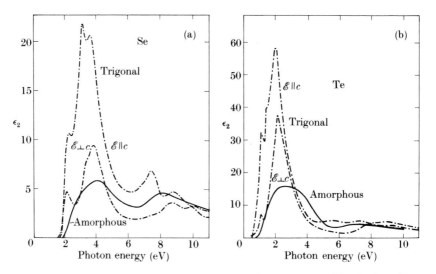

Fig. 10.10. ε_2 spectra of amorphous and crystalline (a) selenium and (b) tellurium. (From Stuke 1970a.)

cases) can be understood on the basis of the calculated band structures (Treusch and Sandrock 1966, Sandrock 1968) shown in Fig. 10.11. The grouping of the bands into three sets of triplets (originating from atomic p states), the upper set being the conduction bands, divides the absorption spectra into two parts. This is evident particularly in the case of Se, the ε_2 spectrum of which exhibits a deep minimum near 6 eV for both of the principal directions of polarization perpendicular to the c axis. This feature is retained in the amorphous forms but most of the fine structure is lost. In addition the strength of the transitions contributing to the first peak is considerably reduced. It is worth mentioning that the electronic band structure in the Δ direction ($\Gamma - Z$) of the Brillouin zone (Fig. 10.10(c)) can also be determined by a tight-binding calculation for an isolated chain, giving similar results (Olechna and Knox 1965). However, the smallest gap occurs in the neighbourhood of the H point, that is in a direction from Γ

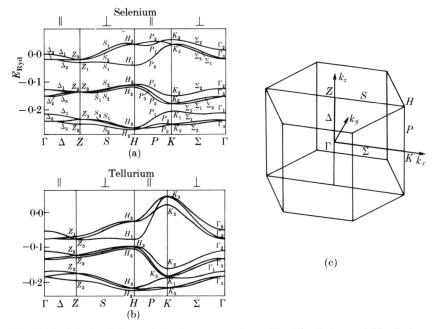

Fig. 10.11. Calculated electronic band structures of crystalline (a) selenium and (b) tellurium. (From Treusch and Sandrock 1966.) (c) Brillouin zone.

corresponding to a crystallographic axis which is neither parallel nor perpendicular to c. The band structure in the direction H–K is determined to a large extent by interaction between chains. Differences in this interaction between the crystalline and amorphous forms probably contribute to the loss of the peak at 2 eV shown for the crystal in Figs. 10.4 and 10.10(a).

Apart from this change at the edge, the spectrum for amorphous Se has been reproduced by Kramer *et al.* (1970) (see also Maschke and Thomas 1970*b*, Kramer 1970*b*), using the convoluted density of states appropriate to trigonal crystalline Se (Fig. 10.12(a)) and relaxing the k-conservation selection rule. In order to account for the relative heights of the two broad maxima seen in Fig. 10.10(a), it was necessary to take a form for the energy variation of matrix elements as shown in Fig. 10.12(b) (solid curve). The large peak in $\{M(\omega)\}^2$ for trigonal Se (broken curve) arises from Umklapp enhancement for transitions near 4 eV which involve the top plane of the Brillouin zone.

A more detailed consideration of the reason for the loss of strength of the first peak in the ε_2 spectrum (which applies to Te as well as Se) has been made by Robertson (1976). Noting that the two broad peaks arise from transitions to the conduction band from the lone-pair and p-bonding states respectively (see Figs. 10.9 and 10.11), Robertson ascribes the reduction in

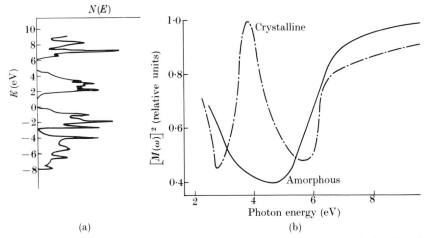

Fig. 10.12. (a) Density of states in trigonal selenium determined from the calculated band structure, part of which was shown in Fig. 10.11. (b) Energy dependence of matrix elements in trigonal and amorphous selenium as calculated by Maschke and Thomas (1970).

height of the first peak to a suppression of intermolecular contributions which in turn arises from the loss of alignment of orbitals between chains in the amorphous form. (The argument parallels that of Martin *et al.* (1976) for the Raman spectra; when coupling between chains is absent, a Raman line at 239 cm^{-1} (see § 6.7.7) reverts to the frequency expected for bond stretching in an isolated chain.) Support for this argument is to be found in the ε_2 spectrum for monoclinic Se (Dalrymple and Spear 1972), the rings in which are relatively isolated and the ε_2 spectrum of which is very similar to that of amorphous Se.

10.3. Electrical properties of amorphous selenium and alloys of selenium

10.3.1. Electrical conductivity

Although amorphous selenium can be obtained by quenching the liquid, most electrical measurements have been made on evaporated films. This is because of the commercial applications of selenium in thin-film form to rectifiers, photocells, the vidicon, and xerography. There is, however, no evidence to suggest significant differences between the properties of films and of bulk samples.

When pure, the conductivity of amorphous selenium is very low ($\sim 10^{-16}\ \Omega^{-1}\,\text{cm}^{-1}$ (Hartke 1962)) at room temperature. This is not, however, out of line with many other amorphous semiconductors of comparable band gaps (>2 eV), and, as shown in Chapter 6, the value of C, if we assume $\sigma = C \exp(-E_\sigma/kT)$, is estimated to be of order

$10^4 \, \Omega^{-1} \, cm^{-1}$. Unfortunately the temperature range over which conductivity measurements can be made is severely limited by the high resistivity at low temperatures and the low crystallization temperature.

Some measurements for the liquid are shown in Fig. 10.13. According to these,

$$E_\sigma \simeq E_S \simeq 1 \cdot 15 \, eV$$

with a positive thermopower. A reasonable assumption, therefore, is that conduction is by hole transport at a mobility edge, as in other liquid chalcogenide compounds (Chapter 5 and § 6.4.4). More recent measurements by Gobrecht, Gawlik, and Mahdjuri (1971), Gobrecht, Mahdjuri, and Gawlik (1971), and Mahdjuri (1975) find, however, that the thermopower is n type in the liquid but changes to p type above 900 K. The n-type behaviour is, we believe, due to impurities and will be discussed in § 10.4.

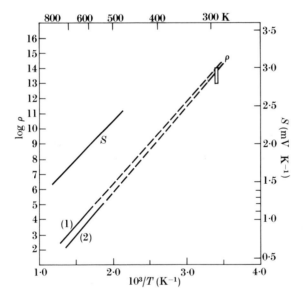

Fig. 10.13. Temperature dependence of resistivity for liquid Se. (1) Henkels and Maczuk 1953 (2) Lizell 1952; and thermoelectric power Henkels and Maczuk 1953. (From Owen 1970.)

Differences between reported values for the d.c. conductivity of amorphous selenium can be attributed in part to its sensitivity to the presence of impurities, particularly oxygen. Results of LaCourse, Twaddell, and Mackenzie (1970) are shown in Fig. 10.14. The room-temperature resistivity of pure deoxygenated selenium (<2 ppm O_2) according to these

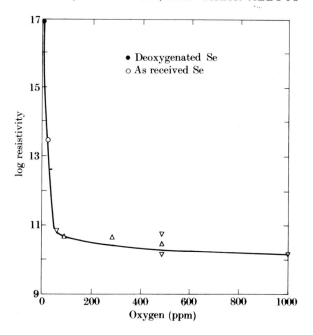

Fig. 10.14. Effects of oxygen on the electrical resistivity of glassy selenium. (From LaCourse *et al.* 1970.)

authors is about $10^{17}\,\Omega\,\text{cm}$ and this drops by more than six orders of magnitude with the presence of approximately 50 ppm O_2. Further addition of oxygen (in the form of SeO_2) has little effect. The addition of As apparently forms AsO_2 and removes the effect of oxygen. Further investigations of the effect of impurities are due to Twaddell, LaCourse, and Mackenzie (1972). These authors find that the room-temperature resistivity of Se of 99·9999% purity is between 10^{17} and $10^{18}\,\Omega\,\text{cm}$ and drops to $10^{11}\,\Omega\,\text{cm}$ through the addition of 20 ppm of O, added as SeO_2 to the melt. Chlorine can have an even greater effect, the resistivity dropping to $10^9\,\Omega\,\text{cm}$ at 500 ppm. Potassium leads to a less marked drop, to $10^{13}\,\Omega\,\text{cm}$. The thermopower of pure amorphous Se and of oxygen-doped Se, according to Twaddell *et al.*, is p type, as is that of liquid Se; these authors report that Cl doping gives an n-type thermopower.

We assume as in § 9.4 that the Fermi energy is pinned by charged dangling bonds. ESR studies in pure Se (Abkowitz 1967) suggest that electrons at the ends of chains are paired, the signal being weak. We suppose, then, that half the chain ends are negative, D^-, and that the other half, D^+, are strongly bonded to an Se atom in a neighbouring chain which becomes three-fold co-ordinated. Fairly direct evidence for dipoles in Se and in $Se + 3\%$ As has been obtained from dielectric measurements by Abkowitz and Pai (1977, 1978).

In order to understand the sensitivity of the conductivity to oxygen, we suppose (Mott 1976d), in common with other investigators (e.g. Mahdjuri 1975), that oxygen is adsorbed to the chain ends, but an additional hypothesis is that it is adsorbed only to *half* of them (the D^-), giving negatively charged oxygen with no unpaired spins. Our hypothesis is that it will not be adsorbed to the positive ends (D^+), which are already strongly bonded to another Se. Thus the D^- centres are destroyed, but not the D^+. According to the considerations of § 9.8, the activation energy E_σ for p-type conduction should then be reduced to two-thirds.

If chlorine is added, the absorption of Cl at a D^- chain end will produce a neutral end to the chain *and* an electron; assuming the strongly bonded D^+ to remain intact, the electron must go to another chlorine and produce an unbonded Cl^-. Since $Se + Cl$ is reported by Twaddell *et al.* to be n type, we suppose that these Cl^- are donors.†

Finally, if potassium is added, the K atom should give up an electron to the D^+ states, producing D^0 or D^-; we suppose that K^+ ions are compensated by D^-, but that, above ~ 100 ppm of potassium, all D^+ disappear. Since the D^- centres can only generate electrons, not holes, the material should be n type. We have no evidence at the time of writing whether this is so.

We now write down the various activation energies. Suppose the band gap is B, with the Fermi energy pinned in pure Se by D^+ and D^- at an energy ε below mid-gap. The conductivity activation energy for pure p-type Se is thus

$$E_\sigma = \tfrac{1}{2}B - \varepsilon \tag{10.1}$$

and in the presence of oxygen it is

$$E_{ox} = \tfrac{2}{3}(\tfrac{1}{2}B - \varepsilon). \tag{10.2}$$

In the presence of potassium, however, the n-type activation energy is

$$E_K = \tfrac{2}{3}(\tfrac{1}{2}B + \varepsilon). \tag{10.3}$$

We now examine the experimental evidence to see (a) if (10.1) and (10.2) are compatible and (b) to deduce ε from (10.3).

Twaddell *et al.* (1972) give activation energies but no details of the temperature range in which the measurements were made, and we prefer

---

† Pfister, Melnyk, and Scharfe (1977) have observed an enhancement of hole drift velocity in As_2Se_3 on adding ~ 0.6 per cent of iodine, but without change in the activation energy or dependence on field. We conjecture that this is due to the disappearance of D^- centres. Possibly here the D^+ centres disappear too.

to suppose that for holes $\sigma = \sigma_0 \exp(-E_\sigma/kT)$ and that σ_0 is independent of impurity content, and to see what value of σ_0 we obtain from the observed conductivities. If $E_\sigma = 1.13$ eV, at room temperature $\exp(-E_\sigma/kT) = 2.3 \times 10^{-20}$; the observed resistivity is $\sim 2 \times 10^{17}$ Ω cm, so σ_0 should be 217 Ω^{-1} cm^{-1}. This is a reasonable value for conduction at a mobility edge; if conduction is by hopping, σ_0 will be less, but so will E_σ, compared with the value extrapolated from the liquid. We shall then take this value of E_σ in our further discussion.

Addition of oxygen, then, should decrease the resistivity by the factor $\exp(\frac{1}{3}E_\sigma/kT) \sim 2 \times 10^7$. The observed drop is from $\sim 2 \times 10^{17}$ to $\sim 2 \times 10^{10}$, and is thus by a factor 10^7. The agreement is excellent.

The drop in resistivity on adding potassium is 10^4. The factor 10^3 (i.e. 10^{7-4}) is, according to eqns (10.2) and (10.3), equal to $\exp\{\frac{4}{3}\varepsilon/kT\}$. Thus, with $T \sim 300$ K,

$$\varepsilon = \tfrac{3}{4}kT \ln 10^3 = 0.13 \text{ eV}.$$

This then gives the displacement of the Fermi level in pure Se from mid-gap.

For Se containing chlorine the observed conductivity corresponds to an activation energy of 0.56 eV. As we stated earlier, we suppose that the Cl$^-$ ions are donors and that the material is an n-type semiconductor. Since there should be some neutral Cl also (not attached to Se chain ends), the material is compensated and 0.56 eV (not twice this) should represent the energy required to remove an electron from Cl$^-$ into the conduction band. The electron affinity of Cl is 3.75 eV, so this would indicate that the conduction-band edge of Se is 3.2 eV below vacuum. Mort (1973b), from photo-emission work due to Mort and Lakatos (1970), gives 5.7 eV for the photo-emission threshold and 2.20 ± 0.25 eV for the gap, so the conduction band is at 3.5 ± 0.25 eV below vacuum, in fair agreement with the result deduced above.

The effect of up to 1 per cent of oxygen on the photoluminescence emission is not marked (Street, Searle, and Austin 1974b, Street 1976). For this we may perhaps deduce that it is the D$^+$ centres that are normally responsible for luminescence.

10.3.2. Drift mobilities

The model of charged dangling bonds thus gives an explanation in broad outline of the effect of impurities on the conductivity of amorphous selenium. Further information can be obtained from the successful use of the transient drift-mobility technique as developed and used on thin material by Spear (1957, 1960), Hartke (1962), Tabak (1970), Pfister (1976), and others. Both hole and electron drift mobilities in thin films have been measured as a function of temperature, pressure, electric field, and sample variables, such as deposition temperature and concentration of additives.

The technique for measurement of drift mobilities has been described in § 6.5.1. Basically it involves measurement of the time required for a thin sheet of charge carriers, produced by photon or electron irradiation at one surface, to drift across the biased sample. Table 10.1 lists the room-

TABLE 10.1

Drift parameters in amorphous selenium

Electrons		Holes		
μ_D	E_D	μ_D	E_D	Reference
$5 \cdot 2 \times 10^{-3}$	0·25	0·135	014	Spear 1957, 1960
$7 \cdot 8 \times 10^{-3}$	0·285	0·165	0·14	Hartke 1962
$4 \cdot 5 \times 10^{-3}$	0·25	0·11	0·20	Grunwald and Blakney 1968
$5 \cdot 8 \times 10^{-3}$	0·33	0·12	0·25	
$6 \cdot 0 \times 10^{-3}$		0·13		
	0·33		0·16	Schottmiller *et al.* 1970
$8 \cdot 3 \times 10^{-3}$		0·16		
		0·13		
			0·23	Tabak 1970
		0·17		
		0·20	0·28	Pfister 1976

μ_D is the drift velocity in $cm^2 \, V^{-1} \, s^{-1}$, E_D is the activation energy in eV.

temperature drift mobilities μ_D and activation energies E_D. In view of the differences in starting materials and conditions of deposition the consistency of some of the earlier data is good, though Pfister's more recent results for holes gives rather larger values for both E_D and μ_D. The spread in the data of Grunwald and Blakney (1968) corresponds to a range in substrate temperatures of 25–58°C (the activation energies showed an almost linear rise with increase in substrate temperature). The substrates in Tabak's experiments were held at 55°C.

There are several possible explanations of an activated drift mobility. One is that the carriers form polarons, which we think improbable; if this were so a substantial part of E_σ (perhaps 0·15 eV) could be identified with the hopping energies W_H (cf. Chapter 3), showing a difference between E_σ and E_S. This does not seem to be observed, at any rate in the liquid. Another is that there is a large range of band-edge localized states due to disorder; but particularly for electrons the large values shown in Table 10.1 make this unlikely. We think it probable that both electrons and holes move at their respective mobility or band edges, the mobility being controlled by traps due to defects which are probably chain ends or impurities.

There are basic differences between hole and electron transport. Some observations and deductions concerning the transport processes in amorphous selenium are as follows.

(a) At room temperature, except at low values of the electric field (<100 volts cm^{-1}), no appreciable loss of carriers during transit through the films is found. However, as will be shown below, much higher fields are necessary to ensure that a large proportion of the carriers leave the generation region. Furthermore, the pulse shapes (ramp-like for voltage, rectangular for current) are not rounded, yielding well-defined transit times for both holes and electrons. These observations indicate virtually no spreading of the thin sheet of charge, such as would be obtained if there was a broad spectrum of release times for carriers trapped during transit. They do not, however, rule out a spectrum of trapping levels with release times much less than the transit time. However, Pfister (1976) finds below 180 K a change to the stochastic process described in § 6.5.1. Also, with his measurements over a much larger temperature range, he finds, as we stated above, a larger activation energy than earlier workers, which is shown in the table. Whether dispersion arises from properties of the band edge, or a range of depths of 'discrete' states† (e.g. D^+ and D^-), is a problem which is not resolved at the time of writing.

(b) The drift mobilities are independent of the magnitude of the applied electric field except at low temperatures (<200 K).

(c) There is no detectable change in the mobilities or activation energies when a hydrostatic pressure of 4·2 kbar is applied (Dolezalek and Spear 1970). This is considered to be strong evidence against hopping transport for injected carriers, at least over the temperature range 230–300 K. In other materials such as sulphur and anthracene, where polaron hopping seems probable, a strong pressure effect is observed.

(d) For hole transport neither the magnitude nor the activation energy of the drift mobility is affected by alloying with up to 2 per cent of As (Hartke 1962) or with S. Alloying with Te reduces the room-temperature drift mobility and increases the activation energy slightly.

(e) For electron transport the activation energy associated with the drift mobility is essentially unchanged by light alloying with As, S, or Te. However, the magnitude of the electron drift mobility falls on light alloying with As and Te but not with S.

† cf. Marshall 1977.

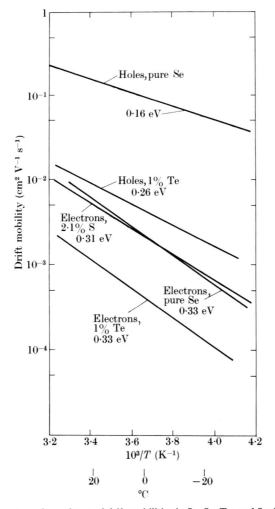

Fig. 10.15. Temperature dependence of drift mobilities in Se, Se–Te, and Se–S alloys. (From Schottmiller *et al.* 1970.)

The alloying effects are shown in Figs. 10.15 and 10.16, plotted from results given by Schottmiller *et al.* (1970) (see also Kolomiets and Lebedev, 1966). For electrons the constancy of E_D suggests that the traps which limit electron motion remain unchanged by alloying with the elements shown. Interpretation of the electron drift mobility data in terms of eqn (6.11), namely

$$\mu_D = \mu_0 \frac{N_c}{N_t} \exp\left(-\frac{E_D}{kT}\right) \qquad (10.4)$$

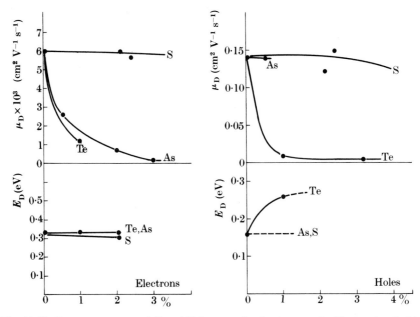

Fig. 10.16. Room-temperature drift mobilities μ_D and activation energies E_D associated with these mobilities in amorphous selenium alloys.

leads, if we suppose that $\mu_0 = 10 \text{ cm}^2 \text{ V}^{-1} \text{ s}^{-1}$ and N_c, the effective density of states at the mobility edge, equal to 10^{20} cm^{-3} in pure Se, that the decrease of μ_D on alloying is then associated with an increase in N_t. On this model, the increase in N_t on alloying with As is found to be approximately equal to the number of As atoms introduced.

For holes, assuming a trap-limited mobility according to eqn (10.4), the trap density would be $2 \times 10^{14} \text{ cm}^{-3}$. However, if conduction were by hopping at the valence-band edge, rather than at a mobility edge, μ_0 could be 10–100 times smaller, and it is not ruled out that the values of N_t are the same as for electrons.

If N_t is indeed the same for electrons as for holes, then the simplest assumption is that the electron and hole traps are charged dangling bonds at the ends of chains. The trapping activation energies would then be those for the processes

$$D^0 \rightarrow \text{electron} + D^+$$

and

$$D^0 \rightarrow \text{hole} + D^-.$$

If this is so, oxygen should remove the hole traps, substituting a deeper trap, and potassium should affect the electron trap. Experimental evidence on this point is lacking.

However, the trapping energies seem small compared with the energies expected, and possibly they may be excited states for carriers captured by D^+ and D^-, that is an electron or a hole bound in a 2s state to the respective charged centre. If so, the (multiphonon) transition probability to the ground state must be small.

10.3.3. Carrier lifetimes and ranges

The *range* for carriers injected into a semiconductor is the distance travelled per unit field before loss by trapping into deep levels or by recombination. It is equal to the product of the drift mobility and the carrier lifetime τ. Use of low electric fields in the drift-mobility experiments described above can lead to transit times of the order of or less than τ. In this case the transit pulses become exponential, and it is possible to obtain values of τ from their shape. Using this method Tabak and Warter (1968) determine the following room-temperature values for hole and electron lifetimes in amorphous selenium:

$$\tau_p = 10\text{--}45 \ \mu s, \qquad \tau_n = 40\text{--}50 \ \mu s.$$

These values give room-temperature ranges of

$$1\cdot3\text{--}6\cdot3 \times 10^{-6} \ \mathrm{cm^2 \ V^{-1}} \quad \text{for holes}$$

and

$$2\cdot4\text{--}3\cdot1 \times 10^{-7} \ \mathrm{cm^2 \ V^{-1}} \quad \text{for electrons.}$$

It should be noted that the hole lifetime and range as determined by this method are 20–100 times larger than previous estimates by Hartke (1962), who used a probably inappropriate Hecht-type analysis of pulse height against electric field.

The effect of light alloying on the carrier lifetimes is interesting. The lifetimes are controlled by deep levels which may or may not be related to the shallow-trap-controlled drift mobilities. Fig. 10.17 shows that the addition of As, even in small quantities, dramatically reduces the hole lifetime but increases the electron lifetime. The hole lifetime is similarly reduced by alloying with S and Tl, whereas it is essentially unchanged with Te and Cl. The electron lifetime is reduced by alloying with Cl and perhaps Te, but is largely unaffected by the addition of Tl or S.

A possible hypothesis already suggested above is that the deep traps are the D^+, D^- centres and the shallow traps are excited states of these, the transition between them being a slow multiphonon process (Chapter 9). If so, removing traces of oxygen would increase the number of D^- centres and thus account for the increase in the lifetime of the holes. Oxygen at the end of a chain produces a negatively charged centre, and thus a shallow

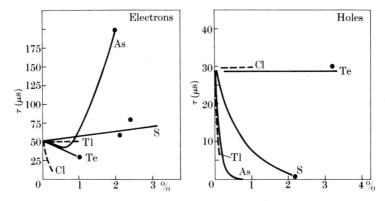

Fig. 10.17. Variation of electron and hole lifetimes in amorphous selenium with light alloy-ing. Data from transient photoconductivity experiments.

trap, but unless its lone-pair orbital (with an electron missing) can bond with a Se lone pair on another chain, it should not produce a deep state.

10.3.4. Photogeneration in amorphous Se; xerography

The discharge of a selenium (or selenium alloy) film in the electrostatic copying process known as xerography (Fig. 10.18) involves the creation of electron–hole pairs by optical absorption in a thin layer at the surface and their subsequent separation under the action of an electric field. Transient photoconductivity experiments (Tabak and Warter 1968, Pai and Ing 1968, Pai and Enck 1975; cf. § 6.5.2) on amorphous selenium have shown that, even for fields sufficiently high that bulk trapping events during transit are negligible, the quantum efficiency of the process is still significantly less than unity. The quantum efficiency, defined as the number of free electron–hole pairs created per absorbed photon, is found to be an increasing function of electric field, temperature, and photon energy. It approaches unity for high values of these parameters.

The dependence of quantum efficiency on photon energy at room temperature and for high values of the electric field has been shown in Fig. 6.36. The displacement of this quantum efficiency edge from the optical absorption edge has been discussed in § 6.5.3 in which a simple model, involving thermalization of the photoexcited carrier pair, was proposed. Fig. 10.19(a,b) show typical field dependencies of the quantum efficiency η at various temperatures and at two different wavelengths, 400 nm ($\approx 3\cdot 10$ eV) and 580 nm ($\approx 2\cdot 14$ eV), corresponding to fairly high and low values of η respectively. In spite of the different shapes of these curves they can, at least above a field of about 10^4 V cm^{-1}, be accounted for by a theory due to Onsager (1938). This theory, originally developed to explain

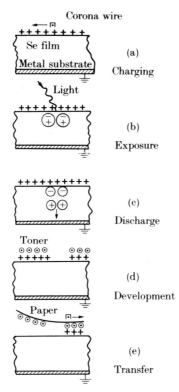

Fig. 10.18. The xerographic process. (a) The photoconducting film is charged positively by a corona discharge induced from a wire, held at a high potential, which is moved parallel to the top surface. (b) The document to be copied is imaged on to the film. Electron–hole pairs are created in the film by strongly absorbed photons reflected from light areas on the document. (c) Under the action of the electric field, the holes drift towards the metal substrate; the electrons move in the opposite direction to neutralize the positive surface charge. (d) Negatively charged 'toner' particles (carbon black dispersed in a low-melting plastic) are cascaded on to the surface, adhering to those areas of the film that have not been discharged. (e) The toner is transferred to paper with the aid of a second corona discharge. The paper is removed and the image made permanent by heating. (For further details of the process see Dessauer and Clark 1965, Mort 1974.)

the departure from ohmic behaviour in weak electrolytes or solid dielectrics, yields an expression for the probability for a pair of oppositely charged carriers to dissociate by Brownian motion in the presence of their Coulomb attraction and a static electric field F, i.e. in a potential of the form $e^2/\kappa r + eFr$ where κ is the dielectric constant of the medium. Apart from the obvious constants, the only parameter in the Onsager theorem is the initial separation r_0. The expression for the probability of escape is complicated but it can be reduced (Pai and Enck 1975) to a convergent series in F.

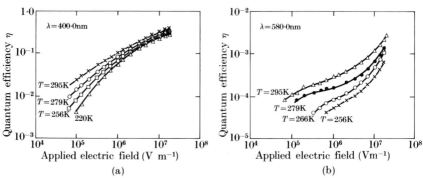

Fig. 10.19. Quantum efficiency of injection of holes *versus* electric field at several temperatures and at an exciting wavelength of (a) 4000 Å and (b) 5800 Å.

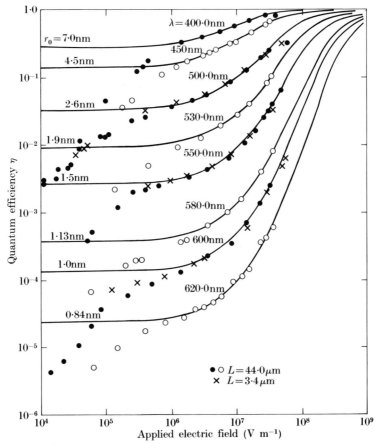

Fig. 10.20. The circles and crosses show the experimental quantum efficiency of photoinjection of holes *versus* applied electric field for different values of the wavelength of exciting radiation. The figure also shows data on films of two different thicknesses. The solid curves are the theoretical Onsager dissociation efficiencies for an initial separation r_0 indicated in the figure. Note that the field is expressed in V m^{-1}.

The theoretical curves used by Pai and Enck to fit the quantum-efficiency data in amorphous Se are shown in Fig. 10.20. Values of r_0 used for each wavelength of excitation are marked on the curves. The departure from theory below about 10^4 V cm^{-1} is attributable to recombination in the region of photoexcitation, as shown conclusively in a novel two-photon experiment by Enck (1973). It is clear that the Onsager mechanism (or for that matter the cruder Poole–Frenkel treatment, § 3.8) must have a low-field cut-off when the potential drop eFr produced by the field is less than about kT at the escape radius ($r_c = e^2/\kappa kT$) (see Warter 1971). For Se at room temperature this low-field cut-off is $\sim 3 \times 10^4$ V cm^{-1}. The behaviour at higher fields *approximates* to the Poole–Frenkel behaviour, for which $\eta \sim \exp(\beta F^{1/2}/kT)$ (Pai and Ing 1968, Pai and Enck 1975), but if interpreted by this expression the values of $\beta (= (4e^3/\kappa)^{1/2})$ are incorrect by a factor of 2 and furthermore depend on temperature.

The Onsager expression yields a temperature dependence of η, for a given r_0, that is approximately exponential in $1/T$. The corresponding activation energies are compared with experimental values as a function of photon energy in Fig. 10.21(a). The dependence is seen to be good except

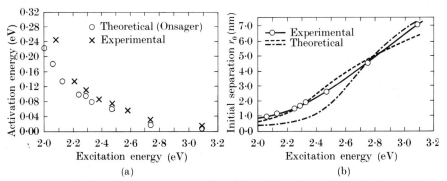

Fig. 10.21. (a) Comparison of the theoretical (Onsager) and experimental values of the activation energy as a function of photon energy at an applied electric field of 7×10^6 V m^{-1}. (b) Experimental initial separation distances as a function of the photon energy. The figure also shows the expected variation of initial distances using a simple model.

for excitation energies close to the absorption edge. Here r_0 is only about 10 Å and the classical assumptions of the theory probably break down. The dependence of r_0 on the excitation energy predicted by the thermalization model developed in § 6.5.3 (see also Knights and Davis 1974) is shown, using appropriate parameters, in Fig. 10.21(b). Although the general behaviour is reproduced, a more detailed theoretical treatment of the initial thermalization process would be desirable.

10.4. Some properties of liquid selenium and Se–Te alloys

Pure liquid selenium has been the subject of many studies, recently reviewed in a book by Cutler (1977). It is thought to consist of a mixture of

chain molecules and eight-membered rings. The theory of bond equilibrium for such liquids is well established (Eisenberg and Tobolsky 1960,
Gee 1952). The concentration of chain ends varies with T as $\exp(-E_d/kT)$
and Eisenberg and Tobolsky, supposing that the viscosity depends on chain
length, find $E_d = 0\cdot54$ eV.

The discussion of this section will seek to apply the model of charged
dangling bonds (§ 9.4) to these liquids, and indeed if much energy was not
released by the bonding of D^+ (a positively charged chain end) to a Se
atom in another chain, the energy of bond breaking would be much greater
than this. We suppose then that $0\cdot54$ eV is half the energy required to form
a D^+ and a D^-. Koningsberger, Van Wolput, and Rieter (1971) have found
that the activation energy to form centres that give an e.s.r. signal is
$0\cdot63$ eV, while Massen, Weijts, and Poulis (1964) find from a study of
magnetic susceptibility that the activation energy is $0\cdot87$ eV. One or other
of these values should be the energy to create D^0, which is greater than the
mean for D^+ and D^-.

Since the concentration of these centres at a given temperature must be
much greater than that of the carriers, for which $E_\sigma \simeq 1$ eV, liquid Se will
never be intrinsic and the Fermi energy will remain pinned. However, the
effect of impurities on the D^+, D^- centres may well be similar to that in the
solid, depending on their concentration and binding energy to chain ends
about which little is known. The results of Gobrecht, Gawlick, and
Mahdjuri (1971), Gobrecht, Mahdjuri, and Gawlick (1971), Mahdjuri
(1975) have been mentioned in § 10.3.1; the most important was that their
liquid Se has an n-type thermopower below 900 K, and we think that this
must be due to an impurity that neutralizes the D^+ chain ends. Above
900 K the concentrations of D^+ and D^- may be such that the impurities do
not saturate them.

We turn now to liquid Se_x–Te_{1-x} which has been investigated by
Perron (1967) and whose results are reproduced in Figs. 10.22 and 10.23.
The first point to note is that these Te-rich alloys become metallic at high
T. If we suppose that metallic behaviour begins when $\sigma \sim 300\ \Omega^{-1}\ cm^{-1}$,
we see for $x = 0\cdot2$ that the thermopower is then of order k/e ($87\mu V^{-1}$ K),
so that the correspondence is reasonable. No direct measurements of
$N(E_F)$ are available, as for instance from the Knight shift. Such measurements do, however, exist for Te (Chapter 5), where the relationship $\sigma \propto$
$\{N(E_F)\}^2$ is obeyed. It seems likely that it would be obeyed by these alloys
too.

However, for σ well below $300\ \Omega^{-1}\ cm^{-1}$ the model of a Fermi energy
pinned by charged dangling bonds should be correct, and it is of interest to
see how this goes over to the pseudogap model when their concentration is
large. In Chapter 5 we quoted the results of Cabane and Friedel (1971)
that in liquid Te, as the temperature rises, three-fold co-ordination

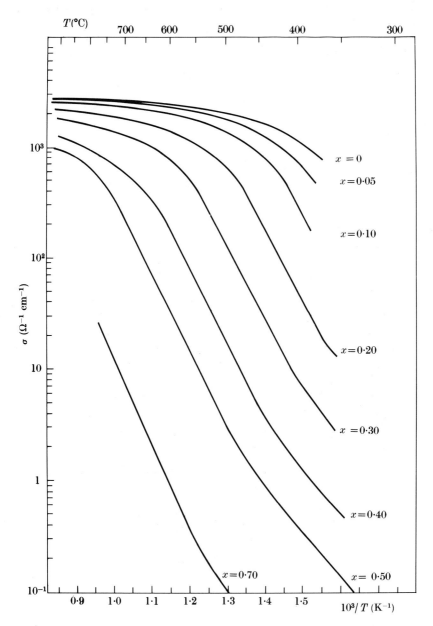

Fig. 10.22. Electrical conductivity of liquid Se_xTe_{1-x} alloys as a function of temperature. (From Perron 1967.)

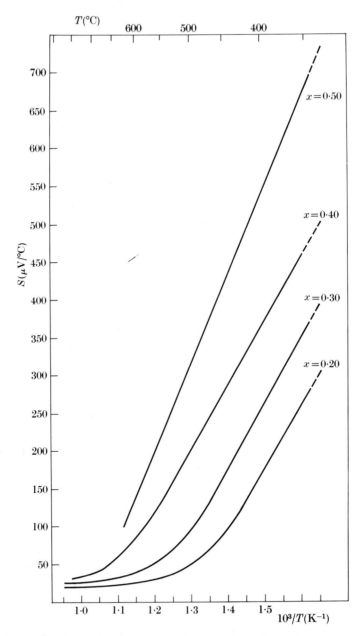

Fig. 10.23. Thermoelectric power S of liquid $Se_x Te_{1-x}$ alloys. (From Perron 1967.)

increases, and in § 9.4 we point out that the D^+ centres imply just this. However, a quantitative treatment is lacking. In Fig. 10.22 it will be noticed that the slope of the plot of $\ln \sigma$ against $1/T$ increases with increasing T; Davis and Mott (1970) supposed that this is not a true activation energy, but occurs because $E_F - E_V$ decreases faster than linearly. Cutler (1977) shows that for $x = 0.5$ the low-temperature part may be of the form $\sigma_{min} \exp(\gamma/k) \exp(-E_0/kT)$ with reasonable values of σ_{min} and γ and E_0, corresponding to the thermopower according to eqn (2.83). To achieve this result he assumes a small contribution from electrons in the conduction band. A problem discussed particularly by Cutler (1977) is why the change of slope in the $\ln \sigma$ versus $1/T$ plot is not reflected in that of the thermopower. We think it possible that the extrapolation of the $\log \sigma$ versus $1/T$ curve without assuming any ambipolar conduction does give a rather small value of σ_0 ($\sim 10 \ \Omega^{-1} \ cm^{-1}$), and that this is characteristic of liquids when the number of carriers at the mobility edge is small. A possible explanation (Cutler, private communication) for p-type liquids is that holes form V_K-type polarons (even if they do not in the solid) and that such polarons behave like a Se_2^+ molecule and move like a heavy ion.

REFERENCES

Many references are to papers presented at international conferences. With the exception of those for which the proceedings have been published in a journal, they are referred to under the title of the place where the meeting took place. In the case of the Leningrad meeting, an indication is made of which of the two volumes is referenced.

Proceedings of the 3rd International Conference on Amorphous and Liquid Semiconductors, held (1969) in Cambridge, U.K., ed. N. F. Mott. (ref: (1970) *J. non-cryst. Solids* **6**.)

Proceedings of the 4th International Conference on Amorphous and Liquid Semiconductors, held (1971) in Ann Arbor, Michigan, eds. M. H. Cohen and G. Lucovsky. (ref: (1972) *J. non-cryst. Solids* **8–10**.)

Proceedings of the 5th International Conference on Amorphous and Liquid Semiconductors, held (1973) in Garmisch-Partenkirchen, FDR, eds. J. Stuke and W. Brenig, Taylor & Francis, London, 1974. (ref: (1974) *Garmisch*.)

Proceedings of the 6th International Conference on Amorphous and Liquid Semiconductors, held (1975) in Leningrad, U.S.S.R., ed. B. T. Kolomiets, Academy of Sciences of U.S.S.R., Nauka, 1976. (refs: (1976) *Leningrad, Electronic Phenomena in Non-Crystalline Semiconductors*; (1976) *Leningrad, Structure and Properties of Non-Crystalline Semiconductors*.)

Proceedings of the 7th International Conference on Amorphous and Liquid Semiconductors, held (1977) in Edinburgh, U.K., ed. W. E. Spear, Centre for Industrial Consultancy and Liaison, University of Edinburgh (ref: (1977) *Edinburgh*.)

Proceedings of an International Conference on Tetrahedrally Bonded Amorphous Semiconductors, held (1974) in Yorktown Heights, N.Y., eds. M. H. Brodsky, S. Kirkpatrick, and D. Weaire, AIP Conference Proceedings No. 20. (ref: (1974) *Yorktown Heights*.)

Proceedings of an International Conference on Structure and Excitations of Amorphous Solids, held (1976) Williamsburg, Va., eds. G. Lucovsky and F. Galeener, AIP Conference Proceedings No. 31. (ref: (1976) *Williamsburg*.)

Proceedings of the 13th Session of the Scottish Universities Summer School in Physics, Electronic and Structural Properties of Amorphous Semiconductors, held (1972) in Aberdeen, U.K., eds. P. G. Le Comber and J. Mort, Academic Press, 1973. (ref: (1973) *Aberdeen*.)

Proceedings of the Symposium on The Structure of Non-Crystalline Materials, held (1976) in Cambridge, U.K., ed. P. H. Gaskell, Taylor & Francis, London, 1977. (ref: (1977) *Cambridge*.)

ABELES, B. and PING SHEN. (1974). *Proc. 13th Int. Conf. on Low Temperature Physics* (eds. K. D. Timmerhaus, W. J. O'Sullivan, and E. F. Hammel), Vol. 3, p. 578. Plenum Press, New York.

—— PINCH, H. L., and GITTLEMAN, J. I. (1975). *Phys. Rev. Lett.* **35**, 247.

—— PING SHEN, COUTTS, M. D., and ARIE, Y. (1975). *Adv. Phys.* **24**, 407.

ABKOWITZ, M. (1967). *J. chem. Phys.* **46**, 4537.

—— and PAI, D. M. (1977). *Phys. Rev. Lett.* **38**, 1412.

—— (1978). *Phys. Rev.* (in press).

—— LE COMBER, P. G., and SPEAR, W. E. (1976). *Commun. Phys.* **1**, 175.

ABOU-CHACRA, R. and THOULESS, D. J. (1974). *J. Phys. C: Solid State Phys.* **7**, 65.

—— ANDERSON, P. W., and THOULESS, D. J. (1973). *J. Phys. C: Solid State Phys.* **6**, 1734.

ABRAM, R. A. (1973). *J. Phys. C: Solid State Phys.* **6**, L379.

—— and EDWARDS, S. F. (1972). *J. Phys. C: Solid State Phys.* **5**, 1183.

ADAMS, P. D. and KRAVITZ, S. (1961). *Internal report*, Department of Metallurgy, Imperial College, London (unpublished).

ADKINS, C. J. (1978). *J. Phys. C: Solid State Phys.* **11**, 851.

—— and HAMILTON, E. M. (1971). *Proc. 2nd Int. Conf. Conduction in Low Mobility Materials* (eds. N. Klein, D. S. Tannhauser, and M. Pollak), p. 229. Taylor & Francis, London.

ADLER, D. (1976). *Williamsburg*, p. 11.

—— and YOFFA, E. J. (1976). *Phys. Rev. Lett.* **36**, 1197.

—— FLORA, L. P., and SENTURIA, S. D. (1973). *Solid State Commun.* **12**, 9.

—— HENISCH, H. K., and MOTT, N. F. (1978). *Rev. mod. Phys.* **50**, 203.

—— COHEN, M. H., FAGEN, E. A., and THOMPSON, J. C. (1970), *J. non-cryst. Solids* **3**, 402.

—— ARNTZ, F. O., FLORA, L. P., MATHUR, B. P., and REINHARD, D. K. (1974). *Garmisch*, p. 859.

AFROMOWITZ, M. A. and REDFIELD, D. (1968). *Proc. 9th Int. Conf. on the Physics of Semiconductors, Moscow*, p. 98. Nauka, Leningrad.

AGARWAL, S. C. (1973). *Phys. Rev. B* **7**, 685.

—— GUHA, S., and NARASIMHAN, K. L. (1975). *J. non-cryst. Solids* **18**, 429.

AL-BERKDAR, F., TAYLOR, P. C., HOLAH, G. D., CROWDER, J. G., and PIDGEON, C. R. (1977). *Edinburgh*, p. 184.

ALBEN, R., CARGILL, G. S., and WENZEL, J. (1976). *Phys. Rev. B* **13**, 835.

—— VON HEIMENDAHL, L., GALISON, P., and LONG, M. (1975). *J. Phys. C: Solid State Phys.* **8**, L468.

—— WEAIRE, D., and STEINHARDT, P. (1973). *J. Phys. C: Solid State Phys.* **6**, L384.

—— —— SMITH, J. E., and BRODSKY, M. H. (1975). *Phys. Rev. B* **11**, 2271.

ALEXANDER, M. N. and HOLCOMB, D. F. (1958). *Rev. mod. Phys.* **40**, 815.

ALLCOCK, G. R. (1956). *Adv. Phys.* **5**, 412.

ALLEN, F. R. (1973). Ph.D. Thesis, University of Cambridge.

—— and ADKINS, C. J. (1972). *Phil. Mag.* **26**, 1027.

—— WALLIS, R. H., and ADKINS, C. J. (1974). *Garmisch*, p. 895.

ALLGAIER, R. S. (1970). *Phys. Rev. B* **2**, 2257.

—— and HOUSTON, B. B. (1962). *Proc. Int. Conf. on the Physics of Semiconductors, Exeter*, p. 172. The Institute of Physics and The Physical Society, London.

—— and SCANLON, W. W. (1958). *Phys. Rev.* **111**, 1029.

ALLISON, J., DAWE, V. R., and ROBSON, P. N. (1972). *J. non-cryst. Solids* **8–10**, 563.

ALTARELLI, M. and DEXTER, D. L. (1972). *Phys. Rev. Lett.* **29**, 1100.

AMBEGAOKAR, V., COCHRAN, S., and KURKIJÄRVI, J. (1973). *Phys. Rev. B* **8**, 3682.

—— HALPERIN, B. I., and LANGER, J. S. (1971). *Phys. Rev. B* **4**, 2612.

AMRHEIN, E. M. and MUELLER, F. H. (1968). *Trans. Faraday Soc.* **64**, 666.

ANDERSON, D. A. and SPEAR, W. E. (1977). *Phil. Mag.* **35**, 1.

—— MOUSTAKAS, T. D., and PAUL, W. (1977). *Edinburgh*, p. 334.

ANDERSON, J. C. (1974). *Phil. Mag.* **30**, 839.

ANDERSON, P. W. (1956). *Phys. Rev.* **102**, 1008.

—— (1958), *Phys. Rev.* **109**, 1492.

—— (1968), *Phys. Rev. Lett.* **21**, 13.

—— (1969). *Phys. Rev.* **181**, 25.

—— (1970). *Comments Solid State Phys.* **2**, 193.

—— (1972*a*). *Proc. Nat. Acad. Sci. USA* **69**, 1097.

—— (1972*b*). *Nature London (Phys. Sci.)* **235**, 163.

—— (1975). *Phys. Rev. Lett.* **34**, 953.

—— (1976). *J. Phys. (Paris)* **C-4**, 339.

ANDERSON, P. W., HALPERIN, P. I., and VARMA, C. M. (1972). *Phil. Mag.* **25**, 1.

ANDREEV, A. A. (1976). *Leningrad, Structure and Properties of Non-Crystalline Semiconductors*, p. 340.

ANDREWS, P. V., WEST, M. B., and ROBESON, C. R. (1969). *Phil. Mag.* **19**, 887.

ANDRIESH, A. M. and KOLOMIETS, B. T. (1965). *Sov. Phys.–Solid State* **6**, 2652.

ANIMALU, A. O. E. and HEINE, V. (1965). *Phil. Mag.* **12**, 1249.

ANSEL'M, A. I. (1953). *Zh. Eksper. Teor. Fiz.* **24**, 83.

AOKI, H. and KAMIMURA, H. (1977). *Solid State Commun.* **21**, 45.

APLING, A. J., LEADBETTER, A. J., and WRIGHT, A. C. (1977). *J. non-cryst. Solids* **23**, 369.

APSLEY, N. and HUGHES, H. P. (1974). *Phil. Mag.* **30**, 963.

—— —— (1975). *Phil. Mag.* **31**, 1327.

—— —— (1976). *Leningrad, Electronic Phenomena in Non-Crystalline Semiconductors*, p. 103.

—— DAVIS, E. A., TROUP, A. P., and YOFFE, A. D. (1977). *Edinburgh*, p. 447.

ARAI, K., ITOH, U., KOMINE, H., and NAMIKAWA, H. (1976). *Leningrad, Structure and Properties of Non-Crystalline Semiconductors*, p. 222.

—— —— BABA, T., SHIMAKAWA, K., and NITTA, S. (1974). *Yorktown Heights*, p. 363.

ARIZUMI, T., YOSHIDA, A., and SAJI, K. (1974). *Garmisch*, p. 1065.

ARNOLD, E. (1974). *Appl. Phys. Lett.* **25**, 705.

ARNOLDUSSEN, T. C., MENEZES, C. A., NAKAGAWA, Y., and BUBE, R. H. (1974). *Phys. Rev. B* **9**, 3377.

ASAF, U. and STEINBERGER, I. T. (1974). *Phys. Rev. B* **10**, 4464.

ASCARELLI, G. and RODRIGUEZ, S. (1961). *Phys. Rev.* **124**, 1321.

ASHCROFT, N. W. and LEKNER, J. (1966). *Phys. Rev.* **145**, 83.

ASHLEY, C. A. and DONIACH, S. (1975). *Phys. Rev. B* **11**, 1279.

AUSTIN, I. G. (1972). *J. Phys. C: Solid State Phys.* **5**, 1687.

—— (1976). In *Linear and non linear electron transport in solids* (ed. J. T. Devreese and V. E. Van Doren), p. 383. Plenum Press, New York.

—— and GARBETT, E. S. (1971). *Phil. Mag.* **23**, 17.

—— —— (1973). *Aberdeen*, p. 393.

—— and MOTT, N. F. (1969). *Adv. Phys.* **18**, 41.

—— and SAYER, M. (1974). *J. Phys. C: Solid State Phys.* **7**, 905.

AXE, J. D., KEATING, D. T., CARGILL, G. S., and ALBEN, R. (1974). *Yorktown Heights*, p. 279.

BABER, W. G. (1937). *Proc. R. Soc. A* **158**, 383.

BAGLEY, B. G. (1970). *Solid State Commun.* **8**, 345.

BAHL, S. K. and BHAGAT, S. M. (1975). *J. non-cryst. Solids* **17**, 409.

—— and CHOPRA, K. L. (1969). *J. appl. Phys.* **40**, 4940.

—— BHAGAT, S. M., and GLOSSER, R. (1973). *Solid State Commun.* **13**, 1159.

—— —— (1974). *Garmisch*, p. 69.

BALLENTINE, L. E. (1965). Ph.D. Thesis, University of Cambridge.

—— (1977). *Proc. 3rd Int. Conf. on Liquid Metals, Bristol* (ed. E. Evans and D. A. Greenwood), p. 188. The Institute of Physics, London.

—— and CHAN, T. (1973). In *The properties of liquid metals* (ed. S. Takeuchi), p. 197. Taylor & Francis, London.

BANERJI, J. and HIRSCH, J. (1974). *Solid State Commun.* **15**, 925.

BANUS, M. D., and REED, T. B. (1970). In *The chemistry of extended defects in non-metallic solids* (ed. LeRoy Eyring and M. O'Keeffe), p. 488. North-Holland, Amsterdam.

BANYAI, L. (1964). In *Physique des Semiconducteurs* (ed. M. Hulin), p. 417. Dunod, Paris.

BARBE, D. F. (1971). *J. Vac. Sci. Technol.* **8**, 102.

BARDASIS, A. and HONE, D. (1967). *Phys. Rev.* **153**, 849.

BARDEEN, J. and SHOCKLEY, W. (1950). *Phys. Rev.* **80**, 72.

BARNA, A., BARNA, P. B., RADNÓCZI, G., SUGAWARA, H., and THOMAS, P. (1976). *Williamsburgh*, p. 199.

BARTRAM, R. H. and STONEHAM, A. M. (1975). *Solid State Commun.* **17**, 1593.

BAUER, R. S. (1974*a*). *Yorktown Heights*, p. 126.

—— (1974*b*). *Garmisch*, p. 595.

—— and GALEENER, F. L. (1972). *Solid State Commun.* **10**, 1171.

BAYM, G. (1964). *Phys. Rev. A* **135**, 1691.

BEAGLEHOLE, D. and ZAVETOVA, M. (1970). *J. non-cryst. Solids* **4**, 272.

BECKER, G. E. and GOBELI, G. W. (1963). *J. chem. Phys.* **38**, 2942.

BEEBY, J. L. (1964). *Proc. R. Soc. A* **279**, 82.

BEEMAN, D. and ALBEN, R. (1977). *Adv. Phys.* **26**, 339. (1977). *Edinburgh*, p. 189.

—— and BOBBS, B. L. (1975). *Phys. Rev. B* **12**, 1399.

BELL, R. J. and DEAN, P. (1968). *Physics and chemistry of glasses*, **9**, 125. (1970). *Discuss. Faraday Soc.* No. 50, p. 55.

—— —— (1972). *Phil. Mag.* **25**, 1381.

—— BIRD, N. F., and DEAN, P. (1968). *J. Phys. C: Solid State Phys.* **1**, 299.

—— —— —— (1974). *J. Phys. C: Solid State Phys.* **7**, 2457

BELLISSENT, P. and TOURAND, G. (1976*a*). *J. Phys. (France)* **37**, 1423. (1976*b*). *Leningrad*, p. 160.

—— —— (1977). *Edinburgh*, p. 98.

BENNETT, A. J. and ROTH, L. M. (1971). *J. Phys. Chem. Solids* **32**, 1251.

BENOÎT, À LA GUILLAUME, C., and VOOS, M. (1973). *Phys. Rev. B* **7**, 1723.

—— —— and SALVAN, F. (1972). *Phys. Rev. B* **5**, 3079.

BEREZIN, A. A., GOLIKOVA, O. A., KAZANIN, M. M., KHOMIDOV, T., MIRLIN, D. N., PETROV, A. V., UMAROV, A. S., and ZAITSEV, V. K. (1974). *J. non-cryst. Solids* **16**, 237.

BERGGREN, K.-F. (1974). *Phil. Mag.* **30**, 1.

—— MARTINO, F., and LINDELL, G. (1974). *Phys. Rev. B* **9**, 4096.

BERGLUND, C. N. and SPICER, W. E. (1964). *Phys. Rev. A* **136**, 1044.

BERRY, M. V: and DOYLE, P. A. (1973). *J. Phys. C: Solid State Phys.* **6**, L6.

BETTS, F., BIENENSTOCK, A., and OVSHINSKY, S. R. (1970). *J. non-cryst. Solids* **4**, 554.

—— —— KEATING, D. T., and DE NEUFVILLE, J. P. (1972). *J. non-cryst. Solids* **7**, 417.

BEYER, W. and STUKE, J. (1974). *Garmisch*, p. 251.

—— —— (1975*a*). *Phys. Status Solidi A* **30**, 511.

—— —— (1975*b*). *Phys. Status Solidi A* **30**, K155.

—— —— and WAGNER, H. (1975). *Phys. Status Solidi A* **30**, 231.

BHATIA, A. B. and KRISHNAN, K. S. (1948). *Proc. R. Soc. A* **194**, 185.

BIENENSTOCK, A. (1973). *J. non-cryst. Solids* **11**, 447.

—— (1974). *Garmisch*, p. 49.

BISHOP, S. G. (1973). *Phys. Lett. A* **44**, 107.

—— and MITCHELL, D. L. (1973). *Phys. Rev. B* **8**, 5696.

—— and SHEVCHIK, N. F. (1974*a*). *Solid State Commun.* **15**, 629.

—— —— (1974*b*). *Proc. 12th Int. Conf. on the Physics of Semiconductors*, Stuttgart (ed. M. H. Pilkuhn), p. 1017. Teubner, Stuttgart.

—— —— (1975). *Phys. Rev. B* **12**, 1567.

—— STROM, U., and GUENZER, C. S. (1974). *Garmisch*, p. 963.

—— —— and TAYLOR, P. C. (1975). *Phys. Rev. Lett.* **34**, 1346.

—— —— (1976*a*). *Phys. Rev. Lett* **36**, 543.

—— —— (1976*b*). *Solid State Commun.* **18**, 573.

—— —— (1977). *Phys. Rev. B* **15**, 2278, Edinburgh, p. 595.

—— TAYLOR, P. C., MITCHELL, D. L., and SLACK, L. H. (1971). *J. non-cryst. Solids* **5**, 351.

BLINOWSKI, J. and MYCIELSKI, J. (1964). *Phys. Rev. A* **136**, 266.

BLOCH, A. N., WEISMAN, R. B., and VARMA, C. M. (1972). *Phys. Rev. Lett.* **28**, 753.

BLOCK, R., SUCK, J. B., GLÄSER, W., FREYLAND, W., and HENSEL, F. (1976). *Ber. Bunsenges Chem.* **80**, 718.

BOGOMOLOV, V. N., KUDINOV, E. K., and FIRSOV, YU. A. (1968). *Sov. Phys.—Solid State* **9**, 2502.

BONCH-BRUEVICH, V. L. (1970*a*). *J. non-cryst. Solids* **4**, 410.

—— (1970*b*). *Phys. Status Solidi* **42**, 35.

—— and ISKRA, V. D. (1975). *Phys. Status Solidi B* **68**, 369.

BORDAS, J. and WEST, J. B. (1976). *Phil. Mag.* **34**, 501.

BORLAND, R. E. (1963). *Proc. R. Soc. A.* **274**, 529.

BOSMAN, A. J. and CREVECOEUR, C. (1966). *Phys. Rev.* **144**, 763.

BÖTTGER, H. and BRYKSIN, V. V. (1977). *Phys. Status Solidi B* **81**, 433.

BOUCHARD, R. J., GILLSON, J. L., and JARRETT, H. S. (1973). *Mater. Res. Bull.* **8**, 489.

BRADLEY, C. C., FABER, T. E., WILSON, E. G., and ZIMAN, J. M. (1962). *Phil. Mag.* **7**, 865.

BRAUN, S. and GRIMMEISS, H. G. (1974). *J. appl. Phys.* **45**, 2658.

BREITLING, G. (1972). *J. non-cryst. Solids* **8–10**, 395.

—— and RICHTER, H. (1969). *Mater. Res. Bull.* **4**, 19.

BRENIG, W., DÖHLER, G. H., and WÖLFLE, P. (1971). *Z. Phys.* **246**, 1.

—— —— —— (1973). *Z. Phys.* **258**, 381.

BRINKMAN, W. F. and RICE, T. M. (1971). *Phys. Rev. B* **4**, 1566.

—— —— (1973). *Phys. Rev. B* **7**, 1508.

BRODSKY, M. H. and TITLE, R. S. (1969). *Phys. Rev. Lett.* **23**, 581.

—— —— (1976). *Williamsburg*, p. 97.

—— CARDONA, M., and CUOMO, J. J. (1977). *Phys. Rev. B* **16**, 3556.

—— KAPLAN, D. M., and ZIEGLER, J. F. (1972*b*). *Proc. 11th Int. Conf. on the Physics of Semiconductors*, *Warsaw*, p. 529. PWN-Polish Scientific Publishers, Warsaw.

—— FRISCH, M. A., ZIEGLER, J. F., and LANFORD, W. A. (1977). *Appl. Phys. Lett.* **30**, 561.

—— GAMBINO, R. J., SMITH, J. E., and YACOBY, Y. (1972*a*). *Phys. Status Solidi B* **52**, 609.

—— TITLE, R. S., WEISER, K., and PETTIT, G. D. (1970). *Phys. Rev. B* **1**, 2632.

BROUERS, F. (1970). *J. non-cryst. Solids* **4**, 428.

—— and BRAUWERS, M. (1975). *J. Phys. (Paris) Lett.* **36**, L-17.

BROWN, D., MOORE, D. S., and SEYMOUR, E. F. W. (1972). *J. non-cryst. Solids* **8–10**, 256.

BROWN, F. C. and RUSTGI, O. P. (1972). *Phys. Rev. Lett.* **28**, 497.

—— BACHRACH, R. Z., and SKIBOWSKI, M. (1977). *Phys. Rev. B* **15**, 4781.

BROWN, G. C. and HOLCOMB, D. F. (1974). *Phys. Rev. B* **10**, 3394.

BRUST, D. (1969*a*). *Phys. Rev. Lett.* **23**, 1232.

—— (1969*b*). *Phys. Rev.* **186**, 768.

BÜCKER, W. (1973). *J. non-cryst. Solids* **12**, 115.

BUCKLEY, W. D. and HOLMBERG, S. H. (1975). *Solid-State Electron.* **18**, 127.

BULLETT, D. W. (1974). *Yorktown Heights*, p. 139.

—— (1975*a*). *J. Phys. C: Solid State Phys.* **8**, L377.

—— (1975*b*). *J. Phys. C: Solid State Phys.* **8**, 2695, 2707.

—— (1975*c*). *Phil. Mag.* **32**, 1063.

—— (1976). *Phys. Rev. B* **14**, 1638.

BUSCH, G. and GÜNTHERODT, H.-J. (1974). *Solid State Phys.* **29**, 235.

—— —— KÜNZI, H. U., and SCHWEIGER, A. (1970). *Phys. Lett A* **33**, 64.

BUTCHER, P. N. (1974*a*). *J. Phys. C: Solid State Phys.* **7**, 2645.

—— (1974*b*). *J. Phys. C: Solid State Phys.* **7**, 3533.

—— (1976*a*). In *Linear and non-linear electron transport in solids* (ed. J. T. Devreese and V. E. van Doren), p. 341. Plenum Press, New York.

—— (1976*b*). *Leningrad, Electronic Phenomena in Non-Crystalline Semiconductors*, p. 89.

—— and FRIEDMAN, L. (1977). *J. Phys. C: Solid State Phys.* **10**, 3803

—— and HAYDEN, K. J. (1977*a*). *Edinburgh*, p. 234.

—— —— (1977*b*). *Phil. Mag.* **36**, 657.

—— and MORYS, P. L. (1973). *J. Phys. C: Solid State Phys.* **6**, 2147.

—— (1974). *Garmisch*, p. 153.

—— and FRIEDEL, J. (1971). *J. Phys. (Paris)* **32**, 73.

CABANE, B. and FROIDEVAUX, C. (1969). *Phys. Lett. A* **29**, 512.

CABRERA, N. and MOTT, N. F. (1948/49). *Rep. Prog. Phys.* **12**, 163.

CALLAERTS, R., DENAYER, M., HASHIMI, F. H., and NAGELS, P. (1970). *Discuss. Faraday Soc.* No. 50, p. 27.

CANNELLA, V., MYDOSH, J. A., and BUDNICK, J. I. (1971). *J. appl. Phys.* **42**, 1689.

CARDONA, M. (1969). *Modulation Spectroscopy. Solid State Phys. Suppl.* 11.

—— and GREENAWAY, D. L. (1964). *Phys. Rev. A* **133**, 1685.

—— GUDAT, W., SONNTAG, B., and YU, P. Y. (1970). *Proc. 10th Int. Conf. on the Physics of Semiconductors, Cambridge, Mass.* (ed. S. P. Keller, J. C. Hensel, and F. Stern), p. 209. United States Atomic Energy Commission, Washington, D.C.

CARE, C. M. and MARCH, N. H. (1975). *Adv. in Phys.* **24**, 101.

CASTELLAN, G. W. and SEITZ, F. (1951). *Proc. Conf. on Semi-conducting Materials, Reading* (ed. H. K. Henisch), p. 8. Butterworths, London.

CASTNER, T. G. and KÄNZIG, W. (1957). *J. Phys. Chem. Solids* **3**, 178.

CATE, R. C., WRIGHT, J. C., and CUSACK, N. E. (1970). *Phys. Lett. A* **32**, 467.

CATTERALL, J. A. and TROTTER, J. (1963). *Phil. Mag.* **8**, 897.

CATTERALL, R. and EDWARDS, P. P. (1975). *J. Phys. Chem.* **39**, 3018.

—— and MOTT, N. F. (1969). *Adv. Phys.* **18**, 665.

CERNOGORA, J., MOLLOT, F., and BENOÎT À LA GUILLAUME, C. (1973). *Phys. Stat. Solidi A* **15**, 401.

—— —— —— (1974). *Proc. 12th Int. Conf. on the Physics of Semiconductors, Stuttgart* (ed. M. H. Pilkuhn), p. 1027. Teubner, Stuttgart.

ČERVINKA, L., HRUBÝ, A., MATYÁŠ, M., ŠIMEČEK, T., ŠKÁCHA, J., ŠTOURAČ, L., TAUC, J., VORLÍČEK, V., and HÖSCHL, P. (1970). *J. non-cryst. Solids* **4**, 258.

CHANG, K. S., SHER, A., PETZINGER, K. G., and WEISZ, G. (1975). *Phys. Rev. B* **12**, 5506.

CHAUDHARI, P. and GRACZYK, J. F. (1974). *Garmisch*, p. 59.

—— —— and CHARBNAU, H. B. (1972). *Phys. Rev. Lett.* **29**, 425.

—— —— and HERD, S. R. (1972). *Phys. Stat. Solidi B* **51**, 801.

—— —— HENDERSON, D., and STEINHARDT, P. (1975). *Phil. Mag.* **31**, 727.

CHEN, J. Y., VETELINO, J. F., and MITRA, S. S. (1976). *Williamsburg*, p. 140.

CHITTICK, R. C. (1970). *J. non-cryst. Solids* **3**, 255.

—— ALEXANDER, J. H., and STERLING, H. F. (1969). *J. electrochem. Soc.* **116**, 77.

CHOO, F. C. and TONG, B. Y. (1976). *Williamsburg*, p. 58.

CHOPRA, K. L. (1969). *Thin film phenomena*. McGraw-Hill, New York.

—— and BAHL, S. K. (1969). *J. appl. Phys.* **40**, 4171.

—— —— (1970). *Phys. Rev. B* **1**, 2545.

—— —— (1972). *Thin solid Films* **11**, 377.

—— RASTOGI, A. C., and PANDYA, D. K. (1974). *Phil. Mag.* **30**, 935.

CHOYKE, W. J., VOSKO, S. H., and O'KEEFFE, T. W. (1971). *Solid State Commun.* **9**, 361.

CLARK, A. H. (1967). *Phys. Rev.* **154**, 750.

—— COHEN, M. M., CAMPI, M., and LANYON, H. P. D. (1974). *J. non-cryst. Solids* **16**, 117.

COCHRAN, W. (1973). *Phys. Rev. B* **8**, 623.

—— (1974). *Yorktown Heights*, p. 177.

COHEN, M. H. (1970*a*). *J. non-cryst. Solids* **4**, 391.

—— (1970*b*). *Proc. 10th Int. Conf. on the Physics of Semiconductors* (ed. S. P. Keller, J. C. Hensel, and F. Stern), p. 645. U.S. Atomic Energy Commission, Washington, D.C.

—— (1973). *Electrons in fluids* (ed. J. Jortner and N. R. Kestner), p. 257. Springer-Verlag.

—— (1977). *Can J. Chem.* **55**, 1906.

—— and JORTNER, J. (1973). *Phys. Rev. Lett.* **30**, 699.

—— —— (1974*a*). *Garmisch*, p. 167.

—— —— (1974*b*). *J. Phys. (Paris) Suppl.* **35**, C4-345.

—— and SAK, J. (1972). *J. non-cryst. Solids* **8–10**, 696.

—— FRITZSCHE, H., and OVSHINSKY, S. R. (1969). *Phys. Rev. Lett.* **22**, 1065.

COLEMAN, M. V., and THOMAS, D. J. D. (1967). *Phys. Status Solidi* **22**, 593; **24**, K111.

—— —— (1968). *Phys. Status Solidi* **25**, 241.

COLES, B. R. and TAYLOR, J. C. (1962). *Proc. R. Soc. A* **267**, 139.

CONNELL, G. A. N., and PAWLIK, J. R. (1976). *Phys. Rev. B* **13**, 787.

—— and PAUL, W. (1972). *J. non-cryst. Solids* **8–10**, 215.

—— and LEWIS, A. (1973). *Phys. Status Solidi B* **60**, 291.

—— and TEMKIN, R. J. (1974*a*). *Phys. Rev. B* **9**, 5323.

—— —— (1974*b*). *Yorktown Heights*, p. 192.

—— CAMPHAUSEN, D. L., and PAUL, W. (1972). *Phil. Mag.* **26**, 541.

—— PAUL, W., and TEMKIN, R. J. (1974). *Garmisch*, p. 541.

—— TEMKIN, R. J., and PAUL, W. (1973). *Adv. Phys.* **22**, 643.

CONWELL, E. M. (1956). *Phys. Rev.* **103**, 51.

CORNET, J. (1977). *Cambridge*, p. 17.

—— and ROSSITER, D. (1973). *J. non-cryst. Solids* **12**, 85.

CREVECOEUR, C. and DE WITT, H. J. (1968). *Solid State Commun.* **6**, 295.

—— —— (1971). *Solid State Commun.* **9**, 445.

CROITORU, N. and VESCAN, L. (1969). *Thin solid Films* **3**, 269.

—— —— POPESCU, C., and LĂZĂRESCU, M. (1970). *J. non-cryst. Solids* **4**, 493.

CROWDER, B. L. and SIENKO, M. J. (1963). *J. chem. Phys.* **38**, 1576.

CUMMING, J. B., KATCOFF, S., PORILE, N. T., TANAKA, S., and WYTTEN-BACH, A. (1964). *Phys. Rev. B* **134**, 1262.

CUSACK, N. E. (1963). *Rep. Prog. Phys.* **26**, 361.

CUTLER, M. (1971). *Phil. Mag.* **24**, 401.

—— (1974). *Phys. Rev. B* **9**, 1762.

—— (1977). *Liquid semiconductors*. Academic Press, New York.

—— and FIELD, M. E. (1968). *Phys. Rev.* **169**, 632.

—— and LEAVY, J. F. (1964). *Phys. Rev. A* **133**, 1153.

—— and MALLON, C. E. (1965). *J. appl. Phys.* **36**, 201.

—— and MOTT, N. F. (1969). *Phys. Rev.* **181**, 1336.

CYROT-LACKMANN, F. and GASPARD, J. P. (1974). *J. Phys. C: Solid State Phys.* **7**, 1829.

DALRYMPLE, R. J. F. and SPEAR, W. E. (1972). *J. Phys. Chem. Solids* **33**, 1071.

D'ALTROY, F. A. and FAN, H. Y. (1956). *Phys. Rev.* **103**, 1671.

DANILOV, A. V. and MOSLI, M. E. (1964). Sov. Phys.—Solid State **5**, 1472.

—— and MYULLER, R. L. (1962). *Zh. Prikl. Khim.* **35**, 2012.

DARBY, J. K. and MARCH, N. H. (1964). *Proc. phys. Soc.* **84**, 591.

DASH, W. C. and NEWMAN, R. (1955). *Phys. Rev.* **99**, 1151.

DA SILVA, J. R. G., PINATTI, D. G., ANDERSON, C. E., and RUDEE, M. L. (1975). *Phil. Mag.* **31**, 713.

DAVER, H., MASSENET, O., and CHAKRAVERTY, B. K. (1974). *Garmisch*, p. 1053.

DAVIES, H. A. and HALL, J. B. (1974). *J. Mater. Sci.* **9**, 707.

DAVIS, E. A. (1970). *J. non-cryst. Solids* **4**, 107.

—— (1973). *Aberdeen*, p. 425.

—— (1974). *Garmisch*, p. 519.

—— and COMPTON, W. D. (1965). *Phys. Rev. A* **140**, 2183.

—— and GREAVES, G. N. (1976). *Leningrad, Electronic Phenomena in Non-Crystalline Semiconductors*, p. 212.

—— and MOTT, N. F. (1970). *Phil. Mag.* **22**, 903.

—— ELLIOTT, S. R., GREAVES, G. N., and JONES, D. P. (1977). *Cambridge*, p. 205.

—— WRIGHT, H., DORAN, N. J., and NEX, C. M. M. (1979). *J. non-cryst. Solids* (in press).

DEBNEY, B. T. (1977). *J. Phys. C: Solid State Phys.* **10**, 4719.

DESSAUER, J. H. and CLARK, H. E. (1965). *Xerography and related processes*. Focal Press, New York.

DE GENNES, P. G. (1960). *Phys. Rev.* **118**, 141.

DE NEUFVILLE, J. P., MOSS, S. C., and OVSHINSKY, S. R. (1974). *J. non-cryst. Solids* **13**, 191.
DEVILLERS, M. A. C. (1974). *J. Phys. F: Metal Phys.* **4**, L236.
—— and ROSS, R. G. (1975). *J. Phys. F: Metal Phys.* **5**, 73.
DE VORE, H. B. (1956). *Phys. Rev.* **102**, 86.
DEVREESE, J. T. (ed.) (1972). *Polarons in ionic crystals and polar semiconductors.* North-Holland, Amsterdam.
DE WIT, H. J. and CREVECOEUR, C. (1972). *J. non-cryst. Solids* **8–10**, 787.
DEXTER, D. L. (1967). *Phys. Rev. Lett.* **19**, 1383.
—— and KNOX, R. S. (1965). *Excitons.* Wiley–Interscience, New York.
—— KLICK, C. C., and RUSSELL, G. A. (1955). *Phys. Rev.* **100**, 603.
DICKENS, P. G., QUILLIAM, R. M. P., and WHITTINGHAM, M. S. (1968). *Mater Res. Bull.* **3**, 941.
DI SALVO, F. J., BAGLEY, B. G., and CLARK, A. H. (1974). *Bull. am. phys. Soc.* **19**, 316.
—— BAGLEY, B. G., and HUTTON, R. S. (1976). *Solid State Commun.* **19**, 97.
DI STEPHANO, T. H., and EASTMAN, D. E. (1971a). *Phys. Rev. Lett.* **27**, 1560.
—— (1971b). *Solid State Commun.* **9**, 2259.
DODELET, J.-P., SHINSAKA, K., and FREEMAN, G. R. (1973). *J. chem. Phys.* **59**, 1293.
DÖHLER, G. H. and BRODSKY, M. H. (1974). *Yorktown Heights*, p. 351.
DOLEZALEK, F. K. and SPEAR, W. E. (1970). *J. non-cryst. Solids* **4**, 97.
DONALLY, J. M. and CUTLER, M. (1968). *Phys. Rev.* **176**, 1003.
—— —— (1972). *J. Phys. Chem. Solids* **33**, 1017.
DONIKOV, L. I. and BORISOVA, Z. U. (1963). *Solid state chemistry* (ed. Z. U. Borisova), p. 100. Transl. Consultants Bureau, New York.
DONOVAN, T. M. (1970). Ph.D. Thesis, Stanford University.
—— (1974). *Yorktown Heights*, p. 1.
—— and HEINEMANN, K. (1971). *Phys. Rev. Lett.* **27**, 1794.
—— and SPICER, W. E. (1968). *Phys. Rev. Lett.* **21**, 1572.
—— ASHLEY, E. J., and SPICER, W. E. (1970). *Phys. Lett A* **32**, 85.
—— SPICER, W. E., and BENNETT, J. M. (1969). *Phys. Rev. Lett.* **22**, 1058.
—— —— —— and ASHLEY, E. J. (1970). *Phys. Rev. B* **2**, 397.
DREWS, R. E., EMERALD, R. L., SLADE, M. L., and ZALLEN, R. (1972). *Solid State Commun.* **10**, 293.
DORDA, G. (1973). *Festkörperprobleme* **13**, 215.
DOUGIER, P. (1975). Thesis, University of Bordeaux.
—— and CASALOT, A. (1970). *J. Solid State Chem.* **2**, 396.
DOUMERC, P. (1974). Thesis, University of Bordeaux.
DOW, J. D. and HOPFIELD, J. J. (1972). *J. non-cryst. Solids* **8–10**, 664.
—— and REDFIELD, D. (1970). *Phys. Rev. B* **1**, 3358.
—— —— (1972). *Phys. Rev. B* **5**, 594.
DRAKE, C. F. and SCANLAN, I. F. (1970). *J. non-cryst. Solids* **4**, 234.
DREIRACH, O., EVANS, R., GÜNTHERODT, H.-J., and KÜNZI, H.-U. (1972). *J. Phys. F: Metal Phys.* **2**, 709.
DREWS, R. E. (1966). *Appl. Phys. Lett.* **9**, 347.
DRUGER, S. D. and KNOX, R. S. (1969). *J. chem. Phys.* **50**, 3143.
DUCKERS, L. J. and ROSS, R. G. (1973). *The properties of liquid metals* (ed. S. Takeuchi), p. 365. Taylor & Francis, London.
DUFFY, M. G., BOUDREAUX, D. S., and POLK, D. E. (1974). *J. non-cryst. Solids* **15**, 435.

DUMAS, J., SCHLENKER, C., and NATOLI, R. C. (1975). *Solid State Commun.* **16**, 493.

DUPREE, B. C., ENDERBY, J. E., NEWPORT, R. J., and VAN ZYTVELD, J. B. (1977). *Proc. 3rd Int. Conf. on Liquid Metals, Bristol* (ed. R. Evans and D. A. Greenwood), p. 337. Institute of Physics, London.

DUWEZ, P. (1976). *Ann. Rev. Mater. Sci.* **6**, 83.

DYNES, R. C., GARNO, J. P., and ROWELL, J. M. (1978). *Phys. Rev. Lett.* **40**, 479.

DYSON, F. J. (1953). *Phys. Rev.* **92**, 1331.

—— (1969*a*). *Phys. Rev.* **178**, 668.

—— (1969*b*). *Phys. Rev.* **181**, 1278.

EAGLES, D. M. (1966). *Phys. Rev.* **145**, 640.

—— (1969*a*). *Phys. Rev.* **178**, 668.

—— (1969*b*). *Phys. Rev.* **181**, 1278.

EASTMAN, D. E. (1971). *Phys. Rev. Lett.* **26**, 1108.

—— and GROBMAN, W. D. (1972*a*). *Phys. Rev. Lett.* **28**, 1378.

—— —— (1972*b*). *Proc. 11th Int. Conf. on the Physics of Semiconductors, Warsaw*, p. 889. PWN-Polish Scientific Publishers, Warsaw.

—— FREEOUF, J. L., and ERBUDAK, M. (1974). *Yorktown Heights*, p. 95.

ECONOMOU, E. N. and ANTONIOU, P. D. (1977). *Solid State Commun.* **21**, 285.

—— and COHEN, M. H. (1970*a*). *Phys. Rev. Lett.* **25**, 1445.

—— —— (1970*b*). *Mater. Res. Bull.* **5**, 577.

—— —— (1970*c*). *Phys. Rev. Lett.* **24**, 218.

—— —— (1972). *Phys. Rev. B* **5**, 2931.

—— and PAPATRIANTAFILLOU, C. (1972). *Solid State Commun.* **11**, 197.

—— —— (1974). *Phys. Rev. Lett.* **32**, 1130.

—— COHEN, M. H., FREED, K. F., and KIRKPATRICK, E. S. (1974). In *Amorphous and liquid semiconductors* (ed. J. Tauc), p. 101. Plenum Press, New York.

EDMOND, J. T. (1966). *Br. J. appl. Phys.* **17**, 979.

—— (1968). *J. non-cryst. Solids* **1**, 39.

—— (1970). *Phys. Status Solidi A* **3**, K129.

EDWARDS, J. T. and THOULESS, D. J. (1972). *J. Phys. C: Solid State Phys.* **5**, 807.

EDWARDS, S. F. (1958). *Phil. Mag.* **3**, 1020.

—— (1961). *Phil. Mag.* **6**, 617.

—— (1962). *Proc. R. Soc. A* **267**, 518.

—— and ANDERSON, P. W. (1975). *J. Phys. F: Metal Phys.* **5**, 965.

EDWARDS, P. P. and SIENKO, M. J. (1978). *Phys. Rev.* (in press).

EFROS, A. L. (1976). *J. Phys. C: Solid State Phys.* **9**, 2021.

—— and SHKLOVSKII, B. I. (1975). *J. Phys. C: Solid State Phys.* **8**, L49.

EGERTON, R. F. (1971). *Appl. Phys. Lett.* **19**, 203.

EISENBERG, A. and TOBOLSKY, A. V. (1960). *J. polymer Sci.* **46**, 19.

EL-HANANY, U. and WARREN, W. W. (1975). *Phys. Rev. Lett.* **34**, 1276.

ELLIOTT, P. J., YOFFE, A. D., and DAVIS, E. A. (1974). *Yorktown Heights*, p. 311.

ELLIOTT, S. R. (1977). *Phil. Mag.* **36**, 1291.

—— (1978). *Phil. Mag. B* **37**, 135, 553.

—— and DAVIS, E. A. (1976). *Williamsburg*, p. 117.

—— —— (1977). *Edinburgh*, p. 637.

—— —— and PITT, G. D. (1977). *Solid State Commun.* **22**, 481.

ELLIS, E., GASKELL, P. M., and JOHNSON, D. W. (1977). *Proc. 4th Int. Conf. on the Physics of Non-Crystalline Solids* (ed. G. H. Frischat), p. 312. Trans. Tech. Publications, Aedermannsdorf, Switzerland.

EMIN, D. (1973*a*). *Adv. Phys.* **22**, 57.

—— (1973*b*). *Aberdeen*, p. 261.

—— (1974). *Phys. Rev. Lett.* **32**, 303.

—— (1975). *Adv. Phys.* **24**, 305.

—— (1977*a*). *Phil. Mag.* **35**, 1189.

—— (1977*b*). *Edinburgh*, p. 249.

—— and HOLSTEIN, T. (1969). *Ann. Phys.* **53**, 439.

—— SEAGER, C. H., and QUINN, R. K. (1972). *Phys. Rev. Lett.* **28**, 813.

ENCK, R. C. (1973). *Phys. Rev. Lett.* **31**, 220.

ENDERBY, J. E. and DUPREE, B. C. (1977). *Phil. Mag.* **35**, 791.

—— and HOWE, R. A. (1968). *Phil. Mag.* **18**, 923.

—— and SIMMONS, C. J. (1970). *Phil. Mag.* **20**, 125.

—— and WALSH, L. (1966). *Phil. Mag.* **14**, 991.

—— NORTH, D. M., and EGELSTAFF, P. A. (1966). *Phil. Mag.* **14**, 961.

ENDO, H., EATAH, A. I., WRIGHT, J. G., and CUSACK, N. E. (1973). *J. phys. Soc. Japan* **34**, 666.

ENDO, S., MITSUI, T. and MIYADAI, T. (1973). *Phys. Lett. A* **46**, 29.

ENGEMANN, D. and FISCHER, R. (1974*a*). *Proc. 12th Int. Conf. on the Physics of Semiconductors, Stuttgart* (ed. M. H. Pilkuhn), p. 1042. Teubner, Stuttgart.

—— —— (1974*b*). *Garmisch*, p. 947.

—— —— (1976). *Williamsburg*, p. 37.

—— —— and MELL, H. (1977). *Edinburgh*, p. 387.

ENGLMAN, R. and JORTNER, J. (1970). *Mol. Phys.* **18**, 145.

ENGSTRÖM, O. and GRIMMEISS, H. G. (1975). *J. appl. Phys.* **46**, 831.

ESSER, B. (1972). *Phys. Status Solidi B* **51**, 735.

EVANS, R. (1970). *J. Phys. C: Solid State Phys.* **3**, S137.

—— GREENWOOD, D. A., and LLOYD, P. (1971). *Phys. Lett. A* **35**, 57.

EVEN, U. and JORTNER, J. (1972). *Phil. Mag.* **25**, 715.

—— —— (1973). *Phys. Rev. B* **8**, 2536.

FABER, T. E. (1967). *Adv. Phys.* **16**, 637.

—— (1972). *Introduction to the theory of liquid metals.* Cambridge University Press.

—— and ZIMAN, J. M. (1965). *Phil. Mag.* **11**, 153.

FAGEN, E. A. and FRITZSCHE, H. (1970). *J. non-cryst. Solids* **2**, 180.

—— HOLMBERG, S. H., SEGUIN, R. W., THOMPSON, J. C., and FRITZSCHE, H. (1970). *Proc. 10th Int. Conf. on the Physics of Semiconductors, Cambridge, Massachusetts* (eds. S. P. Keller, J. C. Hensel, and F. Stern), p. 672. U.S. Atomic Energy Commission, Washington, D.C.

FAN, H. Y. (1951). *Phys. Rev.* **82**, 900.

FANG, F. F. and FOWLER, A. B. (1968). *Phys. Rev.* **169**, 619.

FEHLNER, F. P. and MOTT, N. F. (1970). *Oxid. Met.* **2**, 59.

FELTY, E. J., LUCOVSKY, G., and MYERS, M. B. (1967). *Solid State Commun.* **5**, 555.

FERRE, D., DUBOIS, H., and BISKUBSKI, G. (1975). *Phys. Status Solidi B* **70**, 81.

FERRIER, R. P. and HERRELL, D. J. (1969). *Phil. Mag.* **19**, 853.

—— —— (1970). *J. non-cryst. Solids* **2**, 278.

FISCHER, J. E. and DONOVAN, T. M. (1972). *J. non-cryst. Solids* **8–10**, 202.

FISCHER, K. H. (1977). *Physica B* **86**, 813.

FISCHER, R. and VORNHOLZ, D. (1975). *Phys. Status Solidi B* **68**, 561.

—— HEIM, U., STERN, F., and WEISER, K. (1971). *Phys. Rev. Lett.* **26**, 1182.

FISCHER, T. E. and ERBUDAK, M. (1971). *Phys. Rev. Lett.* **27**, 1220.

FISHER, F. D., MARSHALL, J. M., and OWEN, A. E. (1976). *Phil. Mag.* **33**, 261.

FISHER, G. B., LINDAU, I., ORLOWSKI, B. A., SPICER, W. E., VERHELLE, Y., and WEAVER, H. E. (1974). *Garmisch*, p. 621.

FLASCHEN, S. S., PEARSON, A. D., and NORTHOVER, W. R. (1959). *J. am. ceram. Soc.* **42**, 450.

FLASCK, R., IZU, M., SAPRU, K., ANDERSON, T., OVSHINSKY, S. R., and FRITZSCHE, H. (1977). *Edinburgh*, p. 524.

FOWLER, A. B., FANG, F. F., HOWARD, W. E., and STILES, P. J. (1966). *Phys. Rev. Lett.* **16**, 901.

FRANZ, W. (1958). *Z. Naturforsch. A* **13**, 484.

FREED, K. F. (1972). *Phys. Rev. B* **5**, 4802.

FREEMAN, L. A., HOWIE, A., MISTRY, A. B., and GASKELL, P. H. (1977). *Cambridge*, p. 245.

FRENKEL, J. (1938). *Phys. Rev.* **54**, 647.

FREYLAND, W. and STEINLEITER, G. (1976). *Ber. Bunsenges. phys. Chem.* **80**, 810.

—— PFEIFER, H. P., and HENSEL, F. (1974). *Garmisch*, p. 1327.

FRIEDEL, J. (1954). *Adv. Phys.* **3**, 446.

FRIEDMAN, L. (1963). *Phys. Rev.* **131**, 2445.

—— (1971). *J. non-cryst. Solids* **6**, 329.

—— (1973). *Aberdeen*, p. 363.

—— (1977). *Phil. Mag.* **36**, 553.

—— and HOLSTEIN, T. (1963). *Ann. Phys.* **21**, 494.

FRISCH, H. L. and LLOYD, S. P. (1960). *Phys. Rev.* **120**, 1175.

FRITZSCHE, G. and CUEVAS, M. (1960). *Phys. Rev.* **119**, 1238.

FRITZSCHE, H. (1958). *J. Phys. Chem. Solids* **6**, 69.

—— (1959). *Phys. Rev.* **115**, 336.

—— (1960). *Phys. Rev.* **119**, 1899.

—— (1962). *Phys. Rev.* **125**, 1552.

—— (1971). *J. non-cryst. Solids* **6**, 49.

—— (1973). *Aberdeen*, p. 55.

—— (1974). In *Amorphous and liquid semiconductors* (ed. J. Tauc), p. 221. Plenum Press, New York.

—— (1976). *Leningrad, Electronic phenomena in non-crystalline semiconductors*, p. 65.

—— (1977). *Edinburgh*, p. 3.

—— and OVSHINSKY, S. R. (1970). *J. non-cryst. Solids* **2**, 148.

FRITZSCHE, S. and LARK-HOROVITZ, K. (1959). *Phys. Rev.* **113**, 999.

FRÖHLICH, H. (1947). *Proc. R. Soc. A* **188**, 521.

—— (1954). *Adv. Phys.* **3**, 325.

—— (1958). *Theory of dielectrics* (2nd ed.). Oxford University Press, London.

FROMHOLD, A. T. and NARATH, A. (1964). *Phys. Rev. A* **136**, 487.

FUCHS, R. (1965). *J. chem. Phys.* **42**, 3781.

FUHS, W., HESSE, H.-J. and LANGER, K. H. (1974). *Garmisch*, p. 79.

—— NIEMANN, K., and STUKE, J. (1974). *Yorktown Heights*, p. 345.

GALEENER, F. L. (1971). *Phys. Rev. Lett.* **27**, 421, 769, 1716.

—— and LUCOVSKY, G. (1976). *Phys. Rev. Lett.* **37**, 1474.

GALLAGHER, T. J. (1975). *Simple dielectric liquids, mobility, conduction, and breakdown.* Clarendon Press, Oxford.

GANDAIS, M., THEYE, M. L., FISSON, S., and BOISSONADE, J. (1973). *Phys. Status Solidi B* **58**, 601.

GARDNER, J. A. and CUTLER, M. (1977). *Edinburgh*, p. 838.

GASKELL, P. H. (1975). *Phil. Mag.* **32**, 211.

—— and HOWIE, A. (1974). *Stuttgart*, p. 1076.

—— GIBSON, J. M., and HOWIE, A. (1977). *Cambridge*, p. 181.

GASPARD, J. P. and CYROT-LACKMANN, F. (1973). *J. Phys. C: Solid State Phys.* **6**, 3077.

GAUTIER, F., KRILL, G., LAPIERRE, M. F., PANISSOD, P., ROBERT, C., CZJZEK, G., FINK, J., SCHMIDT, H. (1975). *Phys. Lett. A* **53**, 31.

GEBALLE, T. H. and HULL, G. W. (1955). *Phys. Rev.* **98**, 940.

GEE, G. (1952). *Trans. Faraday Soc.* **48**, 515.

GERSHENZON, E. M., IL'IN, V. A., KURILENKO, I. N., and LITVAK-GORSKAYA, L. B. (1973). *Sov. Phys.—Semicond.* **6**, 1457.

GESCHWIND, S., ROMESTAIN, R., and DEVLIN, G. E. (1976). *J. Phys. (Paris) Colloque* **C4**, 313.

GHOSH, P. K. and SPEAR, W. E. (1968). *J. Phys. C: Solid State Phys.* **1**, 1347.

GILMAN, J. J. (1975). *J. appl. Phys.* **46**, 1625.

GLAZOV, V. M., CHIZHEVSKAYA, S. N., and GLAGOLEVA, N. N. (1969). *Liquid semiconductors*. Plenum Press, New York.

GOBRECHT, H., GAWLIK, D., and MAHDJURI, F. (1971). *Phys. Condens. Matter* **13**, 156.

—— MAHDJURI, F., and GAWLIK, D. (1971). *J. Phys. C: Solid State Phys.* **4**, 2247.

GOGOLIN, A. A., MEL'NIKOV, V. I., and RASHBA, É. I. (1975). *Sov. Phys.— JETP* **42**, 168.

GOLIN, S. (1963). *Phys. Rev.* **132**, 178.

GOODENOUGH, J. B. (1972). *Phys. Rev. B* **5**, 2764.

GOODMAN, A. M. (1967). *Phys. Rev.* **164**, 1145.

GRACZYK, J. F. and CHAUDHARI, P. (1973*a*). *Phys. Status Solidi B* **58**, 163.

—— —— (1973*b*). *Phys. Status Solidi B* **58**, 501.

GRANT, A. J. and DAVIS, E. A. (1974). *Solid State Commun.* **15**, 563.

—— MOUSTAKAS, T. D., PENNEY, T., and WEISNER, K. (1974). *Garmisch*, p. 325.

GREAVES, G. N. (1973). *J. non-cryst. Solids* **11**, 427.

—— (1975). Ph.D. Thesis, University of Cambridge.

—— (1976). *Williamsburg*, p. 136.

GREAVES, G. N. and DAVIS, E. A. (1974*a*). *Phil. Mag.* **29**, 1201.

—— —— (1974*b*). *Proc. 12th Int. Conf. on the Physics of Semiconductors* (ed. M. H. Pilkuhn), p. 1047. Teubner, Stuttgart.

—— —— and BORDAS, J. (1976*a*). *Phil Mag.* **34**, 265.

—— KNIGHTS, J. C., and DAVIS, E. A. (1974). *Garmisch*, p. 369.

—— DAVIS, E. A., BORDAS, J., KELLY, M. J., and BULLETT, D. W. (1976*b*). *Leningrad, Structure and Properties of Non-Crystalline Semiconductors*, p. 264.

—— ELLIOTT, S. R., and DAVIS, E. A. (1979). *Adv. Phys.* (in press).

GREENE, M. P. and KOHN, W. (1965). *Phys. Rev.* **137**, 513.

GREENFIELD, A. J. (1966). *Phys. Rev. Lett.* **16**, 6.

GREENWOOD, D. A. (1958). *Proc. phys. Soc.* **71**, 585.

—— (1973). *Aberdeen*, p. 127.

GREINER, J. D., SHANKS, H. R., and WALLACE, D. C. (1962). *J. chem. Phys.* **36**, 772.

GRIGOROVICI, R. (1968). *Mater. Res. Bull.* **3**, 13.

—— (1973). *Aberdeen*, p. 191.

—— (1974). In *Amorphous and liquid semiconductors* (ed. J. Tauc), p. 45. Plenum Press, New York.

GRIGOROVICI, R. and BALU, A. (1972). *Proc. 11th Int. Conf. on the Physics of Semiconductors, Warsaw*, p. 453. PWN-Polish Scientific Publishers, Warsaw.

—— and MĂNĂILĂ, R. (1969). *J. non-cryst. Solids* **1**, 371.

—— —— (1967). *Thin solid Films* **1**, 343.

—— and VANCU, A. (1968). *Thin solid Films* **2**, 105.

—— CROITORU, N., and DÉVÉNYI, A. (1967). *Phys. Status Solidi* **23**, 621.

GRIMMEISS, H. G. (1977). *Ann. Rev. Mater. Sci.* **7**, 100.

—— LEDEBO, L. Å., OVREN, C., and MORGAN, T. N. (1974). *Proc. 12th Int. Conf. on the Physics of Semiconductors* (ed. M. H. Pilkuhn), p. 386. Teubner, Stuttgart.

GROSSE, P. (1969). *Die Festkörpereigenschaften von Tellur—Springer Tracts modern Phys.* **48**. Springer, Berlin.

GRUNWALD, H. P. and BLAKNEY, R. M. (1968). *Phys. Rev.* **165**, 1006.

GUBANOV, A. I. (1963). *Quantum electron theory of amorphous conductors.* Consultants Bureau, New York, 1965.

GUINER, A., FOURNET, G., WALKER, C. B., and YODOWITCH, K. L. (1955). *Small angle scattering of X-rays.* Wiley, New York.

GUMMEL, H. and LAX, M. (1957). *Ann. Phys. N.Y.* **2**, 28.

GÜNTHERODT, H.-J. and KÜNZI, H. U. (1973). *Phys. kondens. Mater.* **16**, 117.

—— HAUSER, E., and KÜNZI, H. U. (1974). *Phys. Lett. A* **48**, 201.

—— —— —— (1977). *Prod 3rd Int. Conf. on Liquid Metals, Bristol* (ed. R. Evans and D. A. Greenwood), p. 324. The Institute of Physics, London.

GURMAN, S. J. and PENDRY, J. B. (1976). *Solid State Commun.* **20**, 287.

GURNEY, R. W. and MOTT, N. F. (1938). *Trans. Faraday Soc.* **34**, 506.

GUTTMAN, L. (1976). *Williamsburg*, p. 268.

HAACK, D. (1977). *Proc. 4th Int. Conf. on the Physics of Non-Crystalline Solids, Clausthal-Zellerfeld* (ed. G. H. Frischat). p. 384. Trans Tech. Publications. Aedermannsdorf, Switzerland.

HAGENMULLER, P. (1971). *Prog. solid state Chem.* **5**, 71.

HAGSTON, W. E. and LOWTHER, J. E. (1973). *Physica* **70**, 40.

HAISTY, R. W. and KREBS, H. (1969a). *J. non-cryst. Solids* **1**, 399.

—— —— (1969b) *J. non-cryst. Solids* **1**, 427.

HALDER, N. C. and WAGNER, C. N. J. (1967). *J. chem. Phys.* **47**, 4385.

HALPERIN, B. I. (1967). *Adv. chem. Phys.* **13**, 123.

—— and LAX, M. (1966). *Phys. Rev.* **148**, 722.

—— and RICE, T. M. (1968). *Rev. mod. Phys.* **40**, 755.

—— LEKNER, J., RICE, S. A., and GOMER, R. (1967). *Phys. Rev.* **156**, 351.

HAMILTON, E. M. (1972). *Phil. Mag.* **26**, 1043.

HANEMAN, D. (1968). *Phys. Rev.* **170**, 705.

HANSON, R. C. and BROWN, F. C. (1960). *J. appl. Phys.* **31**, 210.

HARRISON, J. P. and MARKO, J. R. (1976). *Phil. Mag.* **34**, 789.

HARRISON, W. A. (1966). *Pseuodopotentials in the theory of metals.* Benjamin, New York.

HARTKE, J. L. (1962). *Phys. Rev.* **125**, 1177.

—— and REGENSBURGER, P. J. (1965). *Phys. Rev. A* **139**, 970.

HARTMAN, R. (1969). *Phys. Rev.* **181**, 1070.

HASEGAWA, A. (1964). *J. phys. Soc. Japan* **19**, 504.

HATTORI, K. (1975). *J. phys. Soc. Japan* **38**, 669.

HAUSER, J. J. (1972). *Phys. Rev. Lett.* **29**, 476.

—— (1973). *Phys. Rev. B* **8**, 3817.

—— (1974*a*). *Phys. Rev. B* **9**, 2623.

—— (1974*b*). *Yorktown Heights*, p. 338.

—— (1975). *Phys. Rev. B* **11**, 738.

—— (1976). *Solid State Commun.* **19**, 1049.

—— and HUTTON, R. S. (1976). *Phys. Rev. Lett.* **37**, 868.

—— and STAUDINGER, A. (1973). *Phys. Rev. B* **8**, 607.

—— DI SALVO, F. J., and HUTTON, R. S. (1977). *Phil. Mag.* **35**, 1557.

HAYDOCK, R., HEINE, V., and KELLY, M. J. (1972). *J. Phys. C: Solid State Phys.* **5**, 2845.

—— —— —— (1975). *J. Phys. C: Solid State Phys.* **8**, 2591.

HAYES, T. M., KNIGHTS, J. C., and MIKKELSEN, J. C. (1977). *Edinburgh*, p. 73.

HAYES, T. M., SEN, P. N., and HUNTER, S. H. (1976). *J. Phys. C: Solid State Phys.* **9**, 4357.

HEIKES, R. R. and URE, R. W. (1961). *Thermoelectricity*, p. 81. Interscience, New York.

HEINE, V. (1970). *Solid State Phys.* **24**, 1.

—— (1971). *J. Phys. C: Solid State Phys.* **4**, L221.

HELMS, C. R., SPICER, W. E., and PERESKOKOV, V. (1974). *Appl. Phys. Lett.* **24**, 318.

HENDERSON, D. (1972). *Computational solid state physics* (eds. F. Herman, N. W. Dalton, and T. R. Koehler), p. 175. Plenum Press, New York.

—— and HERMAN, F. (1972). *J. non-cryst. Solids* **8–10**, 359.

—— and ORTENBURGER, I. B. (1973). *J. Phys. C: Solid State Phys.* **6**, 631.

—— —— and ORTENBURGER, I. B. (1974). *Garmisch*, p. 991.

HENISCH, H. K. (1969). *Sci. Am.* **221**(Nov.), 30.

—— and PRYOR, R. W. (1971). *Solid-State Electron.* **14**, 765.

—— SMITH, W. R., and WIHL, W. (1974). *Garmisch*, p. 567.

HENKELS, H. W. and MACZUK, J. (1953). *J. appl. Phys.* **24**, 1056.

HENRY, C. H. and LANG, D. V. (1977). *Phys. Rev. B* **15**, 989.

HENSEL, F. (1970). *Phys. Lett. A* **31**, 88.

—— (1976). *Ber. Bunsenges. phys. Chem.* **80**, 786.

—— and FRANCK, E. U. (1966). *Ber. Bunzenges. phys. Chem.* **70**, 1154.

—— —— (1968). *Rev. mod. Phys.* **40**, 697.

HENSEL, J. C., PHILLIPS, T. G., and RICE, T. M. (1973). *Phys. Rev. Lett.* **30**, 227.

—— —— and THOMAS, G. A. (1977). *Solid State Phys.* **32**, 87.

HERBERT, D. C. and JONES, R. (1971). *J. Phys. C: Solid State Phys.* **4**, 1145.

HERD, S. R. and CHAUDHARI, P. (1974). *Phys. Status Solidi A* **26**, 627.

HERMAN, F. and VAN DYKE, J. P. (1968). *Phys. Rev. Lett.* **21**, 1575.

—— KORTUM, R. L., KUGLIN, C. C., and SHAY, J. L. (1967). In *II–VI semiconducting compounds* (ed. D. G. Thomas), p. 503. Benjamin, New York.

HIGGINS, J. K., TEMPLE, B. K., and LEWIS, J. E. (1977). *J. non-cryst. Solids* **23**, 187.

HILL, R. M. (1969). *Proc. R. Soc. A* **309**, 377.

—— (1971). *Phil. Mag.* **23**, 59.

HILTON, A. R. (1970). *J. non-cryst. Solids* **2**, 28.

—— and BRAU, M. (1963). *Infrared Phys.* **3**, 69.

HINDLEY, N. K. (1970). *J. non-cryst. Solids* **5**, 17, 31.

HIROSE, M., TARIGUCHI, M., and OSAKA, Y. (1976). *Japan J. appl. Phys.* **15**, 175.

HODGKINSON, R. J. (1974). *J. Phys. C: Solid State Phys.* **7**, L9.

HODGSON, J. N. (1963). *Phil. Mag.* **8**, 735.

HÖHNE, M. and STASIW, M. (1968). *Phys. Status Solidi* **28**, 247.

HOLLAND, L. (1963). *Vacuum deposition of thin films.* Chapman and Hall, London.

HOLSTEIN, T. (1959). *Ann. Phys.* **8**, 343.

—— (1961). *Phys. Rev.* **124**, 1329.

—— (1973). *Phil. Mag.* **27**, 225.

—— and FRIEDMAN, L. (1968). *Phys. Rev.* **165**, 1019.

HOPFIELD, J. J. (1968). *Comments Solid State Phys.* **1**, 16.

HORI, J. (1968). *Spectral properties of disordered chains and lattices.* Pergamon Press, Oxford.

HOSHINO, H., SCHMUTZLER, R. W., and HENSEL, F. (1977). *Proc. 3rd Int. Conf. on Liquid Metals, Bristol* (ed. R. Evans, and D. A. Greenwood), p. 404. The Institute of Physics, London.

HOSHINO, K. and WATABE, M. (1977). *J. Phys. Soc. Japan* 43, 1337.

HOWE, S. H., LE COMBER, P. G., and SPEAR, W. E. (1971). *Solid State Commun.* **9**, 65.

HOWIE, A., KRIVANEK, O. L., and RUDEE, M. L. (1973). *Phil. Mag.* **27**, 235.

HUANG, K. and RHYS, A. (1951). *Proc. R. Soc. A* **204**, 406.

HUBBARD, J. (1964). *Proc. R. Soc. A* **277**, 237.

HUDGENS, S. J. (1973). *Phys. Rev. B* **7**, 2481.

—— (1976). *Phys. Rev. B* **14**, 1547.

—— KASTNER, M., FORBERG, R. R., CANDEA, R. M., and CLAHASEY, C. J. (1976). *Williamsburg*, p. 130.

HUGHES, R. C. (1973). *Phys. Rev. Lett.* **30**, 1333.

—— (1975). *Appl. Phys. Lett.* **26**, 436.

—— (1977). *Phys. Rev. B* **15**, 2012.

HULTGREN, R., GINGRICH, N. S., and WARREN, B. E. (1935). *J. Chem. Phys.* **3**, 351.

HUNG, C. S. and GLEISSMAN, J. R. (1950). *Phys. Rev.* **79**, 726.

HUNTER, S. H., BIENENSTOCK, A., and HAYES, T. M. (1977*a*). *Cambridge*, p. 73.

—— —— (1977*b*). *Edinburgh*, p. 78.

HURST, C. H., and DAVIS, E. A. (1974). *Garmisch*, p. 349. *J. non-cryst. Solids* **16**, 343, 355.

HUTCHISSON, E. (1930). *Phys. Rev.* **36**, 410.

IEDA, M., SAWA, G., and KATO, S. (1971). *J. appl. Phys.* **42**, 3737.

IKEHATA, S., SASAKI, W., and KOBAYASHI, S. (1975). *J. phys. Soc. Japan* **39**, 1492.

ILSCHNER, B. R. and WAGNER, C. (1958). *Acta metall.* **6**, 712.

IOFFE, A. F. and REGEL, A. R. (1960). *Prog. Semicond.* **4**, 237.

IVKIN, E. B. and KOLOMIETS, B. T. (1970). *J. non-cryst. Solids* **3**, 41.

JAMES, H. M. and GINZBARG, A. S. (1953). *J. phys. Chem.* **57**, 840.

JARRETT, H. S., BOUCHARD, R. J., GILLSON, J. L., JONES, G. A., MARCUS, S. M., and WEIHER, J. F. (1973). *Mater. Res. Bull.* **8**, 877.

JENKINS, T. E., LEVY, A. W., HODBY, J. W. (1976). *Phil. Mag.* **34**, 397.

JOANNOPOULOS, J. D. (1976). *Williamsburg*, p. 108.

—— and COHEN, M. L. (1973*a*). *Phys. Rev. B* **7**, 2644.

—— —— (1973*b*). *Phys. Rev. B* **8**, 2733.

—— —— (1976). *Solid State Phys.* **31**, 71.

—— and POLLARD, W. B. (1976). *Solid State Commun.* **20**, 947.

564 REFERENCES

—— and YNDURAIN, F. (1974). *Phys. Rev. B* **10**, 5164.
—— SCHLÜTER, M., and COHEN, M. L. (1975). *Phys. Rev. B* **11**, 2186.
JONES, D. I., LE COMBER, P. G., and SPEAR, W. E. (1977). *Phil. Mag.* **36**, 541.
—— SPEAR, W. E., and LE COMBER, P. G. (1976). *J. non-cryst. Solids* **20**, 259.
JONES, D. P., THOMAS, N., and PHILLIPS, W. A. (1978). *Phil. Mag. B* **38** (in press).
JONES, H. (1934). *Proc. R. Soc. A* **144**, 225.
JONES, R. (1974). *Garmisch*, p. 969.
—— and SCHAICH, W. (1972). *J. Phys. C: Solid State Phys.* **5**, 43.
JONSCHER, A. K. (1975). *Nature (London)* **256**, 566.
JORTNER, J. (1959). *J. chem. Phys.* **30**, 838.
—— (1974). *Proc. 4th Int. Conf. on Vacuum Ultraviolet Radiation Physics* (ed. E.-E. Koch, R. Haensel, and C. Kunz), p. 263. Pergamon, Oxford; Vieweg, Braunschweig.
—— RICE, S. A., and HOCHSTRASSER, R. M. (1969). *Adv. Photochem.* **7**, 149.
—— KESTNER, N. R., RICE, S. A., and COHEN, M. H. (1965). *J. chem. Phys.* **43**, 2614.
JULLIEN, R. and JEROME, D. (1971). *J. Phys. Chem. Solids* **32**, 257.
JUNGK, G. (1971). *Phys. Status Solidi B* **44**, 239.
KAGAN, YU and ZHERNOV, A. P. (1966). *Sov. Phys.—JETP* **23**, 737.
KAMIMURA, H. and MOTT, N. F. (1976). *J. phys. soc. Japan* **40**, 1351.
KANE, E. O. (1963). *Phys. Rev.* **131**, 79.
KANEYOSHI, T. (1972). *J. Phys. C: Solid State Phys.* **5**, L247.
—— (1974). *Phys. Status Solidi B* **66**, K1.
—— (1976). *Phil. Mag.* **33**, 11.
KANZAKI, H. and SAKURAGI, S. (1973). *Photogr. Sci. Eng.* **17**, 69.
—— —— and SAKAMOTO, K. (1971). *Solid State Commun.* **9**, 999.
KÄNZIG, W. (1955). *Phys. Rev.* **99**, 1890.
KAPLAN, D., SOLOMON, I., and MOTT, N. F. (1978). *J. Phys. (Paris)* **39**, L-51.
—— LÉPINE, D., PETROFF, Y., and THIRRY, P. (1975). *Phys. Rev. Lett.* **35**, 1376.
KAPLOW, R., ROWE, T. A., and AVERBACH, B. L. (1968). *Phys. Rev.* **168**, 1068.
KASEN, M. B. (1970). *Phil. Mag.* **21**, 599.
KASTNER, M. (1972). *Phys. Rev. Lett.* **28**, 355.
—— (1977). *Edinburgh*, p. 504.
—— and FRITZSCHE, H. (1970). *Mater. Res. Bull.* **5**, 631.
—— —— (1978). *Phil. Mag.* **37**, 199.
—— ADLER, D., and FRITZSCHE, H. (1976). *Phys. Rev. Lett.* **37**, 1504.
KASUYA, T. and YANASE, A. (1968). *Rev. mod. Phys.* **40**, 684.
—— —— and TAKEDA, T. (1970). *Solid State Commun.* **8**, 1551.
KATZ, M. J. (1965). *Phys. Rev. A* **140**, 1323.
—— KOENIG, S. H., and LOPEZ, A. A. (1965). *Phys. Rev. Lett.* **15**, 828.
KAWAJI, S. and WAKABAYASHI, J. (1977). *Solid State Commun.* **22**, 87.
KEATING, P. N. (1966). *Phys. Rev.* **145**, 637.
KEEM, J. E., HONIG, J. M., and VAN ZANDT, L. L. (1978) *Phil. Mag. B* **37**, 537.
KEIL, T. H. (1966). *Phys. Rev.* **144**, 582.
KELDYSH, L. V. and KOPAEV, YU. V. (1965). *Sov. Phys.—Solid State* **6**, 2219.
KELLER, H. and STUKE, J. (1965). *Phys. Status Solidi* **8**, 831.
KELLER, J. (1971). *J. Phys. C: Solid State Phys.* **4**, 3143.
—— FRITZ, J., and GARRITZ, A. (1974). *J. Phys. (Paris)* **35**, C4-379.
KELLY, M. J. (1974). *Yorktown Heights*, p. 174.
—— and BULLETT, D. W. (1976a). *J. non-cryst. Solids* **21**, 155.
—— —— (1976b). *Solid State Commun.* **18**, 593.
KENNEDY, T. N. and MACKENZIE, J. D. (1967). *Phys. Chem. Glasses* **8**, 169.

KHANNA, S. N. and JAIN, A. (1975). *Phys. Rev. B* **11**, 3662.

KIEL, A. (1964). *Proc. 3rd Int. Conf. on Quantum Electronics, Paris* (ed. P. Grivet and N. Bloembergen), p. 765. Columbia University Press, New York.

KIKUCHI, M. (1970). *J. phys. Soc. Japan* **29**, 296.

—— (1974). *J. phys. Soc. Japan* **37**, 904.

KILLIAS, H. R. (1966). *Phys. Lett.* **20**, 5.

KIMURA, T. and FREEMAN, G. R. (1974). *Can. J. Phys.* **52**, 2220.

KING, C. N., PHILLIPS, W. A., and DE NEUFVILLE, J. P. (1974). *Phys. Rev. Lett.* **32**, 538.

KIRBY, P. B. and DAVIS, E. A. (1979). *Proc. 14th Int. Conf. on the Physics of Semiconductors* (in press).

KIRKPATRICK, S. (1971). *Phys. Rev. Lett.* **27**, 1722.

—— (1973*a*). *Solid State Commun.* **12**, 1279.

—— (1973*b*). *Rev. mod. Phys.* **45**, 574.

—— (1974). *Garmisch*, p. 183.

KISHIMOTO, N., MORIGAKI, K., SHIMIZU, A., and HIRAKI, A. (1976). *Solid State Commun.* **20**, 31.

KLAFTER, J. and JORTNER, J. (1977). *Chem. Phys. Lett.* **49**, 410.

KLINGER, M. I. (1976). *J. Phys. C: Solid State Phys.* **9**, 3955.

KNEPPO, I., LUBY, Š., and ČERVENÁK, J. (1973). *Thin solid Films* **17**, 43.

KNIGHTS, J. C. (1975). *Solid State Commun.* **16**, 515.

—— (1976*a*). *Phil. Mag.* **34**, 663.

—— (1976*b*). *Williamsburg*, p. 296.

—— (1977). *Edinburgh*, p. 435.

—— and DAVIS, E. A. (1974). *J. Phys. Chem. Solids* **35**, 543.

—— and MAHAN, J. E. (1977). *Solid State Commun.* **21**, 983.

—— BIEGELSEN, D. K., and SOLOMON, I. (1977). *Solid State Comm.* **22**, 133.

—— LUCOVSKY, G., and NEMANICH, R. J. (1978). *Phil. Mag. B* **37**, 467.

KNOTEK, M. L. (1974). *Yorktown Heights*, p. 297.

——·(1975*a*). *J. Vac. Sci. Technol.* **12**, 217.

—— (1975*b*). *Solid State Commun.* **17**, 1431.

—— (1977). *Phys. Rev. B* **16**, 2629.

KNOTEK, M. L. and DONOVAN, T. M. (1973). *Phys. Rev. Lett.* **30**, 652.

—— and POLLAK, M. (1974). *Phys. Rev. B* **9**, 664.

—— —— (1976). *Phil. Mag.* **35**, 1133.

—— —— and DONOVAN, T. M. (1974). *Garmisch*, p. 225.

—— —— —— and KURTZMAN, H. (1973). *Phys. Rev. Lett.* **30**, 853.

KNOX, R. S. (1963). *Theory of excitons—Solid State Phys. Suppl.* **5**.

KOBAYASHI, N., IKEHATA, S., KOBAYASHI, S., and SASAKI, W. (1977). *Solid State Commun.* **24**, 9.

KOC, S., ZÁVĚTOVA, M., and ZEMEK, J. (1972). *Thin solid Films* **10**, 165.

KOČKA, J. (1976). *Czech. J. of Phys.* **B26**, 807.

—— TŘÍSKA, A., and STOURAC, L. (1976). *Leningrad*, p. 249.

KOHN, W. (1967). *Phys. Rev. Lett.* **19**, 439.

KOLOMIETS, B. T. (1964). *Phys. Status Solidi* **7**, 359, 713.

—— and LEBEDEV, É. A. (1966). *Sov. Phys.—Solid State* **8**, 905.

—— —— (1967). *Sov. Phys.—Semicond.* **1**, 244.

—— and LYUBIN, V. M. (1973), *Phys. Status Solidi A* **17**, 11.

—— LEBEDEV, É. A., ROGACHEV, N. A., and SHPUNT, V. KH. (1972). *Sov. Phys.—Semicond.* **6**, 167.

—— MAMONTOVA, T. N., and BABAEV, A. A. (1970). *J. non-cryst. Solids* **4**, 289.

——— (1972). *J. non-cryst. Solids* **8–10**, 1004.

—— MAZETS, T. F., ÉFENDIEV, SH. M. (1970). *Sov. Phys.—Semicond.* **4**, 934.

—— MAMONTOVA, T. N., and NEGRESKUL, V. V. (1968). *Phys. Status Solidi* **27**, K15.

——— SMORGONSKAYA, E. A., and BABAEV, A. A. (1972). *Phys. Status Solidi A* **11**, 441.

—— RUKHLYADEV, YU. V., and SHILO, V. P. (1971). *J. non-cryst. Solids* **5**, 402.

——— —— and VASSILYEV, V. A. (1974). *Garmisch*, p. 939.

—— MAZETS, T. F., ÉFENDIEV, SH. M., and ANDRIESH, A. M. (1970). *J. non-cryst. Solids* **4**, 45.

KONINGSBERGER, D. C., VAN WOLPUT, J. H. M. C., and RIETER, P. C. U. (1971). *Chem. Phys. Lett.* **8**, 145.

KOO, J., WALKER, L. R., and GESCHWIND, S. (1975). *Phys. Rev. Lett.* **35**, 1669.

KOPELMAN, R., MONBERG, E. M., OCHS, F. W., and PRASAD, P. N. (1975). *J. chem. Phys.* **62**, 292.

KOSAREV, V. V. (1975). *Sov. Phys.—Semicond.* **8**, 897.

KOSEK, F. and TAUC, J. (1970). *Czech. J. Phys. B* **20**, 94.

KRAMER, B. (1970*a*). *Phys. Status Solidi* **41**, 649.

—— (1970*b*). *Phys. Status Solidi* **41**, 725.

—— (1971). *Phys. Status Solidi B* **47**, 501.

—— and TREUSCH, J. (1974). *Yorktown Heights*, p. 133.

—— MASCHKE, K., THOMAS, P., and TREUSCH, J. (1970). *Phys. Rev. Lett* **25**, 1020.

KREBS, H. (1968). *Fundamentals of inorganic crystal chemistry.* McGraw-Hill, New York.

—— (1969). *J. non-cryst. Solids* **1**, 455.

—— and FISCHER, P. (1971). *Discuss. Faraday Soc.* No. 50, p. 35.

—— and GRUBER, H. U. (1967). *Z. Naturforschg* **22a**, 96.

—— and SCHULTZE-GEBHARDT, F. (1955). *Z. anorg. allg. Chem.* **283**, 263.

—— and STEFFEN, R. (1964). *Z. anorg. allg. Chem.* **327**, 224.

—— HOLTZ, W., and WORMS, K. H. (1957). *Chem. Ber.* **90**, 1031.

KRIEGER, J. B. and MEEKS, T. (1973). *Phys. Rev. B* **8**, 2780.

—— and STRAUSS, S. (1968). *Phys. Rev.* **169**, 674.

KRISHNAN, K. S. and BHATIA, A. B. (1945). *Nature (London)* **156**, 503.

KRIVANEK, O. L., GASKELL, P. H., and HOWIE, A. (1976). *Nature (London)* **262**, 454.

KROLL, D. M. (1974). *Phys. Rev. B* **9**, 1669.

—— and COHEN, M. H. (1972). *J. non-cryst. Solids* **8–10**, 544.

KUBELIK, J. and TRISKA, A. (1973). *Czech. J. Phys. B* **23**, 115.

KUBO, R. (1952). *Phys. Rev.* **86**, 929.

—— (1956). *Can J. Phys.* **34**, 1274.

—— and TOYOZAWA, Y. (1955). *Prog. theor. Phys.* **13**, 160.

KUMEDA, M., JINNO, Y., SUSUKI, M., and SHIMIZU, T. (1976*a*). *Japan J. appl. Phys.* **15**, 201.

—— KOBAYASHI, N., MARUYAMA, E., and SHIMIZU, T. (1976*b*). *Phys. Status Solidi B* **73**, K19.

KURKIJÄRVI, J. (1973). *Phys. Rev. B* **8**, 922.

—— (1974). *Phys. Rev. B* **9**, 770.

KUROSAWA, T. and SUGIMOTO, H. (1975). *Prog. theor. Phys. Suppl.* No. 57, 217.

LA COURSE, W. A., TWADDELL, V. A., and MACKENZIE, J. D. (1970). *J. non-cryst. Solids* **3**, 234.

LAKATOS, A. I. and ABKOWITZ, M. (1971). *Phys. Rev. B* **3**, 1791.

LANDAU, L. D. (1933). *Phys. Z. Sowjetunion* **3**, 664.

LANDAUER, R. (1952). *J. appl. Phys.* **23**, 779.

—— (1970). *Phil. Mag.* **21**, 863.

—— and HELLAND, J. C. (1954). *J. chem. Phys.* **22**, 1655.

LANNIN, J. S. (1976). *Williamsburg*, p. 123.

—— (1977). *Phys. Rev. B* **15**, 3863.

—— ENO, H. F., and LUO, H. L. (1978). *Solid State Commun.* **25**, 81.

LANYON, H. P. D. (1963). *Phys. Rev.* **130**, 134.

LASJAUNIAS, J. C., RAVEX, A., VANDORPE, M., and HUNKLINGER, S. (1975). *Solid State Commun.* **17**, 1045.

LAST, B. J. and THOULESS, D. J. (1971). *Phys. Rev. Lett.* **27**, 1719.

LAUDE, L. D., WILLIS, R., and FITTON, B. (1974). *Garmisch*, p. 277.

LAX, M. (1951). *Rev. mod. Phys.* **23**, 287.

—— (1952). *Phys. Rev.* **85**, 621.

—— and PHILLIPS, J. C. (1958). *Phys. Rev.* **110**, 41.

LAYNE, C. B., LOWDERMILK, W. H., and WEBER, M. J. (1977). *Phys. Rev. B* **16**, 10.

LEADBETTER, A. J., APLING, A. J., and DANIEL, M. F. (1976). *J. non-cryst. Solids* **21**, 47.

—— SMITH, P. M., and SEYFERT, P. (1976). *Phil. Mag.* **33**, 441.

—— APLING, A. J., DANIEL, M. F., WRIGHT, A. C., and SINCLAIR, R. N. (1977). *Cambridge*, p. 23.

LE COMBER, P. G. and SPEAR, W. E. (1970). *Phys. Rev. Lett.* **25**, 509.

—— —— (1976). *Williamsburg*, p. 284.

—— JONES, D. I., and SPEAR, W. E. (1977). *Phil. Mag.* **35**, 1173.

—— MADAN, A., and SPEAR, W. E. (1972). *J. non-cryst. Solids* **11**, 219.

—— —— (1973). *Aberdeen*, p. 373.

—— LOVELAND, R. J., SPEAR, W. E., and VAUGHAN, R. A. (1974). *Garmisch*, p. 245.

LEE, P. A. and BENI, G. (1977). *Phys. Rev. B* **15**, 2862.

—— and PENDRY, J. B. (1975). *Phys. Rev. B* **11**, 2795.

LEE, S. H. (1972). *Appl. Phys. Lett.* **21**, 544.

LEKNER, J. (1967). *Phys. Rev.* **158**, 130.

—— (1968). *Phys. Lett. A* **27**, 341.

LEKNER, J. and BISHOP, A. R. (1972). *Phil. Mag.* **27**, 297.

LEMKE, B. P. and HANEMAN, D. (1975). *Phys. Rev. Lett.* **35**, 1379.

LEPINE, D. J. (1972). *Phys. Rev.* **6**, 436.

LEPINE, D., GRAZHULIS, V. A., and KAPLAN, D. (1976). *Proc. 13th Int. Conf. on the Physics of Semiconductors, Rome* (ed. F. G. Fumi), p. 1081. Tipografia Marves, Rome.

LEVY, A. W., GREEN, M., and GEE, W. (1974). *J. Phys. C: Solid State Phys.* **7**, 352.

LEWIS, A. (1972). *Phys. Rev. Lett.* **29**, 1555.

LEWIS, A. J. (1976). *Leningrad, Electronic Phenomena in Non-Crystalline Semiconductors*, p. 299. *Phys. Rev. B* **14**, 658.

—— CONNELL, G. A. N., PAUL, W., PAWLIK, J. R., and TEMKIN, R. J. (1974). *Yorktown Heights*, p. 27.

LEY, L., KOWALCZYK, S., POLLAK, R., and SHIRLEY, D. A. (1972). *Phys. Rev. Lett.* **29**, 1088.

—— POLLAK, R. A., KOWALCZYK, S., MCFEELY, R., and SHIRLEY, D. A. (1973). *Phys. Rev. B* **8**, 641.

LIANG, K. S., BIENENSTOCK, A., and BATES, C. W. (1974). *Phys. Rev. B* **10**, 1528.

LIANG, W. Y. and BEAL, A. R. (1976). *J. Phys. C: Solid State Phys.* **9**, 2823.

LICCIARDELLO, D. C. and THOULESS, D. J. (1975). *J. Phys. C: Solid State Phys.* **8**, 4157.

LIEB, E. H. and MATTIS, D. C. (1966). *Mathematical physics in one dimension.* Academic Press, New York.

LIEN, S. Y. and SIVERSTEN, J. M. (1969). *Phil. Mag.* **20**, 759.

LIFSHITZ, I. M. (1964). *Adv. Phys.* **13**, 483.

LIGHTSEY, P. A. (1973). *Phys. Rev. B* **8**, 3586.

—— LILIENFELD, D. A., and HOLCOMB, D. F. (1976). *Phys. Rev. B* **14**, 4730.

LIZELL, B. (1952). *J. chem. Phys.* **20**, 672.

LLOYD, P. (1969). *J. Phys. C: Solid State Phys.* **2**, 1717.

LONG, M., GALISON, P., ALBEN, R., and CONNELL, G. A. N. (1976). *Phys. Rev. B* **13**, 1821.

LOVELAND, R. J., SPEAR, W. E., and AL-SHARBATY, A. (1973/74). *J. non-cryst. Solids* **13**, 55.

—— —— —— (1972). *Phys. Status Solidi B* **49**, 633.

LUCOVSKY, G. (1974). *Garmisch*, p. 1099.

—— and GALEENER, F. L. (1976). *Leningrad, Structure and Properties of Non-Crystalline Semiconductors*, p. 207.

—— and KNIGHTS, J. C. (1974). *Phys. Rev. B* **10**, 4324.

—— GALEENER, F. L., GEILS, R. H., and KEEZER, R. C. (1977). *Cambridge*, p. 127.

—— —— KEEZER, R. C., GEILS, R. H., and SIX, H. A. (1974). *Phys. Rev. B* **10**, 5134.

—— MOORADIAN, A., TAYLOR, W., WRIGHT, G. B., and KEEZER, R. C. (1967). *Solid State Commun.* **5**, 113.

LUEDER, H. and SPENKE, E. (1935). *Z. tech. Phys.* **16**, 11.

LUKES, T. (1972). *J. non-cryst. Solids* **8–10**, 470.

—— (1975). *Phil. Mag.* **32**, 1033.

—— and SOMARATNA, K. T. S. (1970). *J. non-cryst. Solids* **4**, 452.

—— NIX, B., and SUPRAPTO, B. (1972). *Phil. Mag.* **26**, 1239.

LYO, S. K. and HOLSTEIN, T. (1973). *Phys. Rev. B* **8**, 682.

MA, W. and ANDERSON, R. M. (1974). *Appl. Phys. Lett.* **25**, 101.

McGILL, T. C. and KLIMA, J. (1972). *Phys. Rev. B* **5**, 1517.

MACKINTOSH, A. R. (1963). *J. chem. Phys.* **38**, 1991.

McWHAN, D. B., RICE, T. M., and SCHMIDT, P. H. (1969). *Phys. Rev.* **177**, 1063.

MADAN, A., LE COMBER, P. G., and SPEAR, W. E. (1976). *J. non-cryst. Solids* **20**, 239.

MADER, S. (1965). *J. Vac. Sci. Technol.* **2**, 35.

—— WIDMER, H., D'HEURLE, F. M., and NOWICK, A. S. (1963). *Appl. Phys. Lett.* **3**, 201.

MAHAN, G. D. (1966). *Phys. Rev.* **145**, 602.

MAHAN, J. E., and BUBE, R. H. (1977). *J. non-cryst. Solids* **24**, 29.

MAHDJURI, F. (1975). *J. Phys. C: Solid State Phys.* **8**, 2248.

MAHR, H. (1963). *Phys. Rev.* **132**, 1880.

MAIN, C. (1974). Ph.D Thesis, University of Edinburgh.

—— and OWEN, A. E. (1973). *Aberdeen*, p. 527.

MAJERNÍKOVÁ, E. (1974a). *Phys. Status Solidi B* **63**, 251.

—— (1974b). *Phys. Lett A* **48**, 266.

—— (1975). *Phys. Status Solidi B* **70**, K47.

MAKINSON, R. E. B. and ROBERTS, A. P. (1962). *Proc. phys. Soc.* **79**, 222.

MALE, J. C. (1967). *Br. J. appl. Phys.* **18**, 1543.

—— (1970). *Electron. Lett.* **6**, 91.

MALHOTRA, A. K. and NEUDECK, G. W. (1974). *Appl. Phys. Lett.* **24**, 557.

—— —— (1975). *J. appl. Phys.* **46**, 2690.

MARCH, N. H. (1968). *Liquid metals.* Pergamon Press, Oxford.

—— (1977). *Can. J. Chem.* **35**, 2165.

MARELLO, V., LEE, T. F., SILVER, R. N., McGILL, T. C., and MAYER, J. W. (1973). *Phys. Rev. Lett.* **31**, 593.

MARK, P. and HARTMAN, T. E. (1968). *J. appl. Phys.* **39**, 2163.

MARSHALL, J., FISHER, F. D., and OWEN, A. E. (1974). *Garmisch*, p. 1305.

MARSHALL, J. M. (1977). *Phil. Mag.* **36**, 959.

—— and MILLER, G. R. (1973). *Phil. Mag.* **27**, 1151.

—— and OWEN, A. E. (1971). *Phil. Mag.* **24**, 1281.

—— —— (1975). *Phil. Mag.* **31**, 1341.

—— —— (1976). *Phil. Mag.* **33**, 457.

MARTIN, B. G. and LERNER, L. S. (1972). *Phys. Rev. B* **6**, 3032.

MARTIN, R. M., LUCOVSKY, G., and HELLIWEL, K. (1976). *Phys. Rev.* **13**, 1383.

MASCHE, K. and THOMAS, P. (1970a). *Phys. Status Solidi* **39**, 453.

—— —— (1970b). *Phys. Status Solidi* **41**, 743.

—— OVERHOF, H., and THOMAS, P. (1974). *Phys. Status Solidi B* **62**, 113.

MASSEN, C. H., WEIJTS, A. G. L. M., and POULIS, J. A. (1964). *Trans. Faraday Soc.* **60**, 317.

MASSENET, O., DAVER, H., and GENESTE, J. (1974). *J Phys. (Paris)* **35**, C4-279.

MATSUBARA, T. and KANEYOSHI, T. (1968). *Prog. theor. Phys.* **40**, 1257.

—— and TOYOZAWA, Y. (1961). *Progr. theor. Phys.* **26**, 739.

MATSUDA, H. and ISHII, K. (1970). *Prog. theor. Phys. Suppl.* **45**, 56.

MATYÁŠ, M. (1976). *Leningrad, Structure and Properties of Non-Crystalline Semiconductors*, p. 13.

MEEK, P. E. (1976a). *Phil. Mag.* **34**, 767.

—— (1976b). *Phil. Mag.* **33**, 897.

—— (1977a). *Proc. 4th Int. Conf. on the Physics of Non-Crystalline Solids, Clausthal-Zellerfeld* (ed. G. H. Frischat), p. 586. Trans. Tech. Publications, Aedermannsdorf, Switzerland.

—— (1977b). *Cambridge*, p. 235.

—— (1977c). *J. Phys. C: Solid State Phys.* **10**, L59.

MEEKS, T. and KREIGER, J. B. (1969). *Phys. Rev.* **185**, 1068.

MEIMARIS, D., KATRIS, J., MARTAKOS, D., and ROILOS, M. (1977). *Phil. Mag.* **35**, 1633.

MELL, H. (1974a). *Garmisch*, p. 203.

—— (1974b). *Yorktown Heights*, p. 357.

—— and STUKE, J. (1970). *J. non-cryst. Solids* **4**, 304.

MELLOR, J. W. (1929). *A comprehensive treatise on inorganic and theoretical chemistry*, IX.

METHFESSEL, S. and MATTIS, D. C. (1968). *Handb. Phys.* **18**, 387.

MIKOSHIBA, N. (1962). *Phys. Rev.* **127**, 1962.

—— (1968). *Rev. mod. Phys.* **40**, 833.

—— and GONDA, S. (1962). *Phys. Rev.* **127**, 1954.

MiLLER, A., and ABRAHAMS, S. (1960). *Phys. Rev.* **120**, 745.

MILLER, L. S., HOWE, S., and SPEAR, W. E. (1968). *Phys. Rev.* **166**, 871.

MILNES, A. G. (1973). *Deep impurities in semiconductors.* Wiley, New York.

MILWARD, R. C. and NEURINGER, L. J. (1965). *Phys. Rev. Lett.* **15**, 664.

MITCHELL, D. L., BISHOP, S. G., and TAYLOR, P. C. (1972). *J. non-cryst. Solids* **8–10**, 231.

MOLE, P. J. (1978), Ph.D. Thesis, University of Cambridge.

MOLLOT, F., CERNOGORA, J., and BENOIT À LA GUILLAUME, C. (1974). *Phys. Status Solidi A* **21**, 281.

MOOIJ, J. H. (1973). *Phys. Status Solidi A* **17**, 521.

MOOKERJEE, A. (1973). *J. Phys. C: Solid State Phys.* **6**, 1340.

MOORE, A. R. (1977). *Appl. Phys. Lett.* **31**, 762.

MOORE, W. J. (1960). *Am. Sci.* **48**, 109.

MOORJANI, K. and FELDMAN, C. (1977). In *Boron and refractory borides* (ed. V. I. Matkovich), p. 581. Springer-Verlag, Vienna.

MOOS, H. W. (1970). *J. Lumin.* **1**, **2**, 106.

MORGAN, M. and WALLEY, P. A. (1971). *Phil. Mag.* **23**, 661.

MORIGAKI, K. and ONDA, M. (1972). *J. phys. Soc. Japan* **33**, 1031.

—— —— (1974). *J. phys. Soc. Japan* **36**, 1049.

MORT, J. (1973*a*). *Aberdeen*, p.475.

—— (1973*b*). *Aberdeen*, p. 493.

—— (1974). *Garmisch*, p. 1361.

—— and LAKATOS, A. I. (1970). *J. non-cryst. Solids* **4**, 117.

MOSS, S. C. (1974). *Garmisch*, p. 17.

—— and GRACZYK, (1969). *Phys. Rev. Lett.* **23**. 1167.

—— —— (1970). *Proc. 10th Int. Conf. on the Physics of Semiconductors, Cambridge, Mass.* (ed. S. P. Keller, J. C. Hensel, and F. Stern), p. 658. United States Atomic Energy Commission, Washington, D.C.

—— FLYNN, P., and BAUER, L.-O. (1973). *Phil. Mag.* **27**, 441.

—— ALBEN, R., ADLER, D., and DE NEUFVILLE, J. P. (1973/74). *J. non-cryst. Solids* **13**, 185.

MOSS, T. S. (1952). *Photoconductivity in the elements.* Butterworths, London.

MOTT, N. F. (1938), *Proc. R. Soc. A* **167**, 384.

—— (1949). *Proc. phys. Soc. A* **62**, 416.

—— (1956). *Can. J. Phys.* **34**, 1356.

—— (1966). *Phil. Mag.* **13**, 989.

—— (1967). *Adv. Phys.* **16**, 49.

—— (1968). *J. non-cryst. Solids* **1**, 1.

—— (1969*a*). *Phil. Mag.* **19**, 835.

—— (1969*b*). *Contemp. Phys.* **10**, 125.

—— (1970). *Phil. Mag.* **22**, 7.

—— (1971). *Phil. Mag.* **24**, 911.

—— (1972*a*). *J. non-cryst. Solids* **8–10**, 1.

—— (1972*b*). *Phil. Mag.* **26**, 505.

—— (1972*c*). *Phil. Mag.* **26**, 1015.

—— (1972*d*). *Phil. Mag.* **26**, 1249.

—— (1972*e*). *Adv. Phys.* **21**, 785.

—— (1973*a*). *Aberdeen*, p. 1.

—— (1973*b*). *Electron. Power* **19**, 321.

—— (1974*a*). *Phil. Mag.* **29**, 59.

—— (1974*b*). *Phil. Mag.* **29**, 613.

—— (1974*c*). *Metal–insulator transitions.* Taylor & Francis, London.

—— (1975*a*). *Phil. Mag.* **32**, 159.

—— (1975b). *J. phys. Chem.* **79**, 2915.

—— (1975c). *Phil. Mag.* **31**, 217.

—— (1975d). *J. Phys. C: Solid State Phys.* **8**, L239.

—— (1976a). *J. Phys. (Paris) Colloq.* No. 4, C-4-301.

—— (1976b). *Phil. Mag.* **34**, 643.

—— (1976c). *Commun. Phys.* **1**, 203.

—— (1976d). *Phil. Mag.* **34**, 1101.

—— (1976e). *Proc. 13th Int. Conf. on the Physics of Semiconductors, Rome* (ed. F. G. Fumi), p. 3. Tipografia Marves, Rome.

—— (1977a). *Phil. Mag.* **35**, 111.

—— (1977b). *Adv. Phys.* **26**, 363.

—— (1977c). *Proc. 4th Int. Conf. on the Physics of Non-Crystalline Solids* (ed. G. H. Frischat), p. 3. Trans. Tech. Publications, Aedermannsdorf, Switzerland.

—— (1977d). *Cambridge*, p. 101.

—— (1977e). *Contemp. Phys.* **18**, 225.

—— (1977f). *Phil. Mag.* **36**, 413.

—— (1978a). *Phil. Mag. B* **37**, 377.

—— (1978b). *Physics of SiO_2 and its interfaces* (ed. S. T. Pantelides), p. 1, Pergamon, New York.

—— and DAVIS, E. A. (1968). *Phil. Mag.* **17**, 1269.

—— and GURNEY, R. W. (1940). *Electronic processes in ionic crystals.* Clarendon Press, Oxford.

—— and JONES, H. (1936). *The theory of the properties of metals and alloys.* Clarendon Press, Oxford.

—— and MASSEY, H. S. W. (1965). *The theory of atomic collisions* (3rd edn.), p. 86. Clarendon Press, Oxford.

—— and STONEHAM, A. M. (1977). *J. Phys. C: Solid State Phys.* **10**, 3391.

—— and STREET, R. A. (1977). *Phil. Mag.* **36**, 33.

—— and TWOSE, W. D. (1961). *Adv. Phys.* **10**, 107.

—— and ZINAMON, Z. (1970). *Rep. Prog. Phys.* **33**, 881.

—— DAVIS, E. A., and STREET, R. A. (1975). *Phil. Mag.* **32**, 961.

—— PEPPER, M., POLLITT, S., WALLIS, R. H., and ADKINS, C. J. (1975). *Proc. R. Soc. A* **345**, 169.

MOUSTAKAS, T. D. (1974). Thesis, Columbia University.

—— and PAUL, W. (1977). *Phys. Rev. B* **16**, 1564.

—— and WEISER, K. (1975). *Phys. Rev. B* **12**, 2448.

MOVAGHAR, B. and SCHWEITZER, L. (1977). *Phys. Status Solidi B* **89**, 491.

MOZZI, R. L. and WARREN, B. E. (1969). *J. appl. Phys.* **2**, 164.

MYTILINEOU, E. and DAVIS, E. A. (1977). *Edinburgh*, p. 632.

—— and ROILOS, M. (1978). *Phil. Mag. B* **37**, 387.

NACHTRIEB, N. H. (1975). *Adv. Chem. Phys.* **31**, 465.

NAGEL, S. R. and TAUC, J. (1975). *Phys. Rev. Lett.* **35**, 380.

—— —— (1977). *Proc. 3rd Int. Conf. on Liquid Metals, Bristol* (ed. R. Evans, and D. A. Greenwood), p. 283. The Institute of Physics, London.

NAGELS, P., CALLAERTS, R., and DENAYER, M. (1974). *Garmisch*, p. 867.

—— —— HASHMI, F. H., and DENAYER, M. (1970). *Phys. Status Solidi* **41**, K39.

NAKAMURA, A. and MORIGAKI, K. (1974). *Solid State Commun.* **14**, 41.

NASHASHIBI, T. S., AUSTIN, I. G., and SEARLE, T. M. (1977a). *Phil. Mag.* **35**, 831.

—— —— —— (1977b). *Edinburgh*, p. 392.

NEMANICH, R. J., LUCOVSKY, G., POLLARD, W., and JOANNOPOULOS, J. D. (1978). *Solid State Commun.* **26**, 137.

—— SOLIN, S. A., and LUCOVSKY, G. (1977). *Solid State Commun.* **21**, 273.

NÉROU, J. P., FILION, A., and GIRARD, P.-E. (1976). *J. Phys. C: Solid State Phys.* **9**, 479.

NEUDECK, G. W. and MALHOTRA, A. K. (1975). *J. appl. Phys.* **46**, 2262.

NEUGEBAUER, C. A. (1970). *Thin solid Films* **6**, 443.

—— and WEBB, M. B. (1962). *J. appl. Phys.* **33**, 74.

NICHOLAS, R. J., STRADLING, R. A., and TIDEY, R. J. (1977). *Solid State Commun.* **23**, 341.

NIELSEN, P. (1972). *Phys. Rev. B* **6**, 3739.

NOOLANDI, J. (1977). *Edinburgh*, p. 224.

NORDHEIM, L. (1931). *Annln Phys.* **9**, 641.

NORTH, D. M., ENDERBY, J. E., and EGELSTAFF, P. A. (1968). *J. Phys. C: Solid State Phys.* **1**, 1075.

NORTON, P. (1976a). *J. appl. Phys.* **47**, 308.

—— (1976b). *Phys. Rev. Lett.* **37**, 164.

NOZIK, A. J. (1972). *Phys. Rev. B* **6**, 453.

OELHAFEN, P. (1975). *Surf. Sci.* **47**, 422.

—— (1976). Diss. No. 5767, Zurich ETH.

OLECHNA, D. J. and KNOX, R. S. (1965). *Phys. Rev. A* **140**, 986.

OLLEY, J. A. (1973). *Solid State Commun.* **13**, 1437.

ONOMICHI, M., ARAI, T., and KUDO, K. (1971). *J. non-cryst. Solids* **6**, 362.

ONSAGER, L. (1938). *Phys. Rev.* **54**, 554.

ORLOWSKI, B. A. and SPICER, W. E. (1972). *Mater. Res. Bull.* **7**, 793.

ORTENBURGER, I. B. and HENDERSON, D. (1972). *Proc. 11th Int. Conf. on the Physics of Semiconductors, Warsaw*, p. 464. PWN-Polish Scientific Publishers, Warsaw.

—— —— (1973). *Phys. Rev. Lett.* **30**, 1047.

—— —— (1974). *Yorktown Heights*, p. 151.

—— RUDGE, W. E., and HERMAN, F. (1972). *J. non-cryst. Solids* **8–10**, 653.

OSMUN, J. W. and FRITZSCHE, H. (1974). *Yorktown Heights*, p. 333.

OVERHOF, H. (1975). *Phys. Status Solidi B* **67**, 709.

—— and THOMAS, P. (1976). *Leningrad, Structure and Properties of Non-Crystalline Semiconductors*, p. 107.

—— UCHTMANN, H., and HENSEL, F. (1976). *J. Phys. F: Met. Phys.* **6**, 523.

OVSHINSKY, S. R. (1968). *Phys. Rev. Lett.* **21**, 1450.

—— (1977). *Edinburgh*, p. 519.

—— and KLOSE, P. H. (1972). *J. non-cryst. Solids* **8–10**, 892.

OWEN, A. E. (1967). *Glass Ind.* **48**, 637, 695.

—— (1970). *Contemp. Phys.* **11**, 227. 257.

—— (1973). *Aberdeen*, p. 161.

—— and ROBERTSON, J. M. (1970). *J. non-cryst. Solids* **2**, 40.

PAESLER, M. A., AGARWAL, S. C., HUDGENS, S. J., and FRITZSCHE, H. (1974). *Yorktown Heights*, p. 37.

PAI, D. M. (1975). *J. appl. Phys.* **46**, 5122.

—— and ENCK, R. C. (1975). *Phys. Rev. B* **11**, 5163.

—— and ING, S. W. (1968). *Phys. Rev.* **173**, 729.

—— and SCHARFE, M. E. (1972). *J. non-cryst. Solids* **8–10**, 752.

PANDYA, D. K., RASTOGI, A. C., and CHOPRA, K. L. (1975). *J. appl. Phys.* **46**, 2966.

PANKOVE, J. I. and CARLSON, D. E. (1976). *Appl. Phys. Lett.* **29**, 620.

—— —— (1977). *Edinburgh*, p. 402.

PANOVA, G. KH., ZHERNOV, A. P., and KUTAĬTSEV, V. I. (1968). *Sov. Phys.—JETP* **26**, 283.

PANTELIDES, S. T. and HARRISON, W. A. (1976). *Phys. Rev. B* **13**, 2667.

PAPATRIANTAFILLOU, C., and ECONOMOU, E. N. (1976). *Phys. Rev. B* **13**, 920.

—— —— and EGGARTER, T. P. (1976). *Phys. Rev. B* **13**, 910.

PARKE, S. and WEBB, R. S. (1973). *J. Phys. Chem. Solids* **34**, 85.

PAUL, W. and CONNELL, G. A. N. (1976). In *Physics of structurally disordered solids* (ed. S. S. Mitra), p. 45. Plenum Press, New York.

—— —— and TEMKIN, R. J. (1973). *Adv. Phys.* **22**, 531.

PAULING, L. (1960). *The nature of the chemical bond* (3rd edn.). Cornell University Press, Ithaca, N.Y.

PEARSON, A. D., DEWALD, J. F., NORTHOVER, W. R., and PECK, W. F. (1962). *Advances in glass technology*, Part I, p. 357. Plenum Press, New York.

PENDRY, J. B. and GURMAN, S. J. (1977). *Cambridge*, p. 61.

PENN, D. R. (1962). *Phys. Rev.* **128**, 2093.

PENNEY, T., HOLTZBERG, F., TAO, L. J., and VON MOLNAR, S. (1973). *AIP Conf. Proc. on Magnetism and Magnetic Materials* (ed. C. D. Graham, and J. J. Rhyne), No. 18, p. 908.

PEPPER, M. (1976). *Commun. Phys.* **1**, 147.

——(1977*a*). *Proc. R. Soc. A* **353**, 225; (1977*b*). *J. Phys. C: Solid State Phys.* **10**, L173.

PEPPER, M., POLLITT, S., and ADKINS, C. J. (1974*a*). *Phys. Lett. A* **48**, 113.

—— —— —— (1974*b*). *J. Phys. C: Solid State Phys.* 7, L273.

—— —— —— and OAKLEY, R. E. (1974). *Phys. Lett. A* **47**, 71.

PERRON, J. C. (1967). *Adv. Phys.* **16**, 657.

—— (1971). *Conductivity in low-mobility materials* (ed. N. Klein, D. S. Tannhauser, and M. Pollak), p. 243. Taylor & Francis, London.

PETERSEN, K. E. and ADLER, D. (1976). *J. appl. Phys.* **47**, 256.

PETERSON, H. and KUNZ, C. (1975). DESY Rep. SR-75/04.

PETTIFER, R. F., MCMILLAN, P. W., and GURMAN, S. J. (1977). *Cambridge*, p. 63.

PFEIFER, H. P., FREYLAND, W., and HENSEL, F. (1976). *Ber. Bunsenges. phys. Chem.* **80**, 716.

PFISTER, G. (1974). *Phys. Rev. Lett.* **33**, 1474.

—— (1976). *Phys. Rev. Lett.* **36**, 271.

—— (1977*a*). *Phil. Mag.* **36**, 1147.

—— (1977*b*). *Edinburgh*, p. 765.

—— and SCHER, H. (1977). *Phys. Rev. B* **15**, 2062.

—— —— (1979). *Adv. Phys.* (in press).

—— MELNYK, A. R., and SCHARFE, M. E. (1977). *Solid State Commun.* **21**, 907.

PHARISEAU, P. and ZIMAN, J. M. (1963). *Phil. Mag.* **8**, 1487.

PHILLIP, H. R. (1966). *Solid State Commun.* **4**, 73.

—— (1971). *J. Phys. Chem. Solids* **32**, 1935.

—— (1972). *J. non-cryst. Solids* **8–10**, 627.

—— and EHRENREICH, H. (1963). *Phys. Rev.* **129**, 1550.

PHILLIPS, J. C. (1971). *Phys. Status Solidi B* **44**, K1.

—— (1976). *Proc. 13th Int. Conf. on the Physics of Semiconductors, Rome* (ed. F. G. Fumi), p. 12. Tipografia Marves, Rome.

PHILLIPS, W. A. (1972). *J. Low Temp. Phys.* **7**, 351.

—— (1976). *Phil. Mag.* **34**, 983.

—— and THOMAS, N. (1977). *Cambridge*, p. 143.

PICHÉ, L., MAYNARD, R., HUNKLINGER, S., and JÄCKLE, J. (1974). *Phys. Rev. Lett.* **32**, 1426.

PIERCE, D. T. and SPICER, W. E. (1971). *Phys. Rev. Lett.* **27**, 1217.

—— —— (1972). *Phys. Rev. B* **5**, 3017.

PIKE, G. E. (1972). *Phys. Rev. B* **6**, 1572.

—— and SEAGER, C. H. (1974). *Phys. Rev. B* **10**, 1421.

PILLER, H., SERAPHIN, B. O., MARKEL, K., and FISCHER, J. E. (1969). *Phys. Rev. Lett.* **23**, 775.

PLATAKIS, N. S., SADAGOPAN, V., and GATOS, H. C. (1969). *J. electrochem. Soc.* **116**, 1436.

POHL, R. O. (1976). In *Phonon scattering in solids* (ed. L. J. Challis, V. W. Rampton, and A. F. G. Wyatt), p. 107. Plenum Press, New York.

—— LOVE, W. F., and STEPHENS, R. B. (1974). *Garmisch*, p. 1121.

POHL, R. W. (1937). *Proc. Phys. Soc.* 49 (extra part), p. 3.

POLK, D. E. (1971). *J. non-cryst. Solids* **5**, 365.

—— and BOUDREAUX, D. S. (1973). *Phys. Rev. Lett.* **31**, 92.

POLLAK, M. (1964). *Phys. Rev. A* **133**, 564.

—— (1971*a*). *Discuss. Faraday Soc.* No. 50, p. 13.

—— (1971*b*). *Phil. Mag.* **23**, 519.

—— (1972). *J. non-cryst. Solids* **11**, 1.

—— (1974). *Phys. Status Solidi B* **66**, 483.

—— (1976). *Leningrad, Electronic Phenomena in Non-Crystalline Semiconductors*, p. 79.

—— (1977). *Phil. Mag.* **36**, 1157.

POLLAK, M. and HAUSER, J. J. (1973). *Phys. Rev. Lett.* **31**, 1304.

—— and GEBALLE, T. H. (1961). *Phys. Rev.* **122**, 1742.

—— and PIKE, G. E. (1972). *Phys. Rev. Lett.* **28**, 1449.

—— and RIESS, I. (1976). *J. Phys. C: Solid State Phys.* **9**, 2339.

—— KNOTEK, M. L., KURTZMAN, H., and GLICK, H. (1973). *Phys. Rev. Lett.* **30**, 856.

POLLITT, S. (1976). *Commun. Phys.* **1**, 207.

POOLE, H. H. (1916). *Phil. Mag.* **32**, 112.

—— (1917). *Phil. Mag.* **34**, 195.

POOLEY, D. (1966). *Proc. phys. Soc.* **87**, 257.

POPIELAWSKI, J. (1972). *J. chem. Phys.* **57**, 929.

—— and GRYKO, J. (1977). *J. chem. Phys.* **66**, 2257.

PROSSER, V. (1961). *Proc. Int. Conf. on Semiconductor Physics, Prague*, p. 993. Czechoslovak Academy of Sciences, Warsaw.

PRYOR, R. W. and HENISCH, H. K. (1971). *Appl. Phys. Lett.* **18**, 324.

QUINN, J. J. and WRIGHT, J. G. (1977). *Proc. 3rd Int. Conf. on Liquid Metals, Bristol* (ed. R. Evans, and D. A. Greenwood), p. 430. The Institute of Physics, London.

QUIRT, J. D. and MARKO, J. R. (1973). *Phys. Rev. B* **7**, 3842.

RAISIN, C., LEVEQUE, G., ROBIN, J., and ROBIN-KANDARE, S. (1974*a*). *Proc. 4th Int. Conf. on Vacuum Ultraviolet Radiation Physics* (ed. E.-E. Koch, R. Haensel, and C. Kunz), p. 502. Pergamon, Oxford; Vieweg, Braunschweig.

—— —— —— —— (1974*b*). *Solid State Commun.* **14**, 723.

RAWSON, H. (1967). *Inorganic glass-forming systems*. Academic Press, New York.

RAZ, B. and JORTNER, J. (1970). *Proc. R. Soc. A* **317**, 113.

—— —— (1973). In *Electrons in fluids* (ed. J. Jortner and N. R. Kestner), p. 413. Springer-Verlag, Vienna.

REDFIELD, D. (1963). *Phys. Rev.* **130**, 916.

—— WITTKE, J. P., and PANKOVE, J. I. (1970). *Phys. Rev. B* **2**, 1830.

REGEL, A. R., ANDREEV, A. A., KOTOV, B. A., MAMADALIEV, M., OKUNEVA, N. M., SMIRNOV, I. A., and SHADRICHEV, E. V. (1970). *J. non-cryst. Solids* **4**, 151.

REHM, W., ENGEMANN, D., FISCHER, R., and STUKE, J. (1976). *Proc. 13th Int. Conf. on the Physics of Semiconductors* (ed. F. G. Fumi). Tipografia Marves, Rome.

REIK, H. G. (1970). *Solid State Commun.* **8**, 1737.

REVESZ, A. G. (1973). *J. non-cryst. Solids* **11**, 309.

RENNINGER, A. L. and AVERBACH, B. L. (1973). *Phys. Rev.* **8**, 1507.

—— RECHTIN, M. D., and AVERBACH, B. L. (1974). *J. non-cryst. Solids* **16**, 1.

RIBBING, C. G., PIERCE, D. T., and SPICER, W. E. (1971). *Phys. Rev. B* **4**, 4417.

RICHTER, H. (1972). *J. non-cryst. Solids* **8–10**, 388.

—— and BREITLING, G. (1958). *Z. Naturforsch. A* **13**, 988.

—— and HERRE, F. (1958). *Z. Naturforsch. A* **13**, 874.

—— and GOMMEL, G. (1957). *Z. Naturforsch. A* **12**, 996.

RIDLEY, B. K. (1963). *Proc. phys. Soc.* **82**, 954.

ROBERTS, G. G. (1973). *Aberdeen*, p. 409.

—— TUTIHASI, S., and KEEZER, R. C. (1968). *Phys. Rev.* **166**, 637.

—— KEATING, B. S., and SHELLEY, A. V. (1974). *J. Phys. C: Solid State Phys.* **7**, 1595.

ROBERTSON, J. (1975). *J. Phys. C: Solid State Phys.* **8**, 3131.

—— (1976). *Phil. Mag.* **34**, 13.

ROBERTSON, J. M. and OWEN, A. E. (1972). *J. non-cryst. Solids* **8–10**, 439.

ROBERTSON, N. and FRIEDMAN, L. (1976). *Phil. Mag.* **33**, 753.

—— —— (1977). *Phil. Mag.* **36**, 1013.

ROBINSON, M. G. and FREEMAN, G. R. (1974). *Can. J. Chem.* **52**, 440.

ROCKSTAD, H. K. (1970.) *J. non-cryst. Solids* **2**, 192.

—— (1971). *Solid State Commun.* **9**, 2233.

—— and DE NEUFVILLE, J. P. (1972). *Proc. 11th Int. Conf. on the Physics of Semiconductors, Warsaw*, p. 542. PWN-Polish Scientific Publishers, Poland.

ROGACHEV, A. A. (1968). *Proc. 9th Int. Conf. on the Physics of Semiconductors, Moscow*, p. 407. Nauka, Leningrad.

ROILOS, M. and MYTILINEOU, E. (1974). *Garmisch*, p. 319.

ROSE, A. (1963). *Concepts in photoconductivity and allied problems.* Wiley–Interscience, New York.

ROULET, B., GAVORET, J., and NOZIÈRES, P. (1969). *Phys. Rev.* **178**, 1072.

ROWE, J. E. (1974). *Appl. Phys. Lett.* **25**, 576.

ROWLAND, S. C., NARASIMHAN, S., and BIENENSTOCK, A. (1972). *J. appl. Phys.* **43**, 2741.

—— RITLAND, F., HAFERBURNS, D., and BIENENSTOCK, A. (1977). *Edinburgh*, p. 135.

RUBIO, J. (1966). *J. Phys. C: Solid State Phys.* **2**, 288.

RUDEE, M. L. (1971). *Phys. Status Solidi B* **46**, K1.

—— (1972). *Thin solid Films* **12**, 207.

—— and HOWIE, A. (1972). *Phil. Mag.* **25**, 1001.

RUFFA, A. R. (1968). *Phys. Status Solidi* **29**, 605.

SALGADO, J., GOMPF, F., and REICHARDT, W. (1974). Progress Rep. Teilinstitut Nukleare Festkorperphysik.

SANDROCK, R. (1968). *Phys. Rev.* **169**, 642.

SASAKI, W. (1976). *J. Phys. (Paris) Colloq.* **C4**, C4-307.

—— IKEHATA, S., and KOBAYASHI, S. (1973). *Phys. Lett. A* **42**, 429.

SAVAGE, J. A. and NIELSEN, S. (1965*a*). *Infrared Phys.* **5**, 195.

—— —— (1965*b*). *Phys. Chem. Glasses* **6**, 90.

SAYER, M., CHEN, R., FLETCHER, R., and MANSINGH, A. (1975). *J. Phys. C: Solid State Phys.* **8**, 2059.

SAYERS, D. E. (1977). *Edinburgh*, p. 61.

—— LYTLE, F. W., and STERN, E. A. (1974). *Garmisch*, p. 403.

—— STERN, E. A., and LYTLE, F. W. (1975). *Phys. Rev. Lett.* **35**, 584.

SCHEIN, L. B., DUKE, C. B., and MCGHIE, A. R. (1978). *Phys. Rev. Lett.* **40**, 197.

SCHER, H. and LAX, M. (1973). *Phys. Rev. B* **7**, 4491, 4502.

—— and MONTROLL, E. W. (1975). *Phys. Rev. B* **12**, 2455.

SCHIFF, L. I. (1955). *Quantum mechanics* (2nd edn.). McGraw-Hill, New York.

SCHLÜTER, M. and CHELIKOWSKY, J. R. (1977). *Solid State Commun.* **21**, 381.

—— JOANNOPOULOS, J. D., and COHEN, M. L. (1974). *Phys. Rev. Lett.* **33**, 89.

SCHMID, A. P. (1968). *J. appl. Phys.* **39**, 3140.

SCHMIDLIN, F. W. (1977). *Phys. Rev. B* **16**, 2362.

SCHMUTZLER, R. W. and HENSEL, F. (1972). *J. non-cryst. Solids* **8–10**, 718.

SCHNAKENBERG, J. (1968). *Phys. Status Solidi* **28**, 623.

SCHNYDERS, H., RICE, S. A., and MEYER, L. (1966). *Phys. Rev.* **150**, 127.

SCHÖNHAMMER, K. (1971). *Phys. Lett. A* **36**, 181.

—— and BRENIG, W. (1973). *Phys. Lett. A* **42**, 447.

SCHOTTMILLER, J., TABAK, M., LUCOVSKY, G., and WARD, A. (1970). *J. non-cryst. Solids* **4**, 80.

SCHRIEFFER, J. R. (1955). *Phys. Rev.* **97**, 641.

—— (1957). In *Semiconductor surfaces* (ed. R. H. Kingston), p. 1. University of Pennsylvania Press, Pittsburgh, Pa.

SCHWIDEFSKY, F. (1973). *Thin solid Films* **18**, 45.

SEAGER, C. H. and PIKE, G. E. (1974). *Phys. Rev. B* **10**, 1435.

—— and QUINN, R. K. (1975). *J. non-cryst. Solids* **17**, 386.

—— EMIN, D., and QUINN, R. K. (1973). *Phys. Rev. B* **8**, 4746.

—— KNOTEK, M. L., and CLARK, A. H. (1974). *Garmisch*, p. 1133.

SEGBAHN, K., NORDLING, C., FAHLMAN, A., NORDBERG, R., HAMRIN, K., HEDMAN, J., JOHANSSON, G., BERGMARK, T., KARLSSON, S. E., LINGREN, I., and LINDBERG, B. J. (1967). ESCA-atomic, molecular and solid state structure by means of electron spectroscopy. *Nova Acta Regiae Soc. sci. Ups.*, Ser. IV, **20**.

SENTURIA, S. D., HEWES, C. R., and ADLER, D. (1970). *J. appl. Phys.* **41**, 430.

SHANFIELD, Z., MONTANO, P. A., and BARRETT, P. H. (1975). *Phys. Rev. Lett.* **35**, 1789.

SHANTE, V. K. S. (1973). *Phys. Lett. A* **43**, 249.

SHARMA, S. K., THEINER, W. A., and GESERICH, H. P. (1974). *Phys. Status Solidi A* **25**, K65.

SHAW, R. F., LIANG, W. Y., and YOFFE, A. D. (1970). *J. non-cryst. Solids* **4**, 29.

SHEVCHIK, N. J. (1974*a*). *Yorktown Heights*, p. 72.

—— (1974*b*). *Phys. Rev. Lett.* **33**, 1572.

—— (1977). *Phil. Mag.* **35**, 261.

—— and BISHOP, S. G. (1975). *Solid State Commun.* **17**, 269.

—— and PAUL, W. (1972). *J. non-cryst. Solids* **8–10**, 381.

—— —— (1973). *J. non-cryst. Solids* **13**, 1.

—— —— (1974). *J. non-cryst. Solids* **16**, 55.

—— TEJEDA, J., and CARDONA, M. (1974*a*). *Garmisch*, p. 609.

—— CARDONA, M., and TEJEDA, J. (1973). *Phys. Rev. B* **8**, 2833.

—————— (1974*b*). *Phys. Rev. B* **9**, 2627.
———— LANGER, D. W., and CARDONA, M. (1973*a*). *Phys. Rev. Lett* **30**, 659.
————————— (1973*b*). *Phys. Status Solidi B* **57**, 245.
SHKLOVSKII, B. I. (1973*a*). *Sov. Phys.—Semicond.* **6**, 1053.
—— (1973*b*). *Sov. Phys.—Semicond.* **6**, 1964.
—— and ÉFROS, A. L. (1971), *Sov. Phys.—JETP* **33**, 468.
—— and SHLIMAK, I. S. (1972). *Sov. Phys.—Semicond.* **6**, 104.
SHLIMAK, I. S. and NIKULIN, E. I. (1972). *JETP Lett.* **15**, 20.
SHOCKLEY, W. (1950). *Electrons and holes in semiconductors.* Van Nostrand, Princeton, N.J.
—— and READ, W. T. (1952). *Phys. Rev.* **87**, 835.
SIEBRAND, W. (1967). *J. chem. Phys.* **46**, 440.
SIEMSEN, K. J. and FENTON, E. W. (1967). *Phys. Rev.* **161**, 632.
SIGEL, G. H. (1973/74). *J. non-cryst. Solids* **13**, 372.
SIK, M. J. and FERRIER, R. P. (1974). *Phil. Mag.* **29**, 877.
SILVER, M. (1977). *Edinburgh*, p. 214.
—— and COHEN, L. (1977). *Phys. Rev. B* **15**, 3276.
SIMMONS, J. G. (1967). *Phys. Rev.* **155**, 657.
—— (1971). *J. Phys. D: Appl. Phys.* **4**, 613.
—— and TAYLOR, G. W. (1974). *J. Phys. C: Solid State Phys.* **7**, 3051.
SLATER, J. C. (1951). *Phys. Rev.* **82**, 538.
—— (1963). *Quantum theory of molecules and solids*, Vol. 1. McGraw-Hill, New York.
SLOWICK, J. H. and BROWN, F. C. (1972). *Phys. Rev. Lett.* **29**, 934.
SMITH, J. E., BRODSKY, M. H., CROWDER, B. L., and NATHAN, M. I. (1972). *J. non-cryst. Solids* **8–10**, 179.
SMITH, P. M., LEADBETTER, A. J., and APLING, A. J. (1975). *Phil. Mag.* **31**, 57.
SMITH, R. E. (1972). *J. non-cryst. Solids* **8–10**, 598.
SNOW, E. H. (1967). *Solid State Commun.* **5**, 813.
SOLIN, S. A. and PAPATHEODOROU, G. N. (1977). *Phys. Rev. B* **15**, 2084.
SOLOMON, I. (1976). *Solid State Commun.* **20**, 215.
—— BIEGELSEN, D., and KNIGHTS, J. C. (1977). *Solid State Commun.* **22**, 505.
SOMMER, A. H. (1966). *J. appl. Phys.* **37**, 2789.
SONG, K. S. (1971). *Can. J. Phys.* **49**, 26.
SOONPAA, H. H. (1976). *Proc. 13th Int. Conf. on the Physics of Semiconductors; Rome* (ed. F. G. Fumi), p. 364. Tipografia Marves, Rome.
SPEAR, W. E. (1957). *Proc. phys. Soc. B* **70**, 669.
—— (1960). *Proc. phys. Soc.* **76**, 826.
—— (1974*a*). *Adv. Phys.* **23**, 523.
—— (1974*b*). *Garmisch*, p. 1.
—— (1977). *Adv. Phys.* **26**, 811.
—— and LE COMBER, P. G. (1972). *J. non-cryst. Solids* **8–10**, 727.
———— (1975). *Solid State Commun.* **17**, 1193.
———— (1976). *Phil. Mag.* **33**, 935.
—— and TANNHAUSER, D. S. (1973). *Phys. Rev. B* **7**, 831.
—— LOVELAND, R. J., and AL-SHARBATY, A. (1974). *J. non-cryst. Solids* **15**, 410.
SPICER, W. E. (1973). *Band structure spectroscopy of metals and alloys* (ed. D. J. Fabian and L. M. Watson), p. 7. Academic Press, New York.
—— (1974). *Garmisch*, p. 499.
—— and DONOVAN, T. M. (1970). *J. non-cryst. Solids* **2**, 66.
———— and FISCHER, J. E. (1972). *J. non-cryst. Solids* **8–10**, 122.

SPRINGTHORPE, A. J., AUSTIN, I. G., and SMITH, B. A. (1965). *Solid State Commun.* **3**, 143.

SRINIVASAN, G. (1971). *Phys. Rev. B* **4**, 2581.

STEINHARDT, P., ALBEN, R., and WEAIRE, D. (1974). *J. non-cryst. Solids* **15**, 199.

STEPHENS, R. B. (1976). *Phys. Rev. B* **13**, 852

STERN, E. A. (1974). *Phys. Rev. B* **10**, 3027.

—— (1976). *Sci. Am.* **234**, 96.

—— SAYERS, D. E., and LYTLE, F. W. (1975). *Phys. Rev. B* **11**, 4836.

STERN, F. (1971). *Phys. Rev. B* **3**, 2636.

—— (1972). *Phys. Rev. B* **5**, 4891.

—— (1974*a*). *Phys. Rev. B* **9**, 2762.

—— (1974*b*). *Crit. Rev. Solid State Sci.* **4**, 499.

STINCHCOMBE, R. B. (1973). *J. Phys. C: Solid State Phys.* **6**, L1.

—— (1974). *J. Phys. C: Solid State Phys.* **7**, 179.

STOCKER, H. J., BARLOW, C. A., and WEIRAUCH, D. F. (1970). *J. non-cryst. Solids* **4**, 523.

STÖHR, H. (1939). *Z. anorg. allg. Chem.* **242**, 138.

STOLEN, R. H. (1970). *Phys. Chem. Glasses* **11**, 83.

STONEHAM, A. M. (1975). *Theory of defects in solids.* Clarendon Press, Oxford.

STREET, R. A. (1976). *Adv. Phys.* **25**, 397.

—— (1977). *Edinburgh*, p. 509. *Solid State Commun.* **24**, 363.

—— (1978). *Phys. Rev. B* **17**, 3984; *Phil. Mag. B* **37**, 35.

—— and MOTT, N. F. (1975). *Phys. Rev. Lett.* **35**, 1293.

—— and YOFFE, A. D. (1972). *J. non-cryst. Solids* **8–10**, 745.

—— AUSTIN, I. G., and SEARLE, T. M. (1975). *J. Phys. C: Solid State Phys.* **8**, 1293.

—— DAVIES, G. R., and YOFFE, A. D. (1971). *J. non-cryst. Solids* **5**, 276.

—— SEARLE, T. M., and AUSTIN, I. G. (1973). *J. Phys. C: Solid State Phys.* **6**, 1830.

—— —— —— (1974*a*). *Garmisch*, p. 953.

—— —— —— (1974*b*). *Proc. 12th Int. Conf. on the Physics of Semiconductors* (ed. M. H. Pilkuhn), p. 1037 Teubner, Stuttgart.

—— —— —— (1974*c*). *Phil. Mag.* **30**, 1181.

—— —— —— (1975). *Phil. Mag.* **32**, 431.

—— AUSTIN, I. G., SEARLE, T. M., and SMITH, B. A. (1974). *J. Phys. C: Solid State Phys.* **7**, 4185.

—— SEARLE, T. M., AUSTIN, I. G., and SUSSMANN, R. S. (1974). *J. Phys. C: Solid State Phys.* **7**, 1582.

STROM, U. and TAYLOR, P. C. (1974). *Garmisch*, p. 375.

—— SCHAFER, D. E., and TAYLOR, P. C. (1977). *Edinburgh*, p. 179.

STROUD, J. S., SCHREUFFER, J. W. H., and TUCKER, R. F. (1965). *Proc. 7th Int. Conf. on Glass*, p. 42. Gordon and Breach, London.

STRUCK, C. W. and FONGER, W. H. (1975). *J. Lumin.* **10**, 1.

STUKE, J. (1970*a*). *J. non-cryst. Solids* **4**, 1.

—— (1970*b*). *Proc. 10th Int. Conf. on the Physics of Semiconductors, Cambridge, Mass.* (ed. S. P. Keller, J. C. Hensel, and F. Stern), p. 14. United States Atomic Energy Commission, Washington, D.C.

—— (1976). *Leningrad, Structure and Properties of Non-Crystalline Semiconductors*, p. 193.

—— (1977). *Edinburgh*, p. 406.

—— and WEISER, G. (1966). *Phys. Status Solidi* **17**, 343.

SUGIURA, T. and MASUDA, Y. (1973). *J. phys. Soc. Japan* **35**, 1254.

SUMI, A. and TOYOZAWA, Y. (1973). *J. phys. Soc. Japan* **35**, 137.

SUSSMANN, R. S., AUSTIN, I. G., and SEARLE, T. M. (1975). *J. Phys. C: Solid State Phys.* **8**, L182.

SUTTON, C. M. (1975). *Solid State Commun.* **16**, 327.

SZEKELY, G. (1951). *J. electrochem. Soc.* **98**, 318.

TABAK, M. D. (1970). *Phys. Rev. B* **2**, 2104.

—— and WARTER, P. J. (1968). *Phys. Rev.* **173**, 899.

TANAKA, K., HAMANAKA, H., and IIZIMA, S. (1977). *Edinburgh*, p. 787.

TANAKA, S. and FAN, H. Y. (1963). *Phys. Rev.* **132**, 1516.

TANGONAN, G. L. (1975). *Phys. Lett A* **54**, 307.

TAUC, J. (1970a). In *The optical properties of solids* (ed. F. Abeles), p. 277. North-Holland, Amsterdam.

——(1970b). *Mater. Res. Bull.* **5**, 721.

—— and MENTH, A. (1972). *J. non-cryst. Solids* **8–10**, 569.

—— and NAGEL, S. R. (1976). *Comments Solid State Phys.* **7**, 69.

—— —— (1977).

—— GRIGOROVICI, R., and VANCU, A. (1966). *Phys. Status Solidi* **15**, 627.

—— MENTH, A., and WOOD, D. L. (1970). *Phys. Rev. Lett.* **25**, 749.

—— ABRAHÁM, A., ZALLEN, R., and SLADE, M. (1970). *J. non-cryst. Solids* **4**, 279.

—— DI SALVO, F. J., PETERSON, G. E., and WOOD, D. L. (1973). In *Amorphous Magnetism* (ed. H. O. Hooper and A. M. de Graaf), p. 119. Plenum Press, New York.

TAYLOR, G. W. and SIMMONS, J. G. (1974). *J. Phys. C: Solid State Phys.* **7**, 3067.

TAYLOR, J. B., BENNETT, S. L., and HEYDING, R. D. (1965). *J. Phys. Chem. Solids* **26**, 69.

TAYLOR, P. C., BISHOP, S. G., and MITCHELL, D. L. (1970). *Solid State Commun.* **8**, 1783.

—— —— —— (1971). *Phys. Rev. Lett.* **27**, 414.

TELNIC, M., VESCAN, L., CROITORU, N., and PEPESCU, C. (1973). *Phys. Status Solidi B* **59**, 699.

TEMKIN, R. J. (1974). *Yorktown Heights*, p. 229.

—— (1978). *J. non-cryst. Solids* **28**, 23.

—— PAUL, W., and CONNELL, G. A. N. (1973). *Adv. Phys.* **22**, 581.

THÈYE, M.-L. (1970). *Opt. Commun.* **2**, 329.

—— (1971). *Mater. Res. Bull.* **6**, 103.

—— (1974). *Garmisch*, p. 479.

THOMAS, C. B., FRAY, A. F., and BOSNELL, J. (1972). *Phil. Mag.* **26**, 617.

THOMAS, P. A. and KAPLAN, D. (1976). *Williamsburg*, p. 85.

THOMPSON, J. C. (1976). *Electrons in liquid ammonia*. Clarendon Press, Oxford.

THORNBER, K. K. and FEYNMAN, R. P. (1970). *Phys. Rev. B* **1**, 4099.

THORPE, M. F. and WEAIRE, D. (1971). *Phys. Rev. B* **4**, 3518.

—— —— (1974). *Garmisch*, p. 917.

THOULESS, D. J. (1970). *J. Phys. C: Solid State Phys.* **3**, 1559.

—— (1974). *Phys. Rev. C* **13**, 93.

—— (1975). *Phil. Mag.* **32**, 877.

TIÈCHE, Y. and ZAREBA, A. (1973). *Phys. kondens. Mater.* **1**, 402.

TITLE, R. S., BRODSKY, M. H., and CUOMO, J. J. (1977). *Edinburgh*, p. 424.

TONG, B. Y. (1974). *Yorktown Heights*, p. 145.

—— SWENSON, J. R., and CHOO, F. C. (1974). *Phys. Rev. B* **10**, 3338.

TORRANCE, J. B., SHAFER, M. W., and MCGUIRE, T. R. (1972). *Phys. Rev. Lett.* **29**, 1168.

TOULOUSE, G. (1975). *C. R. Acad. Sci. Paris* **280**, 629.

—— and PFEUTY, P. (1975). *C. R. Acad. Sci. Paris* **280**, 33.

TOYOTOMI, Y. (1974). Thesis, Tokyo.

TOYOZAWA, Y. (1959*a*). *Prog. theor. Phys. Suppl.* **12**, 111.

—— (1959*b*). *Prog. theor. Phys.* **22**, 455.

—— (1961). *Prog. theor. Phys.* **26**, 29.

—— (1962). *J. phys. Soc. Japan* **17**, 986.

—— (1964). *Tech. Rep. Solid State Phys.* (*Univ. Tokyo*) Ser. A, No. 119.

TREUSCH, J. and SANDROCK, R. (1966). *Phys. Status Solidi* **16**, 487.

TRONC, P., BENSOUSSAN, M., and BRENAC, A. (1973). *Phys. Rev. B* **8**, 5947.

TSAI, C. C., FRITZSCHE, H., TANIELIAN, M. H., GACZI, P. J., PERSANS, P. D., and VESAGHI, M. A. (1977). *Edinburgh*, p. 339.

TSCHIRNER, H.-U., POPP, K., SOBOTTKA, A., and WOBST, M. (1976). *Leningrad, Structure and Properties of Non-Crystalline Semiconductors*, p. 410.

TSU, R., HOWARD, W. E., and ESAKI, L. (1970). *J. non-cryst. Solids* **4**, 322.

TSUI, D. C. and ALLEN, S. J. (1974). *Phys. Rev. Lett.* **32**, 1200.

—— —— (1975). *Phys. Rev. Lett.* **34**, 1293.

TUNSTALL, D. P. (1975). *Phys. Rev. B* **11**, 2821.

TURNBULL, D. (1969). *Contemp. Phys.* **10**, 473.

—— and POLK, D. E. (1972). *J. non-cryst. Solids* **8–10**, 19.

TUTIHASI, S., ROBERTS, G. G., KEEZER, R. C., and DREWS, R. E. (1969). *Phys. Rev.* **177**, 1143.

TWADDELL, V. A., LA COURSE, W. C., and MACKENZIE, J. D. (1972). *J. non-cryst. Solids* **8–10**, 831.

TWOSE, W. D. (1959). Ph.D. Thesis, University of Cambridge.

UDA, T. and YAMADA, E. (1975).

UE, H. and MAEKAWA, S. (1971). *Phys. Rev. B* **3**, 4232.

UPHOFF, H. L. and HEALY, J. H. (1961). *J. appl. Phys.* **32**, 950.

URBACH, F. (1953). *Phys. Rev.* **92**, 1324.

URBAIN, G. and ÜBELACKER, E. (1966). *C. R. Acad. Sci Paris Sér. C* **262**, 699.

VALIANT, J. C. and FABER, T. E. (1974). *Phil. Mag.* **29**, 571.

VALLANCE, R. H. (1938). *A textbook of inorganic chemistry* (Griffin, London, 1938) Part IV, p. 27.

VAN SANTEN, J. H. and JONKER, G. H. (1950). *Physica* **16**, 599.

VAN VECHTEN, J. A. (1969). *Phys. Rev.* **182**, 891.

VASHISHTA, P., BHATTACHARYYA, P., and SINGWI, K. S. (1973). *Phys. Rev. Lett* **30**, 1248.

VENGEL, T. N. and KOLOMIETS, B. T. (1957). *Sov. Phys.—Tech. Phys.* **27.2**, 2314.

VEST, R. W., GRIFFEL, M., and SMITH, J. F. (1958). *J. chem. Phys.* **28**, 293.

VEZZOLI, G. C., WALSH, P. J., KISATSKY, P. J., and DOREMUS, L. W. (1974). *J. appl. Phys.* **45**, 4534.

VOGET-GROTE, U., STUKE, J., and WAGNER, H. (1976). *Williamsburg*, p. 91.

VON MOLNAR, S. (1970). *IBM J. Res. Dev.* **14**, 269.

VUČIČ, Z., ETLINGER, B., and KUNSTELJ, D. (1976). *J. non-cryst. Solids* **20**, 451.

VUL, B. M. (1974). *Suppl. J. Japan Soc. appl. Phys.* **43**, 183.

—— KOTEL'NIKOVA, N. V., ZAVARITSKAYA, E. I., and VORONOVA, I. D. (1976). *Sov. Phys.—Semicond.* **10**, 1351.

WAGNER, C. N. J. (1969). *J. Vac. Sci. Technol.* **6**, 650.

WAGNER, L. F. and SPICER, W. E. (1972). *Phys. Rev. Lett.* **28**, 1381.

WALES, J., LOVITT, G. J., and HILL, R. A. (1967). *Thin solid Films* **1**, 137.

WALLEY, P. A. (1968*a*). Ph.D. Thesis, University of London.

—— (1968*b*). *Thin solid Films* **2**, 327.

—— and JONSCHER, A. K. (1968). *Thin solid Films* **1**, 367.

WALLIS, R. H. (1973). Ph.D. Thesis, University of Nottingham.

WARD, A. T. (1972). *Adv. Chem.* **110**, 163.

WARREN, A. C. (1970). *J. non-cryst. Solids* **4**, 613.

WARREN, W. W. (1970*a*). *J. non-cryst. Solids* **4**, 168.

—— (1970*b*). *Solid State Commun.* **8**, 1269.

—— (1972*a*). *Phys. Rev.* **6**, 2522.

—— (1972*b*). *J. non-cryst. Solids* **8–10**, 241.

—— and BRENNERT, G. F. (1974). *Garmisch*, p. 1047.

—— BRENNERT, G. F., BUEHLER, E., and WERNICK, J. H. (1974). *J. Phys. Chem. Solids* **35**, 1153.

WARTER, P. J. (1971). *Proc. 3rd Int. Conf. on Photoconductivity* (ed. E. M. Pell), p. 311. Pergamon Press, Oxford.

WATANABE, I., INAGAKI, Y., and SHIMIZU, T. (1976). *J. non-cryst. Solids* **22**, 109.

WATKINS, G. D. and CORBETT, J. W. (1965). *Phys. Rev. A* **138**, 543.

WEAIRE, D. (1971). *Phys. Rev. Lett.* **26**, 1541.

—— and THORPE, M. F. (1971). *Phys. Rev. B* **4**, 2508.

—— —— (1973). In *Computational methods for large molecules and localized states in solids* (ed. F. Herman, A. D. McLean, and R. K. Nesbet), p. 295. Plenum Press, New York.

WEBER, W. (1974). *Phys. Rev. Lett.* **33**, 371.

WEBMAN, I., JORTNER, J., and COHEN, M. H. (1975). *Phys. Rev. B* **11**, 2885.

WEEKS, J. D., ANDERSON, P. W., and DAVIDSON, A. G. H. (1973). *J. chem. Phys.* **58**, 1388.

WEGNER, F. J. (1976). *Z. Phys. B* **25**, 327.

WEINBERG, Z. A. and POLLAK, R. A. (1975). *Appl. Phys. Lett.* **27**, 254.

WEINSTEIN, F. C. (1974). *Garmisch*, p. 95.

—— and DAVIS, E. A. (1973/74). *J. non-cryst. Solids* **13**, 153.

WEISER, K. (1972). *J. non-cryst. Solids* **8–10**, 922.

—— and BRODSKY, M. H. (1970). *Phys. Rev. B* **1**, 791.

—— and STUKE, J. (1969). *Phys. Status Solidi* **35**, 747.

—— FISCHER, R., and BRODSKY, M. H. (1970). *Proc. 10th Int. Conf. on the Physics of Semiconductors* (ed. S. P. Keller, J. C. Hensel, and F. Stern), p. 667. United States Atomic Energy Commission, Washington, D. C.

—— GRANT, A. J., and MOUSTAKAS, T. D. (1974). *Garmisch*, p. 335.

WENGER, L. E., AMAYA, K., and KUKKONEN, C. A. (1976). *Phys. Rev. B* **14**, 1327.

WHITE, R. M. (1974*a*). *J. non-cryst. Solids* **16**, 387.

—— (1974*b*). *Phys. Rev. B* **10**, 3426.

—— and ANDERSON, P. W. (1972). *Phil. Mag.* **25**, 737.

—— and WOOLSEY, R. B. (1968). *Phys. Rev.* **176**, 908.

WIGNER, E. (1938). *Trans. Faraday Soc.* **34**, 678.

WIHL, M., STILES, P. J., TAUC, J. (1972). *Proc. 11th Int. Conf. on the Physics of Semiconductors*, Warsaw, p. 484. PWN-Polish Scientific Publishers, Warsaw.

WILLIAMS, R. (1965). *Phys. Rev. A* **104**, 569.

WISER, N. and GREENFIELD, A. J. (1966). *Phys. Rev. Lett.* **17**, 586.

WONG, J. and ANGELL, C. A. (1977). *Glass structure by spectroscopy.* Marcell Dekker, New York.

WONG, P. T. T. and WHALLEY, E. (1970). *Discuss. Faraday Soc.* No. 50, p. 94.

WOOD, D. L. and TAUC, J. (1972). *Phys. Rev. B* **5**, 3144.

WRIGHT, A. C. (1974). *Adv. Struct. Res. Diffr. Methods* **5**, 1.

—— and LEADBETTER, A. J. (1976). *Phys. Chem. Glasses* **17**, 122.

WRIGHT, J. G. (1977). *Proc. 3rd Int. Conf. on Liquid Metals, Bristol* (ed. R. Evans and D. A. Greenwood), p. 251. The Institute of Physics, London.

WU, C. T. and LUO, H. L. (1973/74). *J. non-cryst. Solids* **13**, 437.

WYCKOFF, R. W. G. (1963). *Crystal structures* (2nd edn.), Vol. 1, pp. 31–32. Wiley–Interscience, New York.

YAMANOUCHI, C. (1965). *J. phys. Soc. Japan* **20**, 1029.

—— MIZUGUCHI, K., and SASAKI, W. (1967). *J. phys. Soc. Japan* **22**, 859.

YASHIHIRO, K., TOKUMOTO, M., and YAMANOUCHI, C. (1974). *Proc. 12th Int. Conf. on the Physics of Semiconductors, Stuttgart* (ed. M. H. Pilkuhn), p. 1128, Teubner, Stuttgart.

YIP, K. L. and FOWLER, W. B. (1975). *Phys. Rev. B* **11**, 2327.

YNDURAIN, F. and JOANNOPOULOS, J. D. (1975). *Phys. Rev. B* **11**, 2957.

—— —— (1976). *Phys. Rev. B* **14**, 3569.

YONOZAWA, F., ISHIDA, Y., MARTINO, F., and ASANO, S. (1977). *Proc. 3rd Int. Conf. on Liquid Metals, Bristol* (ed. R. Evans, and D. A. Greenwood), p. 385. The Institute of Physics, London.

YOSHINO, S. and OKAZAKI, M. (1977). *J. Phys. Soc. Japan* **43**, 415.

ZACHARIASEN, W. H. (1932). *J. Am. Chem. Soc.* **54**, 3841.

ZALLEN, R. (1968). *Phys. Rev.* **173**, 824.

—— SLADE, M. L., and WARD, A. T. (1971*b*). *Phys. Rev. B* **3**, 4257.

—— DREWS, R. E., EMERALD, R. L., and SLADE, M. L. (1971*a*). *Phys. Rev. Lett.* **26**, 1564.

ZELLER, R. C. and POHL, R. O. (1971). *Phys. Rev. B* **4**, 2029.

ZEPPENFELD, K. and RAETHER, H. (1966). *Z. Phys.* **193**, 471.

ZIMAN, J. M. (1967). *Adv. Phys.* **16**, 421.

—— (1960). *Electrons and phonons.* Oxford University Press, London.

—— (1961). *Phil. Mag.* **6**, 1013.

—— (1969). *J. Phys. C: Solid State Phys.* **2**, 1230, 1704.

—— (1970). *J. non-cryst. Solids* **4**, 426.

—— (1971). *J. Phys. C: Solid State Phys.* **4**, 3129.

ZITTARTZ, J. and LANGER, J. S. (1966). *Phys. Rev.* **148**, 741.

ZLATKIN, L. B. and MARKOV, YU. F. (1971). *Phys. Status Solidi A* **4**, 391.

ZUMSTEG, F. C. (1976). *Phys. Rev. B* **14**, 1406.

ZVYAGIN, I. P. (1973). *Phys. Status Solidi B* **58**, 443.

INDEX

a.c. conductivity, 59ff., 117ff., 223ff.
  in amorphous
    arsenic, 423
    chalcogenides, 83, 231ff., 464, 467
    germanium, 357ff.
    silicon, 374, 375
alloys, chalcogenide, *see* chalcogenides
  , liquid, 170ff., 543ff.
  , of group V materials, 440, 441
  , silicon–arsenic, 372, 373
  , with germanium, 132ff.
alternating currents, *see* a.c. conductivity
amorphons, 325
amorphous metals, *see* metallic glasses
Anderson
  localization, 9ff., 15ff.
  transitions, 37ff., 98ff.
antiferromagnetic metal, 108
antiferromagnetism, amorphous, 56
antimony
  amorphous, alloys with, 440, 441
    , d.c. conductivity of, 439
    , RDF of, 418
arsenic,
  amorphous, 408ff.
    , a.c. conductivity of, 423
    , CRN of, 414ff.
    , density of electron states in, 432ff.
    , dihedral angles in, 415
    , electrical properties of, 419ff.
    , e.s.r. in, 436, 437
    , Hall effect in, 422, 423
    , infrared spectra of, 305
    , optical absorption edges in, 279, 427, 428
    , optical properties of, 426ff.
    , photoconductivity in, 426, 428, 429
    , photoluminescence in, 435, 436
    , preparation of, 408, 409
    , pressure dependence of conductivity in, 424
    , RDF of, 412ff.
    , specific heat of, 309, 310, 436
    , states in the gap of, 421, 425, 434ff.
    , structure factor of, 412, 418
    , structure of, 410ff.
    , thermopower of, 422, 433
    , X-ray absorption in, 497
  , crystalline forms of, 408, 409
arsenic
  trisulphide, *see* chalcogenides
  triselenide, *see* chalcogenides
  tritelluride, *see* chalcogenides

Baber scattering, 164
band-crossing transitions, 101ff.
band structure,
  of crystalline
    germanium, 376
    selenium and tellurium, 529
    silicon, 386
bipolarons, 490
  in a.c. conduction, 231
bismuth, amorphous, 408, 441
  , metal-insulator transitions in, 102

caesium, liquid, 189, 190
cerium sulphide, 146ff.
CFO model, 48, 211
chalcogenides,
  amorphous, 442ff.
    , a.c. conductivity of, 231ff., 464, 467
    , bonding in, 446ff., 493, 494
    , charged defects in, 49, 88, 89, 214, 215, 231, 464ff., 488ff.
    , core-level spectra of, 495ff.
    , d.c. conductivity of, 224, 241, 454ff.
    , density of electron states in, 491ff.
    , drift mobility in, 250, 254, 470ff.
    , effect of alloying on the conductivity of, 484ff.
    , electroabsorption in, 285ff.
    , electrical properties of, 272, 452ff.
    , e.s.r. in, 318, 319, 461, 467, 482, 504
    , field effect in, 246, 469, 470
    , free-carrier absorption in, 297ff., 506
    , glass-forming regions in, 202, 443, 444
    , Hall effect in, 240ff., 272, 459ff.
    , infrared spectra of, 234, 301ff., 446ff.
    , magnetoresistance in, 243
    , negative Hubbard U in, 466
    , non-ohmic conduction in, 490
    , optical absorption edges in, 265, 279, 290, 292, 298, 474, 476, 480, 481, 497ff.
    , optical properties of, 491ff., 497ff.
    , photoconductivity in, 258ff., 268, 483, 484
    , photodarkening in, 468
    , photogeneration in, 265ff.
    , photoluminescence in, 474ff., 503, 504
    , pinning of the Fermi level in, 46, 212, 461ff., 485ff.
    , preparation of, 442ff.
    , Raman spectra of, 446ff.
    , RDF of, 449ff.

chalcogenides, amorphous – *cont.*
, recombination lifetimes in, 261ff., 483, 484
, screening lengths in, 468, 469
, specific heats of, 305ff.
, states in the gap of, 460ff., 488ff.
, structure of, 203, 445ff.
, switching in, 507ff.
, thermal conductivity of, 306ff.
, thermopower of, 272, 458
, valence-alternation pairs in, 466
, voids in, 443
, X-ray absorption in 495, 497
, crystalline, 481, 482, 491ff., 488, 489, 513
, liquid, 224, 237, 241, 298, 543ff.
charged defects
in amorphous
arsenic, 434ff.
chalcogenides, 49, 88, 89, 214, 215, 231, 464ff., 488ff.
selenium, 532ff.
chromium sulphide, 155
classification of amorphous materials, 200ff.
conduction at a mobility edge versus hopping, 270ff.
continuous random networks, *see* CRN
co-ordination number
in amorphous materials, 44, 203ff., 208, 325, 408, 445, 461, 518
, relation to slope of Urbach edge, 283, 284
core levels, 314, 495ff.
correlation, 103, 104ff., 110ff., 119ff.
, effects on hopping, 37
CRN, 43
of arsenic, 414ff.
of chalcogenides, 451
of germanium and silicon, *frontispiece*, 326ff.
of selenium, 519, 420
crystallization, 200, 203
Curie paramagnetism, 315, 464

D⁺D⁻ model, *see* charged defects
dangling bands, *frontispiece*, *see also* defects charged defects
defects
in amorphous
arsenic, 434ff.
chalcogenides, 49, 88, 89, 214, 215, 231, 461, 464ff., 488ff.
germanium and silicon, 48, 49, 88, 211ff., 333ff., 369, 370
selenium, 532ff.
silicon dioxide, 515

d.c. conductivity
in amorphous
arsenic, 419ff.
antimony, 439
chalcogenides, 224, 241, 454ff.
germanium, 345ff.
selenium, 530ff.
semiconductors, temperature dependence of, 219ff., 224, 270ff.
silicon, 362ff.
deformation potential, 67
degenerate gas of polarons, 94, 95, 145
density of electron states, 1, 7ff., 51, 52, 62
in amorphous
arsenic, 432ff.
chalcogenides, 491ff.
germanium, 396ff.
selenium, 296, 524ff.
silicon, 312, 396ff.
silicon dioxide, 514
tellurium, 524ff.
in liquid metals, 173, 175
Dexter, Klick, and Russell, mechanism of recombination, 87, 300, 483
devitrification, *see* crystallization
differential thermal analysis (DTA), 201
diffraction function, 205
for amorphous
arsenic, 412
chalcogenides, 418
germanium, 333, 334
dihedral angles
in amorphous
arsenic, 415
germanium, 325, 330
selenium, 521, 526ff.
silicon, 324, 328, 330
divacancy in silicon, 213, 369, 370
doped crystalline semiconductors, 111ff.
doping
of amorphous
chalcogenides, 484ff.
semiconductors, 44ff.
silicon, 322, 370ff.
double injection, 103, 507ff.
DOVS, *see* density of electron states
Dow and Redfield theory, 276ff.
drift mobility, 93, 247ff., 270ff.
, dispersive transits in, 250ff., 491
in amorphous
chalcogenides, 250, 254, 470ff.
selenium, 250ff., 531ff.
silicon, 250, 256, 366ff.
silicon dioxide, 512, 513
in liquid xenon, 180, 181
Drude formula, 15, 224, 298

electrical conductivity, *see* d.c. conductivity, electrical properties
electrical properties
  of non-crystalline semiconductors, 209ff.
  of amorphous
    arsenic, 419ff.
    chalcogenides, 452ff.
    germanium, 345ff.
    selenium, 250ff., 530ff.
    silicon, 362ff.
electroabsorption
  of amorphous
    chalcogenides, 285ff.
    germanium, 287
    selenium, 286
electroluminescence, 395, 510
electron-hole droplets, 102, 103
electron spin resonance, *see* e.s.r.
energy distribution curves (EDC), *see* photoemission
ensemble average, 9
e.s.r.,
  in impurity conduction, 123
  in amorphous
    arsenic, 436, 437
    chalcogenides, 318, 319, 461, 467, 482, 504
    germanium, 314ff.
    selenium, 532
    silicon, 36, 314ff.
europium oxide and sulphide, 153, 154
EXAFS, 205, 207
  , spectra of selenium, 207
excitons in amorphous semiconductors, 268, 280
  , bound, interacting with lattice vibrations, 275
  , electric-field broadening of, 275, 276, 279ff.
  in magnesium-bismuth alloys, 280
  in recombination mechanisms, 82ff.
  in ruby, 26
  in silicon dioxide, 513ff.
excitonic phase, 102
extended-state conduction, 25, 219
extended X-ray absorption fine structure, see EXAFS

fatiguing effects in luminescence, 438, 475, 479
F-centre, 267
Fermi glasses, 9, 98ff.
Fermi level, pinning of, 46, 194, 212 ff., 435, 461, 467, 485ff.
field effect
  in amorphous
    arsenic, 435

field effect in amorphous—*cont.*
    chalcogenides, 246, 469, 470
    germanium, 246, 359
    semiconductors, 243ff.
    silicon, 246, 368, 369, 374
Frank–Condon principle, 215
Franz–Keldysh effect, 276, 277
free-carrier absorption
  in amorphous
    chalcogenides, 297ff., 506
    semiconductors, 297ff.
frequency dependence of conductivity, *see* a.c. conductivity

gadolinium sulphide, 157
gallium, amorphous, 441
gallium–tellurium liquid alloys, 193, 194, 197ff.
geminate recombination, 266ff., 392
germanium,
  amorphous, 320ff.
    , a.c. conductivity of, 357ff.
    , alloys with iron and silicon, 132ff.
    , CRN of, *frontispiece*, 326ff.
    , d.c. conductivity of, 345ff.
    , defects in, 48, 49, 88, 211ff., 333ff.
    , density of, 321, 380
    , density of electron states in, 396ff.
    , diamagnetic susceptibility of, 338
    , diffraction functions of, 333, 334
    , dihedral angles in, 325, 330
    , electrical properties of, 345ff.
    , electron micrographs of, 337
    , e.s.r. in, 314ff.
    , field effect in, 246, 359
    , Hall effect in, 349
    , hydrogenated, 340ff.
    , infrared spectra of, 305, 340, 341
    , ion bombardment of, 354, 355
    , magnetoresistance of, 243, 357, 360ff.
    , microcrystalline models of, 331ff.
    , non-ohmic conduction in, 355, 356
    , optical absorption edges in, 283, 292, 340, 375ff., 399
    , optical properties of, 293ff., 339, 375ff.
    , oxygen in, 339, 379
    , preparation of, 320ff.
    , polytetrahedral model of, 332
    , porosity of, 338
    , RDF of, 206, 323ff., 329ff.
    , small-angle X-ray scattering in, 334, 335, 380
    , spin density in, 214, 314ff., 338, 341ff.
    , specific heat of, 310
    , states in the gap of, 352ff.
    , structure of, 322ff.

germanium, armorphous—*cont.*
, thermopower of, 239, 345ff.
, variable-range hopping in, 346ff., 351ff.
, voids impurities and defects in, 333ff.
, crystalline, 325, 376
, impurity conduction in 111ff.
germanium chalcogenides, *see* chalcogenides
glass-forming materials, 202, 203, 443, 444
glass transition (transformation) temperature, 200, 201
glassy metals, *see* metallic glasses
glow-discharge deposition, 321, 322, 409
grain boundaries, resistance of, 178
granular films, 157ff.

Hall effect, 56ff., 240ff., 270ff.
due to polarons, 92
in amorphous
arsenic, 422, 423
chalcogenides, 240ff., 272, 459ff.
germanium, 240, 349
silicon, 373
and impurity conduction, 116, 121ff.
in liquid
mercury, 186, 187
metals, 172ff.
semiconductors, 184
in tungsten bronzes, 150, 151
heterojunctions between crystalline and
amorphous silicon, 365
hopping conduction, 33ff., 27, 28, 32ff., 59ff.,
75ff.
at band edges, 47, 216ff., 270ff., 366ff.
at Fermi level, *see* variable-range hopping
due to polarons, 92ff.
in alloys, 132ff.
in granular films, 158, 159
in impurity conduction, 111ff.
in oxides, 150ff.
Hubbard bands, 104ff., 111ff., 119ff.
Hubbard U, 105, 212, 214, 466, 467

impurity conduction, 111ff.
indium antimonide, impurity conduction in,
129
indium phosphide, arsenide, and antimonide,
amorphous, core levels in, 314
infrared spectra
of amorphous
arsenic, 305
chalcogenides, 234, 301ff., 446ff.
germanium, 305, 340, 341
selenium, 303, 305
semiconductors, 233ff., 301ff.
silicon, 304

interband absorption
in amorphous semiconductors 287ff., *see
also* optical properties
intraband absorption
in amorphous semiconductors, 297ff., *see
also* free-carrier absorption
Ioffe–Regel
criterion for mean free path, 3, 8
rule for co-ordination number, 203

k-conservation selection rule, 2, 288, 30,
Knight shift, 174, 175, 189, 196ff.
Kondo effect, 125
Kubo–Greenwood formula, 11ff.

lanthanum–strontium vanadate 144, 145
lateral disorder, 21, 24, 124
lead chalcogenides, liquid, 165
lead iodide, effect of bombardment on, 280,
281
$LiNbO_3$, polarons in, 92, 94
liquid alloys, 170ff.
caesium, 189, 190
chalcogenides, 224, 237, 241, 297, 458
metals, 161ff., 222, 223
mercury, 185ff.
rare gases, 179ff.
salts, 185
selenium, 531, 543ff.
selenium-tellurium alloys, 543ff.
semimetals and semiconductors, 181ff.
tellurium and alloys, 194ff., 543ff.
localization, due to magnetic fields, 129, 139,
see also Anderson localization
luminescence, *see* electroluminescence, photoluminescence

magnesium–bismuth alloys, 45, 130ff., 280
magnetic susceptibility
of amorphous semiconductors, 318, *see
also* Curie paramagnetism
of electrons in impurity bands, 106
of tellurium, 195
magnetoresistance
in amorphous semiconductors, 242, 243,
357, 360ff., 365
in impurity conduction, 129, 130
mean free path, 2, 13ff.
mercury,
, liquid, 172, 185ff.
, liquid alloys, 170ff.
metal–ammonia system, 93, 109
metallic glasses, 177ff., 203
metal–rare gas systems, 126, 127
microcrystallite models, 208, 331ff.

minimum metallic conductivity, 4, 28ff., 289, 299
  as pre-exponential in semiconductors, 47, 219, 419, 453
  in granular films, 157ff.
  in impurity conduction, 126ff.
  in two-dimensional conduction, 136, 137
manganese oxide, 92
mobility edge, 4, 23, 39ff., 46ff., 211, 215ff.
  in amorphous silicon, 366ff.
mobility gap, 211
monomolecular and bimolecular recombination, 254ff., 392
Mott transition, 104ff., 109
multicomponent glasses, 203, 442
multiphonon processes, 78ff.
multiple-scattering theory, 162, 405
napthalene, drift mobility in, 93
Néel point, 104ff., 110, 156, 315
negative Hubbard U, 214, 316, 318, 466
nickel oxide, 90ff., 140, 141
non-bridging oxygen, 515, 516
non-ohmic conduction, 35, 36, 95ff.
  in amorphous
    chalcogenides, 490
    germanium, 355, 356
nuclear magnetic resonance, 195, 208

one-dimensional problems, 62ff.
Onsager
  escape radius, 263, 269
  theory of dissociation, 264, 542, 543
optical absorption edges
  in amorphous
    and crystalline semiconductors, 273ff., 287ff.
    arsenic, 279, 427, 428
    chalcogenides, 265, 279, 290, 292, 297, 474, 476, 480, 481, 497ff.
    germanium, 292, 340, 375ff., 399
    selenium, 267, 279, 521ff.
    silicon, 290, 344, 384ff.
    silicon dioxide, 513
    tellurium, 279, 521
  , effect of pressure on, 284
  , electric-field broadening of, 276ff.
optical properties
  of amorphous
    arsenic, 426ff.
    chalcogenides, 491ff., 497ff.
    germanium, 293ff., 339, 375ff.
    selenium, 521ff.
    semiconductors, 272ff.
    silicon, 384ff.
    silicon dioxide, 513ff.
    tellurium, 521ff.
oxide glasses, 512ff.

Pauli susceptibility of liquids, 192, 193
Penn gap, model, 296, 382, 383, 431
percolation edges, 39ff.
  theory of, 35, 149, 158, 159, 189, 210, 218, 356
phonon spectra, see infrared spectra
phosphorus, amorphous, 418, 439
photoconductivity
  in amorphous
    arsenic, 426, 428, 429
    chalcogenides, 258ff., 268, 483, 484
    semiconductors, 254ff.
    silicon, 255ff., 263, 264, 316, 374, 389, 390
photodarkening, 468
photoemission
  in amorphous
    arsenic, 432
    chalcogenides, 491ff.
    germanium, 312ff., 396ff.
    selenium, 523, 524
    semiconductors, 311ff.
    silicon, 312, 313, 396
    tellurium, 524
  liquids, 175, 176
photogeneration, 265ff., 540ff.
photoluminescence
  in amorphous
    arsenic, 435, 436
    chalcogenides, 474ff., 503, 504
    selenium, 482, 534
    semiconductors, 300, 301
    silicon, 300, 301, 391ff.
plasma frequency
  in arsenic, 430ff.
  in germanium, 381ff.
polarons, 5, 65ff., 231, 270, 271, 453, 459, 460, 513
Polk model, 326ff.
Poole–Frenkel effect, 97, 269, 490
potential fluctuations in amorphous semiconductors, see spatial fluctuations
pressure,
  effect on drift mobility, 250
    optical absorption edge, 284
    resistivity of arsenic, 424, 425
pseudogaps, 50, 51, 109, 130, 182ff., 185ff., 190ff.
pseudopotentials, 166ff., 403
pyrolytic carbon, 145, 146

quantum efficiency, 265ff., 540ff.

radial distribution function, see RDF

Raman spectroscopy
of amorphous
chalcogenides, 446ff.
selenium, 303ff.
silicon, 304
random electric fields in amorphous semiconductors, 39, 209, 210, 213, 280ff.,
287, 477, 478
rare-earth liquid metals, 177
rare gases, 179ff.
Rayleigh scattering, 499
RDF, 205ff.
of amorphous
antimony, 418
arsenic, 412ff.
chalcogenides, 499ff.
germanium, 206, 323ff., 329ff.
phosphorus, 418
selenium, 518, 519
silicon, 327, 328
recombination, 76ff.
edges, 475
lifetime
in amorphous semiconductors, 261ff.,
483, 484
in amorphous silicon, 257
in threshold switch, 507, 510, 511
RKKY interaction, 133
rubidium, liquid, 169, 170

Seebeck coefficient, see thermopower
Scher–Montroll theory of dispersive transits,
252ff.
screening length, 245, 468, 469
selenium,
amorphous, 517ff.
, alloys with, 537, 538, 443ff.
, carrier lifetimes in, 539, 540
, CRN of, 519, 520
, density of electron states in, 296, 524ff.
, dihedral angles in, 521, 526ff.
, drift mobility in, 250ff., 531ff.
, electrical properties of, 250ff., 530ff.
, electroabsorption in, 285ff.
, e.s.r. in, 532
, EXAFS spectra of, 207
, infrared spectra of, 303, 305
, optical absorption edges in 267, 279,
521ff.
, optical properties of, 521ff.
, oxygen in, 531ff.
, photogeneration in, 266ff., 540ff.
, photoluminescence in, 482, 534
, Raman spectra of, 302ff.
, RDF of, 518, 519
, states in the gap of, 532ff.

selenium, amorphous—cont.
, structure of, 517ff.
, xerographic process in, 530, 540, 541
, X-ray absorption in, 296, 497
, crystalline, 517, 518, 529
, liquid, 531, 543ff.
–tellurium alloys, 543ff.
semiconductors, amorphous, preparation of,
200ff.
semimetals, 50, 181ff.
short-range order, 204
Shubnikov–de Haas oscillations, 135
silica, see silicon dioxide
silicon–arsenic alloys, 372, 373
silicon,
amorphous, 321ff.
, a.c. conductivity of, 374, 375
, alloyed with arsenic, 372, 373
, CRN of, frontispiece, 326ff.
, d.c. conductivity of, 362ff.
, defects in, 48, 49, 88, 211ff., 333ff.
, density of electron states in, 312, 313,
396ff.
, dihedral angles in, 325, 328, 330
, doping of, 212, 322, 370ff., 393ff.
, drift mobility in 250, 256, 366ff.
, electrical properties of, 362ff.
, e.s.r. in, 36, 314ff.
, field effect in, 368, 369, 374
, glow-discharge-deposited, 316, 317,
321, 322, 344, 365ff., 386ff.
, Hall effect in, 373
, hydrogen in, 322, 387
, infrared spectra of, 304
, magnetoresistance of, 243, 365
, optical absorption edges in, 290, 344,
384ff.
, optical properties of, 384ff.
, oxygen in, 342ff., 362
, photoconductivity in, 255ff, 263, 264,
317, 374, 389, 390
, photoemission in, 312, 313, 396
, photoluminescence in, 300, 301, 391ff.
, preparation of, 320ff.
, recombination lifetime in, 257
, Raman spectra of, 304
, RDF of, 327, 328
, spin-dependent photoconductivity in,
316, 317
, states in the gap of, 255, 363, 364, 368,
369
, structure of, 322ff.
, thermopower of, 239, 362ff., 374
, variable-range hopping in, 363, 364
, voids, imperfections, and defects in,
333ff, 369, 370
, X-ray absorption in, 406

silicon, amorphous – *cont.*
, crystalline,
band structure of, 386
, divacancy in, 370
, impurity conduction via, 111ff.
silicon dioxide, 209, 250, 306ff., 456, 512ff.
small-angle X-ray scattering, 334ff., 443
soft modes in amorphous semiconductors, 309, 310
solvated electrons, 93
spatial fluctuations in the band gap of amorphous semiconductors, 39, 209, 210, 213
specific heat,
electronic, in impurity band, 125
of glasses, 230, 305ff, 436
spin-dependent photoconductivity in doped silicon, 316, 317
spin glasses, 110
spin polarons, 107, 155, 156
splat cooling, 178, 200, 320, 443
stabilization of network by defects, 333, 352
states in the gap
of amorphous
arsenic, 421, 425, 434ff.
chalcogenides, 460ff., 488ff.
germanium, 352ff.
selenium, 532ff.
semiconductors, 43, 48ff., 210ff.
silicon, 255, 363, 364, 368, 369
Stokes shift, 83, 300, 391, 439, 475, 481, 489
structure factor, 165ff., 181, *see also* diffraction function
structure, determination of, 204ff.
structure of amorphous arsenic, 410ff.
chalcogenides, 203, 445ff.
germanium, 322ff.
selenium, 517ff.
silicon, 322ff.
silicon dioxide, 512
tellurium, 517ff.

$T^{-1/4}$ behaviour, *see* variable-range hopping at Fermi level
tellurium,
amorphous, 517ff.
, density of electron states in 524ff.
, optical properties of, 279, 521ff.
, structure of, 517ff.
, crystalline, 521
, liquid 194ff.
– selenium aloys, 543ff.
ternary glass systems, 202, 203, 443, 444
thermal conductivity of non-crystalline materials, 305ff.
thermalization of electron-hole pair, 269, 475
thermoelectric power, *see* thermopower

thermopower, 52ff., 235ff., 270ff.
due to polarons, 91, 92, 94, 141
in amorphous
arsenic, 422, 433
chalcogenides, 237, 272, 458
germanium, 239, 345ff.
magnesium–bismuth alloys, 132
silicon, 239, 362ff., 374
in lanthanum-strontium vanadate alloys, 145
in liquid chalcogenides, 237, 458
metals and alloys, 172, 190, 191, 193
selenium and selenium–tellurium alloys, 544, 546
semimetals and semiconductors, 183
in impurity conduction, 116, 141
in vanadium and titanium oxides, 153
threshold switch, 464, 507ff.
tight-binding approximation, 16, 401, 494, 526, 527
titanium dioxide, 90
transit time, in threshold switch, 511, *see also* drift mobility
transition metals, liquids, 176, 177
transition-metal oxide glasses, 142ff.
two-dimensional behaviour, 22, 25, 31, 37, 38
in inversion layers, 135ff.
in amorphous
germanium, 352ff.
silicon, 363, 364
tungsten bronzes, 148ff.

Umklapp processes, 294, 529
UPS (ultraviolet photoemission spectroscopy), *see* photoemission
Urbach's rule, 273ff., 297, 380, 427, 497ff., 523

valence-alternation pairs, 466
vanadium monoxide and VO_x, 150ff.
vanadium-phosphate glasses, 96, 142ff.
variable-range hopping
at band edge, 216ff.
at Fermi level 32ff., 122, 123, 127, 128, 132ff., 136ff., 143, 152, 153, 221, 345ff., 362ff., 420, 425, 439ff., 467
vibrational spectra, *see* infrared spectra
virtual bound state, 9
viscosity of liquid selenium, 544
V_k centre, 71, 93
voids
in amorphous
chalcogenides, 433
silicon and germanium, 208, 209, 210, 213, 333ff., 379

Wigner crystallization, 109

xerographic process, 266, 530, 540, 541
XPS (X-ray photoemission spectroscopy), *see*
    photoemission
X-ray absorption
  in amorphous arsenic, 497

x-ray absorption—*cont.*
  chalcogenides, 495, 497
  selenium, 296, 497
  silicon, 406

Ziman theory of resistivity of liquid metals,
    161, 165ff.